# Antibiotics: Role of Actinobacteria and Myxobacteria

# Antibiotics: Role of Actinobacteria and Myxobacteria

Edited by John Durham

New York

Hayle Medical,
750 Third Avenue, 9th Floor,
New York, NY 10017, USA

Visit us on the World Wide Web at:
www.haylemedical.com



ISBN: 978-1-64647-529-2



**Cataloging-in-Publication Data**

Antibiotics : role of actinobacteria and myxobacteria / edited by John Durham.
p. cm.
Includes bibliographical references and index.
ISBN 978-1-64647-529-2
1. Antibiotics. 2. Actinobacteria. 3. Myxobacterales. 4. Allelopathic agents. I. Durham, John.
RM267 .A58 2023
615.792 2--dc23

# Table of Contents

# Preface

Antibiotics are medicines that are used to fight bacterial infections, and are used for the treatment and prevention of infections caused by bacteria. They can be derived from various sources such as bacteria, fungi and plant extracts. Two of such bacteria are Actinobacteria and Myxobacteria. They are primarily found in the soil. Actinobacteria are a group of gram-positive bacteria, which play critical roles in humus production and decomposition. Myxobacteria are gram-negative bacteria, which live primarily in soil and feed on insoluble organic matter. These bacteria generate a variety of biomedically and industrially useful chemicals including antibiotics, which they export outside the cell. They also generate gephyronic acid, which is an inhibitor of eukaryotic protein synthesis and a potential cancer chemotherapeutic drug. Myxobacteria are also useful models for studying multicellularity in bacterial environment. This book contains some path-breaking studies related to the role of actinobacteria and myxobacteria in the production of antibiotics. Those in search of information to further their knowledge will be greatly assisted by it.

The information contained in this book is the result of intensive hard work done by researchers in this field. All due efforts have been made to make this book serve as a complete guiding source for students and researchers. The topics in this book have been comprehensively explained to help readers understand the growing trends in the field.

I would like to thank the entire group of writers who made sincere efforts in this book and my family who supported me in my efforts of working on this book. I take this opportunity to thank all those who have been a guiding force throughout my life.

**Editor**

# Marine Actinobacteria: Screening for Predation Leads to the Discovery of Potential New Drugs against Multidrug-Resistant Bacteria

**Manar Ibrahimi [1,2,3,*], Wassila Korichi [1,3], Mohamed Hafidi [1,4], Laurent Lemee [2], Yedir Ouhdouch [1,4] and Souad Loqman [3]**

1 Laboratory of Microbial Biotechnologies, Agrosciences and Environment (BioMAgE), Faculty of Sciences Semlalia, Cadi Ayyad University, PO Box 2390, Marrakesh, Morocco; wassila.korichi@edu.uca.ac.ma (W.K.); hafidi@uca.ma (M.H.); ouhdouch@uca.ac.ma (Y.O.)
2 Institut de Chimie des Milieux et Matériaux de Poitiers (IC2MP - CNRS UMR 7285), Université de Poitiers, 4 rue Michel Brunet – TSA 51106, 86073 Poitiers Cedex 9, France; laurent.lemee@univ-poitiers.fr
3 Laboratory of Microbiology and Virology, Faculty of Medicine and Pharmacy, Cadi Ayyad University, PO Box 7010, Marrakesh, Morocco; s.loqman@uca.ma
4 Agro Bio Sciences Program, Mohammed VI Polytechnic University (UM6P), Benguerir, 43150, Morocco
* Correspondence: manar.ibrahimi@univ-poitiers.fr

**Abstract:** Predatory bacteria constitute a heterogeneous group of prokaryotes able to lyse and feed on the cellular constituents of other bacteria in conditions of nutrient scarcity. In this study, we describe the isolation of Actinobacteria predator of other bacteria from the marine water of the Moroccan Atlantic coast. Only 4 Actinobacteria isolates showing strong predation capability against native or multidrug-resistant Gram-positive or Gram-negative bacteria were identified among 142 isolated potential predatory bacteria. These actinobacterial predators were shown to belong to the *Streptomyces* genus and to inhibit the growth of various native or multidrug-resistant micro-organisms, including *Micrococcus luteus*, *Staphylococcus aureus* (native and methicillin-resistant), and *Escherichia coli* (native and ampicillin-resistant). Even if no clear correlation could be established between the antibacterial activities of the selected predator Actinobacteria and their predatory activity, we cannot exclude that some specific bio-active secondary metabolites were produced in this context and contributed to the killing and lysis of the bacteria. Indeed, the co-cultivation of Actinobacteria with other bacteria is known to lead to the production of compounds that are not produced in monoculture. Furthermore, the production of specific antibiotics is linked to the composition of the growth media that, in our co-culture conditions, exclusively consisted of the components of the prey living cells. Interestingly, our strategy led to the isolation of bacteria with interesting inhibitory activity against methicillin-resistant *S. aureus* (MRSA) as well as against Gram-negative bacteria.

**Keywords:** marine habitats; isolation; screening; predator Actinobacteria; antibiotic; multidrug-resistant bacteria

---

## 1. Introduction

In the late 1960s, the discovery of penicillin as the first antibiotic led to a significant reduction of mortality and infectious diseases caused by bacteria [1,2]. However, the discovery of new classes of antibiotics was limited in the late 1960s [3], and bacteria rapidly developed a varied array of antibiotic resistance mechanisms [4,5]. Consequently, the prevalence of multidrug-resistant (MDR) bacteria became a growing problem worldwide [6]. According to 2016–2017 WHO's new Global Antimicrobial Surveillance System (GLASS) report, antibiotic resistance is a serious health threat worldwide.

Nowadays, very few antibiotics are active against multidrug-resistant bacteria [7] and classical pathogenic Gram-negative bacteria. To face this major global healthcare threat and thus combat and stop the spreading of antibiotic-resistant pathogens, it is imperative to discover new and efficient antibiotics with novel mechanisms of action [8,9]. Most of the known natural antibiotics, active against pathogenic microorganisms, are produced by Actinobacteria [10,11]. Actinobacteria could reduce the growth (bacteriostatic activity) or kill (bactericidal activity) other microorganisms such as bacteria, fungi, and even other Actinobacteria thanks to the production of secondary metabolites [12]. These bacteria are thus the most efficient producers of most marketed antibiotics [13,14]. Recent whole-genome sequencing programs have revealed that the biosynthetic potential of Actinobacteria has been greatly under-explored and thus under-exploited [15,16]. Cryptic gene clusters in Actinobacteria are now regarded as an untapped source of bacterial secondary metabolites [17]. Consequently, nowadays, various approaches are being tested to stimulate the production of microbial secondary metabolites by Actinobacteria [13,18–21].

Co-cultivation is one of these strategies [22–24] and has led to the isolation of 33 new secondary metabolites from 12 Actinobacteria [22].

The exploration of the biosynthetic abilities of predatory bacteria is another approach used to discover novel secondary metabolites [25,26]. Predatory bacteria constitute a heterogeneous group of prokaryotes that share the ability to feed on the cellular constituents of other bacteria [27]. They include *Ensifer adhaerens* [28], *Cupriavidus necator* [29,30], *Lysobacter spp.* [31,32], *Bdellovibrionales* (d-BALOs) [26–28], Myxobacteria [20,29,30], and Actinobacteria such as *Streptomyces atrovirens* [33]. The predatory activity of these bacteria involves the secretion of small bioactive molecules acting as predatory weapons and of lytic enzymes [34,35]. Chemical investigations of these bacteria led to the discovery of many new natural products with antibacterial activity, including gulmirecins [36], myxopyronins [37,38], corallopyronins [39], and althiomycin [26,40]. These secondary metabolites are thought to contribute to the killing of prey organisms, since in their absence, the predatory performance of these bacteria is severely affected [25,26]. The ecology and physiology of predator Actinobacteria has been poorly studied. Only a few *Streptomyces* species were reported to manifest predation behavior [41,42]. This might be due to the fact that they are facultative predators, so their predatory activity might not be detected under standard culture conditions. Indeed, predation is likely to be triggered in specific conditions such as reduced nutrients availability [42,43]. It would be interesting to determine whether the Actinobacteria cryptic secondary metabolites biosynthetic pathways are activated in the course of the predation process.

Banning (2010) [44] reported that predation could be extremely beneficial for bacteria living in oligotrophic marine environments, and marine Actinobacteria were shown to be highly metabolically active [45,46]. Morocco, because of its specific geographic position, bears a unique marine environment characterized by important temperature variations throughout the year [47]. These are influenced by the connection between the Gulf Stream, via the Azores Current and the Canary Current, and the North Equatorial Current [48]. Since the antagonistic potential of Actinobacteria from Moroccan marine environment has been poorly explored, we screened for new non-obligate predatory Actinobacteria from Moroccan marine water able to feed on other bacteria cellular constituents in conditions of nutritional stress. Furthermore, we demonstrated the production of secondary metabolites by these bacteria, but the putative role of these molecules in the predation process could not be firmly demonstrated.

## 2. Results

### 2.1. Physico-Chemical Parameters

The sampling sites and the various physicochemical properties of the collected marine water are provided in Table S1. The sampling sites were characterized by rather high salinity (ranging from 35.1 mg/L to 35.8 mg/L) and conductivity (ranging from 53.4 mS/cm to 54.2 mS/cm). The pH of the water varied between 7.76 and 8.08.

*2.2. Isolation and Screening of Predatory Actinobacteria*

In total, 142 morphologically different predatory bacteria were isolated from 9 water marine samples collected from different locations of the Atlantic Ocean along the Moroccan coast. After 14 to 21 days of incubation at 28 °C, predatory bacteria colony-forming units (CFU) appeared on agar-agar that contained just 1 mL of live washed cells of *Micrococcus luteus* at the concentration $10^{12}$ CFU/mL as the sole nutrient source (Figure 1d). The water samples were diluted up to $10^{-6}$ before plating. The isolation of predatory Actinobacteria was performed by three different methods described in Materials and Methods. *M. luteus* was chosen as the prey because it is the only reported prey for *Streptomyces* species with predation behavior, and the yellow color of its CFU facilitates its experimental monitoring. The appearance of predatorial CFU from marine water samples on agar-agar containing the prey indicated the presence of at least one predator (Figure 1d), whereas, in controls samples, no CFU were observed (Figure 1a,b). A control sample consisting of agar-agar flooded with nutrient broth confirmed the viability of the prey, as shown in Figure 1c.

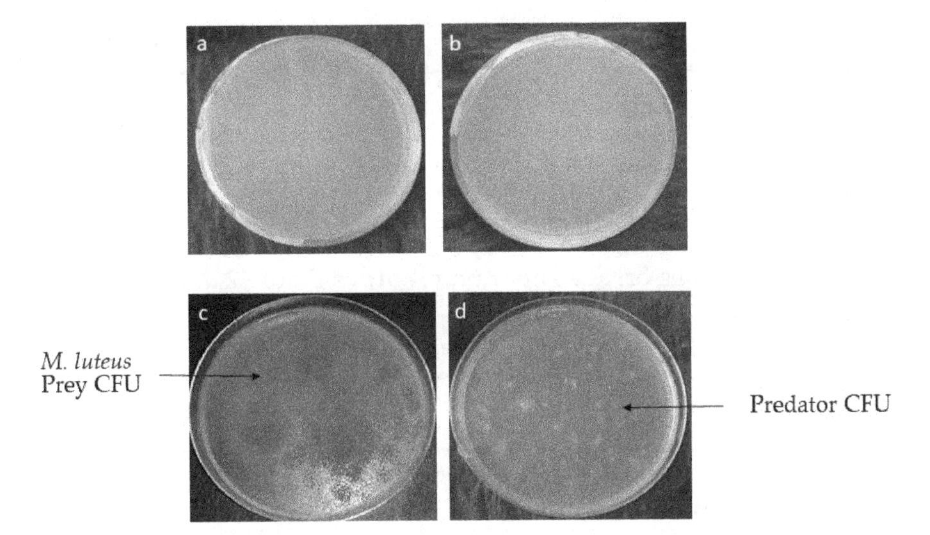


**Figure 1.** Predator isolation procedures and controls. (**a**) Plating of 1 mL of *Micrococcus luteus* at the concentration $10^{12}$ colony-forming units (CFU)/mL embedded in 20 mL of molten agar-agar at 45 °C: no growth detected; (**b**) plating of 100 µL of a water marine sample dilution on the surface of 20 mL of agar-agar: no growth detected; (**c**) growth of 1 mL of *M. luteus* at the concentration $10^{12}$ CFU/mL embedded in 20 mL of molten agar-agar at 45 °C flooded by 1 mL of nutrient broth; (**d**) growth of potential predators able to use *M. luteus* as a nutritional source.

The obtained CFU were purified in the same isolation medium to obtain pure cultures of the predatory bacteria after 72 to 96 h of incubation at 28 °C. These cultures showed filamentous morphology characteristics of Actinobacteria under light microscopy (G X10) as well as specific features described by the International *Streptomyces* Project (ISP) [49]. After purification, the isolates were stored in 20% sterile glycerol and frozen at −80 °C. These results confirmed that *M. luteus* can be used as a nutritional source by predatory Actinobacteria.

In total, morphologically different predatory bacteria were isolated from marine water samples originating from nine different locations of the Atlantic Ocean along the Moroccan coast.

*2.3. Selection of Actinobacteria with Predatory Ability and Assessement of the Putative Prey Specificity of the Selected Actinobacteria Predators*

After this primary screening, 142 bacteria isolates were obtained as potential predators. To assess their predatory ability, we tested each pure isolate individually for its aptitude to grow on agar-agar with live prey cells as the sole source of nutrients. Out of the all isolates, only four different morphotypes of predatory Actinobacteria, designed EMM111, EMM112, EMM183, and EMM194 were selected

(Table 1). The selection was based on the following criteria: growth after 72–96 h of incubation, consistency, pigmentation of substrate and aerial mycelium, microscopic observation, and specific features defined by the ISP.

**Table 1.** Origin and number of predatory Actinobacteria isolates.

| Samples | Number of Predatory Bacteria | Number of Predatory Actinobacteria | Method of Isolation | Code |
|---|---|---|---|---|
| 1 | 23 | 2 | 2 | EMM111 |
| | | | 1 | EMM112 |
| 2 | 12 | 1 | 1 | EMM111 |
| 3 | 17 | 1 | 1 | EMM111 |
| 4 | 6 | 1 | 1 | EMM111 |
| 5 | 11 | 0 | - | - |
| 6 | 5 | 1 | 1 | EMM111 |
| 7 | 18 | 1 | 1 | EMM111 |
| 8 | 30 | 1 | 2 | EMM183 |
| 9 | 20 | 1 | 2 | EMM194 |

In order to determine the possible prey specificity of the four selected isolates, the latter were grown on agar-agar inoculated with the following Gram-positive and Gram-negative bacteria belonging to different taxonomic groups, *M. luteus* (ML), *Staphylococcus aureus* (SA), methicillin-resistant *S. aureus* (MRSA), *Escherichia coli* (EC), and ampicillin-resistant *E. coli* (AREC). The four strains tested showed clear macroscopic expanding zones on agar-agar on all preys (Table 2 and Figure 2). This indicated the total lack of specificity of this predation process.

**Table 2.** Comparison of the predatory activity and the antibacterial activity on rich medium of four predatory Actinobacteria isolates.

| Strains | Predatory Activity | | | | | Antibacterial Activity in a Rich Medium (Bennett) | | | | |
|---|---|---|---|---|---|---|---|---|---|---|
| | ML | SA | EC | MRSA | AREC | ML | SA | EC | MRSA | AREC |
| EMM111 | + | + | + | + | + | 22.16 ± 0.7 | 17.5 ± 0.5 | 0 | 16.3 | 0 |
| EMM112 | + | + | + | + | + | 45 ±1 | 35.4 ± 0.9 | 23 ± 1 | 34.6 ± 0.5 | 15 ± 1 |
| EMM183 | + | + | + | + | + | 0 | 35 ± 1 | 0 | 23.26 ± 3.2 | 0 |
| EMM194 | + | + | + | + | + | 0 | 0 | 4.4 ± 0.5 | 0 | 0 |

Predatory activity is expressed by the presence (+) or absence (0). The experiments were repeated three times (n = 3) with each independent assay. Antibacterial activity in a rich medium (Bennett) is expressed by inhibition diameter (mm). Each value represents the mean ± SD of three replicates.

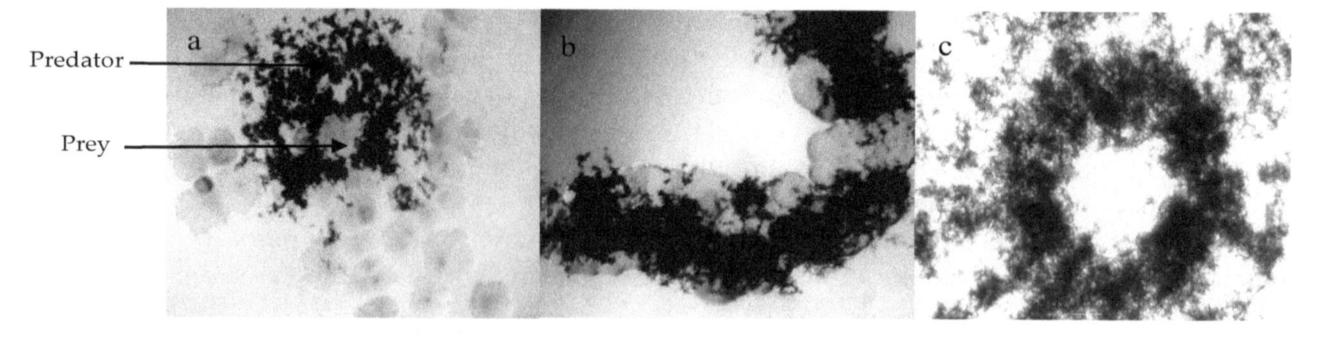


**Figure 2.** Images of Actinobacteria EMM111 predation of *M. luteus*. (**a**) and (**b**) Predation of *M. luteus* by Actinobacteria observed by phase-contrast microscopy at magnitude of 400x; (**c**) total elimination of prey cells after 15 days of incubation. The scale bar is 10 μm.

## 2.4. *Impact of Actinobacteria on M. Luteus Viability*

In an attempt to determine whether the predatory process involves the production of diffusible killing substances by Actinobacteria, the latter was deposited in the center of a plate of agar-agar containing *M. luteus* as a prey. Clearing zones signaling *M. luteus* lysis were clearly seen around the Actinobacteria deposit. *M. luteus* present at various distances (indicated with numbers from 1, i.e., proximal to 6, i.e., distal) from the Actinobacteria clearing zones was transplanted on nutrient agar (NA). After 24 h (Figure 3b) or 48 h (Figure 3c) of incubation, it was obvious that the growth of *M. luteus* originating from zones 1 and 2 was either absent or not as dense as that of *M. luteus* originating from zones 5 or 6 or from the control's zone 7 incubated separately in the absence of predator (Figure 3b). This indicated that Actinobacteria had a clear negative impact on *M. luteus* viability, but we could not determine whether this was due to the predatory activity of Actinobacteria or to its ability to produce bactericidal substances.

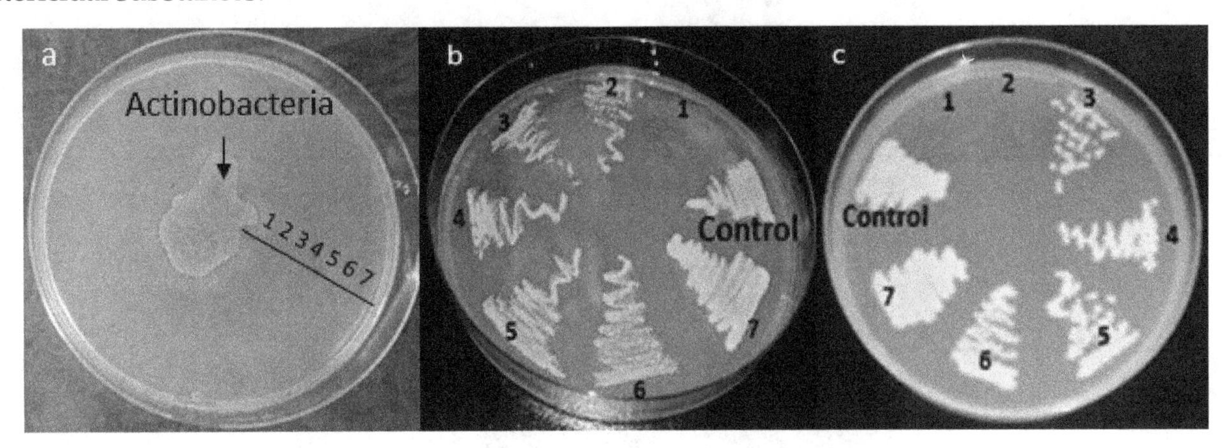


**Figure 3.** Impact of Actinobacteria on *M. luteus* viability. Pure culture growth of predatory Actinobacteria in the presence of washed live cells of *M. luteus* as the sole source of nutrients on agar-agar medium (**a**). Transplanted viable *M. luteus* prey cells present at various distances from Actinobacteria after 24 h (**b**) and 48 h (**c**) of incubation.

## 2.5. *Production of Antimicrobial Compounds by Actinobacteria Grown on Bennett Medium*

Since the clear negative impact that Actinobacteria had a on *M. luteus* viability could be due to the production of antimicrobial substances, the production of antibacterial substances by the four predatory Actinobacteria isolates was assessed using Bennett medium (Table 2). Two Actinobacteria (EMM 111 and 112) produced substances limiting the growth of *M. luteus*, but EMM183 and 194 did not. Otherwise, most of the Actinobacteria showed a significant degree of antagonistic activities against all tested microorganisms (Figure 4A–C). Among the selected isolates, EMM111 had significant inhibitory activity against Gram-positive SA, MRSA, and ML. The predation activity of EMM111 towards MRSA was studied in more detail by monitoring the increase and decrease of fatty acid biomarkers characteristic of the predator and the prey, respectively [50]. EMM194 showed a significant antibacterial activity against Gram-negative bacteria (EC), and EMM183 showed a significant antagonistic activity against SA and MRSA. However, the most promising isolate against MRSA was EMM112, whereas the commercial antibiotics ceftriaxone (30 µg) and cefoxitin (30 µg) had no inhibitory impact on this multidrug-resistant bacteria. EMM112 also exhibited the highest antibacterial activity against both Gram-positive and Gram-negative negative bacteria and showed the largest inhibition zones against *M. luteus*.

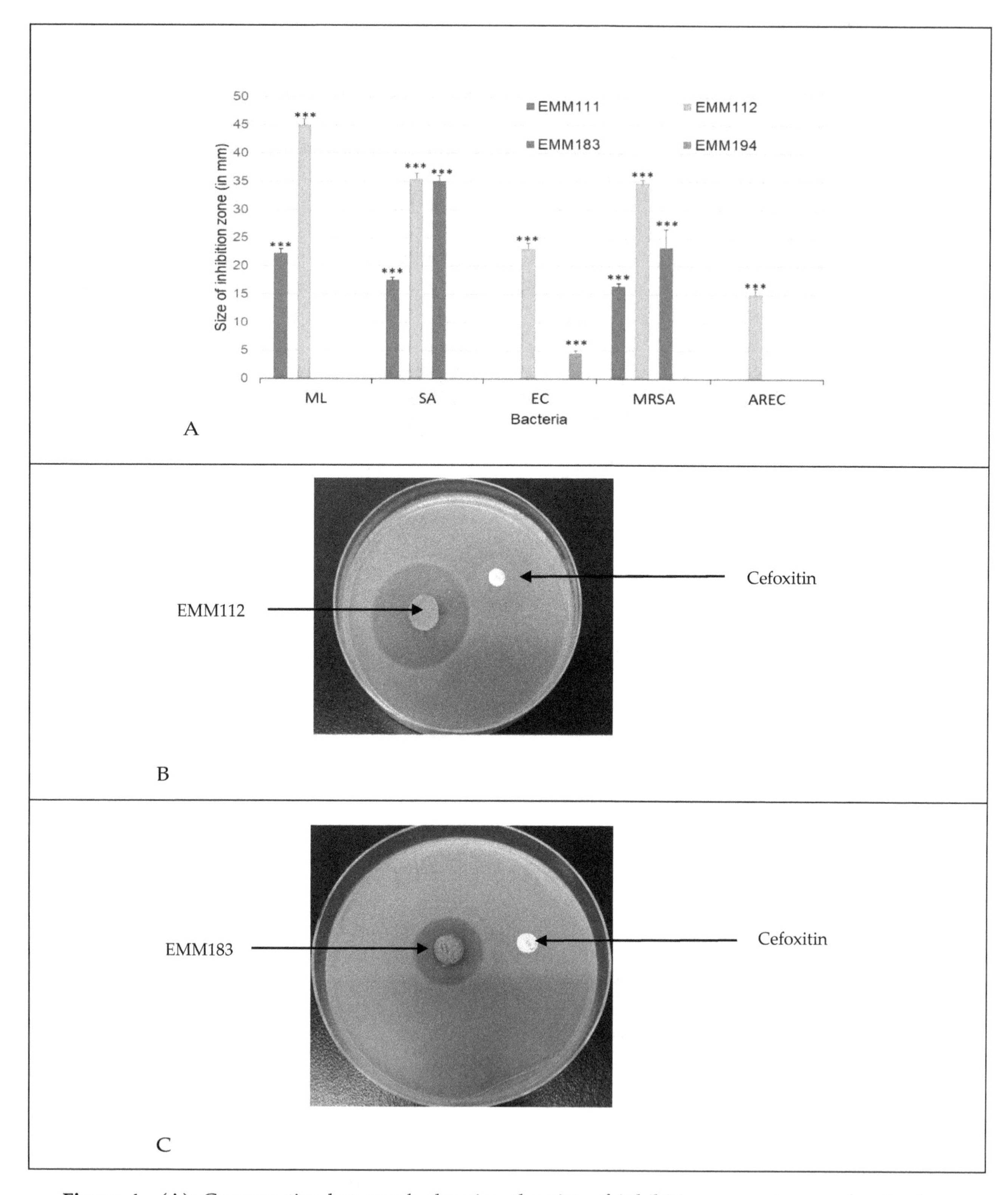


**Figure 4.** (**A**): Comparative bar graph showing the size of inhibition zones in mm for the four non-obligate predatory Actinobacteria using Bennett medium against different bacteria. Each value represents the mean ± SD of three replicates; ***$p < 0.001$ indicates significant differences compared the agar cylinders of Bennett medium using two-way analysis of variance (ANOVA). ML: *M. luteus*, S: *Staphylococcus aureus*, EC: *Escherichia coli*, MRSA: methicillin-resistant *S. aureus*, AREC: ampicillin-resistant *E. coli*; (**B**): Antibacterial activity of EMM112 using Bennett medium against MRSA; (**C**): Antibacterial activity of EMM183 using Bennett medium against MRSA.

*2.6. Morphological, Physiological, and Molecular Characterization of the Selected Facultative Predator Actinobacteria Isolates*

Morphological observation revealed that the four predatory Actinobacteria isolates were different (Table 3). The color of aerial and substrate mycelium of the strains varied. EMM111 produced a variety of pigments according to the medium composition, while the other isolates were not pigmented, except EMM183, which produced a brown pigment in ISP2. None of the four isolates produced melanin on peptone yeast extract agar (ISP 6). On the basis of the sequence of their 16S RNA, the four selected isolates were shown to belong to the genus *Streptomyces* (Table 4). The level of identity of EMM112 and EMM 183 with *Streptomyces coelicoflavus* and *Streptomyces mutabilis*, respectively, was 100%, whereas that of EMM111 and EMM194 with *Streptomyces griseoflavus* and *Streptomyces champavatii*, respectively, was 99%.

**Table 3.** Morphological and physiological characteristics of four selected predatory Actinobacteria isolates.

| Properties | EMM111 | EMM112 | EMM183 | EMM194 |
|---|---|---|---|---|
| **ISP2** | | | | |
| **Color of aerial mycelium** | beige | orange | white | white |
| **Color of substrate mycelium** | white | brown | beige | white |
| **Diffusible pigment produced** | orange | - | brown | - |
| **ISP6** | | | | |
| **Color of aerial mycelium** | beige | grey | white | white |
| **Color of substrate mycelium** | beige | black | white | white |
| **Diffusible pigment produced** | - | - | - | - |
| **ISP7** | | | | |
| **Color of aerial mycelium** | beige | grey | pink | white |
| **Color of substrate mycelium** | white | black | pink | brown |
| **Diffusible pigment produced** | grey | - | - | - |

The sign "-" means negative.

**Table 4.** Comparison of percent similarities between our 16S rRNA gene sequence and sequences present in the genomic database banks using NCBI BLAST.

| Strains | Percentage of Sequence Identity (%) | Actinobacteria Strains | Accession |
|---|---|---|---|
| **EMM111** | 99 | *Streptomyces griseoflavus* | NR_042291.1 |
| **EMM112** | 100 | *Streptomyces coelicoflavus* | NR_041175.1 |
| **EMM183** | 100 | *Streptomyces mutabilis* | NR_044139.1 |
| **EMM194** | 99 | *Streptomyces champavatii* | NR_115669.1 |

## 3. Discussion

Our study constitutes the first report confirming that oligotrophic environments like marine water might be a promising source of predatory Actinobacteria. Previous investigations demonstrated that in the absence of nutrients sources, diverse bacteria become predator and consume a variety of prey microorganisms as a nutritional resource [28,50–52]. Predatory bacteria are found in a wide variety of habitats, including seawater [27,53,54], and were reported to produce important therapeutic molecules active against human, animal, or plant pathogenic bacteria [55–58].

In the course of our study, we selected four different predatory Actinobacteria from Moroccan marine water (Figure 1 and Table 1). Our results demonstrated that predatory Actinobacteria could use Gram-positive and Gram-negative bacteria as feeding resources. They were all identified as belonging to the *Streptomyces* genus. Our results are consistent with previous reports showing that *Streptomyces* spp. are non-obligatory predators [33,42,59]. To date, only a few *Streptomyces* spp. predators of bacteria have been described. The first reported predatory Actinobacteria was a *Streptomyces* [41]. Then, the predation of *M. luteus* by *Streptoverticillium* was proposed [60]. Next, Kumbhar and coworkers [42] showed that

*Streptomyces* are non-obligate epibiotic predators of a variety of microorganisms like *S. aureus*, *E. coli*, *Bacillus* spp., *Pseudomonas aeruginosa*, and *Klebsiella* spp. Recently, a marine *S. atrovirens* demonstrated a good predatory activity on a range of preys [33]. Finally, our study confirmed the existence of predatory activity in *Streptomyces* species. *S. griseoflavus* EMM111, *S. coelicoflavus* EMM112, and *S. mutabilis* EMM183 were already known as a terrestrial strain of *Streptomyces* [61–63], whereas *S. champavatii* EMM194 was previously isolated from a marine environment [64]. Some marine Actinobacteria can indeed have a terrestrial origin [65], being transported from a terrestrial to a marine environment by rivers, rainfall, wind, and spores leaching [65,66].

The results of the present study are potentially promising, since they indicate that predatory Actinobacteria can be easily isolated by a simple strategy requiring nutrient resources scarcity to induce the predatory behavior. In this study, marine water samples were simply plated on agar-agar plates by three methods. The first method used marine water dilution and direct spread of the prey directly onto the surface of the agar-agar plates. In the second method, the prey cells and molten agar-agar at 45 °C were shaken on sterile Scott bottles and poured into Petri dishes. Then, marine water dilutions were spread directly onto the surface. For the last technique, each marine water sample dilution and the prey were first transferred into sterile Petri dishes and mixed with molten agar-agar at 45 °C. In such condition, a great number of predatory bacteria could be isolated. The absence of growth of strains isolates on agar-agar control and their growth on agar-agar in the presence of *M. luteus* confirmed the predation. All four selected Actinobacteria isolates showed ability to grow on different types of bacterial cells (Gram-positive, Gram-negative, and multidrug-resistant bacteria), demonstrating the lack of specificity of the predation process that requires the excretion of lytic enzymes, in line with previous reports for other predatory bacteria [44]. Most of these species also produced anti-microbial molecules on Bennett medium. In condition of predation, these strains reduced the viability of *M. luteus*, but whether this is due to the predatory/lytic effect of these strains or to their ability to produce killing molecules is not known. Indeed, for instance, *S. champavatii* EMM194, which did not produce any molecule inhibiting the growth of the tested bacteria (except *E. coli*) when grown on Bennett medium, was still able to use all the bacteria tested as a prey. This suggests that predation and antibiotic-producing activities might not be linked. However, we cannot exclude that EMM194 produces specific bio-active molecules exclusively in the presence of its preys. Indeed, the production of bio-active secondary metabolites is often linked to the composition of the growth media that, in our condition, was exclusively constituted of components of living cells; in addition, the co-cultivation of Actinobacteria with other bacteria is known to lead to the production of compounds that are not produced in monoculture [22–24].

Since the four actinobacterial strains isolated were able to grow in the absence of prey on a variety of rich media, they should be considered as non-obligate predators. These bacteria are also able to produce anti-bacterial molecules, but the connection of these productions with the predatory process remains to be elucidated. The purification and elucidation of the structure of the antibacterial compounds produced by the selected isolates during the predation process or in monocultures are in progress.

## 4. Conclusions

This study constitutes the first report of the isolation predatory Actinobacteria from Moroccan marine water. The obtained data suggest that this oligotrophic environment is a promising reservoir of facultative predatory Actinobacteria. Four different facultative predatory Actinobacteria were isolated and identified. The four selected Actinobacteria showed a strong predatory activity spectrum and produced antibacterial substances on Bennett medium. It is possible that the antibacterial activity detected contributed to the killing and lysis of the bacteria, even if this could not be demonstrated firmly. Genome sequencing of the four predatory Actinobacteria is a crucial step to identify genes expressed during predation and specifically those directing the biosynthesis of potential killing compounds.

## 5. Materials and Methods

### *5.1. Samples and Microbial Strains*

#### 5.1.1. Sample Collection

Marine water samples were collected from nine different Moroccan locations of the Atlantic Ocean in April 2016. Sterile bottles were used to collect marine water samples. pH, temperature, and conductivity of the samples were measured at the moment of sampling, using a hand-held multiparameter device HI98194, Germany. The collected samples were transferred into sterile polythene bottles and stored at 4 °C for 24 h before further study.

#### 5.1.2. Bacterial Strains

Prey cell strains used during investigations were *M. luteus* (ML) (ATCC381), *S. aureus* (SA) (ATCC 25923), methicillin-resistant *S. aureus* (MRSA) (NCTC 12493), *E. coli* (EC) (ATCC 8739), and ampicillin-resistant *E. coli* (AREC) (ATCC 35218). Bacterial inocula were prepared by growing cells on nutrient agar plates (3 g beef extract; 5 g peptone; 8 g sodium chloride; 15 g agar; 1000 mL distilled water; final pH 6.8 +/− 0.2) for 24 h at 37 °C.

### *5.2. Preparation of the Prey Cells and Screening of Predatory Actinobacteria Potential*

We designed a new technique for the primary screening of predatory Actinobacteria isolates using *M. luteus* as a prey. The latter was cultivated on nutrient plates for 24 h at 37 °C, then collected, centrifuged at 4000 rpm for 10 min at 10 °C, and washed three times using sterile physiological water (NaCl 9 g/L). The final prey cell suspension was at a concentration of $10^{12}$ CFU/mL. Isolation of predatory Actinobacteria was performed from all water samples diluted up to $10^{-6}$ by three different methods. In the first method, 1 mL of *M. luteus* at the concentration $10^{12}$ CFU/mL and 100 μL of marine water dilution were plated directly on the surface of 20 mL of agar-agar plates containing 15 g of agar-agar per liter of sterile water. In the second method, 1 mL of *M. luteus* at the concentration $10^{12}$ CFU/mL was added to 20 mL of molten agar-agar at 45 °C, homogenized, and poured into Petri dishes, and 100 μL of marine water dilutions was spread directly onto the surface of the plates. In the last technique, 1 mL of each marine water sample dilution and M. *luteus* at the concentration $10^{12}$ CFU/mL was transferred into empty sterile Petri dishes, and 20 mL of molten agar-agar (45 °C) was poured on this mixture. The plates were incubated at 37 °C for 21 days. Each sample was analyzed separately and in triplicates, and several control experiments were carried out. The first control consisted of plates of agar-agar to test sterility. For the second control, 100 μL of water sample dilutions were plated on agar-agar. In the last control, agar-agar was inoculated with 1 mL of *M. luteus* washed live cells and incubated for 24 h at 37 °C. Since *M. luteus* could not grow on this medium, the plates were then flooded with 1 mL of nutrient broth to assess viability and abundance of the prey.

After 21 days of incubation, the growth of predatory Actinobacteria isolates was evaluated. Predatory Actinobacteria colonies were recognized on the basis of their morphological characteristics using light microscopy (G X10). The colony-forming units surrounded by zones of hydrolysis were selected as potential predators and purified. Most Actinobacteria showed vegetative and aerial mycelium, while others showed only substrate mycelium. Isolates were maintained on nutrient agar medium and stored at +4 °C for 2 months. Alternatively, cultures were re-suspended in 20% sterile glycerol and frozen at −80 °C.

### *5.3. Selection of Actinobacteria with Predatory Ability and Assessement of the Putative Prey Specificity of the Selected Actinobacteria Predators*

The selection and the assessment of the putative prey specificity of t the selected purified isolates was tested using washed cell of *M. luteus, S. aureus, E.a coli,* methicillin-resistant *S. aureus,* and ampicillin-resistant *E. coli*. In this experiment, 1 mL of prey cells at $10^{12}$ CFU/mL were first plated on agar-agar plates, then 10 µL of predatory Actinobacteria isolates at $10^6$ CFU/mL was inoculated in the center of the plates.

### 5.4. Impact of Actinobacteria on M. Luteus Viability

We investigated the impact of Actinobacteria on *M. luteus* viability. For this, 1 mL of washed prey cells of *Mi. luteus* at $10^{12}$ CFU/mL was spread on agar-agar plates. Then, 10 µL of predatory Actinobacteria isolate at $10^6$ CFU/mL was inoculated in the center. In order to assess the prey viability at various distances of the Actinobacteria predator, samples taken at various distances were sub-cultured on nutrient agar plates for 24 h and 48 h at 37 °C. Control subcultures of the prey were also incubated in the absence of the predator.

### 5.5. Production of Antimicrobial Compounds by Actinobacteria Grown on Bennett Medium

The antibacterial activity was tested by the plate-diffusion method [67]. The four selected predatory Actinobacteria were grown on Bennett agar medium at 28 °C for 21 day. Agar cylinders (6 mm in diameter) were cut and placed on Mueller–Hinton agar, previously seeded by the test microorganisms. The plates were kept during 4 h at 4 °C for a good diffusion of the secondary metabolites produced, then incubated at 37 °C. Inhibition zones were determined after 24 hours. Antibacterial activity was evaluated in vitro using the five strains (ML, SA, EC, MRSA, and AREC) mentioned above. Antibiotics such as ceftriaxone (30 µg) and cefoxitin (30 µg) were used to reveal MDR bacteria according to Clinical and Laboratory Standards Institute guideline [68]. All tests were carried out in triplicate.

### 5.6. Statistical Analysis

The data obtained from the antibacterial activity were subjected to two-way analysis of variance (ANOVA). Significance of the differences between group means was assessed with the Tukey test; $p < 0.05$ was considered statistically significant. The statistical analysis was carried out using the software XLSTAT Version 2016.02.27444.

### 5.7. Morphological, Physiological, and Molecular Characterization of the Selected Facultative Predator Actinobacteria Isolates

The cultural, morphological, and physiological characterizations of the selected predatory Actinobacteria isolates were carried out according to the ISP [49]. For the molecular identification, the DNA of selected Actinobacteria was sequenced as described by Hopwood et al. 1985 [69]. The 16S rDNA was amplified by PCR with Taq DNA polymerase and primers 27F (5-AGAGTTTGATCCTGGCTCAG-3) and 1492R (5-TACGGYTACCTTGTTACGACTT-3). The conditions for thermal cycling were as follows: denaturation of the target DNA at 98 °C for 3 min followed by 30 cycles at 94 °C for 1 min, primer annealing at 52 °C for 1 min, and primer extension at 72 °C for 5 min. At the end of the cycling, the reaction mixture was held at 72 °C for 5 min and then cooled to 4 °C. The PCR product obtained was sequenced using an automated sequencer and the same primers used for sequence determination (Macrogen Inc., Seoul, Korea). The sequence was compared for similarity with the reference sequences of bacteria species contained in genomic database banks, using the NCBI Blast available at http://www.ncbi.nlm.nih.gov/

**Author Contributions:** Conceptualization: Y.O., M.I; Methodology: M.I., Y.O., S.L., L.L., W.K.; Validation: Y.O., S.L., L.L.; Formal analysis: M.I., Y.O., S.L., L.L.; Investigation: M.I, W.K.; Resources: Y.O., S.L., L.L.; Data curation: M.I.,Y.O., S.L., L.L.,; Writing – original draft: M.I.; Writing – review & editing: M.I.,Y.O., S.L., L.L., M.H.; Visualization: M.I.,Y.O., S.L., L.L., M.H., W.K; Supervision: Y.O., S.L., L.L.; Project administration: Y.O., L.L; Funding acquisition: Y.O., S.L., L.L. All authors have read and agreed to the published version of the manuscript

**Acknowledgments:** We would like to thank the French ministries of Europe and foreign affairs (MEAE), of higher education, research, and innovation (MESRI) and the Moroccan ministry of higher education, research, and managerial training via the partnership Hubert Curien (PHC) Toubkal n° 17/48–Campus France n° 36856WA who financially supported this work.

## References

1. Avorn, J.L.; Barrett, J.F.; Davey, P.G.; McEwen, S.A.; O'Brien, T.F.; Levy, S.B. Response, World Health Organization Dept of Epidemic and Pandemic Alert and Alliance for the Prudent Use of Antibiotic. In *Antibiotic Resistance: Synthesis of Recommendations by Expert Policy Group*; World Health Organization: Geneva, Switzerland, 2001.
2. Cragg, G.M.; Grothaus, P.G.; Newman, D.J. Impact of natural products on developing new anti-cancer agents. *Chem. Rev.* **2009**, *109*, 3012–3043. [CrossRef]
3. Coates, A.R.; Halls, G.; Hu, Y. Novel classes of antibiotics or more of the same? *Br. J. Pharm.* **2011**, *163*, 184–194. [CrossRef]
4. Demerec, M. Origin of Bacterial Resistance to Antibiotics. *J. Bacteriol.* **1948**, *56*, 63. [CrossRef]
5. Lieberman, J.M. Appropriate antibiotic use and why it is important: The challenges of bacterial resistance. *Pediatri. Infect. Dis. J.* **2003**, *22*, 1143–1151. [CrossRef]
6. Lorcy, A.; Dubé, E. Les enjeux des bactéries multi-résistantes à l'hôpital. Innovations technologiques, politiques publiques et expériences du personnel. *Anthropologie & Santé. Revue Internationale Francophone D'anthropologie de la Santé* **2018**. [CrossRef]
7. Furusawa, C.; Horinouchi, T.; Maeda, T. Toward prediction and control of antibiotic-resistance evolution. *Curr. Opin. Biotechnol.* **2018**, *54*, 45–49. [CrossRef] [PubMed]
8. Sarkar, R.; Paul, R.; Roy, D.; Thakur, I.; Ray, J.; Sau, T.J.; Haldar, K.; Mondal, J. Rising levels of antibiotic resistance in bacteria: A cause for concern. *J. Assoc. Physicians India* **2017**, *65*, 107. [PubMed]
9. Van der Meij, A.; Worsley, S.F.; Hutchings, M.I.; Van Wezel, G.P. Chemical ecology of antibiotic production by actinomycetes. *FEMS Microbiol. Rev.* **2017**, *41*, 392–416. [CrossRef]
10. Kavitha, A.; Vijayalakshmi, M.; Sudhakar, P.; Narasimha, G. Screening of actinomycetes strains for the production of antifungal metabolites. *AJMR* **2010**, *4*, 027–032.
11. Ouchari, L.; Boukeskasse, A.; Bouizgarne, B.; Ouhdouch, Y. Antimicrobial potential of actinomycetes isolated from the unexplored hot Merzouga desert and their taxonomic diversity. *Biol. Open* **2019**, *8*, bio035410. [CrossRef]
12. Ortiz-Ortiz, L.; Bojalil, L.F.; Yakoleff, V. *Biological, Biochemical, and Biomedical Aspects of Actinomycetes*; Elsevier: Amsterdam, The Netherlands, 2013; ISBN 978-1-4832-7369-3.
13. Abdelmohsen, U.R.; Grkovic, T.; Balasubramanian, S.; Kamel, M.S.; Quinn, R.J.; Hentschel, U. Elicitation of secondary metabolism in actinomycetes. *Biotechnol. Adv.* **2015**, *33*, 798–811. [CrossRef]
14. Arul Jose, P.; Jebakumar, S.R.D. Non-*streptomycete* actinomycetes nourish the current microbial antibiotic drug discovery. *Front. Microbiol.* **2013**, *4*, 240.
15. Baltz, R.H. Natural product drug discovery in the genomic era: Realities, conjectures, misconceptions, and opportunities. *J. Ind. Microbiol. Biotechnol.* **2019**, *46*, 281–299. [CrossRef] [PubMed]
16. Ikeda, H.; Shin-ya, K.; Omura, S. Genome mining of the *Streptomyces avermitilis* genome and development of genome-minimized hosts for heterologous expression of biosynthetic gene clusters. *J. Ind. Microbiol. Biotechnol.* **2014**, *41*, 233–250. [CrossRef]
17. Olano, C.; García, I.; González, A.; Rodriguez, M.; Rozas, D.; Rubio, J.; Sánchez-Hidalgo, M.; Braña, A.F.; Méndez, C.; Salas, J.A. Activation and identification of five clusters for secondary metabolites in *Streptomyces albus* J1074. *Microb. Biotechnol.* **2014**, *7*, 242–256. [CrossRef] [PubMed]
18. Scherlach, K.; Hertweck, C. Triggering cryptic natural product biosynthesis in microorganisms. *Org. Biomol. Chem.* **2009**, *7*, 1753–1760. [CrossRef]
19. Dashti, Y.; Grkovic, T.; Abdelmohsen, U.; Hentschel, U.; Quinn, R. Production of induced secondary metabolites by a co-culture of sponge-associated actinomycetes, *Actinokineospora* sp. EG49 and *Nocardiopsis* sp. RV163. *Mar. Drugs* **2014**, *12*, 3046–3059. [CrossRef]

20. Sung, A.; Gromek, S.; Balunas, M. Upregulation and identification of antibiotic activity of a marine-derived *Streptomyces* sp. via co-cultures with human pathogens. *Mar. Drugs* **2017**, *15*, 250. [CrossRef] [PubMed]
21. Vikeli, E.; Widdick, D.A.; Batey, S.F.; Heine, D.; Holmes, N.A.; Bibb, M.J.; Martins, D.J.; Pierce, N.E.; Hutchings, M.I.; Wilkinson, B. In situ activation and heterologous production of a cryptic lantibiotic from a plant-ant derived *Saccharopolyspora* species. *Appl. Environ. Microbiol.* **2019**. [CrossRef]
22. Hoshino, S.; Onaka, H.; Abe, I. Activation of silent biosynthetic pathways and discovery of novel secondary metabolites in actinomycetes by co-culture with mycolic acid-containing bacteria. *J. Ind. Microbiol. Biotechnol.* **2019**, *46*, 363–374. [CrossRef]
23. Pettit, R.K. Mixed fermentation for natural product drug discovery. *Appl. Microbiol. Biotechnol.* **2009**, *83*, 19–25. [CrossRef] [PubMed]
24. Ueda, K.; Beppu, T. Antibiotics in microbial coculture. *J. Antibiot.* **2017**, *70*, 361. [CrossRef] [PubMed]
25. Xiao, Y.; Wei, X.; Ebright, R.; Wall, D. Antibiotic production by myxobacteria plays a role in predation. *J. Bacteriol.* **2011**, *193*, 4626–4633. [CrossRef]
26. Korp, J.; Gurovic, M.S.V.; Nett, M. Antibiotics from predatory bacteria. *Beilstein J. Org. Chem.* **2016**, *12*, 594–607. [CrossRef]
27. Jurkevitch, E. Predatory behaviors in bacteria-diversity and transitions. *Microbe Am. Soc. Microbiol.* **2007**, *2*, 67. [CrossRef]
28. Casida, L.E., Jr. *Ensifer adhaerens* gen. nov., sp. nov.: A bacterial predator of bacteria in soil. *Int. J. Syst. Evol. Microbiol.* **1982**, *32*, 339–345. [CrossRef]
29. Kreutzer, M.F.; Kage, H.; Nett, M. Structure and biosynthetic assembly of cupriachelin, a photoreactive siderophore from the bioplastic producer *Cupriavidus necator* H16. *J. Am. Chem. Soc.* **2012**, *134*, 5415–5422. [CrossRef]
30. Makkar, N.S.; Casida, L.E., Jr. *Cupriavidus necator* gen. nov., sp. nov.; a Nonobligate Bacterial Predator of Bacteria in Soil. *Int. J. Syst. Evol. Microbiol.* **1987**, *37*, 323–326. [CrossRef]
31. Li, S.; Jochum, C.C.; Yu, F.; Zaleta-Rivera, K.; Du, L.; Harris, S.D.; Yuen, G.Y. An antibiotic complex from *Lysobacter* enzymogenes strain C3: Antimicrobial activity and role in plant disease control. *Phytopathology* **2008**, *98*, 695–701. [CrossRef]
32. Seccareccia, I.; Kost, C.; Nett, M. Quantitative Analysis of *Lysobacter* Predation. *Appl. Environ. Microbiol.* **2015**, *81*, 7098–7105. [CrossRef]
33. Baig, U.I.; Pund, A.; Holkar, K.; Watve, M. Predator prey and the third beneficiary. *bioRxiv* **2019**, 730895. [CrossRef]
34. Berleman, J.E.; Kirby, J.R. Deciphering the hunting strategy of a bacterial wolfpack. *FEMS Microbiol. Rev.* **2009**, *33*, 942–957. [CrossRef] [PubMed]
35. Reichenbach, H.; Höfle, G. Biologically active secondary metabolites from myxobacteria. *Biotechnol. Adv.* **1993**, *11*, 219–277. [CrossRef]
36. Schieferdecker, S.; König, S.; Weigel, C.; Dahse, H.-M.; Werz, O.; Nett, M. Structure and Biosynthetic Assembly of Gulmirecins, Macrolide Antibiotics from the Predatory Bacterium *Pyxidicoccus fallax*. *Chem. Eur. J.* **2014**, *20*, 15933–15940. [CrossRef] [PubMed]
37. Kohl, W.; Irschik, H.; Reichenbach, H.; Höfle, G. Antibiotika aus Gleitenden Bakterien, XVII. Myxopyronin A und B—Zwei neue Antibiotika aus *Myxococcus fulvus* Stamm Mx f50. *Liebigs Ann. Chem.* **1983**, *1983*, 1656–1667. [CrossRef]
38. Irschik, H.; Jansen, R.; Höfle, G.; Gerth, K.; Reichenbach, H. The corallopyronins, new inhibitors of bacterial RNA synthesis from Myxobacteria. *J. Antibiot.* **1985**, *38*, 145–152. [CrossRef]
39. Schäberle, T.F.; Schmitz, A.; Zocher, G.; Schiefer, A.; Kehraus, S.; Neu, E.; Roth, M.; Vassylyev, D.G.; Stehle, T.; Bierbaum, G.; et al. Insights into Structure–Activity Relationships of Bacterial RNA Polymerase Inhibiting Corallopyronin Derivatives. *J. Nat. Prod.* **2015**, *78*, 2505–2509. [CrossRef]
40. Kunze, B.; Reichenbach, H.; Augustiniak, H.; Höfle, G. Isolation and identification of althiomycin from *Cystobacter fuscus* (Myxobacterales). *J. Antibiot.* **1982**, *35*, 635–636. [CrossRef]
41. Waksman, S.A.; Woodruff, H.B. *Actinomyces antibioticus*, a new soil organism antagonistic to pathogenic and non-pathogenic bacteria. *J. Bacteriol.* **1941**, *42*, 231. [CrossRef]
42. Kumbhar, C.; Mudliar, P.; Bhatia, L.; Kshirsagar, A.; Watve, M. Widespread predatory abilities in the genus *Streptomyces*. *Arch. Microbiol.* **2014**, *196*, 235–248. [CrossRef]

43. Banning, E.C.; Casciotti, K.L.; Kujawinski, E.B. Novel strains isolated from a coastal aquifer suggest a predatory role for *Flavobacteria*. *FEMS Microbiol. Ecol.* **2010**, *73*, 254–270. [CrossRef] [PubMed]
44. Banning, E.C. *Biology and Potential Biogeochemical Impacts of Novel Predatory Flavobacteria*; DTIC Document; Massachusetts Institute of Technology: Cambridge, MA, USA, 2010.
45. Subramani, R.; Aalbersberg, W. Marine actinomycetes: An ongoing source of novel bioactive metabolites. *Microbiol. Res.* **2012**, *167*, 571–580. [CrossRef] [PubMed]
46. Subramani, R.; Sipkema, D. Marine rare actinomycetes: A promising source of structurally diverse and unique novel natural products. *Mar. Drugs* **2019**, *17*, 249. [CrossRef] [PubMed]
47. Rouhi, A.; Sif, J.; Ferssiwi, A.; Gillet, P.; Deutch, B. Reproduction and population dynamics of *Perinereis cultrifera* (*Polychaeta*: *Nereididae*) of the Atlantic coast, El Jadida, Morocco. *Cahiers de Biologie Marine* **2008**, *49*, 151.
48. Benchoucha, S.; Berraho, A.; Bazairi, H.; Katara, I.; Benchrifi, S.; Valavanis, V.D. Salinity and temperature as factors controlling the spawning and catch of *Parapenaeus longirostris* along the Moroccan Atlantic Ocean. In *Essential Fish Habitat Mapping in the Mediterranean*; Springer: Berlin/Heidelberg, Germany, 2008; pp. 109–123.
49. Shirling, E.T.; Gottlieb, D. Methods for characterization of *Streptomyces* species. *Int. J. Syst. Evol. Microbiol.* **1966**, *16*, 313–340. [CrossRef]
50. Ibrahimi, M.; Korichi, W.; Loqman, S.; Hafidi, M.; Ouhdouch, Y.; Lemee, L. Thermochemolysis–GC-MS as a tool for chemotaxonomy and predation monitoring of a predatory *Actinobacteria* against a multidrug resistant bacteria. *J. Anal. Appl. Pyrolysis* **2019**, *145*, 104740. [CrossRef]
51. Singh, B.N. Myxobacteria in soils and composts; their distribution, number and lytic action on bacteria. *Microbiology* **1947**, *1*, 1–10. [CrossRef]
52. Sangkhobol, V.; Skerman, V.B. *Saprospira* species—Natural predators. *Curr. Microbiol.* **1981**, *5*, 169–174. [CrossRef]
53. McBride, M.J.; Zusman, D.R. Behavioral analysis of single cells of *Myxococcus xanthus* in response to prey cells of *Escherichia coli*. *FEMS Microbiol. Lett.* **1996**, *137*, 227–231. [CrossRef]
54. Jurkevitch, E. *Predatory Prokaryotes: Biology, Ecology and Evolution*; Springer Science & Business Media: Berlin, Germany, 2006; Volume 4.
55. Markelova, N.Y. Predacious bacteria, *Bdellovibrio* with potential for biocontrol. *Int. J. Hyg. Environ. Health* **2010**, *213*, 428–431. [CrossRef]
56. Fratamico, P.M.; Cooke, P.H. Isolation of *Bdellovibrios* that prey on *Escherichia coli* O157: H7 and *Salmonella* species and application for removal of prey from stainless steel surfaces. *J. Food Saf.* **1996**, *16*, 161–173. [CrossRef]
57. Sockett, R.E.; Lambert, C. *Bdellovibrio* as therapeutic agents: A predatory renaissance? *Nat. Rev. Microbiol.* **2004**, *2*, 669–675. [CrossRef] [PubMed]
58. Schisler, D.A.; Khan, N.I.; Boehm, M.J.; Slininger, P.J. Greenhouse and field evaluation of biological control of *Fusarium* head blight on durum wheat. *Plant Dis.* **2002**, *86*, 1350–1356. [CrossRef] [PubMed]
59. Casida, L.E. Minireview: Nonobligate bacterial predation of bacteria in soil. *Microb. Ecol.* **1988**, *15*, 1–8. [CrossRef]
60. Casida, L.E. Bacterial predators of Micrococcus luteus in soil. *Appl. Environ. Microbiol.* **1980**, *39*, 1035–1041. [CrossRef]
61. Othman, B.A.; Askora, A.; Awny, N.M.; Abo-Senna, A.S.M. Characterization of virulent bacteriophages for *Streptomyces griseoflavus* isolated from soil. *Pak. J. Biotechnol.* **2008**, *5*, 109–119.
62. Hassan, R.; Shaaban, M.I.; Abdel Bar, F.M.; El-Mahdy, A.M.; Shokralla, S. Quorum sensing inhibiting activity of *Streptomyces coelicoflavus* isolated from soil. *Front. Microbiol.* **2016**, *7*, 659. [CrossRef]
63. Belghit, S.; Driche, E.H.; Bijani, C.; Zitouni, A.; Sabaou, N.; Badji, B.; Mathieu, F. Activity of 2,4-Di-tert-butylphenol produced by a strain of *Streptomyces mutabilis* isolated from a Saharan soil against *Candida albicans* and other pathogenic fungi. *J. Mycol. Médicale* **2016**, *26*, 160–169. [CrossRef]
64. Pesic, A.; Baumann, H.; Kleinschmidt, K.; Ensle, P.; Wiese, J.; Süssmuth, R.; Imhoff, J. Champacyclin, a new cyclic octapeptide from *Streptomyces* strain C42 isolated from the Baltic Sea. *Mar. Drugs* **2013**, *11*, 4834–4857. [CrossRef]
65. Okazai, T. Actinomycetes tolerant to increased NaCl concentration and their metabolites. *J. Ferment. Technol.* **1975**, *53*, 833–840.

66. Goodfellow, M.; Williams, S.T. Ecology of actinomycetes. *Annu. Rev. Microbiol.* **1983**, *37*, 189–216. [CrossRef] [PubMed]
67. Shomura, T.; Yoshida, J.; Amano, S.; Kojima, M.; Inouye, S.; Niida, T. Studies on Actinomycetales producing antibiotics only on agar culture. *J. Antibiot.* **1979**, *32*, 427–435. [CrossRef] [PubMed]
68. Wayne, P.A. *Performance Standards for Antimicrobial Susceptibility Testing*; Twenty-First Informational Supplement, CLSI Document M100-S21; Clinical and Laboratory Standards Institute: Wayne, PA, USA, 2011.
69. Hopwood, D.A. *Genetic Manipulation of Streptomyces: A Laboratory Manual*; John Innes Foundation: Norwich, UK; Cold Spring Harbour Laboratory: New York, NY, USA, 1985.

# 2

# *Corynebacterium glutamicum* CrtR and Its Orthologs in *Actinobacteria*: Conserved Function and Application as Genetically Encoded Biosensor for Detection of Geranylgeranyl Pyrophosphate

**Nadja A. Henke [1], Sophie Austermeier [1,2], Isabell L. Grothaus [1,3], Susanne Götker [1], Marcus Persicke [4], Petra Peters-Wendisch [1] and Volker F. Wendisch [1,*]**

1 Faculty of Biology & CeBiTec, Bielefeld University, 33615 Bielefeld, Germany; n.henke@uni-bielefeld.de (N.A.H.); sophie.austermeier@leibniz-hki.de (S.A.); grothaus@uni-bremen.de (I.L.G.); sgoetker@cebitec.uni-bielefeld.de (S.G.); petra.peters-wendisch@uni-bielefeld.de (P.P.-W.)

2 Department of Microbial Pathogenicity Mechanisms, Leibniz Institute for Natural Product Research and Infection Biology (HKI), 07745 Jena, Germany

3 Faculty of Production Engineering, Bremen University, 28359 Bremen, Germany

4 Faculty of CeBiTec, Bielefeld University, 33615 Bielefeld, Germany; marcusp@cebitec.uni-bielefeld.de

* Correspondence: Volker.wendisch@uni-bielefeld.de.

**Abstract:** Carotenoid biosynthesis in *Corynebacterium glutamicum* is controlled by the MarR-type regulator CrtR, which represses transcription of the promoter of the *crt* operon (P*crtE*) and of its own gene (P*crtR*). Geranylgeranyl pyrophosphate (GGPP), and to a lesser extent other isoprenoid pyrophosphates, interfere with the binding of CrtR to its target DNA in vitro, suggesting they act as inducers of carotenoid biosynthesis. CrtR homologs are encoded in the genomes of many other actinobacteria. In order to determine if and to what extent the function of CrtR, as a metabolite-dependent transcriptional repressor of carotenoid biosynthesis genes responding to GGPP, is conserved among *actinobacteria*, five CrtR orthologs were characterized in more detail. EMSA assays showed that the CrtR orthologs from *Corynebacterium callunae, Acidipropionibacterium jensenii, Paenarthrobacter nicotinovorans, Micrococcus luteus* and *Pseudarthrobacter chlorophenolicus* bound to the intergenic region between their own gene and the divergently oriented gene, and that GGPP inhibited these interactions. In turn, the CrtR protein from *C. glutamicum* bound to DNA regions upstream of the orthologous *crtR* genes that contained a 15 bp DNA sequence motif conserved between the tested bacteria. Moreover, the CrtR orthologs functioned in *C. glutamicum* in vivo at least partially, as they complemented the defects in the pigmentation and expression of a P*crtE_gfp*uv transcriptional fusion that were observed in a *crtR* deletion mutant to varying degrees. Subsequently, the utility of the P*crtE_gfp*uv transcriptional fusion and chromosomally encoded CrtR from *C. glutamicum* as genetically encoded biosensor for GGPP was studied. Combined FACS and LC-MS analysis demonstrated a correlation between the sensor fluorescent signal and the intracellular GGPP concentration, and allowed us to monitor intracellular GGPP concentrations during growth and differentiate between strains engineered to accumulate GGPP at different concentrations.

**Keywords:** *C. glutamicum*; regulation of carotenogenesis; GGPP; biosensor

---

## 1. Introduction

Microbial single-cell biosensors have become valuable tools in metabolic engineering due to their easy detection through a fluorescent output signal, their single-cell resolution and their compatibility with viable cells [1,2]. Biosensors facilitate the screening or selection process for a desired product by sensing the presence of the inconspicuous molecule coupled to a conspicuous reporter [2]. These reporters are often based on transcription regulators or riboswitches, as these molecules undergo a conformational change triggered by binding of the analyte, which is directly linked to transcriptional control of the reporter gene [1,2]. Natural biosensors are rare because of the lack of known regulators that are specific for the detection of a desired metabolite. Therefore, intense efforts have been undertaken in biosensor identification and characterization in order to monitor intracellular concentrations of industrially relevant compounds. Rational strain engineering for industrial purposes is often limited by the high complexity of metabolic networks. On the other side, classical strain development based on random mutagenesis is typically limited by the screening capacity, and often lacks an easy to manage readout system to judge the performance of the generated mutants [3]. Biosensors allow high-throughput screening for the monitoring of the production performance, at least for products accumulating intracellularly, while production performance for secreted products can only be deduced indirectly [1,4,5]. Biosensors have been shown to augment and accelerate metabolic engineering based on a new build-test-learn cycle [5].

Isoprenoid pyrophosphates such as GGPP are typically present in low concentrations in the cell. Therefore, an effective biosensor is sought for terpenoid process/strain optimization. Isoprenoid pyrophosphates are building blocks for the synthesis of many high-value terpenoids, including the carotenoid astaxanthin or the sesquiterpenoid patchoulol [6–8]. These important secondary metabolites find various applications in the food, feed and cosmetic industries, either as additives or as high-performance ingredients in the health industry [9,10]. Chemical synthesis as well as isolation from natural sources is expensive and/or results in insufficient amounts, and thus the microbial production of terpenoids is receiving increasing attention [11,12]. As an example, the high-value astaxanthin is produced with the microalgae *Haematococcus pluvialis* [13], the red yeast *Xanthophyllomyces dendrorhous* [14] and the bacterium *Paracoccus carotinifaciens* [15]. Microbial carotenoid production by engineered strains is on the rise, as much higher production titers, for example of 6.5 g/L β-carotene with optimized *Yarrowia lipolytica*, can be achieved [16]; however, the industrial production of carotenoids by engineered organisms is currently rare. Besides the intensive engineering of the precursor supply and the optimization of terminal carotenoid biosynthesis, central carbon fluxes as well as cofactor-regeneration might be promising targets [16] in order to achieve carotenoid production titers of >10 g/L. There are two engineered biosensors that exist pertaining to the detection of mevalonate, an intermediate of the mevalonate pathway of isopentenyl pyrophosphate (IPP) biosynthesis [17,18], which cannot be used to monitor IPP biosynthesis via the MEP pathway. Direct IPP sensing has been achieved by a synthetic fusion of the IPP-binding isopentenyl pyrophosphate:dimethylallyl pyrophosphate isomerase Idi with the DNA-binding domain of AraC [19]. This fluorescence reporter responded to the extracellular addition of mevalonate to an *E. coli* strain equipped with the mevalonate pathway; however, no direct evidence was observed that intracellular IPP was sensed by the synthetic biosensor [19]. Such orthogonal biosensors are supposed to interact with the endogenous cellular network less commonly, which might be favorable for scoring production [19], but might be a disadvantage when native regulatory mechanisms are examined.

Here, a biosensor based on the metabolite-dependent MarR-type transcriptional repressor CrtR from *C. glutamicum* is described. This soil bacterium with GRAS status has been used safely in industrial amino acids production for over 60 years [20] since its discovery as a natural glutamate producer [21]. *C. glutamicum* has been metabolically engineered for the sustainable production of various, mostly nitrogenous, compounds [22]. Notably, *C. glutamicum* is a natural carotenoid producer, and its yellow pigmentation is due to the unusual C50 carotenoid decaprenoxanthin [23,24]. The carotenoid precursor GGPP is synthesized from IPP and DMAPP (dimethylallyl pyrophosphate), which are generated from pyruvate and GAP in the MEP pathway [23–25], primarily by the prenyltransferase IdsA [26]. The genes that are necessary for the conversion of GGPP to the final yellow decaprenoxanthin are organized in a single operon (*crtE, cg0722, crtB, crtI,* $crtY_e$, $crtY_f$, *crtEb*) transcribed from P*crtE* [24]. The knowledge about carotenogenesis in *C. glutamicum* has guided metabolic engineering in such a way as to enhance production of the native decaprenoxanthin [23], and to enable the production of nonnative C40 and C50 carotenoids [27–29], including the industrially relevant astaxanthin [30,31]. Several metabolic engineering approaches have been used to improve the production of terpenoids by *C. glutamicum*. First, *dxs* encoding the committed enzyme in the MEP pathway was overexpressed [32,33]. Second, balancing the DMAPP to IPP ratio via the overexpression of *idi* improved patchoulol production when combined with *dxs* overexpression [8]. Third, the overproduction of the two endogenous GGPP synthases IdsA and CrtE enhanced decaprenoxanthin production due to increased synthesis of GGPP [26]. The overexpression of these genes from IPTG-inducible promoters was orthogonal and independent from endogenous transcription regulatory feedback. Recently, a membrane-fusion protein comprising CrtZ and CrtW was published, and it was shown that the additional overexpression of precursor biosynthesis genes enhanced astaxanthin product formation [31]. In this regard, precursor-dependent transcriptional regulation may be beneficial, for example in the on-demand conversion of GGPP to the chosen target terpenoid. A biosensor system for the detection of GGPP would represent a powerful tool for strain development, in particular with regard to investigations into the MEP pathway concerning efficient precursor supply. *C. glutamicum* possesses the transcriptional repressor CrtR for the control of decaprenoxanthin biosynthesis [23]. Like most MarR-type regulators, *C. glutamicum* CrtR represses gene transcription by binding to the intergenic region between its own gene and the divergently oriented *crt* operon [23,34]. In vitro analysis showed that isoprenoid pyrophosphates act as inducers of CrtR in *C. glutamicum*. CrtR binding to P*crtE* was inhibited by GGPP, and to lesser extents by FPP, GPP, DMAPP and IPP [23]. Thus, in *C. glutamicum*, GGPP leads to the derepression of the *crt* operon and its own gene by CrtR in a metabolite-dependent feed forward mechanism [23].

CrtR has also been associated with the light-dependent regulation of carotenogenesis, although the mechanism remains unclear [35]. CrtR homologs are found mainly in actinobacteria, and *crtR* genes often cluster with carotenoid biosynthetic genes and/or a *mmpL*-like transporter gene [23]. This suggested a conserved regulatory function of the CrtR orthologs with respect to transcriptional control of carotenoid biosynthetic genes and/or a *mmpL*-like transporter gene in other actinobacteria. MmpL (mycobacterial membrane protein large) proteins often export hydrophobic or lipid-like substances across the cell membrane in mycobacteria [36]. Here, we studied if, and to what extent, the function of CrtR as a metabolite-dependent transcriptional repressor of carotenoid biosynthesis genes responding to GGPP is conserved among actinobacteria. Moreover, we developed the first genetically encoded biosensor system for the detection of intracellular GGPP based on CrtR from *C. glutamicum*.

## 2. Results

### 2.1. CrtR Orthologs from Actinobacteria Showed Binding to Their Own Promoters and Derepression by GGPP

Previously, the MarR-type regulator CrtR was identified as a GGPP-dependent repressor of the carotenogenic gene cluster *crtE-mmpl-crtBIY$_{e/f}$Eb*, and was shown to auto-regulate its own expression [23]. Moreover, 94 CrtR homologs with at least a 25% amino acid identity with CrtR from *C. glutamicum* were identified [23]. In order to study if the function of these MarR-type regulators as GGPP-dependent transcriptional repressors of carotenogenesis is conserved, the orthologs of five actinobacteria, with increasing phylogenetic distances between themselves and *C. glutamicum*, were selected for further analysis. The *crtR* genes from *C. callunae*, *A. jensenii* and *P. nicotinovorans*, as well as *C. glutamicum*, have in common that they are co-localized with genes of carotenoid biosynthesis [23] (Figure 1). The phylogenetically more distant *crtR* genes from *P. chlorophenolicus* and *M. luteus* co-localize only with a *mmpL* gene. The genome of *P. chlorophenolicus* lacks carotenogenesis genes except *crtR*, whereas the carotenogenic genes of *M. luteus* are encoded in loci distant from *crtR* (Figure 1).

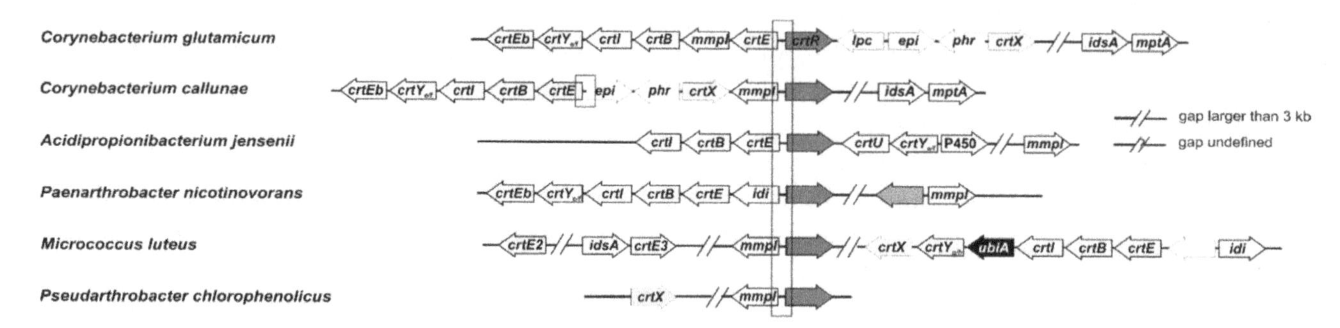


**Figure 1.** Genomic organization of *crtR* from *C. glutamicum* ATCC 13,032 and the *crtR* orthologs from *C. callunae* DSM 20,147, *A. jensenii* DSM 20,535, *P. nicotinovorans* Hce-1, *M. luteus* NCTC 2665 and *P. chlorophenolicus* A6. Boxed areas highlight the putative promoter regions tested in bandshift assays. The *crtR* orthologs (given in red) are transcribed divergently either to *crt* genes (given in yellow) or to *mmpL* genes. *P. nicotinovorans* contains in addition a *crtR* paralog (given in pink), which is transcribed divergently to gene *mmpL*. Other carotenoid associated genes (given in white) and genes for prenylation (given in black) in the close vicinity of *crtR* genes are included.

The CrtR orthologs from *C. callunae* ($CrtR_{Cc}$), *A. jensenii* ($CrtR_{Aj}$), *P. nicotinovorans* ($CrtR_{Pn}$), *P. chlorophenolicus* ($CrtR_{Pc}$) and *M. luteus* ($CrtR_{Ml}$) were fused with an N-terminal His-tag, and the proteins were purified by Ni-NTA affinity chromatography. In order to test if *crtR* autoregulation—as observed for CrtR from *C. glutamicum*—is conserved, each protein was tested for binding to the intergenic DNA sequences between its own gene and the divergently transcribed gene (Figure 2). Indeed, all CrtR orthologs analyzed bound to the DNA sequences upstream of their own *crtR* gene (Figure 2). This indicated *crtR* autoregulation and/or regulation of the respective divergently transcribed genes (*mmpl* for *C. callunae, M. luteus* and *P. chlorophenolicus*, *crtE* for *A. jensenii* and *idi* for *P. nicotinovorans*) (Figure 1). The finding that *A. jensenii* CrtR, *P. nicotinovorans* CrtR and *M. luteus* CrtR did not shift all target DNA may either indicate that the CrtR protein–target DNA interaction is less tight in these bacteria, or it may be due to technical reasons, e.g., due to the purification of the tagged proteins that may differ between the five CrtR proteins analyzed.

Since GGPP inhibits the binding of CrtR from *C. glutamicum* to its target promoter [23], it was tested if GGPP could also inhibit the binding of the other CrtR orthologs to their respective target DNA. Indeed, GGPP inhibited the interaction between the tested CrtR proteins and their target DNA (Figure 2A).

Thus, these in vitro results revealed that the binding of CrtR orthologs, from several actinobacteria, to their own upstream DNA sequences, and the inhibitory effect of GGPP, are conserved.

For *C. callunae*, a second putative target DNA sequence was tested, namely the intergenic region between its carotenogenic gene cluster *crtEBIY$_{e/f}$Eb*, which is located a few genes upstream of *crtR* in this bacterium, and the divergently transcribed gene *epi* (Figure 1). This intergenic DNA sequence from *C. callunae* was bound by a CrtR protein from *C. callunae* ($CrtR_{Cc}$) unless GGPP was added (Figure 2B). Thus, the GGPP-dependent regulation of carotenoid biosynthesis genes by CrtR is conserved at least in the closely related *C. callunae*, and possibly in other actinobacteria.

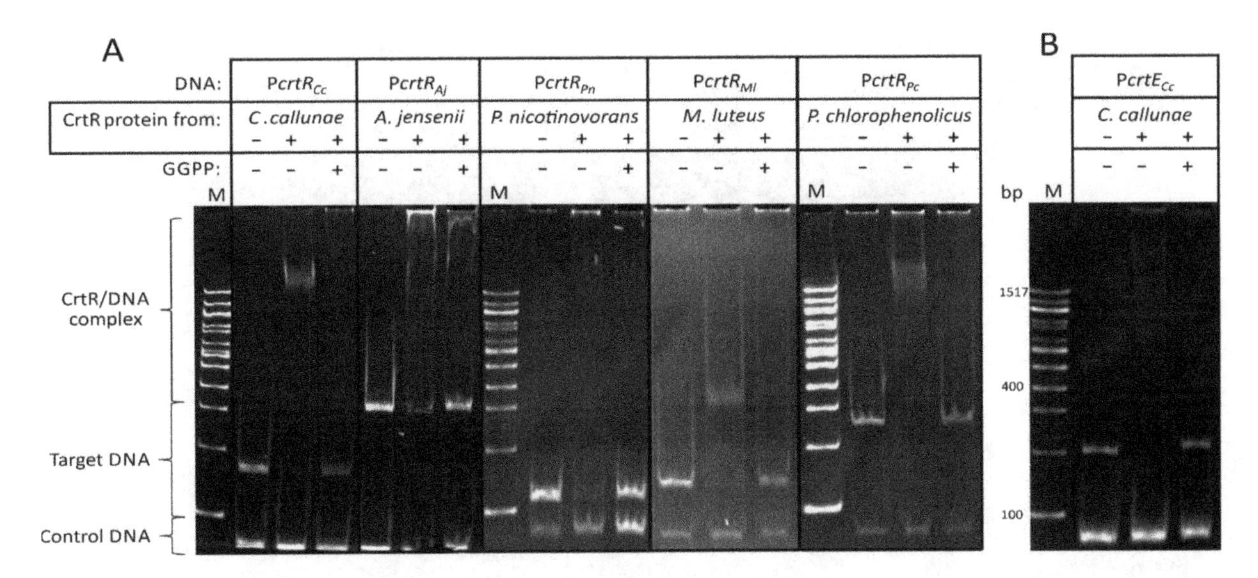


**Figure 2.** In vitro characterization of CrtR orthologs. (**A**) Bandshift assays of CrtR orthologs from *C. callunae, A. jensenii, P. nicotinovorans, M. luteus* and *P. chlorophenolicus* with their respective own putative promoter region and the inhibition of the binding by GGPP. (**B**) Bandshift assays of CrtR from *C. callunae* with a putative *crtE* promoter region and the inhibition of the binding by GGPP. The presense or absence of CrtR protein and GGPP are indicated by "+" and "−", respectively.

## 2.2. *CrtR from C. glutamicum Binds to Heterologous crtR Promoter DNA Sequences*

The binding of the *C. glutamicum* CrtR protein to heterologous *crtR* promoter DNA sequences was studied in order to (i) test if this specific DNA binding is conserved across the actinobacteria analyzed, and, if it is, to (ii) identify the putative DNA sequence motif. The intergenic DNA sequences between the five orthologous *crtR* genes and the respective divergently transcribed genes were used in a bandshift assay with His-tagged CrtR proteins from *C. glutamicum* (Figure 3A). The strong binding of $CrtR_{Cg}$ to the *crtR* promoter sequences from *C. callunae* and *P. nicotinovorans* was detected, whereas the interactions between $CrtR_{Cg}$ and the *crtR* promoter sequences from *A. jensenii, M. luteus* and *P. chlorophenolicus* were weak (Figure 3A).

Previously, we narrowed down the target DNA sequence, to which CrtR from *C. glutamicum* binds, to 19 bp (5′-CCCATGAG**AATT**TATTTTT-3′), and mutational analysis revealed that exchanging the central four nucleotides (**TTAA**) simultaneously interfered with binding [23]. Inspection of the DNA sequences upstream of the *crtR* genes from *C. callunae, A. jensenii, P. nicotinovorans, M. luteus* and *P. chlorophenolicus* revealed that this motif was present in the intergenic DNA regions in these species (Figure 3B), and conserved to some extent (Figure 3C). It is evident that conservation of the central **TTAA** sequence is not sufficient to explain the observed binding preferences of $CrtR_{Cg}$ and the variuos *crtR* promoter sequences studied. It remains to be elucidated how specific nucleotides of the 15 bp motif (other than the central **TTAA**) affect the binding of $CrtR_{Cg}$ protein to DNA.

The derived consensus DNA binding motif of CrtR proteins from the studied species is depicted in Figure 3D, with sequence conservation and relative frequency for each nucleotide position.

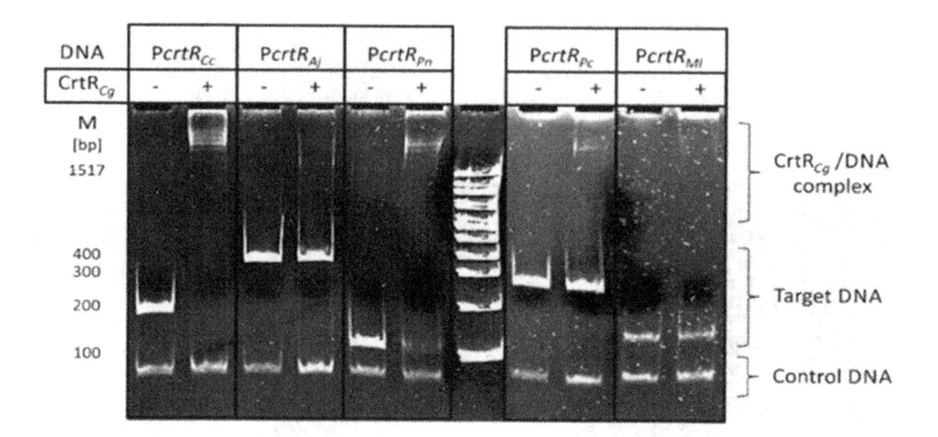


B

***Corynebacterium glutamicum* – PcrtR**

CATTCGAGTATCACACGGCCAGTTATCTCGCAAAAATTCCCAATCGTTGTATATGGCGCTTTATTTTGATGAAGTACAGAAAGTGTGAATTTGGGTCCATAAAAA
TAATGTGCCTA**C**AAGAAATTTATAGTATCCCATGAGTTAATATTTTTAAAAATAAACTTTATCTGACTTTGTAGAAAAAGGTGATTACTATGCTGAAT**A**TG

***Corynebacterium callunae* - PcrtR**

CGCAAGATGCCGGGAAATACGCTCTGAAGAGGGGTTTTCATGGCCTTAAACCTTTATATATGCAAATTGTTAAATTATCCAACAGTGTAAGGTATCCCATTTTCT
TAAAAAAAGGGTTAATATCCATCTATGTCCGATCCGCAA

***Corynebacterium callunae* - PcrtE**

TTGATCCTTAAGATGAGGGAAAACCTGGAAGGTCAGCAGTCAATTCACTTTGGGTAAATTTTGAAGACGCAGTGTTATCTTATGGTGTAGTATATCTGATGTGAT
AGTAGACGGATGTTGTTTTCCAAAATGATGGAGAGAAAGATTTGAGTCTTGGAACTCCACAGGAGACGCCGAAACGATG

***Paenarthrobacter nicotinovorans* - PcrtR**

CATACTGCGCCTTCCTCCTGCAACCTGAGTCTTTACAGCAGCCCGAATATTGCTAGACAAGCTAGAAGAAAGGCTACCATGTAGGAAGTTAGGTGTAACATACTC
AGCATG

***Pseudarthrobacter chlorophenolicus* - PcrtR**

TGCCTTCCATGCGGATGGTCCACGCGGGACAAAGCGGAGTAGGATTGGTTCCACATTAGACTATCTCCATGATAGAGATAAGTGCCAGTCAAGCTTTTTCAACGG
GGGCCTCCGCCGGGGGCTTCCACGGCTGAGCCAGGTAATGCCAAGGCAGCTGAAGTAATATCCCTCAGGGCCTGTGCGCAAAGACGGACGGCCCCACGAACTCCA
GGAGGTGACCATGAACGGCAACAATCCGGGCA

***Acidipropionibacterium jensenii* - PcrtR**

TGCAGATGGGCGCTGTCATCGCTGCCCCGCAGTGCGTGGGAGTGACGACTGGGCAGCTGTGAGACTGTCGTTCATCAGTGATCCCTCCAGAAGTAAGTGGCTTCC
TTGAGACTTTAATAGCCAGGCTAGCAATATGTAAAGGATGTATCTGACTGTGCTGATAATCTTTCCCCGTGAGTGAAGACCGCGATGCGCAGACGAGTCGCCTCA
GTCTCTACGACGTCGAGTCAAGTGACCTGGTGGACCGGTCCTCGCTGTCGGACCCCGACATCGCCCAGATCAACACCCTGATGGGCGCCTTCGCGCGGT

***Micrococcus luteus* - PcrtR**

GGGTGCGGAAGGCGGACAGGTCGGGCGGGGTGGGCATGCCCCCGAGCGTAGGGGATGGTTCGATGATCAAACAACCATGGGGCTGGCTAGACTCCCGGCCATGAC
CACGCAGCCCCCCTCCGCC

C

| | |
|---|---|
| ***C. glutamicum* PcrtR:** | TGAG**TTAA**TATTTTT |
| ***C. callunae* PcrtR:** | AGGG**TTAA**TATCCAT |
| ***P. nicotinovorans* PcrtR:** | AGGTG**TAA**CATACTC |
| ***P. chlorophenolicus* PcrtR:** | TGAAG**TAA**TATCCCT |
| ***A. jensenii* PcrtR:** | TGTGC**T**G**A**TAATCTT |
| ***M. luteus* PcrtR:** | TGATCA**AA**CAACCAT |
| ***C. callunae* PcrtE:** | TGGTG**TA**GTATATCT |

D

**Figure 3.** Characterization of CrtR from *C. glutamicum* in vitro. (**A**) Bandshift assays of His-tagged CrtR protein from *C. glutamicum* (CrtR$_{Cg}$) and the intergenic DNA sequences between the *crtR* orthologs from *C. callunae, A. jensenii, P. nicotinovorans, M. luteus* and *P. chlorophenolicus,* and the respective divergently transcribed genes. (**B**) Putative -10 and -35 promoter DNA sequences (underlined), translation start codons (italics) and the putative conserved CrtR binding sequences (boxed). The mapped transcriptional start sites of *C. glutamicum crtR* and *crtE* are given in bold. (**C**) Putative conserved CrtR binding sequences (conserved nucleotides are given in yellow; the **TTAA** sequence that was shown previously to be required for *C. glutamicum* CrtR binding by mutational analysis is depicted in bold face). (**D**) The graphical representation of the derived consensus DNA binding motif of the CrtR proteins from *C. glutamicum, C. callunae, A. jensenii, P. nicotinovorans, M. luteus* and *P. chlorophenolicus* (designed using WebLogo).

### 2.3. CrtR Orthologs from Actinobacteria Affected Carotenogenesis and Expression of a CrtE Transcriptional Fusion in C. glutamicum In Vivo

To monitor the promoter activity of the carotenogenic gene cluster of *C. glutamicum* in vivo, the promoter probe vector pEPR1 was used [37]. This reporter system was employed to determine the promoter activity of the carotenogenic promoter (P*crtE*) from *C. glutamicum* in the absence of endogenous chromosomally encoded CrtR, but in the presence of CrtR orthologs from other actinobacteria. The CrtR orthologs were at different phylogenetic distances compared to CrtR from *C. glutamicum*, and therefore different protein identities: *C. callunae* (62% identity), *A. jensenii* (57% identity), *P. nicotinovorans* (53% identity), *M. luteus* (35% identity) and *P. chlorophenolicus* (35% identity). To this end, the *crtR* orthologs from *C. callunae*, *A. jensenii*, *P. nicotinovorans*, *M. luteus* and *P. chlorophenolicus*, as well as the *crtR* from *C. glutamicum* as reference, were expressed from the strong, constitutive promoter P*gap* in divergent orientation to the P*crtE*_*gfp*uv transcriptional fusion. The respective vectors were named pTEST_CrtR$_{Cg}$, pTEST_CrtR$_{Cc}$, pTEST_CrtR$_{Aj}$, pTEST_CrtR$_{Pn}$, pTEST_CrtR$_{Ml}$ and pTEST_CrtR$_{Pc}$ (Figure S1). First, maximal repression using the vector pTEST-*crtR*$_{Cg}$ was determined and compared to a two-vector system, in which the expression of *crtR* and the reporter gene fusion P*crtE*_*gfp*uv were decoupled; *crtR* was expressed by IPTG-inducible pEC-XT_*crtR*$_{Cg}$, while pEPR1_P*crtE* contained the reporter gene fusion P*crtE*_*gfp*uv (Figure S1). In the strains carrying the P*crtE*_*gfp*uv fusion, but lacking *crtR* (WTΔ*crtR*(pTEST) and WTΔ*crtR*(pEPR1_P*crtE*), pigmentation due to decaprenoxanthin accumulation (3.2–3.3 mg/g CDW) and GFPuv fluorescence (1.3 normalized MFI) were high (Figure S2). The IPTG-inducible expression of *crtR* in the two-vector system reduced decaprenoxanthin accumulation and GFPuv fluorescence 38- and 32-fold, respectively (Figure S2). Reduction of decaprenoxanthin accumulation and GFPuv fluorescence using pTEST_*crtR*$_{Cg}$ (200- and 55-fold, respectively; Figure S2) was even higher (*crtR*$_{Cg}$ is transcribed from the strong and constitutive P*gap*, shown above). Next, the in vivo effects of the CrtR orthologs were tested. Constitutive overexpression of the *crtR* orthologs from the *gap* promoter revealed differential effects on carotenogenesis and expression of the P*crtE*_*gfp*uv transcriptional fusion (Figure 4). WTΔ*crtR* carrying the empty vector pTEST showed the expected intense yellow pigmentation due to derepression of the chromosomal carotenoid biosynthesis genes (Figure 4A), as well as the derepressed expression of the P*crtE*_*gfp*uv transcriptional fusion (Figure 4B). Upon plasmid-borne expression of CrtR repressor genes from *C. glutamicum*, *C. callunae*, *A. jensenii* and *P. nicotinovorans*, pigmentation was strongly reduced to less than 1 mg/g CDW (Figure 4A), which corresponded to the strongly reduced expression of the P*crtE*_*gfp*uv transcriptional fusion (Figure 4B). Repression by CrtR from *C. callunae* was nearly as tight as that by endogenous CrtR, leading to a relative GFPuv signal of less than 0.1 (Figure 4B). CrtR orthologs from *A. jensenii* and *P. nicotinorovans* also repressed the *crtE* promoter fusion very efficiently, resulting in a GFPuv signal of less than 0.2 (Figure 4B). Upon expression of *M. luteus crtR*, carotenoid biosynthesis was reduced to a much lesser extent (Figure 4A), while the expression of the P*crtE*_*gfp*uv transcriptional fusion was as high as in the empty vector control (Figure 4B). CrtR from *P. chlorophenolicus* reduced pigmentation, but expression of the P*crtE*_*gfp*uv transcriptional fusion was unaffected (Figure 4A,B).

Taken together, the repression of carotenogenesis, and of the expression of *crtE* transcriptional fusion, in *C. glutamicum* in vivo by CrtR orthologs from other actinobacteria was possible. The efficacy was highest for closely related species, and it decreased with phylogenetic distance.

### 2.4. Construction and Analysis of a GGPP Biosensor

The expression control of P*crtE*_*gfp*uv transcriptional fusion by CrtR from *C. glutamicum* has been described previously [19], and the results from above prompted us to consider its use in combination with chromosomally encoded *crtR* as a GGPP biosensor, i.e., using *C. glutamicum* WT(pEPR1_P*crtE)* as a biosensor strain. Targeted metabolic engineering typically addresses four modules: carotenoid biosynthesis, precursor supply, central carbon metabolism and redox cofactor regeneration. A CrtR-based biosensor may allow us to simultaneously optimize the latter three modules via fluorescence reporter output. While GGPP inhibits the DNA binding of CrtR [23], no feeding regimen is known to predictably alter the intracellular GGPP concentration. Therefore, a genetic approach was chosen, and two *C. glutamicum* strains, expected to accumulate GGPP to intracellular concentrations different from the wild type, were constructed (Figure 5).

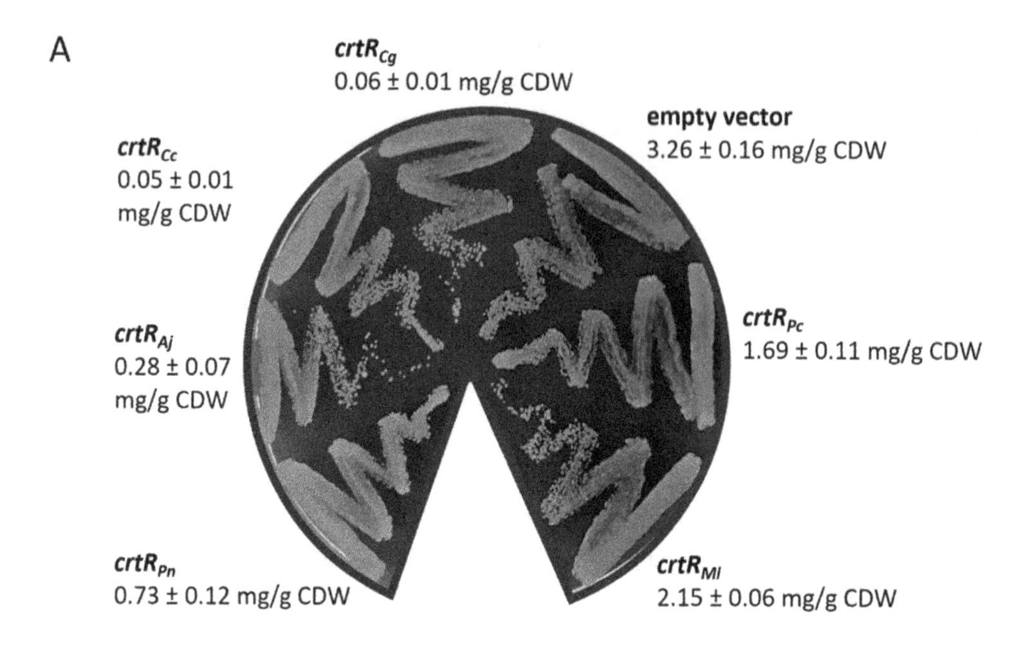


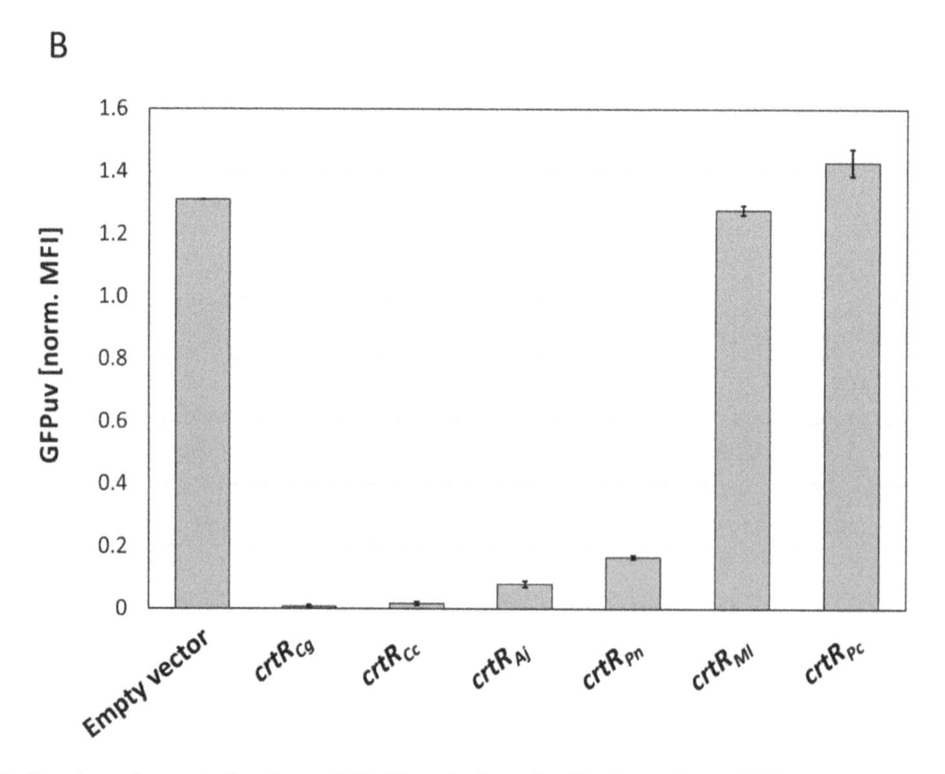


**Figure 4.** In vivo characterization of CrtR orthologs in *C. glutamicum* WTΔ*crtR*. (**A**) Phenotypes on LB plates after incubation at 30 °C for 24 h and carotenoid concentration in mg/g CDW (β-carotene equivalents) of WTΔ*crtR* strains harboring pTEST derivatives expressing *crtR* genes from the indicated bacteria. (**B**) Flow cytometry analysis of the strains depicted in (A) during exponential growth in LB. Mean fluorescence intensities (MFI) of GFPuv signals were normalized to autofluorescence and shown for at least two biological replicates.

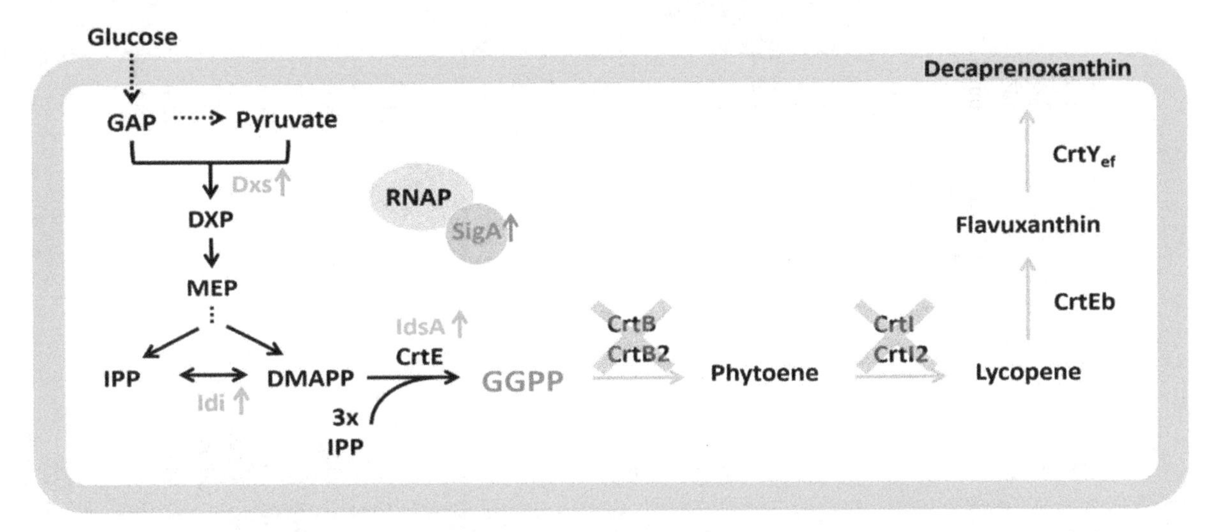


**Figure 5.** GGPP and decaprenoxanthin biosynthesis pathway in *C. glutamicum*. GAP: glyceraldehyde 3-phosphate, DXP: 1-deoxy-1-xylulose-5-phosphate; MEP: methylerythritol phosphate; IPP: isopentenyl pyrophosphate; DMAPP: dimethylallyl pyrophosphate; GGPP: geranylgeranyl pyrophosphate; RNAP: RNA-Polymerase core enzyme; SigA: housekeeping primary sigma factor A; Dxs: 1-deoxy-1-xylulose5-phosphate synthase; Idi: isopentenyl pyrophosphate isomerase; IdsA/CrtE: GGPP synthase; CrtB/CrtB2: phytoene synthase; CrtI/I2: phytoene desaturase; CrtEb: lycopene elongase; CrtY$_{ef}$: C50 ε-cyclase; genes overexpressed in strains GGPPA and GGPPB are shown in green, genes deleted on both strains are indicated by red crosses; blue shows genes overexpressed only in strain GGPPB.

First, the conversion of GGPP by the endogenous phytoene synthases CrtB [24] was prevented through the deletion of *crtB* and *crtB2I'I2* from the chromosome of the *C. glutamicum* WT yielding strain WTΔ*crtB*Δ*crtB2I'I2* (Table 1). Second, the supply of the precursor molecules DMAPP and IPP was increased by overexpression of the genes encoding the MEP pathway enzymes 1-deoxy-D-xylulose 5-phosphate synthase (Dxs) and Idi. It is known for *C. glutamicum* [28] and other organisms that the first enzymatic step in the MEP pathway strongly limits the flux [38–40]. Dxs is supposed to be feedback-regulated by isoprenoid pyrophosphates [38], and the overexpression of *dxs* was shown to increase the flux towards terpenoid biosynthesis [28,30,33]. In addition, the ratio of DMAPP to IPP was shown to be important for optimized isoprenoid production, and *idi* overexpression equilibrates intracellular concentrations of DMAPP and IPP [28,33,39]. Third, GGPP synthesis was improved by plasmid-driven overexpression of the major GGPP synthase gene *idsA* from *C. glutamicum* [26] (pEC-XT_*idsA*; Table 1). Combining the three strategies resulted in strain GGPPA (WTΔ*crtB*Δ*crtB2I'I2* (pEKEx3-*dxs_idi*) (pECXT-*idsA*)) (Table 1) (Figure 5). Since engineering of the RNA polymerase sigma factor A improved isoprenoid carotenoid production [41], *sigA* from *C. glutamicum* was overexpressed in a synthetic operon with *idsA* (pEC-XT_*idsA_sigA*) (Table 1). The resulting strain GGPPB (WTΔ*crtB*Δ*crtB2I'I2* (pEKEx3-*dxs_idi*) (pECXT-*idsA_sigA*)) differs from strain GGPPA only by the additional overexpression of *sigA* (Table 1) (Figure 5). The intracellular GGPP concentrations of the *C. glutamicum* strains WT, GGPPA and GGPPB were expected to differ.

After transformation with the biosensor plasmid pEPR1_P*crtE*, the strains were grown in CGXII minimal medium with glucose. Samples were taken during exponential growth 12 h after inoculation and analyzed by LC-MS (Figure 6). As expected, WT (pEPR1_P*crtE*) accumulated the lowest GGPP concentration, with less than 0.1 mM GGPP (Figure 6). In comparison, GGPPA (pEPR1_P*crtE*) accumulated about 23-fold more GGPP (2.3 ± 0.5 mM; Figure 6). With 4.0 ± 0.4 mM, strain GGPPB (pEPR1_P*crtE*) exhibited the highest concentration of GGPP, i.e., about 1.7-fold higher than GGPPA (pEPR1_P*crtE*). Thus, it was confirmed that the genetic approach altered the intracellular GGPP concentrations as anticipated.

**Table 1.** Strains, genomic DNA and plasmids used in this study.

| Strain, gDNA or Plasmid | Relevant Characteristics or Sequence | Reference |
|---|---|---|
| | *E. coli* strains | |
| *E.coli* DH5α | F-*thi*-1 *endA1 hsdr17*(r-, m-) *supE44* Δ*lacU169* (Φ80*lacZ*ΔM15) *recA1 gyrA96* | [42] |
| S17-1 | *recA pro hsdR* RP4-2-Tc::Mu-Km::Tn7 integrated into the chromosome | [43] |
| *E.coli* BL21 (DE3) | *F– ompT gal dcm lon hsdSB(rB–mB–) λ(DE3 [lacI lacUV5-T7p07 ind1 sam7 nin5]) [malB+]K-12(λS)* | [44] |
| *E.coli* BL21 (DE3) (pLysS) | *F– ompT gal dcm lon hsdSB(rB–mB–)λ(DE3 [lacI lacUV5-T7p07 ind1 sam7 nin5]) [malB+]K-12(λS) pLysS[T7p20 orip15A](CmR)* | Promega |
| | ***C. glutamicum strains*** | |
| *C. glutamicum* WT | ATCC 13032, wild type | [45] |
| WTΔ*crtR* | ATCC 13,032 with deletion of *crtR* (cg0725) | [23] |
| WTΔ*crtB*Δ*crtB2I'I2* | ATCC 13,032 with deletion of *crtB* (cg0721) and *crtB2I'I2* (OP_cg2672) | this work |
| GGPPA | WTΔ*crtB*Δ*crtB2I'I2* derivative with plasmid-driven IPTG-inducible expression of MEP pathway genes *dxs* (cg2083) and *idi* (cg2531) from pEKEx3 and the GGPP synthase gene *idsA* (cg2384) from pEC-XT. | this work |
| GGPPB | WTΔ*crtB*Δ*crtB2I'I2* derivative with plasmid-driven IPTG-inducible expression of MEP pathway genes *dxs* (cg2083) and *idi* (cg2531) from pEKEx3 and the GGPP synthase gene *idsA* (2384) and primary sigma factor gene *sigA* (cg2092) from pEC-XT. | this work |
| | **Genomic DNA** | |
| *Acidipropionibacterium jensenii* | Wild type, DSM 20535, ATCC 4868 | [46], DSMZ |
| *Corynebacterium callunae* | Wild type, DSM 20147, ATCC 15991 | [47] |
| *Micrococcus luteus* | Wild type, DSM 20030, ATCC 4698 | [48], DSMZ |
| *Paenarthrobacter nicotinovorans* | Wild type, DSM 420, ATCC 49919 | [49], DSMZ |
| *Pseudarthrobacter chlorophenolicus* | Wild type, DSM 12829, ATCC 700700 | [50], DSMZ |
| | **Plasmids** | |
| pEPR1 | $Km^R$, pCG1 $oriV_{CG}$, *gfp*uv, promoterless, *C. glutamicum/E.coli* shuttle promoter-probe vector | [37] |
| pEPR1_P*crtE* | pEPR1 derivate containing the promoter of *crtE* (P*crtE*) | [23] |
| pTEST | pEPR1_P*crtE* derivate containing an additional expression cassette for expression of *crtR* orthologs from the *gap* promoter | this work |
| pTEST_$crtR_{Cg}$ | pTEST derivate for expression of the *crtR* from *C. glutamicum* | this work |
| pTEST_$crtR_{Cc}$ | pTEST derivate for heterologous expression of the *crtR* orthologs from *C. callunae* | this work |
| pTEST_$crtR_{Aj}$ | pTEST derivate for heterologous expression of the *crtR* ortholog from *A. jensenii* | this work |
| pTEST_$crtR_{Pn}$ | pTEST derivate for heterologous expression of the *crtR* ortholog from *P. nicotinovorans* | this work |
| pTEST_$crtR_{Pc}$ | pTEST derivate for heterologous expression of the *crtR* ortholog from *P. chlorophenolicus* | this work |
| pTEST_$crtR_{Ml}$ | pTEST derivate for heterologous expression of the *crtR* ortholog from *M. luteus* | this work |
| pET16b | Expression plasmid for production of His-tagged proteins | Novagen |
| pET16b_$crtR_{Cg}$ | pET16b derivate for production of His-tagged CrtR from *C. glutamicum* | this work |
| pET16b_$crtR_{Cc}$ | pET16b derivate for production of His-tagged CrtR *C. callunae* | this work |
| pET16b_$crtR_{Aj}$ | pET16b derivate for production of His-tagged CrtR *A. jensenii* | this work |

**Table 1.** *Cont.*

| Strain, gDNA or Plasmid | Relevant Characteristics or Sequence | Reference |
|---|---|---|
| | **Plasmids** | |
| pET16b_*crtR*$_{Pn}$ | pET16b derivate for expression of the *crtR* from *P. nicotinovorans* | this work |
| pET16b_*crtR*$_{Pc}$ | pET16b derivate for production of His-tagged CrtR *P. chlorophenolicus* | this work |
| pET16b_*crtR*$_{Ml}$ | pET16b derivate for production of His-tagged CrtR *M. luteus* | this work |
| pEKEx3 | Spec$^R$, P*tac lacI*$^q$, pBL1 *oriVCg*, *C. glutamicum/E. coli* expression shuttle vector | [51] |
| pEKEx3_*dxs_idi* | pEKEx3 derivate for IPTG-inducible expression of *dxs* and *idi* from *C. glutamicum* containing an artificial ribosome binding site | this work |
| pEC-XT99A | Tet$^R$, P*trc lacI*$^q$, pGA1 *oriVCg*, *C. glutamicum/E. coli* expression shuttle vector | [52] |
| pEC-XT_*idsA* | pEC-XT99A derivate for IPTG-inducible expression of *idsA* from *C. glutamicum* containing an artificial ribosome binding site | this work |
| pEC-XT_*idsA_sigA* | pEC-XT99A derivate for IPTG-inducible expression of *idsA* and *sigA* from *C. glutamicum* containing an artificial ribosome binding site | this work |

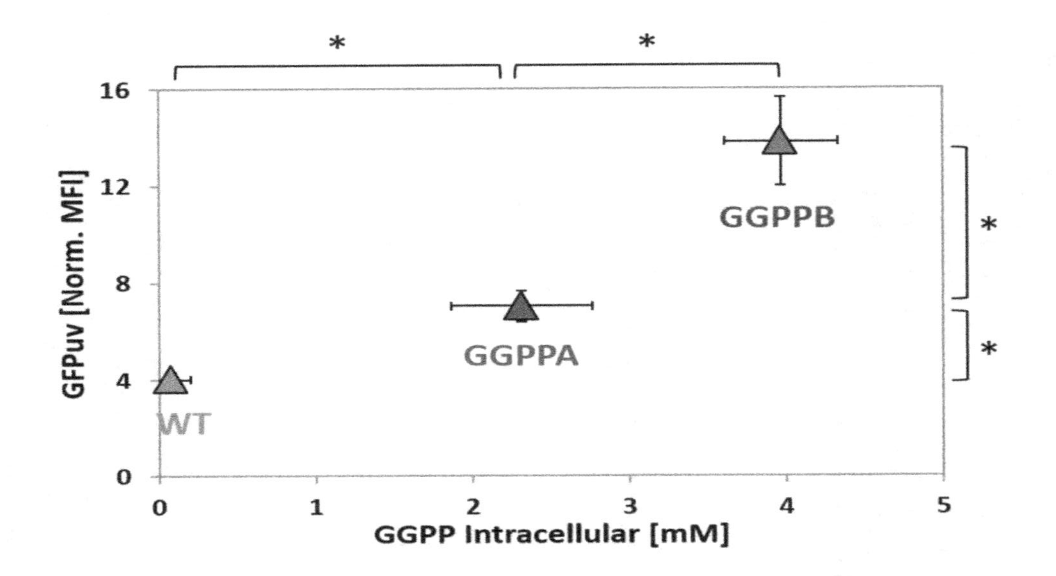


**Figure 6.** Biosensor-based differentiation between strains accumulating different GGPP concentrations. The intracellular GGPP concentrations are given in mM, and the GFPuv signals in mean fluorescence intensities were normalized to autofluorescence. Strains WT (pEPR1_P*crtE*), GGPPA (pEPR1_P*crtE*) and GGPPB (pEPR1_P*crtE*) were cultivated in CGXII (100 mM Gluc + 100 µM IPTG) and data were taken after 12 h. Statistical significance was calculated with paired Student t-test (two-tailed); *: $p$-value $< 0.05$.

Since the strains harbored plasmid pEPR1_*PcrtE*, the sensing of intracellular GGPP by chromosomally encoded CrtR could be tested. The normalized mean fluorescence intensity (MFI) of the *PcrtE_gfp*uv fusion observed after 12 h growth in glucose minimal medium was lowest for WT (pEPR1_*PcrtE*) (MFI of 4.0 ± 0.3). The biosensor signal of GGPPB (pEPR1_*PcrtE*), of 14.0 ± 1.8, was about two-fold higher than that of GGPPA (pEPR1_*PcrtE*) (MFI of 7.0 ± 0.7; Figure 6). Thus, biosensor signal output correlated with intracellular GGPP concentration.

In order to determine if the *PcrtE_gfp*uv fusion can be used to monitor variations in the GGPP concentration during growth, strain GGPPB (pEPR1_*PcrtE*) was analyzed in a time-course experiment in CGXII minimal medium supplemented with 100 mM glucose and 100 μM IPTG (Figure 7). Samples were taken every 1.5 h for determination of the intracellular GGPP concentration and flow cytometry analysis (Figure 7). Both the intracellular GGPP concentration and the GFPuv signal strongly increased in the first 7.5–10.5 h after inoculation. An offset between the GGPP concentration and the GFPuv signal was observed, with the increase in the latter being delayed by about 4 h (Figure 7). This offset may be explained by the time required to synthesize GFPuv after CrtR has sensed an increased GGPP concentration and *PcrtE_gfp*uv has been derepressed. The intracellular GGPP concentration reached its maximum approximately after 7.5 h, with about 3 mM GGPP, and decreased afterwards, while the GFPuv reached its maximum after 10.5 h of cultivation (Figure 7).

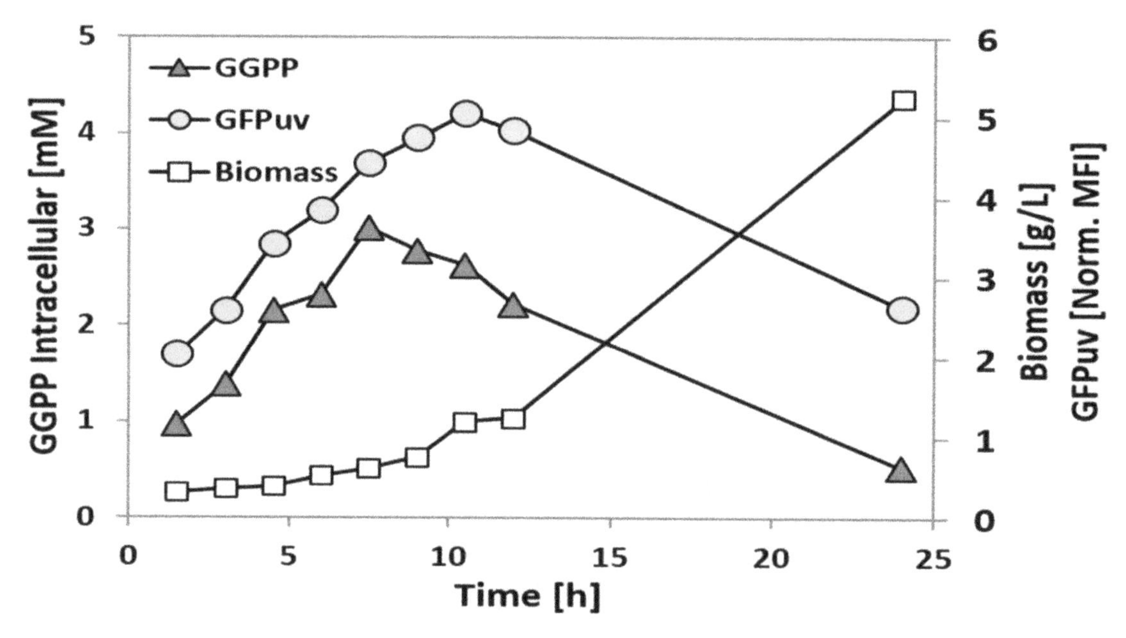


**Figure 7.** GGPP concentration and GFPuv fluorescence during growth of GGPP accumulating *C. glutamicum* strain GGPPB (pEPR1_*PcrtE*). Intracellular GGPP concentration (blue triangles; in mM) and GFPuv signal (green circles; mean fluorescence intensities normalized to autofluorescence) were monitored during growth in CGXII (100 mM Gluc + 100 μM IPTG). Biomass concentrations are given in gCDW/L (empty squares).

## 3. Discussion

This study revealed the conserved functions of CrtR orthologs in six actinobacteria with respect to GGPP-dependent regulation. The repression of carotenogenesis, and of the expression of a *crtE* transcriptional fusion, in *C. glutamicum* in vivo was highest for closely related species, and it decreased with phylogenetic distance. The *PcrtE_gfp*uv transcriptional fusion was suitable for monitoring intracellular GGPP concentrations in a strain with chromosomally encoded *crtR* as a genetically encoded biosensor system.

The conserved role of CrtR orthologs as GGPP-dependent transcriptional regulators suggested that they are relevant for the control of carotenogenesis and/or *mmpL* genes. Although not tested, it is tempting to speculate that these proteins may be used to monitor intracellular GGPP concentrations in their native hosts. *C. glutamicum* and *C. callunae* are close relatives, synthesizing the C50 carotenoid decaprenoxanthin and its glycosides as pigments [25,27]. *P. nicotinovorans* is pigmented most probably due to the accumulation of carotenoids [35]. *M. luteus* is a yellow-pigmented bacterium due to the accumulation of the C50 carotenoid sarcinaxanthin and its glycosides [53]. By contrast, *A. jensenii* does not synthesize a carotenoid, but the polyene pigment granadaene [54], while *P. chlorophenolicus* is a non-pigmented soil bacterium [55]. Interestingly, relatives of *P. chlorophenolicus*, such as *Arthrobacter arilaitensis*, are pigmented most probably due to the accumulation of C50 carotenoids, and are found on the surface of smear-ripened cheeses [55,56]. Thus, carotenogenesis is not conserved in actinobacteria possessing CrtR orthologs that control the expression of their own gene and/or the divergently oriented gene(s) in a GGPP-dependent manner. Besides carotenogenesis genes, *mmpL* genes are also transcribed divergently to *crtR*, and are presumably controlled by CrtR. MmpL transporters are considered candidate targets for the development of anti-tuberculosis drugs [57], as they couple lipid synthesis and the export of bulky, hydrophobic substrates [58]. Thus, MmpL proteins are essential for the cell envelope, and support the infectivity and persistence of *M. tuberculosis* in its host [59]. Moreover, MmpL proteins of *M. tuberculosis* are involved in the oxidative stress response [60].

MarR-type transcriptional regulators, including CrtR and its orthologs, are found in all bacteria, and are natural sensors that allow adaptation to environmental stresses such as ROS, toxic compounds or antibiotics (hence the name multiple antibiotic resistance regulators) [61]. As shown for the CrtR orthologs studied here (see Figures 1 and 2), MarR-type regulators typically repress genes in the close vicinity [34]. MarR-type regulators typically bind and respond to low-molecular-weight compounds [34,62], such as GGPP for the CrtR orthologs. *C. glutamicum* possesses nine MarR-type regulators, eight of which have been characterized in some detail. RosR [63], CosR [64], OhsR [65] and OsmC [66] play roles in the response to ROS stress, whereas CarR [67], MalR [68] and PhdR [69] deal with other environmental stresses, such as toxic compounds or cell-membrane associated stress. The GGPP-dependent control of carotenogenesis by CrtR from *C. glutamicum* and its orthologs studied here may be considered a stress response, since the antioxidative properties of carotenoids counteract oxidative stress. This function would be in line with the CrtR-mediated control of *mmpL* genes that are involved in the oxidative stress response (see above).

MarR-type regulators typically bind to a palindromic 16–20 bp target site that overlaps with the -10 or -35 promoter regions for steric inhibition of RNA polymerase binding [34]. Previously, a 19 bp DNA sequence with a central TTAA motif was shown to be essential for the binding of CrtR from *C. glutamium* to its DNA target site [23]. Here, we showed that CrtR from *C. glutamicum* bound to DNA sequences upstream of *crtR* from other actinobacteria, and inspection of the DNA sequences revealed a conserved 15 bp binding motif, including the central TTAA base pairs (see Figure 3). The binding of $CrtR_{Cg}$ decreased with increasing deviation from the consensus motif (see Figure 3). The CrtR binding motif is typical for the MarR-type family of transcriptional regulators [34]. The observed graded effect of $CrtR_{Cg}$ binding to promoter sequences with increasing deviation from the consensus motif in vitro was congruent with the in vivo finding that different CrtR orthologs affected the pigmentation and expression of the P*crtE_gfp*uv transcriptional fusion more weakly when their phylogenetic distance to *C. glutamicum* was greater (see Figure 4). This is in line with the phylogenetic analysis of CrtR orthologs [23].

The application of biosensors has become a prominent tool in strain development over the last few years [5,70]. In this study, CrtR from *C. glutamicum* was demonstrated to be the first genetically encoded biosensor for the detection of GGPP that allows one to distinguish between *C. glutamicum* strains that have accumulated GGPP to different intracellular concentrations (Figure 6), and to monitor GGPP accumulation over time during growth (Figure 7). Since the CrtR-based biosensor system was suitable for the detection of intracellular GGPP concentrations between 0.1 and at least 4 mM (Figure 6), it is plausible that the described system is applicable to the screening of mutants accumulating GGPP well above wild type levels, and their enrichment/isolation, by flow cytometry. As an alternative application, the on-demand expression control of GGPP converting enzymes in response to intracellular GGPP concentration can be envisioned for strain optimization with respect to the production of GGPP-derived diterpenoids and/or carotenoids. On-demand production may be established by transcriptional fusion of the P*crtE* to the gene of interest, e.g., a diterpenoid synthase. This approach may improve production as the terminal biosynthesis pathway is initiated only in the presence of high concentrations of the precursor GGPP, which may prevent the accumulation of toxic GGPP concentrations. Biosensor approaches to on-demand expression control have been successfully applied, e.g., to improve lysine production by *C. glutamicum* [71,72].

The central role of GGPP as a terpenoid and carotenoid precursor suggests a wide application range for the GGPP-based biosensor developed here [10], since the tens of thousands of terpenoids derived from GGPP represent one of the biggest sources of valuable natural products for human use [73].

## 4. Materials and Methods

### 4.1. Bacterial Strains, Media and Growth Conditions

Strains and plasmids that were used in this study are listed in Table 1. *C. glutamicum* ATCC 13,032 [45] served as the wild type and was used as the basic strain for genetic engineering. Modifications aimed at higher production levels of GGPP and the establishment of the biosensor system. Precultures of *C. glutamicum* were performed in LB/BHI medium with 50 mM glucose as carbon and energy source [74] supplemented with the appropriate antibiotic at 30 °C and 120 rpm. The main cultures of *C. glutamicum* consisted of 50 mL CGXII medium with 100 mM glucose and 100 μM IPTG and were inoculated to an initial optical density ($OD_{600}$) of 1. The $OD_{600}$ of the cultures was measured with the Shimadzu UV-1202 spectrophotometer (Duisburg, Germany).

### 4.2. Recombinant DNA Work and Gene Expression

Cloning of plasmids was done in *E. coli* DH5α using PCR-generated fragments that were purified using the NucleoSpin kit (Macherey-Nagel, Düren, Germany). Oligonucleotides were ordered from Metabion GmbH (Planegg/Steinkirchen, Germany) (Table 2). For plasmid construction standard PCR, restriction and dephosphorylation reactions [75] were performed as well as Gibson Assembly [76]. Transformation of *E. coli* was performed via the RbCl method [42]. Cloned DNA insert fragments were verified by sequencing. Transformation of *C. glutamicum* was performed via electroporation using a Gene Pulser Xcell™ (Bio-Rad Laboratories GmbH, Munich, Germany) at 2.5 kV, 200 Ω and 25 μF [74]. For expression of CrtR in the pTEST vector (NA, Tables 1 and 2) and the production of His-tagged CrtR from various organisms in the pET16b vector (HN, Tables 1 and 2), the respective genes were amplified from chromosomal DNA using primer pairs NA25/26 and HN83/HN84 for *C. glutamicum*, NA27/28 and HN85/HN86 for *C. callunae*, NA31/32 and HN87/HN88 for *A. jensenii*, NA33/34 and HN89/HN90 for *P. nicotinovorans*, NA39/40 and HN93/HN94 for *M. luteus* and NA41/42 and HN95/HN96 for *P. chlorophenolicus*, respectively. The purified PCR products were cloned into pTEST restricted with *BamHI* and pET16b restricted with *Nde*I using Gibson assembly [76], respectively. Chromosomal DNA was extracted from DSMZ (see Table 1).

**Table 2.** Oligonucleotides used in this study.

| | Oligonucleotide (5′→3′) |
|---|---|
| NH45 | CATGCCTGCAGGTCGACTCTAGAGGAAAGGAGGCCCTTCAGATGGGAATTCTGAACAGTATTTCAA |
| NH46 | GTTCGTGTGGCAGTTTTATTCCCCGAACAGGGAATC |
| NH47 | AACTGCCACACGAACGAAAGGAGGCCCTTCAGATGTCTAAGCTTAGGGGCATG |
| NH48 | ATTCGAGCTCGGTACCCGGGGATCTTACTCTGCGTCAAACGCTTC |
| NH49 | ATGGAATTCGAGCTCGGTACCCGGGGAAAGGAGGCCCTTCAGATGGCTTACTCCGCTATGGCTA |
| NH50 | GCATGCCTGCAGGTCGACTCTAGAGGATCTTAGTTCTGGCGGAAAGCAA |
| NH51 | GTTCGTGTGGCAGTTTTAGTTCTGGCGGAAAGCAA |
| NH52 | ATGGAATTCGAGCTCGGTACCCGGGGAAAGGAGGCCCTTCAGATGGACTTTCCGCAGCAACTCG |
| NH53 | GCATGCCTGCAGGTCGACTCTAGAGGATCTTATTTATTACGCTGGATGATGTAGTCC |
| NH54 | GTTCGTGTGGCAGTTTTATTTATTACGCTGGATGATGTAGTCC |

**Table 2.** *Cont.*

| | Oligonucleotide (5′→3′) |
|---|---|
| NH55 | ATGGAATTCGAGCTCGGTACCCGGGGAAAGGAGGCCCTTCAGATGAGCAGTTTCGATGCCCA |
| NH56 | GCATGCCTGCAGGTCGACTCTAGAGGATCTTACATCCGACGTTCGGTTGA |
| NH57 | GTTCGTGTGGCAGTTTTACATCCGACGTTCGGTTGA |
| NH58 | ATGGAATTCGAGCTCGGTACCCGGGGAAAGGAGGCCCTTCAGATGGTAGAAAACAACGTAGCAA |
| NH59 | GCATGCCTGCAGGTCGACTCTAGAGGATCTTAGTCCAGGTAGTCGCGAAG |
| NH60 | AACTGCCACACGAACGAAAGGAGGCCCTTCAGATGGTAGAAAACAACGTAGCAA |
| NH63 | GCAAAGTTGTTGTCGTAGTC |
| NH64 | ATGAAAACGTTGTTGCCAT |
| NH65 | ATGAAGACGCCACTGAC |
| NH66 | CGGTGAGCTCGGCATCT |
| NH67 | GTGCCTTGCGAGCTGTCT |
| TH17 | CTGTTGATGACGACGAGGAG |
| pE-CXT fw | AATACGCAAACCGCCTCTCC |
| pE-CXT rv | TACTGCCGCCAGGCAAATTC |
| crtE-E | GTGACCATGAGGGCGAAAGC |
| crtE-F | TCACATAGTCCGGCGTTTGC |
| idsA-E | GCAGCTTCGCCAGAGTGTAT |
| idsA-F | CAATGCGGACAATGCTCCAG |
| 581 | CATCATAACGGTTCTGGC |
| 582 | ATCTTCTCTCATCCGCCA |
| P*gap* fw | TGGCCTTTTGCTGGCCTTTTGCTCACTGCGAAATCTTTGTTTCCCCG |
| P*gap* rv | GGATCCGTTGTGTCTCCTCTAAAGATT |
| *term* fw | AATCTTTAGAGGAGACACAACGGATCCTTTTGGCGGATGAGAGAA |
| *term* rv | AATCAGGGGATAACGCAGGAAAGAACAAAAGAGTTTGTAGAA |
| NA25- Cg fw | TACAATCTTTAGAGGAGACACAACGGAAAGGAGGCCCTTCAGATGCTGAATATGCAGGAACCA |
| NA26- Cg rv | AAAATCTTCTCTCATCCGCCAAAAGTTACTCCGTGTTGAGCCATGG |
| NA27- Cc fw | TACAATCTTTAGAGGAGACACAACGGAAAGGAGGCCCTTCAGATGTCCGATCCGCAAGAACC |
| NA28- Cc rv | AAAATCTTCTCTCATCCGCCAAAAGTTAATGTGAGGAAGACTCGAAC |
| NA31- Aj fw | TACAATCTTTAGAGGAGACACAACGGAAAGGAGGCCCTTCAGATGAGTGAAGACCGCGATG |
| NA32- Aj rv | AAAATCTTCTCTCATCCGCCAAAAGTTACCGCGGGTGGCGC |
| NA33-An fw | TACAATCTTTAGAGGAGACACAACGGAAAGGAGGCCCTTCAGATGTCCAGTCTTGAAGAAATGC |
| NA34-An rv | AAAATCTTCTCTCATCCGCCAAAAGTTAGCGTGGAGCCGCAG |
| NA39- Ml fw | TACAATCTTTAGAGGAGACACAACGGAAAGGAGGCCCTTCAGATGACCACGCAGCCCC |
| NA40- Ml rv | AAAATCTTCTCTCATCCGCCAAAAGTTACGGGTCCTCCGGGG |
| NA41- Pc fw | TACAATCTTTAGAGGAGACACAACGGAAAGGAGGCCCTTCAGATGAACGGCAACAATCCG |
| NA42- Pc rv | AAAATCTTCTCTCATCCGCCAAAAGTTACCCGGCTGGACGC |
| HN83-Cg-fw | GCGGCCATATCGAAGGTCGTCATCTGAATATGCAGGAACCAG |
| HN84-Cg-rv | TAGCAGCCGGATCCTCGAGCATTACTCCGTGTTGAGCCATG |
| HN85-Cc-fw | GCGGCCATATCGAAGGTCGTCATTCCGATCCGCAAGAACCCC |
| HN86-Cc-rv | TAGCAGCCGGATCCTCGAGCATTAATGTGAGGAAGACTCGAAC |
| HN87-Aj-fw | GCGGCCATATCGAAGGTCGTCATAGTGAAGACCGCGATGC |
| HN88-Aj-rv | TAGCAGCCGGATCCTCGAGCATTACCGCGGGTGGCGC |
| HN89-Pn-fw | GCGGCCATATCGAAGGTCGTCATTCCAGTCTTGAAGAAATGCC |
| HN90-Pn-rv | TAGCAGCCGGATCCTCGAGCATTAGCGTGGAGCCGCAG |
| HN93-Ml-fw | GCGGCCATATCGAAGGTCGTCATACCACGCAGCCCCCC |
| HN94-Ml-rv | TAGCAGCCGGATCCTCGAGCATTACGGGTCCTCCGGGG |
| HN95-Pc-fw | GCGGCCATATCGAAGGTCGTCATAACGGCAACAATCCGGGC |
| HN96-Pc-rv | TAGCAGCCGGATCCTCGAGCATTACCCGGCTGGACGC |
| Pc-PcrtR-fw | TGCCTTCCATGCGGATGGTC |
| Pc-PcrtR-rv | TGCCCGGATTGTTGCCGTTC |

### 4.3. *Extraction of Carotenoids from Bacterial Cells and HPLC Analysis*

The carotenoid extraction from *C. glutamicum* was performed as described previously [30] using 1 mL of the cell cultures. Pigments were isolated from the cell pellets with a methanol:acetone mixture (7:3) at 60 °C for 15 min with shaking at 500 rpm. The clear supernatant was used for HPLC analysis after centrifugation of the extract for 10 min at 13,000× *g*. The carotenoid concentration of cell extracts was determined through absorbance at 471 nm by high performance liquid chromatography (HPLC) analysis, performed on an Agilent 1200 series HPLC system (Agilent Technologies Sales & Services GmbH & Co. KG, Waldbronn, Germany), including a diode array detector (DAD) for UV/visible (Vis) spectrum recording. Separation of the carotenoids was performed by application of a column system consisting of a precolumn (LiChrospher 100 RP18 EC-5, 40 × 4 mm, CS-Chromatographie, Langerwehe, Germany) and a main column (LiChrospher 100 RP18 EC-5, 125 × 4 mm, CS-Chromatographie, Langerwehe, Germany) with methanol/water (9:1) (A) and methanol (B) as the mobile phase.

The following gradient was used at a flow rate of 1.5 mL/min: 0 min B—0%; 10 min B—100%; 32.5 min B—100%. The quantification of decaprenoxanthin was calculated based on a β-carotene standard (Merck, Darmstadt, Germany) and reported as β-carotene equivalents.

### *4.4. Analysis of Fluorescence via Flow Cytometry*

Cell cultures were analyzed regarding their fluorescent intensity. Samples were diluted to a final $OD_{600}$ of 0.1 with pure CGXII medium and immediately analyzed with the FACS Gallios$^{TM}$ (Beckman Coulter GmbH, Krefeld, Germany). Alternatively, samples for fluorescence analysis were harvested and stored at 4 °C. *C. glutamicum* (pEPR1) was used as the autofluorescence reference. The GFPuv signal was measured with a blue solid-state laser at 405 nm excitation and fluorescence was detected using a 525/50 nm band-pass filter.

### *4.5. Overproduction and Purification of the Transcriptional Regulator CrtR*

After transformation of the pET16b derivatives in *E. coli* BL21(DE3) or *E. coli* BL21(DE3) (pLysS) transformants carrying the respective plasmids pET16b-$crtR_{Cg}$, pET16b-$crtR_{Cc}$, pET16b-$crtR_{Aj}$, pET16b-$crtR_{Pn}$, pET16b-$crtR_{Ml}$ and pET16b-$crtR_{Pc}$ were grown at 37 °C in 500 mL LB medium with 10 μg/mL ampicillin to an $OD_{600}$ of 0.5 before adding IPTG (0.5 mM) for induction of the gene expression. After induction, cells were cultivated at 21 °C for an additional 4 h and were harvested by centrifugation. Pellets were stored at −20 °C. Crude extract preparation and protein purification via Ni-NTA chromatography was performed as described elsewhere [23]. The purified regulator proteins were used for EMSA experiments without removing the N-terminal His-tag.

### *4.6. Electrophoretic Mobility Shift Assay (EMSA)*

To analyze the physical protein–DNA interaction between the different CrtR proteins and their putative native target DNA, bandshift assays were performed [77]. The His-tagged CrtR proteins were mixed in varying molar excess with 30–90 ng of PCR amplified and purified promoter fragments of the target genes in bandshift (BS) buffer (50 mM Tris–HCl, 10% (*v*/*v*) glycerol, 50 mM KCl, 10 mM $MgCl_2$, 0.5 mM EDTA, pH 7.5) in a total volume of 20 μL. The 5′ UTR of *crtR* genes were PCR-amplified and purified with NucleoSpin kit (MACHEREY-NAGEL GmbH & Co. KG, Düren, Germany). Promoter fragments were amplified using the respective oligonucleotide pairs (Table 2). A 78 bp-fragment of the upstream region of cg2228 was added in every sample as a negative control using oligonucleotides cg2228_fw and cg2228_rv. After 30 min of incubation at room temperature, gel shift samples were separated on a native 6% (*w*/*v*) polyacrylamide. Additionally, the binding affinity in the presence of 100–650 μM GGPP as effector was analyzed by incubation of the protein with the effector under buffered conditions for 15 min at room temperature prior to the addition of the promoter. Subsequently, the gel shift samples were separated on a 6% DNA retardation gel (Life Technologies GmbH, Darmstadt, Germany) at 100 V buffered in 44.5 mM Tris, 44.5 mM boric acid and 1 mM EDTA at pH 8.3. Staining of the DNA was achieved with ethidium bromide.

### *4.7. Extraction of GGPP and LC-MS Analysis*

For isolation of GGPP, 10 mL of culture were harvested at 4000 rpm and 15 min. The supernatant was removed and the cells stored till further use (−80 °C). The cell pellet was defrosted on ice and resuspended in 600 μL acidified methanol (pH 5). Pyrophosphates were extracted by 3 × 30 s shaking in silamat (Ivoclar Vivadent AG, Schaan, Liechtenstein) in the presence of 300 μL silica beads. The clear supernatant was used for LC-MS analysis after subsequent centrifugation for 10 min at 13,000× *g*. LC-MS measurement was performed on a LaChrom ULTRA system (San Jose, CA, USA) using a SeQuant Zic-pHILIC column (5 μm 150 × 2.1 mm) (Merck Millipore, Darmstadt, Germany). As a buffer system, 10 mM ammonium bicarbonat pH 9.3 (A) and acetonitrile (B) was used with a flow rate of 0.2 mL/min; 0–5 min 5% A (const.), 5–20 min 35% A (gradient), 20–25 min 5% A (gradient), 25–35 min 5% A (const.); pre-run 15 min. with 2 μL. Isoprenoid pyrophosphates were identified using a micrOTOFQ (Bruker Daltonics, Billerica, MA, USA) according to their masses (GPP 313.0601; FPP 381.1227; GGPP 449.1853 [M-H]$^{-}$) and elution time in accordance to a standard (Sigma-Aldrich, Merck, Darmstadt, Germany).

**Author Contributions:** N.A.H., P.P.-W. and V.F.W. planned and designed the experiments. N.A.H., S.A., I.L.G., S.G. and M.P. performed the experiments. N.A.H., M.P. and P.P.-W. analyzed the data. N.A.H. and P.P.-W. drafted the manuscript. V.F.W. coordinated the study and finalized the manuscript. All authors read and approved the final manuscript.

**Acknowledgments:** We thank Tim Treis for optimization of the GGPP extraction protocol.

## References

1. Mahr, R.; Frunzke, J. Transcription factor-based biosensors in biotechnology: Current state and future prospects. *Appl. Microbiol. Biotechnol.* **2016**, *100*, 79–90. [CrossRef] [PubMed]
2. Rogers, J.K.; Church, G.M. Genetically encoded sensors enable real-time observation of metabolite production. *Proc. Natl. Acad. Sci. USA* **2016**, *113*, 2388–2393. [CrossRef] [PubMed]
3. Liu, Y.; Liu, Y.; Wang, M. Design, Optimization and Application of Small Molecule Biosensor in Metabolic Engineering. *Front. Microbiol.* **2017**, *8*, 2012. [CrossRef] [PubMed]
4. Zhang, J.; Jensen, M.K.; Keasling, J.D. Development of biosensors and their application in metabolic engineering. *Curr. Opin. Chem. Biol.* **2015**, *28*, 1–8. [CrossRef] [PubMed]
5. De Paepe, B.; Peters, G.; Coussement, P.; Maertens, J.; De Mey, M. Tailor-made transcriptional biosensors for optimizing microbial cell factories. *J. Ind. Microbiol. Biotechnol.* **2017**, *44*, 623–645. [CrossRef]
6. Rohmer, M.; Seemann, M.; Horbach, H.; Bringer-Meyer, S.; Sahm, H. Glyceraldehyde 3-phosphate and pyruvate as precursors of isoprenic units in an alternative non-mevalonate pathway for terpenoid biosynthesis. *J. Am. Chem. Soc.* **1996**, *118*, 2564–2566. [CrossRef]
7. Britton, L.-J. Pfander. In *Carotenoids Handbook*; Birkhauser Verlag: Basel, Switzerland, 2004. [CrossRef]
8. Henke, N.A.; Wichmann, J.; Baier, T.; Frohwitter, J.; Lauersen, K.J.; Risse, J.M.; Peters-Wendisch, P.; Kruse, O.; Wendisch, V.F. Patchoulol Production with Metabolically Engineered *Corynebacterium glutamicum*. *Genes (Basel)* **2018**, *9*. [CrossRef]
9. Schempp, F.M.; Drummond, L.; Buchhaupt, M.; Schrader, J. Microbial cell factories for the production of terpenoid flavor and fragrance compounds. *J. Agric. Food Chem.* **2017**. [CrossRef]
10. Schrader, J.; Bohlmann, J. *Biotechnology of Isoprenoids*; Springer International Publishing: New York, NY, USA, 2015. [CrossRef]
11. Misawa, N. Pathway engineering for functional isoprenoids. *Curr. Opin. Biotechnol.* **2011**, *22*, 627–633. [CrossRef]
12. George, K.W.; Alonso-Gutierrez, J.; Keasling, J.D.; Lee, T.S. Isoprenoid drugs, biofuels, and chemicals–artemisinin, farnesene, and beyond. *Adv. Biochem. Eng. Biotechnol.* **2015**, *148*, 355–389. [CrossRef]
13. Novoveská, L.; Ross, M.E.; Stanley, M.S.; Pradelles, R.; Wasiolek, V.; Sassi, J.-F. Microalgal Carotenoids: A Review of Production, Current Markets, Regulations, and Future Direction. *Mar. Drugs* **2019**, *17*, 640. [CrossRef] [PubMed]
14. Rodriguez-Saiz, M.; de la Fuente, J.L.; Barredo, J.L. *Xanthophyllomyces dendrorhous* for the industrial production of astaxanthin. *Appl. Microbiol. Biotechnol.* **2010**, *88*, 645–658. [CrossRef] [PubMed]
15. Tsubokura, A.; Yoneda, H.; Mizuta, H. *Paracoccus carotinifaciens* sp. nov., a new aerobic gram-negative astaxanthin-producing bacterium. *Int. J. Syst. Bacteriol.* **1999**, *49* (Pt. 1), 277–282. [CrossRef] [PubMed]
16. Li, C.; Swofford, C.A.; Sinskey, A.J. Modular engineering for microbial production of carotenoids. *Metab. Eng. Commun.* **2020**, *10*, e00118. [CrossRef] [PubMed]
17. Pfleger, B.F.; Pitera, D.J.; Newman, J.D.; Martin, V.J.; Keasling, J.D. Microbial sensors for small molecules: Development of a mevalonate biosensor. *Metab. Eng.* **2007**, *9*, 30–38. [CrossRef] [PubMed]
18. Tang, S.Y.; Cirino, P.C. Design and application of a mevalonate-responsive regulatory protein. *Angew. Chem. Int. Ed. Engl.* **2011**, *50*, 1084–1086. [CrossRef]
19. Chou, H.H.; Keasling, J.D. Programming adaptive control to evolve increased metabolite production. *Nat. Commun.* **2013**, *4*, 2595. [CrossRef]

20. Lee, J.H.; Wendisch, V.F. Production of amino acids—Genetic and metabolic engineering approaches. *Bioresour. Technol.* **2017**, *245*, 1575–1587. [CrossRef]
21. Kinoshita, S.; Udaka, S.; Shimono, M. Studies on the amino acid fermentation. Production of L-glutamic acid by various microorganisms. *J. Gen. Appl. Microbiol.* **1957**, *3*, 193–205. [CrossRef]
22. Wendisch, V.F. Metabolic engineering advances and prospects for amino acid production. *Metab. Eng.* **2020**, *58*, 17–34. [CrossRef]
23. Henke, N.A.; Heider, S.A.E.; Hannibal, S.; Wendisch, V.F.; Peters-Wendisch, P. Isoprenoid Pyrophosphate-Dependent Transcriptional Regulation of Carotenogenesis in *Corynebacterium glutamicum*. *Front. Microbiol.* **2017**, *8*, 633. [CrossRef] [PubMed]
24. Heider, S.A.; Peters-Wendisch, P.; Wendisch, V.F. Carotenoid biosynthesis and overproduction in *Corynebacterium glutamicum*. *BMC Microbiol.* **2012**, *12*, 198. [CrossRef] [PubMed]
25. Krubasik, P.; Takaichi, S.; Maoka, T.; Kobayashi, M.; Masamoto, K.; Sandmann, G. Detailed biosynthetic pathway to decaprenoxanthin diglucoside in *Corynebacterium glutamicum* and identification of novel intermediates. *Arch. Microbiol.* **2001**, *176*, 217–223. [CrossRef] [PubMed]
26. Heider, S.A.; Peters-Wendisch, P.; Beekwilder, J.; Wendisch, V.F. IdsA is the major geranylgeranyl pyrophosphate synthase involved in carotenogenesis in *Corynebacterium glutamicum*. *FEBS J.* **2014**, *281*, 4906–4920. [CrossRef] [PubMed]
27. Heider, S.A.; Peters-Wendisch, P.; Wendisch, V.F.; Beekwilder, J.; Brautaset, T. Metabolic engineering for the microbial production of carotenoids and related products with a focus on the rare C50 carotenoids. *Appl. Microbiol. Biotechnol.* **2014**, *98*, 4355–4368. [CrossRef] [PubMed]
28. Heider, S.A.; Wolf, N.; Hofemeier, A.; Peters-Wendisch, P.; Wendisch, V.F. Optimization of the IPP precursor supply for the production of lycopene, decaprenoxanthin and astaxanthin by *Corynebacterium glutamicum*. *Front. Bioeng. Biotechnol.* **2014**, *2*, 28. [CrossRef] [PubMed]
29. Heider, S.A.; Peters-Wendisch, P.; Netzer, R.; Stafnes, M.; Brautaset, T.; Wendisch, V.F. Production and glucosylation of C50 and C40 carotenoids by metabolically engineered *Corynebacterium glutamicum*. *Appl. Microbiol. Biotechnol.* **2014**, *98*, 1223–1235. [CrossRef]
30. Henke, N.A.; Heider, S.A.; Peters-Wendisch, P.; Wendisch, V.F. Production of the Marine Carotenoid Astaxanthin by Metabolically Engineered *Corynebacterium glutamicum*. *Mar. Drugs* **2016**, *14*, 124. [CrossRef]
31. Henke, N.A.; Wendisch, V.F. Improved Astaxanthin Production with *Corynebacterium glutamicum* by Application of a Membrane Fusion Protein. *Mar. Drugs* **2019**, *17*, 621. [CrossRef]
32. Sprenger, G.A.; Schorken, U.; Wiegert, T.; Grolle, S.; de Graaf, A.A.; Taylor, S.V.; Begley, T.P.; Bringer-Meyer, S.; Sahm, H. Identification of a thiamin-dependent synthase in *Escherichia coli* required for the formation of the 1-deoxy-D-xylulose 5-phosphate precursor to isoprenoids, thiamin, and pyridoxol. *Proc. Natl. Acad. Sci. USA* **1997**, *94*, 12857–12862. [CrossRef]
33. Frohwitter, J.; Heider, S.A.; Peters-Wendisch, P.; Beekwilder, J.; Wendisch, V.F. Production of the sesquiterpene (+)-valencene by metabolically engineered *Corynebacterium glutamicum*. *J. Biotechnol.* **2014**, *191*, 205–213. [CrossRef] [PubMed]
34. Wilkinson, S.P.; Grove, A. Ligand-responsive transcriptional regulation by members of the MarR family of winged helix proteins. *Curr. Issues Mol. Boil.* **2006**, *8*, 51–62.
35. Sumi, S.; Suzuki, Y.; Matsuki, T.; Yamamoto, T.; Tsuruta, Y.; Mise, K.; Kawamura, T.; Ito, Y.; Shimada, Y.; Watanabe, E.; et al. Light-inducible carotenoid production controlled by a MarR-type regulator in *Corynebacterium glutamicum*. *Sci. Rep.* **2019**, *9*, 13136. [CrossRef] [PubMed]
36. Viljoen, A.; Dubois, V.; Girard-Misguich, F.; Blaise, M.; Herrmann, J.L.; Kremer, L. The diverse family of MmpL transporters in mycobacteria: From regulation to antimicrobial developments. *Mol. Microbiol.* **2017**, *104*, 889–904. [CrossRef]
37. Knoppova, M.; Phensaijai, M.; Vesely, M.; Zemanova, M.; Nesvera, J.; Patek, M. Plasmid vectors for testing in vivo promoter activities in *Corynebacterium glutamicum* and *Rhodococcus erythropolis*. *Curr. Microbiol.* **2007**, *55*, 234–239. [CrossRef]
38. Banerjee, A.; Wu, Y.; Banerjee, R.; Li, Y.; Yan, H.; Sharkey, T.D. Feedback inhibition of deoxy-D-xylulose-5-phosphate synthase regulates the methylerythritol 4-phosphate pathway. *J. Biol. Chem.* **2013**, *288*, 16926–16936. [CrossRef]
39. Lv, X.; Xu, H.; Yu, H. Significantly enhanced production of isoprene by ordered coexpression of genes *dxs*, *dxr*, and *idi* in *Escherichia coli*. *Appl. Microbiol. Biotechnol.* **2013**, *97*, 2357–2365. [CrossRef]

40. Lv, X.; Gu, J.; Wang, F.; Xie, W.; Liu, M.; Ye, L.; Yu, H. Combinatorial pathway optimization in *Escherichia coli* by directed co-evolution of rate-limiting enzymes and modular pathway engineering. *Biotechnol. Bioeng.* **2016**, *113*, 2661–2669. [CrossRef]
41. Taniguchi, H.; Henke, N.A.; Heider, S.A.E.; Wendisch, V.F. Overexpression of the primary sigma factor gene *sigA* improved carotenoid production by *Corynebacterium glutamicum*: Application to production of β-carotene and the non-native linear C50 carotenoid bisanhydrobacterioruberin. *Metab. Eng. Commun.* **2017**, *4*, 1–11. [CrossRef]
42. Hanahan, D. Studies on transformation of *Escherichia coli* with plasmids. *J. Mol. Biol* **1983**, *166*, 557–580. [CrossRef]
43. Simon, R.; Priefer, U.; Puhler, A. A Broad Host Range Mobilization System for In Vivo Genetic Engineering: Transposon Mutagenesis in Gram Negative Bacteria. *Nat. Biotech.* **1983**, *1*, 784–791. [CrossRef]
44. Studier, F.W.; Moffatt, B.A. Use of bacteriophage T7 RNA polymerase to direct selective high-level expression of cloned genes. *J. Mol. Biol.* **1986**, *189*, 113–130. [CrossRef]
45. Abe, S.; Takayarna, K.; Kinoshita, S. Taxonomical studies on glutamic acid producing bacteria. *J. Gener. Appl. Microbial.* **1967**, *13*, 279–301. [CrossRef]
46. Johnson, J.L.; Cummins, C.S. Cell wall composition and deoxyribonucleic acid similarities among the anaerobic coryneforms, classical propionibacteria, and strains of *Arachnia propionica*. *J. Bacteriol.* **1972**, *109*, 1047–1066. [CrossRef]
47. Yamada, K.; Komagata, K. Taxonomic Studies on *Coryneform* Bacteria. *J. Gen. Appl. Microbial.* **1970**, *16*, 103–113. [CrossRef]
48. Schleifer, K.H.; Kloos, W.E.; Moore, A. Toxonomic Status of Micrococcus luteus (Schroeter 1872) Cohn 1872: Correlation Between Peptidoglycan Type and Genetic Compatibility. *Int. J. Syst. Evolut. Microbiol.* **1972**, *22*, 224–227.
49. Kodama, Y.; Yamamoto, H.; Amano, N.; Amachi, T. Reclassification of two strains of *Arthrobacter oxydans* and proposal of *Arthrobacter nicotinovorans* sp. nov. *Int. J. Syst. Bacteriol.* **1992**, *42*, 234–239. [CrossRef]
50. Westerberg, K.; Elvang, A.M.; Stackebrandt, E.; Jansson, J.K. *Arthrobacter chlorophenolicus* sp. nov., a new species capable of degrading high concentrations of 4-chlorophenol. *Int. J. Syst. Evol. Microbiol.* **2000**, *50* (Pt. 6), 2083–2092. [CrossRef]
51. Stansen, C.; Uy, D.; Delaunay, S.; Eggeling, L.; Goergen, J.L.; Wendisch, V.F. Characterization of a *Corynebacterium glutamicum* lactate utilization operon induced during temperature-triggered glutamate production. *Appl. Environ. Microbiol.* **2005**, *71*, 5920–5928. [CrossRef]
52. Kirchner, O.; Tauch, A. Tools for genetic engineering in the amino acid-producing bacterium *Corynebacterium glutamicum*. *J. Biotechnol.* **2003**, *104*, 287–299. [CrossRef]
53. Netzer, R.; Stafsnes, M.H.; Andreassen, T.; Goksoyr, A.; Bruheim, P.; Brautaset, T. Biosynthetic pathway for gamma-cyclic sarcinaxanthin in *Micrococcus luteus*: Heterologous expression and evidence for diverse and multiple catalytic functions of C(50) carotenoid cyclases. *J. Bacteriol.* **2010**, *192*, 5688–5699. [CrossRef] [PubMed]
54. Vanberg, C.; Lutnaes, B.F.; Langsrud, T.; Nes, I.F.; Holo, H. *Propionibacterium jensenii* produces the polyene pigment granadaene and has hemolytic properties similar to those of *Streptococcus agalactiae*. *Appl. Environ. Microbial.* **2007**, *73*, 5501–5506. [CrossRef] [PubMed]
55. Monnet, C.; Loux, V.; Gibrat, J.-F.; Spinnler, E.; Barbe, V.; Vacherie, B.; Gavory, F.; Gourbeyre, E.; Siguier, P.; Chandler, M.; et al. The *arthrobacter arilaitensis* Re117 genome sequence reveals its genetic adaptation to the surface of cheese. *PLoS ONE* **2010**, *5*, e15489. [CrossRef] [PubMed]
56. Sutthiwong, N.; Dufossé, L. Production of carotenoids by *Arthrobacter arilaitensis* strains isolated from smear-ripened cheeses. *FEMS Microbiol. Lett.* **2014**, *360*, 174–181. [CrossRef] [PubMed]
57. Ma, S.; Huang, Y.; Xie, F.; Gong, Z.; Zhang, Y.; Stojkoska, A.; Xie, J. Transport mechanism of *Mycobacterium tuberculosis* MmpL/S family proteins and implications in pharmaceutical targeting. *Biol. Chem.* **2020**, *401*, 331–348. [CrossRef] [PubMed]
58. Chalut, C. MmpL transporter-mediated export of cell-wall associated lipids and siderophores in mycobacteria. *Tuberc. (Edinb. Scotl.)* **2016**, *100*, 32–45. [CrossRef]
59. Melly, G.; Purdy, G.E. MmpL Proteins in Physiology and Pathogenesis of *M. tuberculosis*. *Microorganisms* **2019**, *7*. [CrossRef] [PubMed]
60. Arumugam, P.; Shankaran, D.; Bothra, A.; Gandotra, S.; Rao, V. The MmpS6-MmpL6 Operon Is an Oxidative

Stress Response System Providing Selective Advantage to *Mycobacterium tuberculosis* in Stress. *J. Infect. Dis.* **2019**, *219*, 459–469. [CrossRef]

61. Schumacher, M.A.; Brennan, R.G. Structural mechanisms of multidrug recognition and regulation by bacterial multidrug transcription factors. *Mol. Microbiol.* **2002**, *45*, 885–893. [CrossRef]
62. Grove, A. Regulation of Metabolic Pathways by MarR Family Transcription Factors. *Comput. Struct. Biotechnol. J.* **2017**, *15*, 366–371. [CrossRef]
63. Bussmann, M.; Baumgart, M.; Bott, M. RosR (Cg1324), a hydrogen peroxide-sensitive MarR-type transcriptional regulator of *Corynebacterium glutamicum*. *J. Biol. Chem.* **2010**, *285*, 29305–29318. [CrossRef] [PubMed]
64. Si, M.; Chen, C.; Su, T.; Che, C.; Yao, S.; Liang, G.; Li, G.; Yang, G. CosR is an oxidative stress sensing a MarR-type transcriptional repressor in *Corynebacterium glutamicum*. *Biochem. J.* **2018**, *475*, 3979–3995. [CrossRef] [PubMed]
65. Si, M.; Su, T.; Chen, C.; Liu, J.; Gong, Z.; Che, C.; Li, G.; Yang, G. OhsR acts as an organic peroxide-sensing transcriptional activator using an S-mycothiolation mechanism in *Corynebacterium glutamicum*. *Microb. Cell Fact.* **2018**, *17*, 200. [CrossRef] [PubMed]
66. Si, M.; Su, T.; Chen, C.; Wei, Z.; Gong, Z.; Li, G. OsmC in *Corynebacterium glutamicum* was a thiol-dependent organic hydroperoxide reductase. *Int. J. Boil. Macromol.* **2019**, *136*, 642–652. [CrossRef]
67. Si, M.; Chen, C.; Wei, Z.; Gong, Z.; Li, G.; Yao, S. CarR, a MarR-family regulator from *Corynebacterium glutamicum*, modulated antibiotic and aromatic compound resistance. *Biochem. J.* **2019**, *476*, 3141–3159. [CrossRef]
68. Hünnefeld, M.; Persicke, M.; Kalinowski, J.; Frunzke, J. The MarR-Type Regulator MalR Is Involved in Stress-Responsive Cell Envelope Remodeling in *Corynebacterium glutamicum*. *Front. Microbiol.* **2019**, *10*, 1039. [CrossRef]
69. Kallscheuer, N.; Vogt, M.; Kappelmann, J.; Krumbach, K.; Noack, S.; Bott, M.; Marienhagen, J. Identification of the phd gene cluster responsible for phenylpropanoid utilization in *Corynebacterium glutamicum*. *Appl. Microbiol. Biotechnol.* **2016**, *100*, 1871–1881. [CrossRef]
70. Ng, C.Y.; Khodayari, A.; Chowdhury, A.; Maranas, C.D. Advances in *de novo* strain design using integrated systems and synthetic biology tools. *Curr. Opin. Chem. Biol.* **2015**, *28*, 105–114. [CrossRef]
71. Zhou, L.B.; Zeng, A.P. Engineering a Lysine-ON Riboswitch for Metabolic Control of Lysine Production in *Corynebacterium glutamicum*. *ACS Synth. Biol.* **2015**, *4*, 1335–1340. [CrossRef]
72. Zhou, L.B.; Zeng, A.P. Exploring lysine riboswitch for metabolic flux control and improvement of L-lysine synthesis in *Corynebacterium glutamicum*. *ACS Synth. Biol.* **2015**, *4*, 729–734. [CrossRef]
73. Tetali, S.D. Terpenes and isoprenoids: A wealth of compounds for global use. *Planta* **2019**, *249*, 1–8. [CrossRef] [PubMed]
74. *Handbook of Corynebacterium Glutamicum*; Eggeling, L.; Bott, M. (Eds.) CRC Press Taylor & Francis Group: Boca Raton, FL, USA, 2005.
75. Sambrook, J.; Russell, D. *Molecular Cloning. A Laboratory Manual*, 3rd ed.; Cold Spring Harbor Laboratoy Press: Cold Spring Harbor, NY, USA, 2001.
76. Gibson, D.G.; Young, L.; Chuang, R.Y.; Venter, J.C.; Hutchison, C.A., 3rd; Smith, H.O. Enzymatic assembly of DNA molecules up to several hundred kilobases. *Nat. Methods* **2009**, *6*, 343–345. [CrossRef] [PubMed]
77. Krause, J.P.; Polen, T.; Youn, J.W.; Emer, D.; Eikmanns, B.J.; Wendisch, V.F. Regulation of the malic enzyme gene *malE* by the transcriptional regulator MalR in *Corynebacterium glutamicum*. *J. Biotechnol.* **2012**, *159*, 204–215. [CrossRef] [PubMed]

3

# Community Structures and Antifungal Activity of Root-Associated Endophytic Actinobacteria of Healthy and Diseased Soybean

**Chongxi Liu [1,2], Xiaoxin Zhuang [1], Zhiyin Yu [1,2], Zhiyan Wang [2], Yongjiang Wang [2], Xiaowei Guo [1,2], Wensheng Xiang [1,3,*] and Shengxiong Huang [2,*]**

1 Heilongjiang Provincial Key Laboratory of Agricultural Microbiology, Northeast Agricultural University, Harbin 150030, China
2 State Key Laboratory of Phytochemistry and Plant Resources in West China, Kunming Institute of Botany, Chinese Academy of Sciences, Kunming 650201, China
3 State Key Laboratory for Biology of Plant Diseases and Insect Pests, Institute of Plant Protection, Chinese Academy of Agricultural Sciences, Beijing 100193, China
* Correspondence: xiangwensheng@neau.edu.cn (W.X.); sxhuang@mail.kib.ac.cn (S.H.)

**Abstract:** The present study was conducted to examine the influence of a pathogen *Sclerotinia sclerotiorum* (Lib.) de Bary on the actinobacterial community associated with the soybean roots. A total of 70 endophytic actinobacteria were isolated from the surface-sterilized roots of either healthy or diseased soybeans, and they were distributed under 14 genera. Some rare genera, including *Rhodococcus*, *Kribbella*, *Glycomyces*, *Saccharothrix*, *Streptosporangium* and *Cellulosimicrobium*, were endemic to the diseased samples, and the actinobacterial community was more diverse in the diseased samples compared with that in the heathy samples. Culture-independent analysis of root-associated actinobacterial community using the high-throughput sequencing approach also showed similar results. Four *Streptomyces* strains that were significantly abundant in the diseased samples exhibited strong antagonistic activity with the inhibition percentage of 54.1–87.6%. A bioactivity-guided approach was then employed to isolate and determine the chemical identity of antifungal constituents derived from the four strains. One new maremycin analogue, together with eight known compounds, were detected. All compounds showed significantly antifungal activity against *S. sclerotiorum* with the 50% inhibition ($EC_{50}$) values of 49.14–0.21 mg/L. The higher actinobacterial diversity and more antifungal strains associated with roots of diseased plants indicate a possible role of the root-associated actinobacteria in natural defense against phytopathogens. Furthermore, these results also suggest that the root of diseased plant may be a potential reservoir of actinobacteria producing new agroactive compounds.

**Keywords:** *Sclerotinia sclerotiorum* (Lib.) de Bary; diseased soybean root; antifungal activity; actinobacterial community; new agroactive compounds

---

## 1. Introduction

Sclerotinia stem rot (SSR) caused by a fungus *Sclerotinia sclerotiorum* (Lib.) de Bary is a highly destructive disease leading to serious economic losses to crops throughout the world. This fungus can infect over 400 plant species, including many economically important crops and vegetables [1–3]. Generally, the development of resistant cultivars is a long-term approach for controlling the disease [2,4]. However, the disease has yet been difficult to control because of the limited resource of the resistant genes. Therefore, fungicides have been used as the auxiliary method for controlling SSR in practice [5]. The benzimidazole and dicarboximide fungicides were the most efficient fungicides in controlling

SSR [6]. However, the continuous use of these fungicides with high concentration can amplify the resistant level of phytopathogens [7–9]. Thus, development of new antifungal agents would be a constant need for controlling the disease.

Endophytic microorganisms residing inside plants have been found in majority of plant species [10]. A growing body of literature recognizes that some of these microorganisms are involved in plant defense against the phytopathogens through a range of mechanisms, including competition for an ecological niche or a substrate, secretion of antibiotics and lytic enzymes, and induction of systemic resistance (ISR) [11,12]. Recent studies on plant-microbe interactions reveal that plants can specifically attract bacteria for their ecological and evolutionary benefit by secreting root exudates [13–15]. It has even been postulated that plants can recruit beneficial microorganisms from soil to counteract pathogen assault [16,17]. For example, it has previously been observed that colonization of the roots of *Arabidopsis* by beneficial rhizobacteria *Bacillus subtilis* FB17 was greatly stimulated when leaves were infected by *Pseudomonas syringae* pv. *tomato* [18].

The phylum *Actinobacteria* consists of a wide range of Gram-positive bacteria with high guanine-plus-cytosine (G + C) content. Actinobacterial species are known to produce a vast diversity of active natural products including antibiotics, antitumor gents, enzymes and immunosuppressive agents, which have been widely used in pharmaceutical, agricultural and other industries [19,20]. Recently, endophytic actinobacteria have attracted significant interest for their capacity to produce abundant bioactive metabolites, which may contribute to their host plants by promoting growth and health [21,22]. A vast majority of endophytic actinobacteria have been isolated from a variety of plants including various crop plants, medicinal plants, and different woody tree species [23–28]. Further, recent cultivation-independent analysis using 16S rRNA gene-based methods revealed that actinobacteria can be specifically enriched in plant roots, and are more abundant in diseased plants than in healthy plants, which may provide probiotic functions for the host plants [29–31]. Thus, it is hypothesized that endophytic actinobacteria from disease plants may be a promising source for the discovery of new antifungal agents against *S. sclerotiorum*.

This prospective study was designed to test the above hypothesis by (i) using culture-independent and dependent methods to compare the generic diversity and antifungal activity of root-associated endophytic actinobacteria of field-growing healthy and diseased plants soybean plants and (ii) identifying antifungal metabolited produced by the outstanding actinobacteria isolated.

## 2. Materials and Methods

### 2.1. Plant Materials

Root samples were collected from soybean (cultivar: Hefeng-50) plants identified as SSR symptomatic (diseased) or asymptomatic (healthy) based on typical symptoms in a heavily infected soybean field in Suihua, Heilongjiang province, North China (46.63° N 126.98° E). The diseased samples showed lesions encircling up to 1/3 of stem diameter referring to a severity scale of three [32]. Each healthy plant and diseased plant located as close neighbors were determined as one group. The samples were brought to the lab in a cooler with ice in July 2017 and were processed immediately.

### 2.2. Isolation of Endophytic Actinobacteria

Three groups of root samples were used for isolation of endophytic actinobacteria. The root samples were air dried for 24 h at room temperature and then washed in water with an ultrasonic step (160 W, 15 min) (KH-160TDV, Hechuang, China) to remove the surface soils and adherent epiphytes completely. After drying, the sample was cut into pieces of 5–10 mm in length and then subjected to a seven-step surface sterilization procedure: A 60-sec wash in sterile tap water containing cycloheximide (100 mg/L) and nalidixic acid (20 mg/L), followed by a wash in sterile water, a 5 min wash in 5% (v/v)

NaOCl, a 10 min wash in 2.5% (w/v) $Na_2S_2O_3$, a 5 min wash in 75% (v/v) ethanol, a wash in sterile water and a final rinse in 10% (w/v) $NaHCO_3$ for 10 min. After being thoroughly dried under sterile conditions, the surface-sterilized samples were subjected to continuous drying at 100 °C for 15 min. The sample was then cut up in a commercial blender and ground with a mortar and pestle, employing 1 mL of 0.5 M potassium phosphate buffer (pH 7.0) per 100 mg tissue. Tissue particles were allowed to settle down at 4 °C for 20–30 min, and an aliquot of 200 μL supernatants were spread on a series of isolation media and incubated at 28 °C for 2–3 weeks. Each isolation medium was supplemented with nalidixic acid (20 mg/L) and cycloheximide (50 mg/L) to inhibit the growth of Gram-negative bacteria and fungi. Five isolation media: Humic acid-vitamin (HV) agar [33], Gause's synthetic agar no. 1 [34], dulcitol-proline agar (DPA) [35], cellulose-proline agar [36], and amino acid agar (serine 0.05%, threonine 0.05%, alanine 0.05%, arginine 0.05%, agar powder 2%, pH 7.2–7.4) were selected for the isolation. After 14 days of aerobic incubation at 28 °C, the actinobacterial colonies were transferred onto oatmeal agar (International *Streptomyces* Project medium 3, ISP3) [37] and repeatedly re-cultured until pure cultures were obtained, and maintained as glycerol suspensions (20%, v/v) at −80 °C.

### 2.3. *Phenotypic and Molecular Characterization of Actinobacterial Isolates*

The purified colonies were cultivated on ISP 3 at 28 °C for two weeks, and then grouped according to their phenotypic characteristics, including the characteristics of colonies on plates, color of aerial and substrate mycelium, spore mass color, spore chain morphology, and production of diffusible pigment. Those colonies with the same characteristics were classified as one species. The number of species was counted to compare the diversity of root-associated endophytic actinobacteria from healthy and diseased soybean.

Different phenotypic isolates were further subjected to 16S rRNA gene sequence analysis for the genus and species identification. The total DNA was extracted using the lysozyme-sodium dodecyl sulfate-phenol/chloroform method [38]. The primers and procedure for PCR amplification were carried out as described by Kim et al. [39]. The PCR products were purified and ligated into the vector pMD19-T (Takara Biomedical Technology, Beijing, China) and sequenced by an Applied Biosystems DNA sequencer (model 3730XL). The almost full-length 16S rRNA gene sequences (~1500 bp) were obtained and aligned with multiple sequences obtained from the GenBank/EMBL/DDBJ databases using CLUSTAL X 1.83 software. Phylogenetic tree was constructed with neighbor-joining method [40] using Molecular Evolutionary Genetics Analysis (MEGA) software version 7.0 [41]. The bootstrap method with 1000 repetitions was using to assess the topology of the phylogenetic tree [42]. Phylogenetic distances were calculated according to the Kimura two-parameter model [43]. The 16S rRNA gene sequence similarities were determined using the EzBiocloud server [44]. The obtained gene sequences have been deposited in the GenBank database.

### 2.4. *Screening for Antagonistic Actinobacteria*

The phytopathogenic *S. sclerotiorum* strain used in this study was kindly provided by the Soybean Research Institute of Northeast Agricultural University (Harbin, China). Antagonistic activity of isolates were evaluated through the dual culture plate assay [45]. The isolates were point-inoculated at the margin of potato dextrose agar (PDA) [46] plates and incubated for three days at 28 °C, after which a fresh mycelial PDA agar plug of the fungus was transferred to the opposite margin of the corresponding plate. After additional days of incubation at 20 °C for seven days, inhibition of hyphal growth of the fungus was scored. The inhibition rates were calculated as follows:

$$\text{Inhibition rate (\%)} = Wi/W \times 100\% \qquad (1)$$

where *Wi* is the width of inhibition and *W* is the width between the pathogen and actinobacteria. Each test was repeated three times and the average was calculated.

### *2.5. Isolation and Characterization of Antifungal Compounds*

The antifungal compounds were isolated using an in vitro antifungal activity-guided method [47]. The active isolates were inoculated into 250 mL flask containing 50 mL of tryptone soy broth (TSB) [48] and cultivated for two days at 28 °C with shaking at 200 rpm. Then, 12.5 mL of the seed culture was transferred into 1 L Erlenmeyer flask containing 250 mL of the fermentation medium (soluble starch 1%, dextrose 2%, tryptone 0.5%, yeast extract 0.5%, NaCl 0.4%, $K_2HPO_4$ $3H_2O$ 0.05%, $MgSO_4$ $7H_2O$ 0.05%, $CaCO_3$ 0.5%, pH 7.2–7.4) and incubated at 28 °C for seven days with shaking at 200 rpm. The fermentation broth (20 L) was centrifuged (4000 rev/min, 20 min), and the supernatant and bacterial biomass were extracted with ethylacetate and methanol, respectively. Both extracts were concentrated by a rotary evaporator under reduced pressure until dry and mixed after dissolving their dried residues with methanol.

The crude extracts were divided into seven fractions using column fractionation packed with silica gel (200−300 mesh, Qingdao Marine Chemical Inc., Qingdao, China) eluting with petroleum ether/ethylacetate (20:1, 10:1, 5:1, 3:1, 2:1, 1:1 and 0:1). The bioactive fractions were then subjected to Sephadex LH-20 (Pharmacia, Uppsala, Sweden) and eluted with methanol to obtain several subfractions. The active subfractions were further separated by semipreparative HPLC (Hitachi-DAD, Tokyo, Japan) using a YMC-Triart $C_{18}$ column (250 × 10 mm i.d., 5 μm) at a flow rate of 3.0 mL/min, and the potent active principles were finally isolated.

Structural determination of the active compounds were made according to spectroscopic analysis. NMR spectra were measured with a Bruker Avance III-600 spectrometer in $CDCl_3$ or $CD_3OD$ using TMS as internal standard. The ESI-MS spectrum was taken on a Waters Xevo TQ-S ultrahigh pressure liquid chromatography triple quadrupole mass spectrometer. The HR-ESI-MS spectrum was acquired with an Agilent G6230 Q-TOF mass instrument. The UV spectrum was recorded in chloroform using a Shimadzu UV-2401PC UV-VIS spectrophotometer. The IR spectrum was obtained using a Bruker Tensor 27 FTIR. Optical rotation was determined in chloroform using a JASCO P-1020 polarimeter. The ECD spectrum was measured on a Chirascan circular dichroism spectrometer (Applied Photophysics Corporation Limited, Leatherhead, UK).

### *2.6. Antifungal Assay of Elucidated Bioactive Compounds*

The active compounds were dissolved in methanol and diluted to different concentrations, which were then added to PDA medium. A fresh fungal plug of the fungus (5 mm in diameter) was placed in the center of the agar plate and incubated at 20 °C. Experiments were performed in triplicate, and the plate with the same amount of methanol was used as control. When the control plate was covered completely with fungal mass, the percentage of inhibition was calculated with the formula as follows:

$$\text{Inhibition (\%)} = (1 - D/D_c) \times 100 \quad (2)$$

where $D$ is average diameter of the treatment and $D_c$ is average diameter of the control. Data were subjected to linear regression analysis, and the effective concentrations required for 50% inhibition ($EC_{50}$) were calculated.

### *2.7. Culture-Independent Community Analysis*

Total community DNA was extracted from three groups of surface-sterilized root samples using FastDNA®SPIN for soil kit (MP Biomedicals, Solon, CA, USA) according to the manufacturers' instructions. The purity and concentration of DNA were detected using NanoPhotometer spectrophotometer (Implen, München, Germany) and Qubit 2.0 Flurometer (Life Technologies, Carlsbad, CA, USA). Bacterial DNA pyrosequencing was based on ~460 bp amplicons generated by the PCR primers: 341F (5′-CCTACGGGNGGCWGCAG-3′) and 805R (5′-GACTACHVGGGTATCTAATCC-3′) with the barcode spanning the hypervariable regions V3-V4 of the 16S rRNA gene The PCR reaction was carried out in 30 μL reactions with 15 μL of Phusion High-Fidelity PCR Master Mix (New England

Biolabs, Ipswich, MA, USA), 0.2 μM of forward and reverse primers, and about 10 ng template DNA. Thermal cycling conditions were 95 °C for 3 min, followed by 25 cycles of 95 °C for 30 s, 55 °C for 30 s, and 72 °C for 30 s, with a final extension at 72 °C for 10 min. PCR products was mixed in equidensity ratios. Then, mixture PCR products was purified with GeneJET Gel Extraction Kit (Thermo Scientific, Fermentas, Germany).

Sequencing libraries were generated using NEB Next Ultra DNA Library Prep Kit for Illumina (New England Biolabs, Ipswich, Massachusetts, USA) following the manufacturer's recommendations and index codes were added. The library quality was assessed on the Qubit@ 2.0 Fluorometer and Agilent Bioanalyzer 2100 system (Agilent Technologies, Palo Alto, CA, USA). At last, the library was sequenced on an Illumina MiSeq platform and 250 bp paired-end reads.

When the sequencing was finished, we needed to filter the raw data to secure the quality, which mainly included the following steps: (1) Cut the polluted adapter, (2) remove low quality reads, specifically reads with average quality less than 19, based on the Phred algorithm, and (3) remove the reads with N base exceeding 5%.

According to overlap of the clean data, we spliced the paired reads by using the PEAR software [49] to merged sequences. The sequences were then removed Chimeras and clustered into operational taxonomic units (OTUs) by UCLUST [50] based on 97% pairwise identity. Taxonomic classification of the representative sequence for each OTU was done using a RDP classifier or QIIME's closed reference strategy against the 16S rRNA gene database [51].

### *2.8. Statistical Analysis*

Statistical analyses were performed via SPSS software, version 18.0. Means were compared via analysis of variance one-way ANOVA using the least significant differences test (LSD, $p < 0.05$). The data were reported as means ± standard deviation.

## 3. Results

### *3.1. Isolation and Distribution of Endophytic Actinobacteria*

A total of 1574 endophytic actinobacterial colonies were successfully isolated. Based on their phenotypic characteristics, the colonies were preliminary classified into 70 species. Among the 70 species, 15 species were from healthy soybean roots, 27 species were from diseased soybean roots, and 28 species were shared by healthy and diseased soybean roots. The diversity of actinobacteria from diseased soybean root were greater than those from healthy soybean root. 16S rRNA gene sequence analysis of the 70 isolates revealed that they were distributed under 14 genera: *Streptomyces*, *Micromonospora*, *Actinomadura*, *Nonomuraea*, *Microbacterium*, *Rhodococcus*, *Promicromonospora*, *Microbispora*, *Kribbella*, *Mycolicibacterium*, *Glycomyces*, *Saccharothrix*, *Streptosporangium* and *Cellulosimicrobium* within the class *Actinobacteria*. *Streptomyces* was the most frequently isolated genus (58%, 41 isolates), followed by *Micromonospora* (5 isolates) and *Nonomuraea* (4 isolates). Some rare genera, including *Rhodococcus*, *Kribbella*, *Glycomyces*, *Saccharothrix*, *Streptosporangium* and *Cellulosimicrobium*, were endemic to the diseased samples (Figure 1). The 16S rRNA gene sequences were deposited in GenBank with accession numbers: MH919371–MH919374 and MN058215–MH058280.

### *3.2. In Vitro Antagonism of S. sclerotiorum and Identification of Bioactive Strains*

Strains DAAG3-11, DGS1-1, DDPA2-14 and DGS3-15 exhibited strong antagonistic activity against *S. sclerotiorum*, with inhibition activity rates ranging from 54.1% to 87.6% (Figure 2). Strains DGS1-1, DDPA2-14 and DGS3-15 were from diseased soybean roots. For strain DAAG3-11, 176 colonies were from diseased soybean roots, whereas 11 colonies were from healthy soybean roots. Based on the 16S rRNA gene sequences, strains DAAG3-11, DGS1-1, DDPA2-14 and DGS3-15 were closely related to *Streptomyces sporoclivatus*, *Streptomyces cavourensis*, *Streptomyces capitiformicae* and *Streptomyces pratensis*, respectively (Table 1).

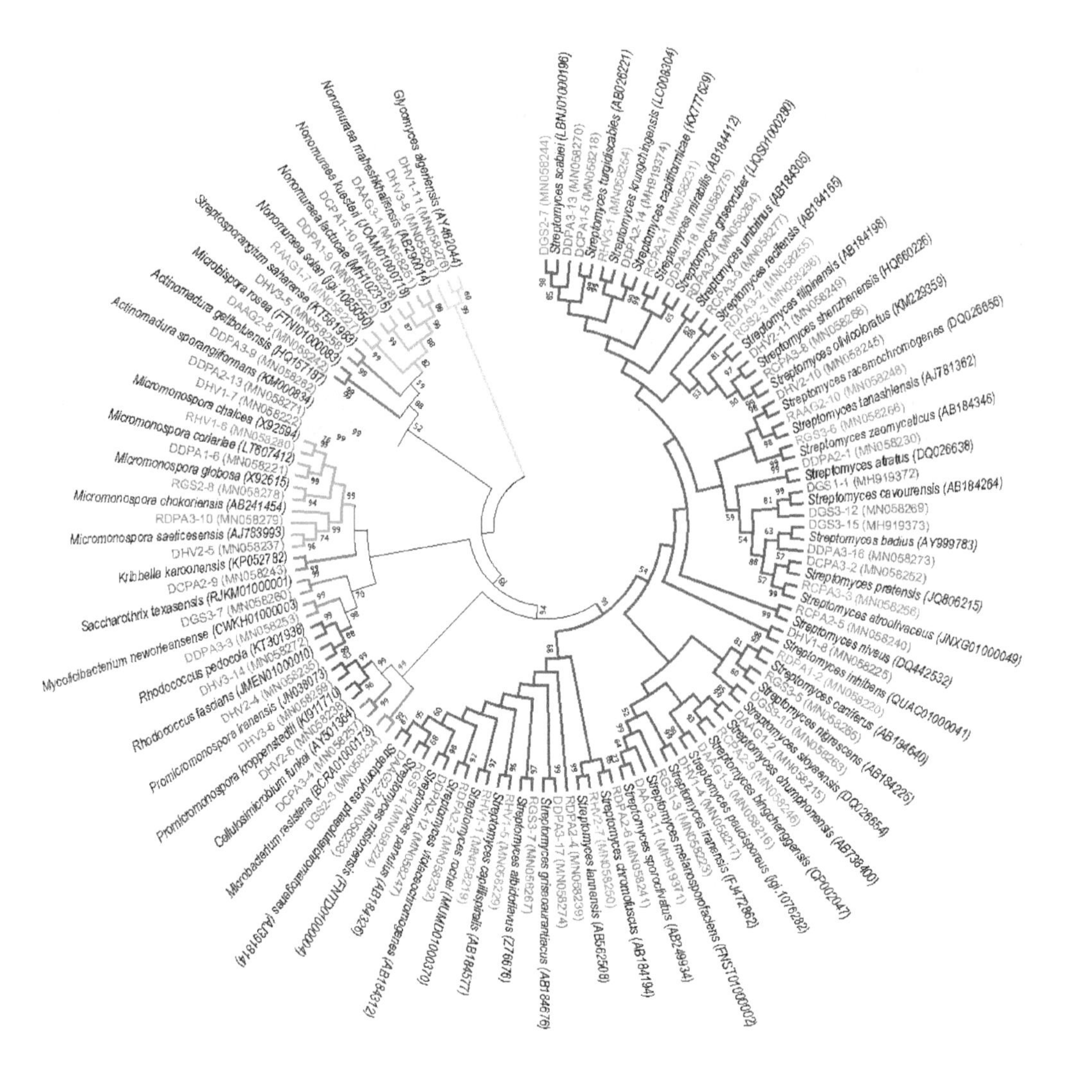

**Figure 1.** Neighbor-joining phylogenetic tree of 16S rRNA gene sequences from 70 endophytic actinobacteria in this study and their phylogenetic neighbors. Numbers at nodes are bootstrap values (percentages of 1000 replications); only values > 50% are shown. GenBank accession numbers of 16S rRNA gene sequences are shown next to isolate names. A branch indicated by the same color belongs to the same genus. Isolates indicated by green and purple are endemic to the healthy and diseased samples, respectively. Isolates shared by healthy and diseased samples are indicated with red.

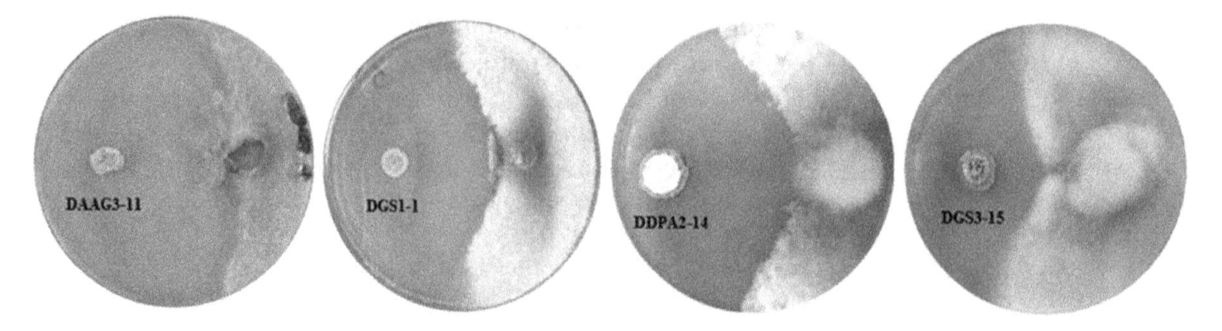


**Figure 2.** Dual culture plate assay between four endophytic actinobacteria against *S. sclerotiorum*.

**Table 1.** Antagonistic potential endophytic actinobacteria isolated from healthy and diseased soybean root, and similarity values for 16S rRNA gene sequences.

| Isolate No. and NCBI Genbank Accesion No. | Closest Type Strain with Accession Number | Similarity | Isolated From | Colony Number | *S. sclerotiorum* Mycelial Growth Inhibition (%) * |
|---|---|---|---|---|---|
| DAAG3-11 (MH919371) | *Streptomyces sporoclivatus* (AB249934) | 100% | Healthy soybean root | 11 | 87.6 ± 1.8 a |
| | | | Diseased soybean root | 176 | |
| DGS1-1 (MH919372) | *Streptomyces cavourensis* (AB184264) | 99.9% | Diseased soybean root | 13 | 78.9 ± 1.9 b |
| DDPA2-14 (MH919374) | *Streptomyces capitiformicae* (KX777629) | 100% | Diseased soybean root | 9 | 68.6 ± 3.4 c |
| DGS3-15 (MH919373) | *Streptomyces pratensis* (JQ806215) | 99.9% | Diseased soybean root | 6 | 54.1 ± 2.2 d |

* Values are the means ± SE ($n = 4$). Data within the same column followed by different letters are significantly different.

### *3.3. Identification and Activity Evaluation of the Antifungal Compounds*

An antifungal activity-guided separation of the components of four active strains against *S. sclerotiorum*, using the in vitro antifungal assay, led to the isolation of nine compounds as their active principles (Figure 3). Out of the nine compounds, compounds **1** and **2** were from strain DAAG3-11, compound **3** was from strain DGS1-1, compounds **4–7** were from strain DGS3-15, and compounds **8** and **9** were from strain DDPA2-14. Compounds **1–8** are known compounds, which structures were elucidated as azalomycins $F_{4a}$ (**1**) [52], azalomycins $F_{5a}$ (**2**) [52], bafilomycin $B_1$ (**3**) [53], actinolactomycin (**4**) [54], dimeric dinactin (**5**) [55], tetranactin (**6**) [56], dinactin (**7**) [56] and maremycin G (**8**) [57] by analysis of their spectroscopic data and comparison with literature values (Figure S1). Compound **9** is a new maremycin analogue.

Compound **9** was obtained as a yellow, amorphous powder. Its molecular formula $C_{22}H_{27}N_3O_4S$ was determined by high resolution electrospray ionization mass spectrometry (HRESIMS) data (*m/z* 468.1555, $[M + Na]^+$, calculated for 468.1564), corresponding to 11 degrees of unsaturation. The IR spectrum indicated the presence of hydroxy (3424 $cm^{-1}$) and carbonyl (1720, 1682 $cm^{-1}$) groups. The $^1H$ NMR data (Table 2) of **9** suggested the presence of one 1, 2-disubstituted benzene system at $\delta_H$ 7.43 (1H, d, $J$ = 7.8 Hz, H-4), 7.12 (1H, td, $J$ = 7.6, 1.0 Hz, H-5), 7.37 (1H, td, $J$ = 7.7, 1.2 Hz, H-6) and 6.89 (1H, d, $J$ = 7.9 Hz, H-7). The $^1H$ NMR data of **9** also revealed the presence of two methyl signals at $\delta_H$ 2.17, (3H, s, H-23) and $\delta_H$ 3.23, (3H, s, H-25). $^{13}C$ NMR spectrum of **9** showed 11 $sp^2$-carbons including three carbonyls at $\delta_C$ 204.84 (C-21), 178.33 (C-2), 168.5 (C-13) and eight aromatic or olefinic carbons at $\delta_C$ 152.27 (C-16), 142.83 (C-9), 130.36 (C-7), 130.01 (C-4), 125.53 (C-5), 123.24 (C-6), 109.08 (C-8), 100.43 (C-17). In the $sp^3$-carbon region, the spectrum showed three methine at $\delta_C$ 42.84 (C-10), 52.94 (C-11), 52.8 (C-14), four methylene at $\delta_C$ 27.04 (C-18), 21.04 (C-19), 38.79(C-20), 38.75(C-22), and three methyl carbons at $\delta_C$ 16.36 (C-23), 8.92 (C-24), 26.65 (C-25).

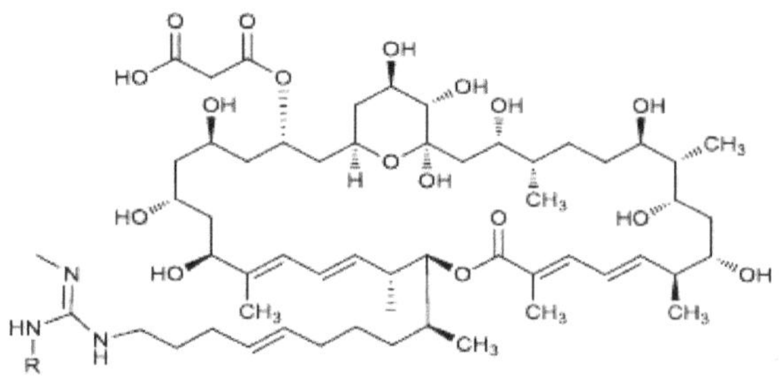

1: R=H
2: R=$CH_3$

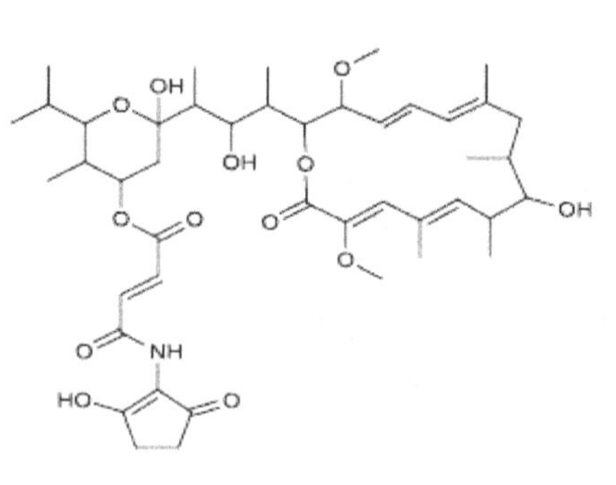

3

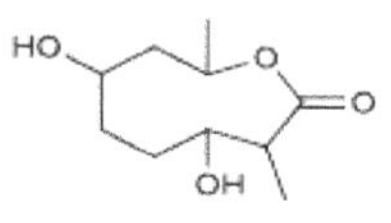

4

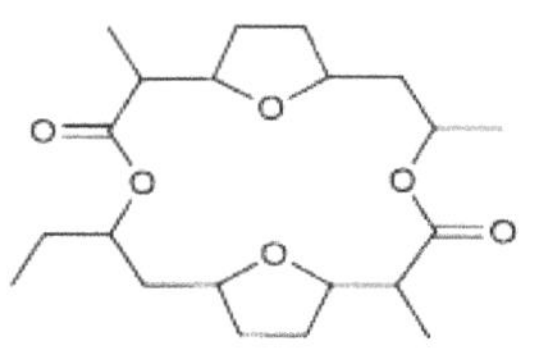

5

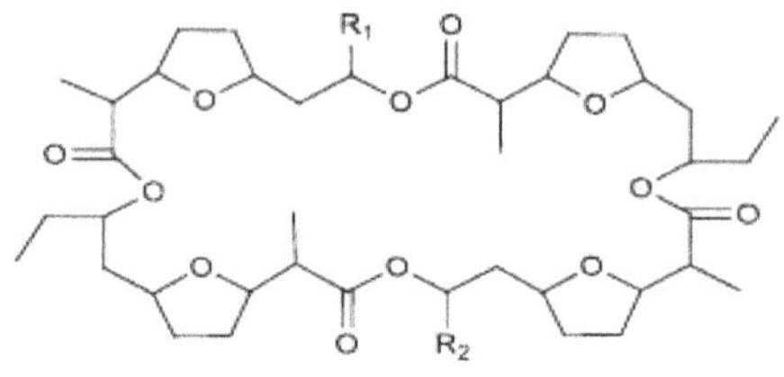

6: $R_1=R_2=CH3$
7: $R_1=R_2=H$

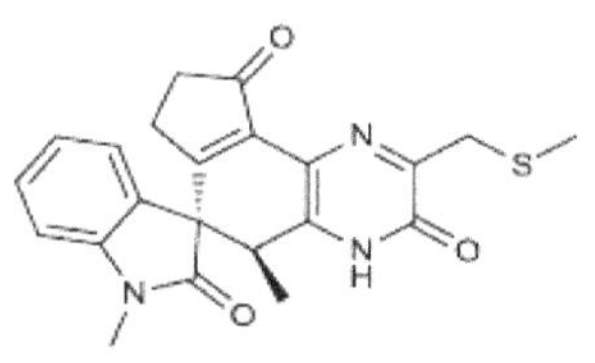

8

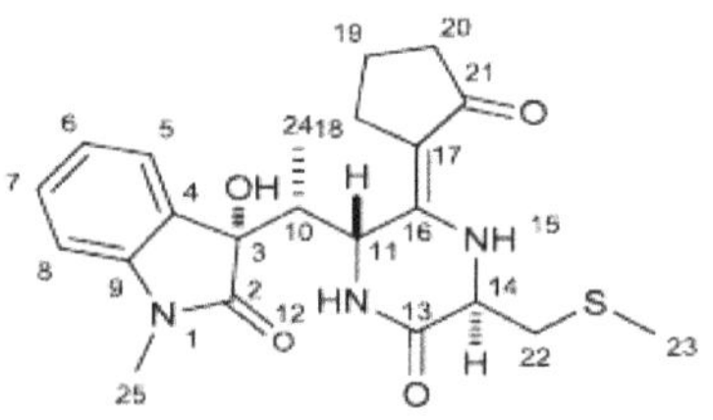

9

**Figure 3.** The structures of compounds **1–9**.

Comparison the NMR data of **9** with FR900452 [58], an indole diketopiperazine motif linked with a cyclopentenone moiety, which was isolated from the fermentation broth of *Streptomyces* sp. B9173, implied that **9** was identified as a reduced form of FR900452 in which the cyclopentenone moiety is hydrogenated to cyclopentanone. As shown in Figure 4, the accurate assignments of all protons and carbons for compound **9** were preformed through the correlations in 2D-NMR spectra ($^1H$–$^1H$ COSY, HSQC and HMBC, Figure S2). The HMBC couplings Me-25/C-9/C-2, H-5/C-3, and H-10/C-2/C-3/C-4, along with $^1H$–$^1H$ COSY correlations of H-5/H-6/H7/H-8, revealed N-methyl-2-oxindole unit. In addition, $^1H$–$^1H$ COSY correlations of H-18/H-19/H-20, as well as the HMBC cross peaks H-18/C-17/C-16, H-20/C-21, demonstrated that oxopiperazinyl moiety was linked to C-16/C-17 on the cyclopentenone moiety. $^1H$–$^1H$ COSY correlations of Me-24/H-10/H-11, together with the HMBC correlations from Me-24/C-3/C-10/C-11, indicated that N-methyl-2-oxindole unit was linked to C-10/C-5 on the oxopiperazinyl moiety. The $^1H$ and $^{13}C$ NMR spectroscopic data of **9** were also indicative of methyl mercaptomethylene moieties [$\delta_C$ 38.75, $\delta_{H\text{-}CH2}$ 3.17 (1H, m), 2.83 (1H, dd, 14.0, 8.2), $\delta_{C\text{-}CH3}$ 16.36, $\delta_{H\text{-}S\text{-}CH3}$ 2.15 (3H, s)]. The HMBC cross peaks Me-23/C22, H-22/C-14/C-13, along with $^1H$–$^1H$ COSY correlations of H-22/H-14, evidenced that methyl mercaptomethylene moieties were linked to C-14 on the oxopiperazinyl moieties, respectively. Therefore, the planar structure of **9** was elucidated as a reduced form of FR900452, depicted in Figure 3.

**Table 2.** $^{1}H$ NMR and $^{13}C$ NMR data of compound **9** in $CDCl_3$.

| Position | $\delta_C$ | $\delta_H$ (*J* in Hz) |
|---|---|---|
| 2 | 178.33 | |
| 3 | 78.69 | |
| 4 | 130.01 | |
| 5 | 125.53 | 7.43 (1H, d, 7.8) |
| 6 | 123.24 | 7.12 (1H, td, 7.6, 1.0) |
| 7 | 130.36 | 7.37 (1H, td, 7.7, 1.2) |
| 8 | 109.08 | 6.89 (1H, d, 7.9) |
| 9 | 142.83 | |
| 10 | 42.84 | 2.06 (1H, m) |
| 11 | 52.94 | 5.36 (1H, s) |
| 12 | | 11.01 (brs) |
| 13 | 168.23 | |
| 14 | 52.8 | 4.14 (1H, dd, 8.2, 3.3) |
| 15 | | 7.09 (brs) |
| 16 | 152.27 | |
| 17 | 100.43 | |
| 18 | 27.04 | 2.24 (1H, s) |
| | | 2.44 (1H, ddd, 13.9, 8.4, 3.8) |
| 19 | 21.04 | 1.83 (2H, m) |
| 20 | 38.79 | 2.28 (2H, m) |
| 21 | 204.84 | |
| 22 | 38.75 | 3.17 (1H, m) |
| 22 | | 2.83 (1H, dd, 14.0, 8.2) |
| 23 | 16.36 | 2.15 (3H, s) |
| 24 | 8.92 | 1.19 (3H, d, 7.0) |
| 25 | 26.65 | 3.23 (3H, s) |

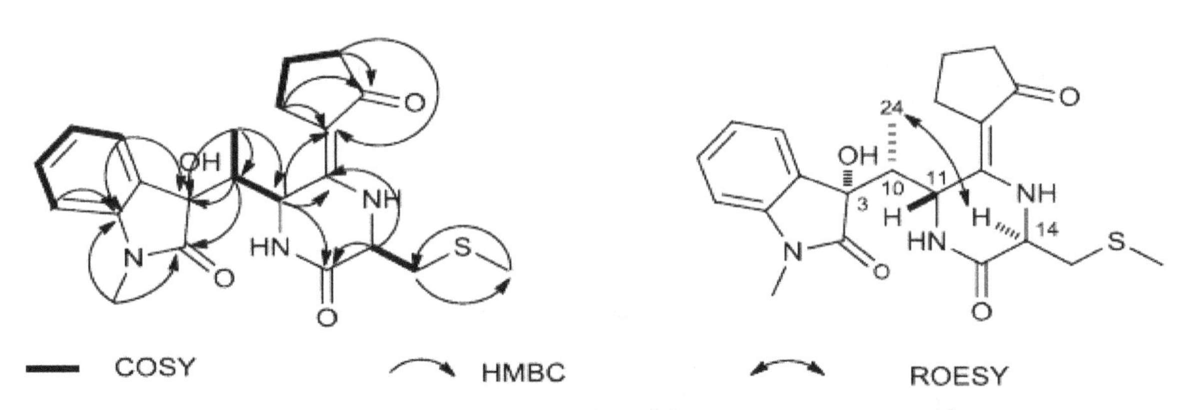


**Figure 4.** Key 1H–1H COSY, HMBC and ROESY correlations of compound **9**.

The relative configuration of compound **9** was determined by interpretation of its ROESY NMR spectrum. The correlations of H-14/Me-24, and H-10/H-11, indicated that H-14 and Me-24 were α-oriented, whereas H-10 and H-11 were β-oriented (Figure 4). Based on the close skeleton, the comparison of ECD spectra between **9** and N-demethylmaremycin B [57], and the largely consistent data supported that the configurations of 3-OH was α-oriented. Ultimately, the absolute configuration of was identified as 3S, 10R, 11R, 14R, resulting from the same trends of cotton effects (CEs) in the experimental ECD spectra of **9** and N-demethylmaremycin B.

The in vitro antifungal activity of compounds **1–9** against *S. sclerotiorum* was determined at various concentrations. All compounds showed significantly antifungal activity against *S. sclerotiorum* with the $EC_{50}$ values ranging from 49.14 mg/L to 0.21 mg/L (Table 3). Thus, it further confirmed that these compounds were the main antifungal constituents produced by the four active strains.

**Table 3.** $EC_{50}$ values of active compounds against *S. sclerotiorum*.

| Compounds | 1 | 2 | 3 | 4 | 5 | 6 | 7 | 8 | 9 |
|---|---|---|---|---|---|---|---|---|---|
| $EC_{50}$ (mg/L) | 4.87 ± 0.16 a | 4.96 ± 0.13 a | 0.21 ± 0.02 b | 49.14 ± 0.82 c | 5.33 ± 0.15 ae | 3.69 ± 0.05 d | 5.60 ± 0.11 e | 3.46 ± 0.12 d | 3.70 ± 0.05 d |

Values are the means ± SE ($n = 9$). Data within the same column followed by different letters are significantly different.

### 3.4. Culture-Independent Communities

A total of 4116 OTUs containing 745708 high-quality reads were detected in the soybean root microbiome. The raw sequencing reads for this project were submitted to the National Center for Biotechnology Information Short Read Archive under accession numbers SRR8056376–SRR8056381. The predominant bacterial phyla were *Proteobacteria*, *Bacteroidetes* and *Actinobacteria* in the soybean roots. To compare the microbial communities obtained in healthy and diseased root samples from each group, the relative abundance of order *Rhizobiales* that can improve rhizobial nodulation and nitrogen fixation was significantly greater in the healthy samples, whereas the order *Actinobacteria* were more abundant in the diseased samples (Figure 5).

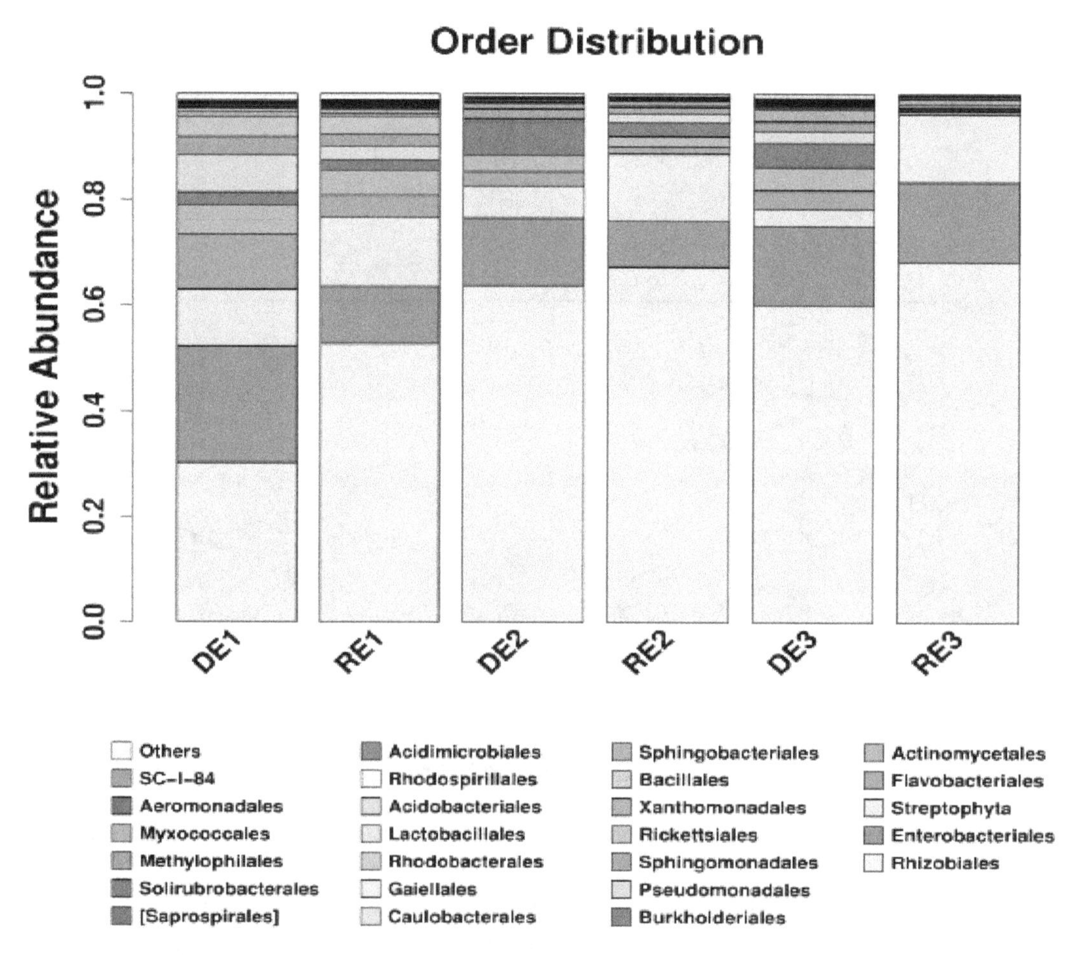


**Figure 5.** Analysis of culture-independent endophytic communities at order level in the soybean roots. DE, diseased sample; RE, healthy sample.

## 4. Discussion

The multifaceted approach adopted in this study, linking culture-independent and culture-dependent analysis, showed that actinobacteria were more abundant or diverse in the diseased soybean roots. This finding was in agreement with the previous study that the phylum *Actinobacteria* was higher in '*Candidatus* Liberibacter asiaticus'-infected citrus samples compared with that in healthy samples [31]. Another similar study also showed that potato plants infected with *Erwinia carotovora* subsp. *atroseptica* increased bacterial diversity [59]. The higher diversity of endophytic actinobacteria in diseased but healthy plants suggests that they may be involved in pathogen defense [60]. Indeed, extensive research has shown that endophytic actinobacteria has the capacity to control plant pathogens [22]. The in vitro antagonism assays demonstrated that four strains showed strong antifungal activity against *S. sclerotiorum*. Among the four antagonistic strains, all colonies of three strains were absolutely from diseased soybean roots, and another strain was also significantly enriched in diseased soybean roots compared to healthy soybean roots. A similar study has also shown that the rhizosphere soil of diseased tomato plant harbored a high percentage of antagonists [61]. Studies over the past few years have provided important information that plants possess a sophisticated defense mechanism by actively recruiting root-associated microbes from soil upon pathogen attack [18,62]. By adjusting the quantity and composition of its root secretion, plants can determine the composition of the root microbiome by affecting microbial diversity, density, and activity [63,64]. Our results seem to be consistent with previous observations. However, those strains with antagonistic activity in vitro may not be simply translated into biocontrol bacteria. Their biocontrol effects are influenced by various factors. For example, the antagonistic strains should reach a certain amount inside the plants to demonstrate a significant biocontrol effect [65,66]. Moreover, their secondary metabolite producing ability inside the plants may be influenced by plant physiological environment. Therefore, further work is required to assess the biocontrol efficiency in vivo and root-colonizing capacity of antagonistic strains by pathogen infection.

To learn more about the chemical nature of the antifungal activity, nine active compounds including six macrolides, two diketopiperazines and one 2-oxonanonoids, were finally obtained. Out of which, bafilomycin $B_1$ (**3**) showed strongest inhibitory activity against *S. sclerotiorum*. Bafilomycin $B_1$ has been reported to be produced by several *Streptomyces* strains and to show inhibitory activity against various fungi in vitro, such as *Rhizoctonia solani*, *Aspergillus fumigatus*, *Botrytis cinerea*, *Penicillium roqueforti*, and so on [67,68]. The antifungal activity of this compound against *S. sclerotiorum* was first reported in this paper. Azalomycins $F_{4a}$ (**1**) and $F_{5a}$ (**2**) were first isolated from the broth of *Streptomyces hygroscopicus* var. *azalomyceticus* [69]. Azalomycins F complex, including azalomycins $F_{3a}$, $F_{4a}$ and $F_{5a}$ showed remarkable antifungal activity against asparagus (*Asparagus officinalis*) pathogens *Fusarium moliniforme* and *Fusarium oxysporumas* as well as powdery mildew pathogen *Botrytis* spp. [70]. The antifungal activity of the pure compound was first demonstrated in our research. Azalomycins possess broad-spectrum antibacterial and antifungal activities, and almost all of them were produced by *Streptomyces*, which were isolated repeatedly from soil and plant roots in the field by our laboratory (data no shown). This emphasizes the possible importance of *Streptomyces* producing azalomycins to protect plants against phytopathogens. A mixture of dinactin (**7**), trinactin and the major component tetranactin (**6**) is known as commercial pesticides polynactin (liuyangmycin in China), which can effectively control spider mites under wet conditions [71]. In addition, tetranactin (**6**) also exhibited significant antifungal activity against plant pathogen *Botrytis cinerea* with a minimum inhibitory concentration (MIC) of 24 $\mu g \cdot mL^{-1}$ [72]. Besides dinactin (**7**) and tetranactin (**6**), the monomer (**4**) and dimer (**5**) of polynactin were also isolated in this study, all of which were active against *S. sclerotiorum*. The antifungal activities of actinolactomycin (**4**), dimeric dinactin (**5**) and dinactin (**7**) have not been reported as yet. The findings reported here shed new light on the application of polynactin. Natural indole diketopiperazines exhibited a wide range of biological activities including antitumor [73], antibacterial [74], antifungal [75] and antiviral activities [76]. FR900452 is sulfur-containing indole diketopiperazines that showed specific and potent inhibitory activity against the platelet aggregation induced by platelet-activating factor [77]. Maremycin G (**8**) and compound **9**, structurally related to FR900452, showed significant antifungal activity against *S. sclerotiorum*. To our knowledge, this is the first report of the antifungal property of maremycins. Further research is needed to confirm the efficacy of in vivo disease control provided by the nine active compounds under laboratory and field conditions.

## 5. Conclusions

In summary, we report that soybean infected by *S. sclerotiorum* (Lib.) de Bary had a higher populations of actinobacteria and enhanced root colonization of antagonistic populations. In addition, eight known compounds and one new compound that exerted significant antifungal activity against *S. sclerotiorum* were obtained. These findings suggest that diseased plant samples could be a potential source for screening novel agroactive compounds, which contribute to a better understanding of plant–microbe interactions and provide new strategies for the development of agricultural antibiotics.

**Author Contributions:** C.L., X.Z., Z.W. and X.G. performed the experiments. Z.Y. and Y.W. analyzed the data. C.L. wrote the paper. W.X. and S.H. designed the experiments and reviewed the manuscript. All authors read and approved the final manuscript.

## References

1. Boland, G.J.; Hall, R. Index of plant hosts of *Sclerotinia sclerotiorum*. *Can. J. Plant. Pathol.* **1994**, *16*, 93–108. [CrossRef]
2. Bardin, S.D.; Huang, H.C. Research on biology and control of *Sclerotinia* diseases in Canada. *Can. J. Plant. Pathol.* **2001**, *23*, 88–98. [CrossRef]
3. Firoz, M.J.; Xiao, X.; Zhu, F.X.; Fu, Y.P.; Jiang, D.H.; Schnabel, G.; Luo, C.X. Exploring mechanisms of resistance to dimethachlone in *Sclerotinia sclerotiorum*. *Pest. Manag. Sci.* **2016**, *72*, 770–779. [CrossRef] [PubMed]
4. Boland, G.J. Stability analysis for evaluating the influence of environment on chemical and biological control of white mold (*Sclerotinia sclerotiorum*) of bean. *Biol. Control.* **1997**, *9*, 7–14. [CrossRef]
5. Lee, Y.H.; Cho, Y.S.; Lee, S.W.; Hong, J.K. Chemical and biological controls of balloon flower stem rots caused by *Rhizoctonia solani* and *Sclerotinia sclerotiorum*. *Plant. Pathol. J.* **2012**, *28*, 156–163. [CrossRef]
6. Hu, S.; Zhang, J.; Zhang, Y.; He, S.; Zhu, F. Baseline sensitivity and toxic actions of boscalid against *Sclerotinia sclerotiorum*. *Crop. Prot.* **2018**, *110*, 83–90. [CrossRef]
7. Ma, H.X.; Feng, X.J.; Chen, Y.; Chen, C.J.; Zhou, M.G. Occurrence and characterization of dimethachlon insensitivity in *Sclerotinia sclerotiorum* in jiangsu province of China. *Plant. Dis.* **2009**, *93*, 36–42. [CrossRef] [PubMed]
8. Kuang, J.; Hou, Y.P.; Wang, J.X.; Zhou, M.G. Sensitivity of *Sclerotinia sclerotiorum* to fludioxonil: in vitro determination of baseline sensitivity and resistance risk. *Crop. Prot.* **2011**, *30*, 876–882. [CrossRef]
9. Zhou, F.; Zhang, X.L.; Li, J.L.; Zhu, F.X. Dimethachlon resistance in *Sclerotinia sclerotiorum* in China. *Plant. Dis.* **2014**, *98*, 1221–1226. [CrossRef]
10. Smith, S.A.; Tank, D.C.; Boulanger, L.A.; Bascom-Slack, C.A.; Eisenman, K.; Kingery, D. Bioactive endophytes warrant intensified exploration and conservation. *PLoS ONE* **2008**, *3*, e3052. [CrossRef]
11. Sturz, A.V.; Christie, B.R.; Nowak, J. Bacterial endophytes: potential role in developing sustainable systems of crop production. *Crit. Rev. Plant. Sci.* **2000**, *19*, 1–30. [CrossRef]
12. Lodewyckx, C.; Vangronsveld, J.; Porteous, F.; Moore, E.R.B.; Taghavi, S.; Mezgeay, M.; van der Lelie, D. Endophytic bacteria and their potential applications. *CRC. Crit. Rev. Plant. Sci.* **2002**, *21*, 583–606. [CrossRef]
13. Bais, H.P.; Weir, T.L.; Perry, L.G.; Gilroy, S.; Vivanco, J.M. The role of root exudates in rhizosphere interactions with plants and other organisms. *Annu. Rev. Plant. Biol.* **2006**, *57*, 233–266. [CrossRef]
14. Schulz, B.; Boyle, C. What are endophytes? In *Microbial Root Endophytes*; Schulz, B.J.E., Boyle, C.J.C., Sieber, T.N., Eds.; Springer: Berlin, Germany; pp. 1–13.
15. Sorensen, J.; Sessitsch, A. Plant-associated bacteria—Lifestyle and molecular interactions. In *Modern Soil Microbiology*; Van Elsas, J.D., Jansson, J.K., Trevors, J.T., Eds.; CRC Press: Boca Raton, FL, USA; pp. 211–236.
16. Mendes, R.; Kruijt, M.; de Bruijn, I.; Dekkers, E.; van der Voort, M.; Schneider, J.H.M.; Piceno, Y.M.; DeSantis, T.Z.; Andersen, G.L.; Bakker, P.A.H.M.; et al. Deciphering the rhizosphere microbiome for disease-suppressive bacteria. *Science* **2011**, *332*, 1097–1100. [CrossRef]
17. Liu, C.X.; Song, J.; Wang, X.J.; Xiang, W.S. Recruitment of defensive microbs and plant protection. *Sci. Sin. Vitae* **2016**, *46*, 1–8.
18. Rudrappa, T.; Czymmek, K.J.; Paré, P.W.; Bais, H.P. Root-secreted malic acid recruits beneficial soil bacteria. *Plant. Physiol.* **2008**, *148*, 1547–1556. [CrossRef]

19. Bèrdy, J. Bioactive microbial metabolites. *J. Antibiot.* **2005**, *58*, 1–26. [CrossRef]
20. Qin, S.; Li, W.J.; Dastager, S.G.; Hozzein, W.N. Editorial: Actinobacteria in special and extreme habitats: Diversity, function roles, and environmental adaptations. *Front. Microbiol.* **2016**, *7*, 1415. [CrossRef]
21. Singh, R.; Dubey, A.K. Diversity and applications of endophytic actinobacteria of plants in special and other ecological niches. *Front. Microbiol.* **2018**, *9*, 1767. [CrossRef]
22. Ek-Ramos, M.J.; Gomez-Flores, R.; Orozco-Flores, A.A.; Rodríguez-Padilla, C.; González-Ochoa, G.; Patricia Tamez-Guerra, P. Bioactive products from plant-endophytic Gram-positive bacteria. *Front. Microbiol.* **2019**, *10*, 463. [CrossRef]
23. Conn, V.M.; Franco, C.M. Isolation and identification of actinobacteria from surface-sterilized wheat roots. *Appl. Environ. Microbiol.* **2003**, *69*, 5603–5608.
24. Tian, X.; Cao, L.; Tan, H.; Han, W.; Chen, M.; Liu, Y.; Zhou, S. Diversity of cultivated and uncultivated actinobacterial endophytes in the stems and roots of rice. *Microb. Ecol.* **2007**, *53*, 700–707. [CrossRef]
25. Zhao, K.; Penttinen, P.; Guan, T.W.; Xiao, J.; Chen, Q.; Xu, J.; Lindström, K.; Zhang, L. The diversity and antimicrobial activity of endophytic actinomycetes isolated from medicinal plants in Panxi plateau, China. *Curr. Microbiol.* **2011**, *62*, 182–190. [CrossRef]
26. Li, J.; Zhao, G.Z.; Huang, H.Y.; Qin, S.; Zhu, W.Y.; Zhao, L.X.; Xu, L.H.; Zhang, S.; Li, W.J.; Strobel, G. Isolation and characterization of culturable endophytic actinobacteria associated with *Artemisia annua* L. *Antonie. Leeuwenhoek.* **2012**, *101*, 515–527. [CrossRef]
27. Miao, G.P.; Zhu, C.S.; Feng, J.T.; Han, L.R.; Zhang, X. Effects of plant stress signal molecules on the production of wilforgine in an endophytic actinomycete isolated from *Tripterygium wilfordii* Hook. f. *Curr. Microbiol.* **2015**, *70*, 571–579. [CrossRef]
28. Wei, W.; Zhou, Y.; Chen, F.; Yan, X.; Lai, Y.; Wei, C.; Chen, X.; Xu, J.; Wang, X. Isolation, diversity, and antimicrobial and immunomodulatory activities of endophytic actinobacteria from tea cultivars Zijuan and Yunkang-10 (*Camellia sinensis* var assamica). *Front. Microbiol.* **2018**, *9*, 1304. [CrossRef]
29. Bulgarelli, D.; Rott, M.; Schlaeppi, K.; Ver Loren van Themaat, E.; Nahal Ahmadinejad, N.; Assenza, F.; Rauf, P.; Huettel, B.; Reinhardt, R.; Schmelzer, E.; et al. Revealing structure and assembly cues for *Arabidopsis* root-inhabiting bacterial microbiota. *Nature* **2012**, *488*, 91–95. [CrossRef]
30. Lundberg, D.S.; Lebeis, S.L.; Paredes, S.H.; Yourstone, S.; Gehring, J.; Malfatti, S.; Tremblay, J.; Engelbrektson, A.; Kunin, V.; del Rio, T.G.; et al. Defining the core *Arabidopsis thaliana* root microbiome. *Nature* **2012**, *488*, 86–90. [CrossRef]
31. Trivedi, P.; He, Z.; Van Nostrand, J.D.; Albrigo, G.; Zhou, J.; Wang, N. Huanglongbing alters the structure and functional diversity of microbial communities associated with citrus rhizosphere. *ISME. J.* **2012**, *6*, 363–383. [CrossRef]
32. Boland, G.J.; Hall, R. Growth room evaluation of soybean cultivars for resistance to *Sclerotinia sclerotiorum*. *Can. J. Plant. Sci.* **1986**, *66*, 559–564. [CrossRef]
33. Hayakawa, M.; Nonomura, H. Humic acid-vitamin agar, a new medium for the selective isolation of soil actinomycetes. *J. Ferment. Technol.* **1987**, *65*, 501–509. [CrossRef]
34. Atlas, R.M. *Handbook of Microbiological Media*; Parks, L.C., Ed.; CRC Press: Boca Raton, FL, USA, 1993.
35. Guan, X.J.; Liu, C.X.; Fang, B.Z.; Zhao, J.W.; Jin, P.J.; Li, J.M. *Baia soyae* gen. nov., sp. nov., a mesophilic representative of the family *Thermoactinomycetaceae*, isolated from soybean root [*Glycine max* (L.) Merr]. *Int. J. Syst. Evol. Microbiol.* **2015**, *65*, 3241–3247. [CrossRef]
36. Qin, S.; Li, J.; Chen, H.H.; Zhao, G.Z.; Zhu, W.Y.; Jiang, C.L.; Xu, L.H.; Li, W.J. Isolation, diversity, and antimicrobial activity of rare actinobacteria from medicinal plants of tropical rain forests in Xishuangbanna, China. *Appl. Environ. Microbiol.* **2009**, *75*, 6176–6186. [CrossRef]
37. Shirling, E.B.; Gottlieb, D. Methods for characterization of *Streptomyces* species. *Int. J. Syst. Bacteriol.* **1966**, *16*, 313–340. [CrossRef]
38. Maniatis, T.; Fritsch, E.F.; Sambrook, J. *Molecular Cloning: A Laboratory Manual*, 2nd ed.; Cold Spring Harbor Laboratory: Cold Spring Harbor, NY, USA, 1982.
39. Kim, S.B.; Brown, R.; Oldfield, C.; Gilbert, S.C.; Iliarionov, S.; Goodfellow, M. *Gordonia amicalis* sp. nov., a novel dibenzothiophene-desulphurizing actinomycete. *Int. J. Syst. Evol. Microbiol.* **2000**, *50*, 2031–2036. [CrossRef]
40. Saitou, N.; Nei, M. The neighbor-joining method: A new method for reconstructing phylogenetic trees. *Mol. Biol. Evol.* **1987**, *4*, 406–425.

41. Kumar, S.; Stecher, G.; Tamura, K. MEGA7: molecular evolutionary genetics analysis version 7.0 for bigger datasets. *Mol. Biol. Evol.* **2016**, *33*, 1870. [CrossRef]
42. Felsenstein, J. Confidence limits on phylogenies: an approach using the bootstrap. *Evolution* **1985**, *39*, 783–791. [CrossRef]
43. Kimura, M. *The Neutral Theory of Molecular Evolution*; Cambridge Universiry Press: Cambridge, UK, 1983.
44. Yoon, S.H.; Ha, S.M.; Kwon, S.; Lim, J.; Kim, Y.; Seo, H.; Chun, J. Introducing EzBioCloud: A taxonomically united database of 16S rRNA gene sequences and whole-genome assemblies. *Int. J. Syst. Evol. Microbiol.* **2017**, *67*, 1613–1617.
45. Hamzah, T.N.T.H.; Lee, S.Y.; Hidayat, A.; Terhem, R.; Faridah-Hanum, I.; Mohamed, R. Diversity and characterization of endophytic fungi isolated from the tropical mangrove species, *Rhizophora mucronata*, and identification of potential antagonists against the soil-borne fungus, *Fusarium solani*. *Front. Microbiol.* **2018**, *9*, 1707. [CrossRef]
46. Zhang, J.; Wang, X.J.; Yan, Y.J.; Jiang, L.; Wang, J.D.; Li, B.J.; Xiang, W.S. Isolation and identification of 5-hydroxyl-5-methyl-2-hexenoic acid from *Actinoplanes* sp. HBDN08 with antifungal activity. *Bioresource. Technol.* **2010**, *101*, 8383–8388. [CrossRef]
47. Liu, C.X.; Zhang, J.; Wang, X.J.; Qian, P.T.; Wang, J.D.; Gao, Y.M.; Yan, Y.J.; Zhang, S.Z.; Xu, P.F.; Li, W.B.; et al. Antifungal activity of borrelidin produced by a *Streptomyces* strain isolated from soybean. *J. Agric. Food. Chem.* **2012**, *60*, 1251–1257. [CrossRef]
48. Zhang, Y.J.; Liu, C.X.; Zhang, J.; Shen, Y.; Li, C.; He, H.R.; Wang, X.J.; Xiang, W.S. *Actinomycetospora atypica* sp. nov., a novel soil actinomycete and emended description of the genus *Actinomycetospora*. *Antonie. Leeuwenhoek.* **2014**, *105*, 891–897. [CrossRef]
49. Zhang, J.; Kobert, K.; Flouri, T.; Stamatakis, A. PEAR: A fast and accurate Illumina Paired-End reAd mergeR. *Bioinformatics* **2014**, *30*, 614–620. [CrossRef]
50. Edgar, R.C. Search and clustering orders of magnitude faster than BLAST. *Bioinformatics* **2010**, *26*, 2460–2461. [CrossRef]
51. Caporaso, J.G.; Kuczynski, J.; Stombaugh, J.; Bittinger, K.; Bushman, F.D.; Costello, E.K.; Fierer, N.; Peña, A.G.; Goodrich, J.K.; Gordon, J.I.; et al. QIIME allows analysis of high-throughput community sequencing data. *Nat. Methods* **2010**, *7*, 335–336. [CrossRef]
52. Yuan, G.; Lin, H.; Wang, C.; Hong, K.; Liu, Y.; Li, J. $^{1}$H and $^{13}$C assignments of two new macrocyclic lactones isolated from *Streptomyces* sp. 211726 and revised assignments of azalomycins $F_{3a}$, $F_{4a}$ and $F_{5a}$. *Magn. Reson. Chem.* **2011**, *49*, 30–37. [CrossRef]
53. Crevelin, E.J.; Canova, S.P.; Melo, I.S.; Zucchi, T.D.; da Silva, R.E.; Moraes, L.A. Isolation and characterization of phytotoxic compounds produced by *Streptomyces* sp. AMC 23 from red mangrove (*Rhizophora mangle*). *Appl. Biochem. Biotechnol.* **2013**, *171*, 1602–1616. [CrossRef]
54. Han, B.; Cui, C.B.; Cai, B.; Ji, X.F.; Yao, X.S. Actinolactomycin, a new 2-oxonanonoidal antitumor antibiotic produced by *Streptomyces flavoretus* 18522, and its inhibitory effect on the proliferation of human cancer cells. *Chin. Chem. Lett.* **2005**, *4*, 471–474.
55. Zhao, P.J.; Fan, L.M.; Li, G.H.; Zhu, N.; Shen, Y.M. Antibacterial and antitumor macrolides from *Streptomyces* sp. Is9131. *Arch. Pharm. Res.* **2005**, *28*, 1228–1232. [CrossRef]
56. Shishlyannikova, T.A.; Kuzmin, A.V.; Fedorova, G.A.; Shishlyannikov, S.M.; Lipko, I.A.; Sukhanova, E.V.; Belkova, N.L. Ionofore antibiotic polynactin produced by *Streptomyces* sp. 156A isolated from Lake Baikal. *Nat. Prod. Res.* **2017**, *31*, 639–644. [CrossRef]
57. Lan, Y.X.; Zou, Y.; Huang, T.T.; Wang, X.Z.; Brock, N.L.; Deng, Z.X.; Lin, S.J. Indole methylation protects diketopiperazine configuration in the maremycin biosynthetic pathway. *Sci. Chin. Chem.* **2016**, *59*, 1224–1228. [CrossRef]

# Biosynthesis of Polyketides in *Streptomyces*

**Chandra Risdian [1,2,*], Tjandrawati Mozef [3] and Joachim Wink [1,*]**

1 Microbial Strain Collection (MISG), Helmholtz Centre for Infection Research (HZI), 38124 Braunschweig, Germany
2 Research Unit for Clean Technology, Indonesian Institute of Sciences (LIPI), Bandung 40135, Indonesia
3 Research Center for Chemistry, Indonesian Institute of Sciences (LIPI), Serpong 15314, Indonesia; tjandrawm@yahoo.com
* Correspondence: chandra.risdian@gmail.com or chandra.risdian@helmholtz-hzi.de (C.R.); Joachim.Wink@helmholtz-hzi.de (J.W.)

**Abstract:** Polyketides are a large group of secondary metabolites that have notable variety in their structure and function. Polyketides exhibit a wide range of bioactivities such as antibacterial, antifungal, anticancer, antiviral, immune-suppressing, anti-cholesterol, and anti-inflammatory activity. Naturally, they are found in bacteria, fungi, plants, protists, insects, mollusks, and sponges. *Streptomyces* is a genus of Gram-positive bacteria that has a filamentous form like fungi. This genus is best known as one of the polyketides producers. Some examples of polyketides produced by *Streptomyces* are rapamycin, oleandomycin, actinorhodin, daunorubicin, and caprazamycin. Biosynthesis of polyketides involves a group of enzyme activities called polyketide synthases (PKSs). There are three types of PKSs (type I, type II, and type III) in *Streptomyces* responsible for producing polyketides. This paper focuses on the biosynthesis of polyketides in *Streptomyces* with three structurally-different types of PKSs.

**Keywords:** *Streptomyces*; polyketides; secondary metabolite; polyketide synthases (PKSs)

---

## 1. Introduction

Polyketides, a large group of secondary metabolites, are known to possess remarkable variety, not only in their structure, and but also in their function [1,2]. Polyketides exhibit a wide range of bioactivities such as antibacterial (e.g., tetracycline), antifungal (e.g., amphotericin B), anticancer (e.g., doxorubicin), antiviral (e.g., balticolid), immune-suppressing (e.g., rapamycin), anti-cholesterol (e.g., lovastatin), and anti-inflammatory activity (e.g., flavonoids) [3–9]. Some organisms can produce polyketides such as bacteria (e.g., tetracycline from *Streptomyces aureofaciens*) [10], fungi (e.g., lovastatin from *Phomopsis vexans*) [11], plants (e.g., emodin from *Rheum palmatum*) [12], protists (e.g., maitotoxin-1 from *Gambierdiscus australes*) [13], insects (e.g., stegobinone from *Stegobium paniceum*) [14], and mollusks (e.g., elysione from *Elysia viridis*) [15]. These organisms could use the polyketides they produce as protective compounds and for pheromonal communication in the case for insects.

Since the beginning of the 1940s, the history of antibiotics has greatly related to microorganisms. One of the groups of bacteria that produce many important antibiotics is Actinobacteria. Actinobacteria are Gram-positive, have high GC content, and comprise various genera known for their secondary metabolite production, such as *Streptomyces*, *Micromonospora*, *Kitasatospora*, *Nocardiopsis*, *Pseudonocardia*, *Nocardia*, *Actinoplanes*, *Saccharopolyspora*, and *Amycolatopsis* [16,17]. Their most important genus is *Streptomyces*, which has a filamentous form like fungi and has become a source of around two-thirds of all known natural antibiotics [18]. Among the antibiotics produced by *Streptomyces*, polyketides are one group of the very important compounds. Some examples of polyketides produced by *Streptomyces* are rapamycin (produced by *Streptomyces hygroscopicus*), oleandomycin (produced by *Streptomyces antibioticus*), actinorhodin (produced by *Streptomyces coelicolor*A3(2)), daunorubicin (produced by *Streptomyces peucetius*) and caprazamycin (produced *by Streptomyces* sp. MK730-62F2) [19–23].

Biosynthesis of polyketides is very complex because the process involves multifunctional enzymes called polyketide synthases (PKSs). The mechanism of PKS is similar to fatty acid synthase (FAS). The process includes many enzymatic reactions with different enzymes such as acyltransferase (AT), which has a role in catalyzing the attachment of the substrate (e.g., acetyl or malonyl) to the acyl carrier protein (ACP), and ketosynthase (KS), which catalyzes the condensation of substrates attached in ACP. After condensation of the substrates, the reaction continues by incorporating ketoreductase (KR), which reduces keto ester, dehydratase (DH), which dehydrates the compound, and enoylreductase (ER), which reduces the carbon-carbon double bond in the molecule (Figure 1). Unlike in FAS, the process catalyzed by KR, DH, and ER is optional in PKSs, which can give the various structures of polyketides with keto groups, hydroxy groups, and/or double bonds in different locations of the molecule [24–26]. In *Streptomyces*, there are three types of PKSs (type I, type II, and type III) [27–29]. This review describes the biosynthesis of polyketides in *Streptomyces* with three distinct types of PKSs. The focus is only on the *Streptomyces* genus because it is one of the most important producers of bioactive compounds and one of the most well-studied microbes in terms of polyketide biosynthesis. To the best of our knowledge, this is the first review that describes the three types of PKSs that are involved in the biosynthesis of polyketides in *Streptomyces*.

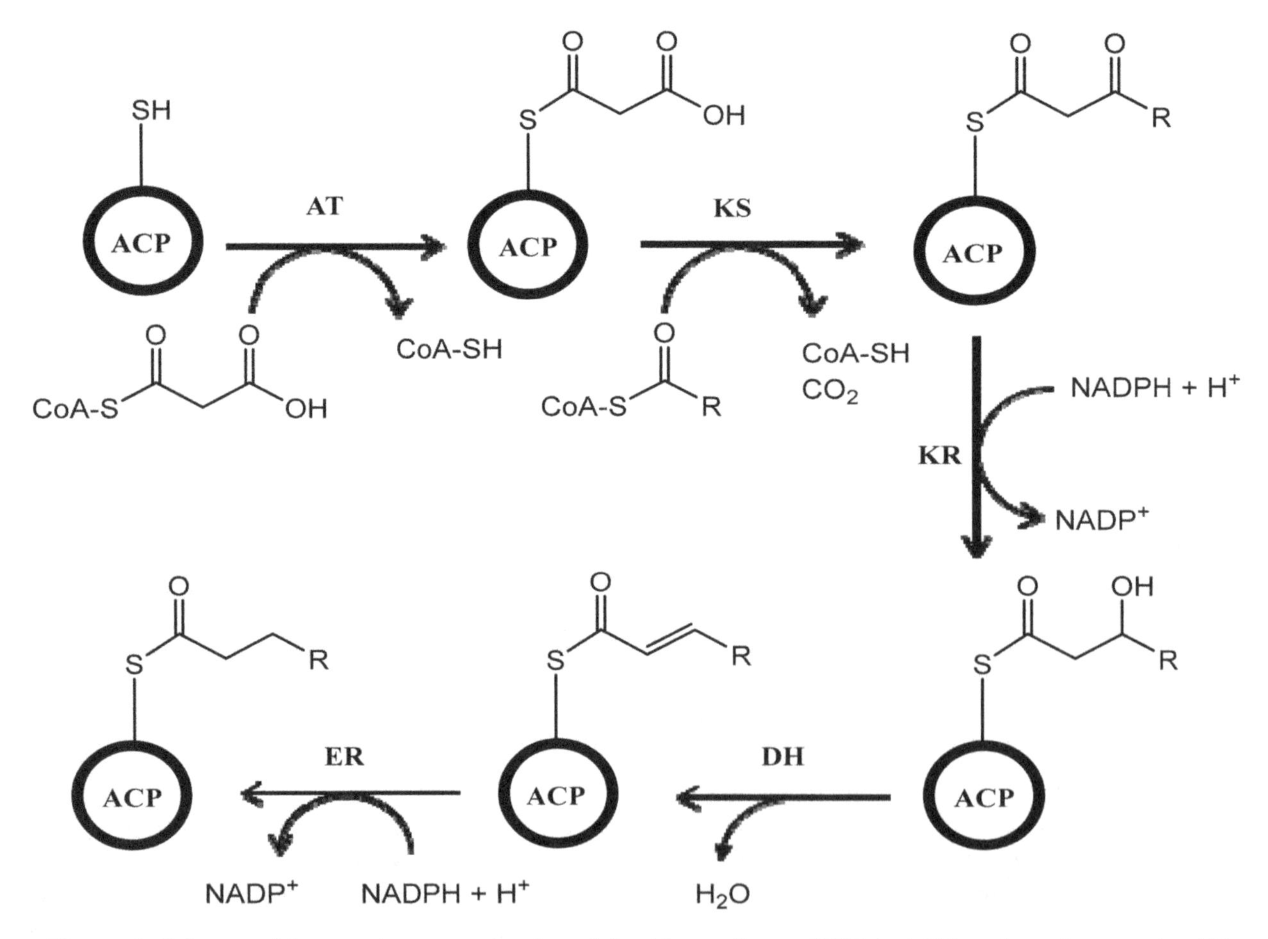


**Figure 1.** Scheme of the reaction occurring in polyketide synthases (PKSs). ACP, acyl carrier protein; AT, acyltransferase; KS, ketosynthase; KR, ketoreductase; DH, dehydratase; ER, enoylreductase. Adapted with permission of Portland Press, from Vance, S.; Tkachenko, O.; Thomas, B.; Bassuni, M.; Hong, H.; Nietlispach, D.; Broadhurst, W. Sticky swinging arm dynamics: studies of an acyl carrier protein domain from the mycolactone polyketide synthase. *Biochem. J.* **2016**, *473*, 1097–1110 [30].

## 2. Polyketide Synthases Type I

The type I polyketide synthases (type I PKSs) involve huge multifunctional proteins that have many modules containing some domains, in which a particular enzymatic reaction occurs (Figure 2). Each module has the responsibility of performing one condensation cycle in a non-iterative way. Because this system works with some modules, it is also called modular PKS. The essential domains existing in each module are acyltransferase (AT), keto synthase (KS), and acyl carrier protein (ACP) which collaborate to produce β-keto ester intermediate. In addition, the other domains that may be present in the module are β-ketoreductase (KR), dehydratase (DH), and enoyl reductase (ER), which are responsible for keto group modification. In the process of producing polyketide, the expanding polyketide chain is transferred from one module to another module until the completed molecule is liberated from the last module by a special enzyme [2,26,31].

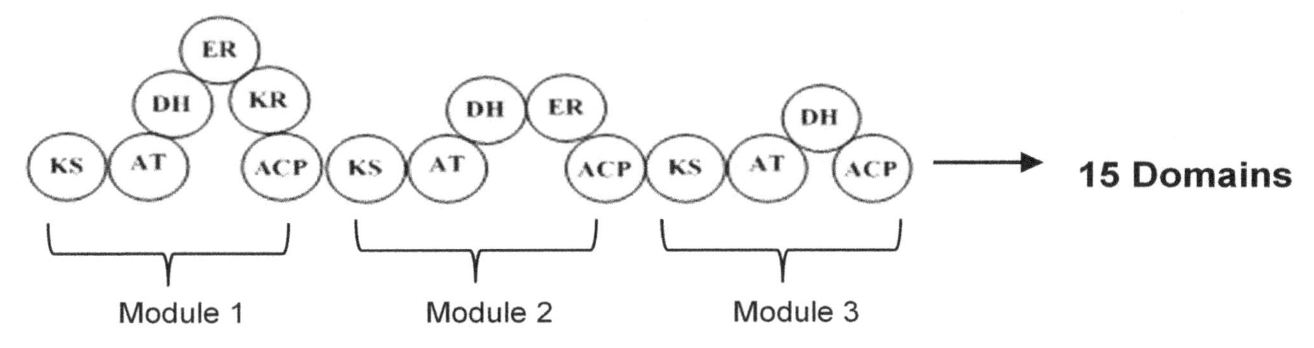


**Figure 2.** Structure of type I PKSs with three modules and 15 domains. ACP, acyl carrier protein; AT, acyltransferase; KS, ketosynthase; KR, ketoreductase; DH, dehydratase; ER, enoylreductase.

Furthermore, type I PKSs are generally responsible for producing macrocyclic polyketides (macrolides), although there was also a study reporting that type I PKSs are also involved in the biosynthesis of linear polyketide tautomycetin [32]. Macrolide belongs to a polyketide compound characterized by a macrocyclic lactone ring, which has various bioactivities such as antibacterial, antifungal, immunosuppressing, and anticancer. As an antibacterial agent, macrolide works by inhibiting protein synthesis by binding to the 50S ribosomal subunit and blocking the translocation steps of protein synthesis [8,27,33]. Some examples of macrolides produced by *Streptomyces* are rapamycin, FK506, spiramycin, avermectin, methymycin, narbomycin, and pikromycin, as shown in Figure 3 [34–37]. These compounds were produced by multifunctional polypeptides encoded by a biosynthetic gene cluster. The list of some polyketides produced by *Streptomyces* with their huge multifunctional proteins can be seen in Table 1.

### 2.1. *Biosynthesis of Rapamycin*

Rapamycin is a 31-membered ring macrolide produced by *Streptomyces hygroscopicus* isolated firstly from the soil of Easter Island (Chile) in the South Pacific Ocean. It is a hydrophobic compound and known as an antifungal compound against *Candida albicans*, *Cryptococcus neoformans*, *Aspergillus fumigatus*, *Fusarium oxysporum*, and some pathogenic species from the genus *Penicillium*. The antifungal mechanism of this compound has been described by diffusing into the cell and binding to intracellular receptor immunophilin FKB12. The FKBP12-rapamycin complexes inhibit enzymes required for signal transduction and cell growth. These enzymes are TOR (target of rapamycin) kinases that are conserved and very important for cell cycle progression. Interestingly, it was also reported that rapamycin has not only antifungal activity, but also anticancer and immunosuppressant activity [8,27,38,39].

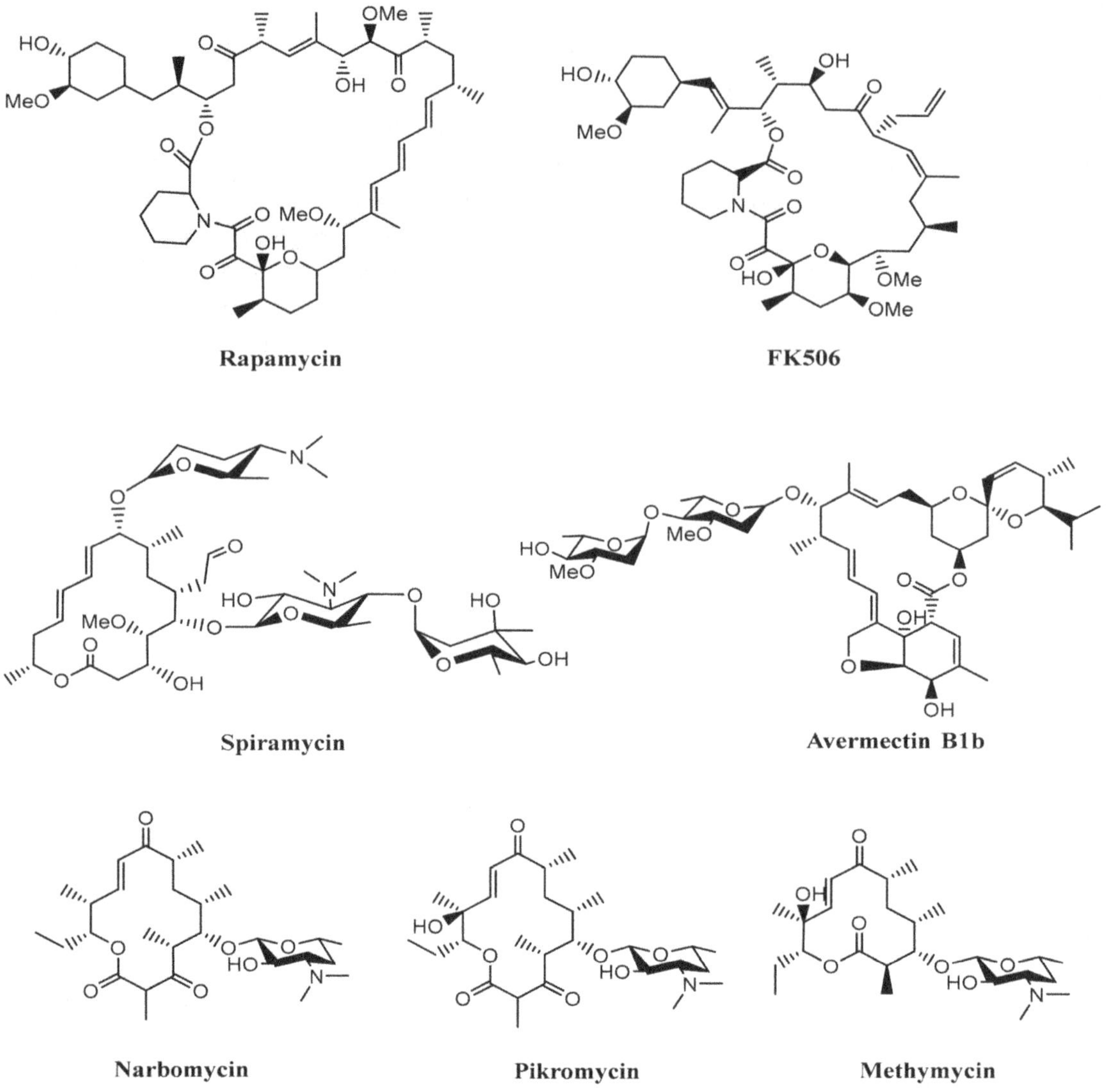

Rapamycin FK506

Spiramycin Avermectin B1b

Narbomycin Pikromycin Methymycin

**Figure 3.** Some of the macrolides produced by *Streptomyces*.

Rapamycin is synthesized by type I PKS rapamycin synthase (RAPS) [40]. The rapamycin-PKS gene cluster (*rapPKS*) is 107.3 kb in size and has three remarkably large ORFs (open reading frames), *rapA*, *rapB*, and *rapC* which encode multifunctional protein RAPS1 (~900 kDa), RAPS2 (~1.07 MDa), and RAPS3 (~660 kDa), respectively. Protein RAPS1 comprises four modules for polyketide chain extension; protein RAPS2 contains six modules responsible for continuing the process of polyketide chain elongation until C-16; and RAPS3 possesses four modules that have a role in completing the polyketide fraction of the rapamycin molecule. Overall, these three giant proteins encompass 70 domains or enzymatic functions, and because of this, rapamycin PKSs are considered as the most complex multienzyme system discovered so far [26,27,34].

In rapamycin PKSs, there is a loading domain (LD) before module 1. In LD, there are three domains, i.e., coenzyme A ligase (CL), enoylreductase (ER), and acyl carrier protein (ACP) domain, which are considered to play a role in activating, reducing a free shikimic-acid-derived moiety starter unit, and finally passing it to the ketosynthase (KS) domain of the first module, respectively. The extender units required for producing rapamycin are malonyl-CoA and methylmalonyl-CoA. The mechanism of transferring from the last domain in rapamycin PKSs and cyclisation of polyketide molecule is assisted by pipecolate-incorporating enzyme (PIE), as depicted in Figure 4. This enzyme (170 kDa) is encoded by gene *rapP*, which is also located in the *rapPKS* gene cluster [26,27,34].

**Table 1.** Some polyketides produced by *Streptomyces* and their type I PKSs.

| Polyketide | Structure | Producer | Type I PKSs | Ref. |
|---|---|---|---|---|
| Avermectin | 16-membered ring macrolide | *Streptomyces avermitilis* | AVES1-4 | [41] |
| Chalcomycin | 16-membered ring macrolide | *Streptomyces bikiniensis* | ChmGI-V | [42] |
| Candicidin | 38-membered ring polyene macrolide | *Streptomyces griseus* IMRU 3570 | CanP1-3, and CanPF | [43,44] |
| FK506 (Tacrolimus) | 23-membered ring macrolide | *Streptomyces tsukubaensis*, *Streptomyces* sp. *MA6858* | FkbABC | [35,45] |
| FK520 (Ascomycin) | 23-membered ring macrolide | *Streptomyces hygroscopicus* var. *ascomyceticus* | FkbABC | [46] |
| Methymycin, Neomethymycin, Narbomycin, Pikromycin | 12-membered ring macrolide, 12-membered ring macrolide, 14-membered ring macrolide, 14- membered ring macrolide | *Streptomyces venezuelae* | PikAI-IV | [37] |
| Pimaricin | 26-membered ring polyene macrolide | *Streptomyces natalensis* | PIMS0 and PIMS1 | [47] |
| Rapamycin | 31- membered ring macrolide | *Streptomyces hygroscopicus* | RAPS1-3 | [34] |
| Spiramycin | 16- membered ring macrolide | *Streptomyces ambofaciens* | SrmGI-V | [36] |
| Tautomycetin | Linear | *Streptomyces* sp. *CK4412* | TmcA and TmcB | [32] |
| Tylosin | 16- membered ring macrolide | *Streptomyces fradiae* | TYLGI-V | [48] |

### 2.2. *Biosynthesis of Avermectin*

Avermectin is a 16-membered ring macrolide and one of the notable anthelmintic compounds produced by *Streptomyces avermitilis* [41,49]. The biosynthesis of avermectin involves type I PKSs (AVES1, AVES2, AVES3, and AVES4). AVES1 (414 kDa) contains one loading domain and two modules; AVES2 (666 kDa) consists of four modules; AVES 3 (575 kDa) comprises three modules; and AVES4 (510 kDa) has three modules. The process of avermectin biosynthesis includes assembling of the polyketide-derived initial aglycon (6, 8a-seco-6, 8a-deoxy-5-oxoavermectin aglycons) by AVES1–4, alteration of the initial aglycon to avermectin aglycons, and, as the last step, the glycosylation of avermectin aglycons to produce avermectins. The starter unit for avermectin biosynthesis is isobutyryl-CoA (derived from valine) or 2-methylbutyryl-CoA (derived from isoleucine), whereas the extender units involved in the production of avermectin are seven malonyl-CoAs (for acetate units) and five methylmalonyl-CoAs (for propionate units). The nucleotide sequence of the avermectin biosynthetic gene cluster comprises 18 ORFs spanning a distance of 82 kb, in which four large ORFs encode the avermectin polyketide synthase (AVES1, AVES2, AVES3, and AVES4) and some of the 14 ORFs encode polypeptides having important roles in avermectin biosynthesis [41].

### 2.3. *Biosynthesis of Candicidin*

Candicidin is a 38-membered ring polyene macrolide produced by *Streptomyces griseus* IMRU 3570 that has antifungal activity. Like the other polyene compounds, the antifungal mechanism of candicidin is also by disrupting the fungal cell membrane. Candicidin has both the amino sugar mycosamine and the aromatic component p-aminoacetophenone in its macrolide structure [43,50].

The candicidin biosynthetic gene cluster (<205 kb) was cloned and partially sequenced. Four genes, canP1, canP3, canP2 (incomplete), and canPF (incomplete), were determined as genes encoding parts of type I PKSs (CanP1, CanP2, CanP3, and CanPF). CanP1 contains one loading domain and one module; CanP2 consists of three modules; and CanP2 comprises six modules. CanPF hypothetically serves as one end of the PKS gene cluster. The starter unit is PABA (p-aminobenzoic acid), and the extender units are four methylmalonyl-CoAs and 17 malonyl-CoAs. At the end of the process in PKS, the molecule is released by thioesterase (CanT). In the next step, the compound is cyclized to become candicidin aglycone, oxidized by P450 monooxygenase (CanC) with aid from ferredoxin (CanF). The last step is glycosylation by adding mycosamine to the structure [43,44].

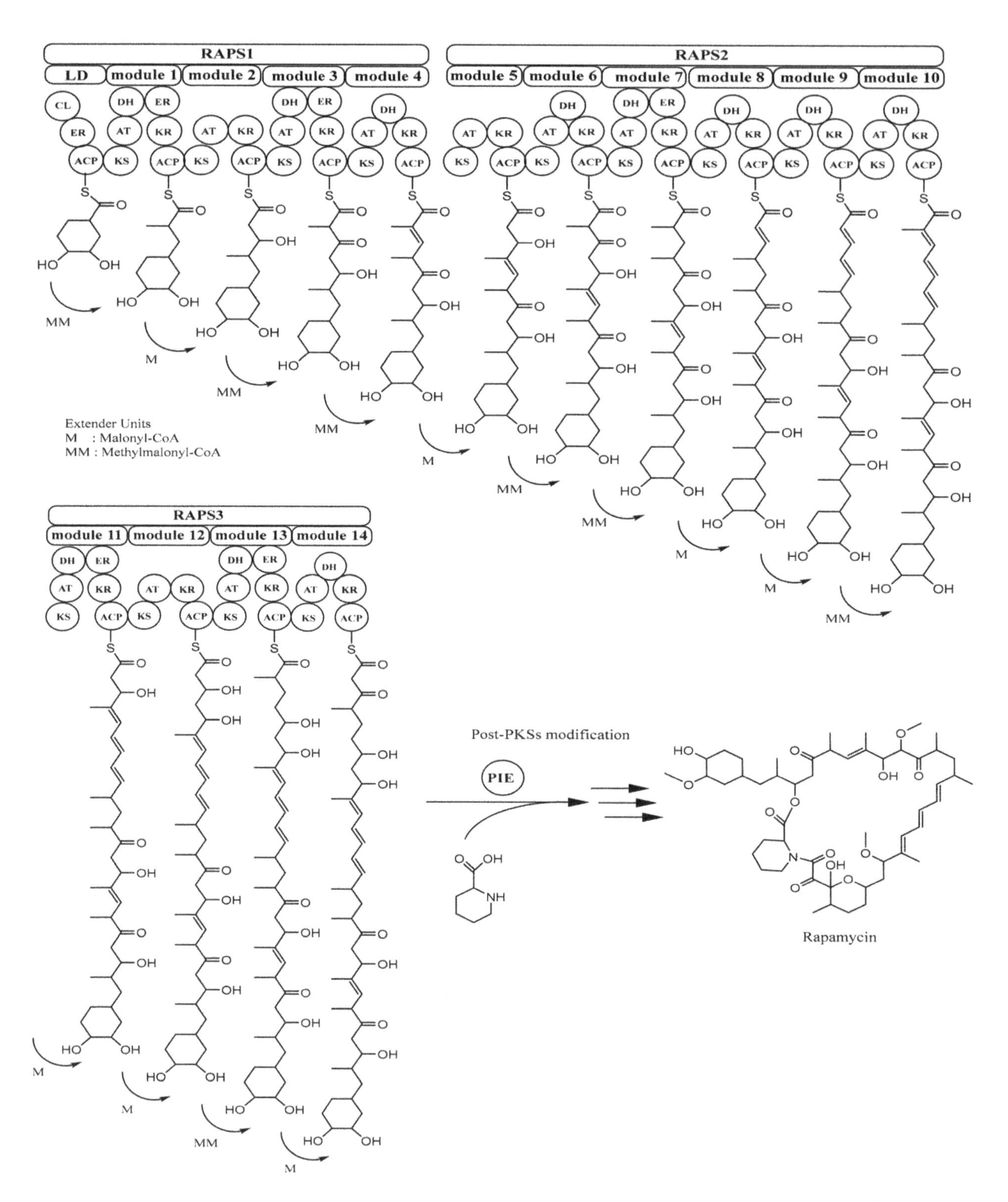


**Figure 4.** Biosynthesis of rapamycin. ACP, acyl carrier protein; AT, acyltransferase; KS, ketosynthase; KR, ketoreductase; DH, dehydratase; ER, enoylreductase; PIE, pipecolate-incorporating enzyme. Adapted with permission from Schwecke, T.; Aparicio, J.F.; Molnár, I.; König, A; Khaw, L.E.; Haydock, S.F.; Oliynyk, M.; Caffrey, P.; Cortés, J.; Lester, J.B. The biosynthetic gene cluster for the polyketide immunosuppressant rapamycin. Proc. Natl. Acad. Sci. USA 1995, 92, 7839–7843, doi:10.1073/pnas.92.17.7839 [34]. Copyright (1995) National Academy of Sciences, U.S.A. Adapted with permission of The Royal Society of Chemistry 2001, from Staunton, J.; Weissman, K.J. Polyketide biosynthesis: A millennium review. *Nat. Prod. Rep.* **2001**, *18*, 380–416 [26]; permission conveyed through Copyright Clearance Center, Inc.

### 2.4. *Biosynthesis of Tautomycetin*

Tautomycetin, firstly isolated from *Streptomyces griseochromogenes* and then from *Streptomyces* sp. CK4412, is an antifungal compound and an activated T cell-specific immunosuppressive compound. The inhibition of T-cells' proliferation is by the apoptosis mechanism. Unlike the other type I polyketide-derived compounds, tautomycetin has a linear structure [32,51].

The tautomycetin (TMC) biosynthetic gene cluster (~70 kb) has two ORFs that encode type I PKSs (Tmc A and TmcB). TmcA has six modules including the loading module, and TmcB has four modules, the TE (thioesterase) domain of which is located in TmcB. TE domain is responsible for releasing the intermediate chain of the compound from the PKS. The biosynthesis of TMC requires malonyl-CoA as a starter unit and the extender units such as 4 malonyl-CoAs, 4 methylmalonyl-CoAs, and 1 ethylmalonyl-CoA. After being released from PKS, the intermediate compound is modified by post-PKS mechanisms such as hydroxylation, decarboxylation, dehydration, and esterification with the cyclic C8 dialkylmaleic anhydride moiety [32].

## 3. Polyketide Synthases Type II

The type II polyketide synthases (type II PKSs) are responsible for producing aromatic polyketide. Based on the polyphenolic ring system and their biosynthetic pathways, the aromatic polyketides produced by type II PKSs generally are classified into seven groups, i.e., anthracyclines, angucyclines, aureolic acids, tetracyclines, tetracenomycins, pradimicin-type polyphenols, and benzoisochromanequinones [52].

Anthracyclines consists of a linear tetracyclic ring system with quinone–hydroquinone groups in rings B and C. Angucyclines have an angular tetracyclic ring system. The aureolic acids have a tricyclic chromophore. Tetracyclines contain a linear tetracyclic ring system without quinone–hydroquinone groups in rings B and C. Tetracenomycins have a linear tetracyclic ring system with the quinone group in ring B. Pradimicin-type polyphenols are considered as extended angucyclines. Benzoisochromanequinones contain a quinone derivative from the isochroman structure [52]. Some examples of aromatic polyketide produced by *Streptomyces* are actinorhodin (benzoisochromanequinones), doxorubicin (anthracyclines), jadomycin B (angucyclines), oxytetracycline (tetracyclines), mithramycin (aureolic acids), tetracenomycin C (tetracenomycins), and benastatin A (pradimicin-type polyphenols) (Figure 5) [28,52–57].

Unlike type I PKSs that involve huge multifunctional proteins that have many modules containing domains and perform the enzymatic reaction in a non-iterative way, the type II PKSs have monofunctional polypeptides and work iteratively to produce aromatic polyketide. However, like the type I PKS, the type II PKSs also comprise the acyl carrier protein (ACP) that functions as an anchor for the nascent polyketide chain. In addition to possessing ACP, the type II PKSs also consists of two ketosynthases units ($KS_\alpha$ and $KS_\beta$) that work cooperatively to produce the poly-β-keto chain. The $KS_\alpha$ unit catalyzes the condensation of the precursors; on the other hand, the role of $KS_\beta$ in the type II PKSs is as a chain length-determining factor. The three major systems (ACP, $KS_\alpha$, and $KS_\beta$) are called "minimal PKS" that work iteratively to produce aromatic polyketide. The other additional enzymes such as ketoreductases, cyclases, and aromatases cooperate to transform the poly-β-keto chain into the aromatic compound core. Furthermore, the post-tailoring process is conducted by oxygenases and glycosyl and methyl transferases [52,58–60]. The list of some aromatic polyketides produced by *Streptomyces* with their type II PKSs can be seen in Table 2.

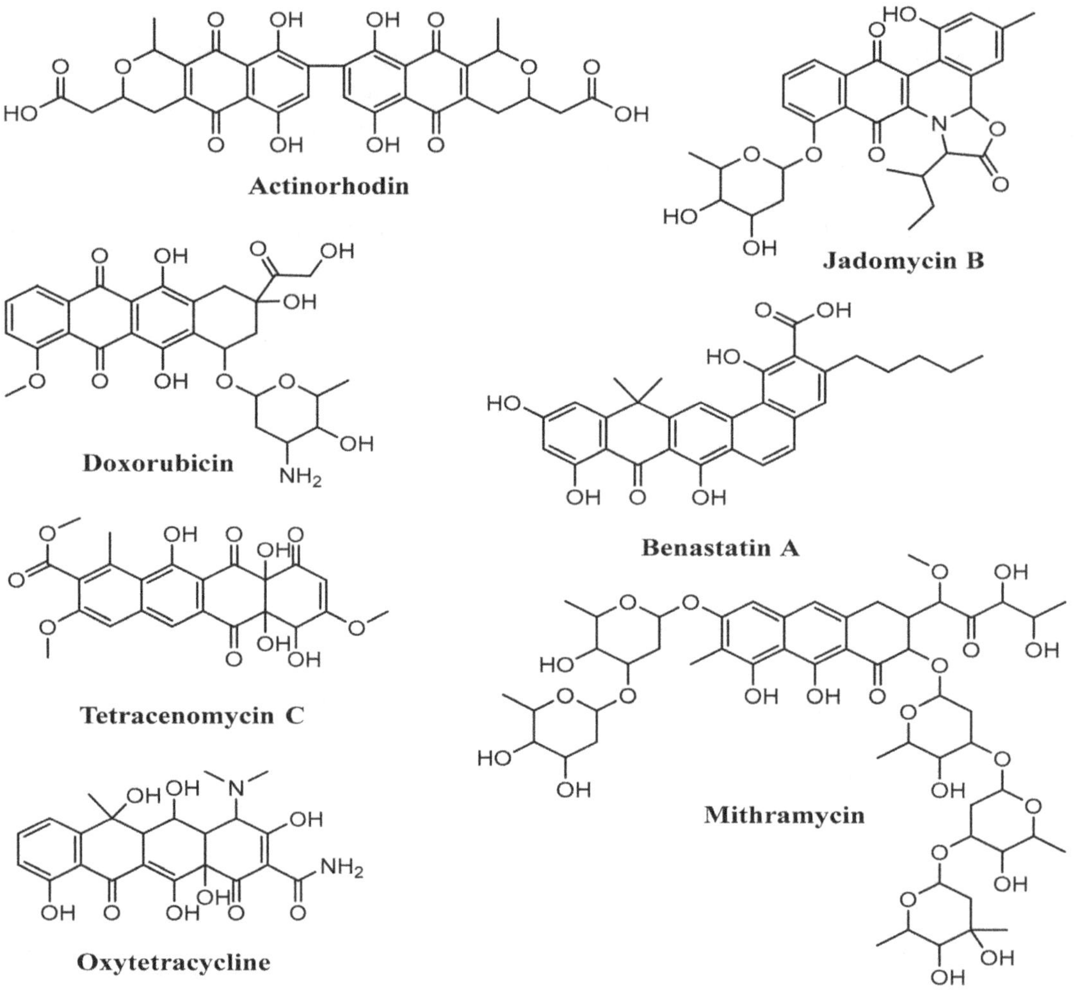

**Figure 5.** Some aromatic polyketides produced by *Streptomyces*.

### *3.1. Biosynthesis of Doxorubicin*

Doxorubicin was isolated from *Streptomyces peucetius* in the early of 1960s. It belongs to anthracyclines that have a tetracyclic ring containing quinone and a hydroquinone group in their structure. Doxorubicin is one of the important drugs for the treatment of cancer such as breast cancer, childhood solid tumors, soft tissue sarcomas, and aggressive lymphomas. There are some proposed mechanisms for how doxorubicin kills the cancer cells: (i) intercalation of DNA and interference of topoisomerase-II-mediated DNA repair; and (ii) formation of free radicals and their deterioration of cell components such as cellular membranes, DNA, and proteins [61–63].

Daunorubicin (DNR)-doxorubicin (DXR) type II PKSs, encoded by *dps* genes in *Streptomyces peucetius*, are involved in the formation of doxorubicin. The biosynthesis of doxorubicin requires one propionyl-CoA as the starter unit and nine malonyl-CoAs as the extender units. The process involves two "minimal PKSs" (DpsC-DpsD-DpsG and DpsA-DpsB-DpsG) to produce a 21-carbon decaketide as an intermediate compound. The repetitive process is conducted by $KS_{\alpha}$ (DpsA), $KS_{\beta}$ (DpsB), and ACP (DpsG). The next process employs several enzymes such as ketoreductase (DpsE), cyclases (DpsF, DpsY, and DnrD), oxygenase (DnrG and DnrF), and methyl transferase (DnrC) to produce ε-rhodomycinone, an important intermediate of doxorubicin biosynthesis. The remaining steps to synthesize doxorubicin utilize glycosyltransferase (DnrS) with the thymidine-diphospho (TDP) derivative of L-daunosamine, methyl esterase (DnrP), oxygenase (DoxA), and methyl transferase (DnrK) (Figure 6) [60,64–68].

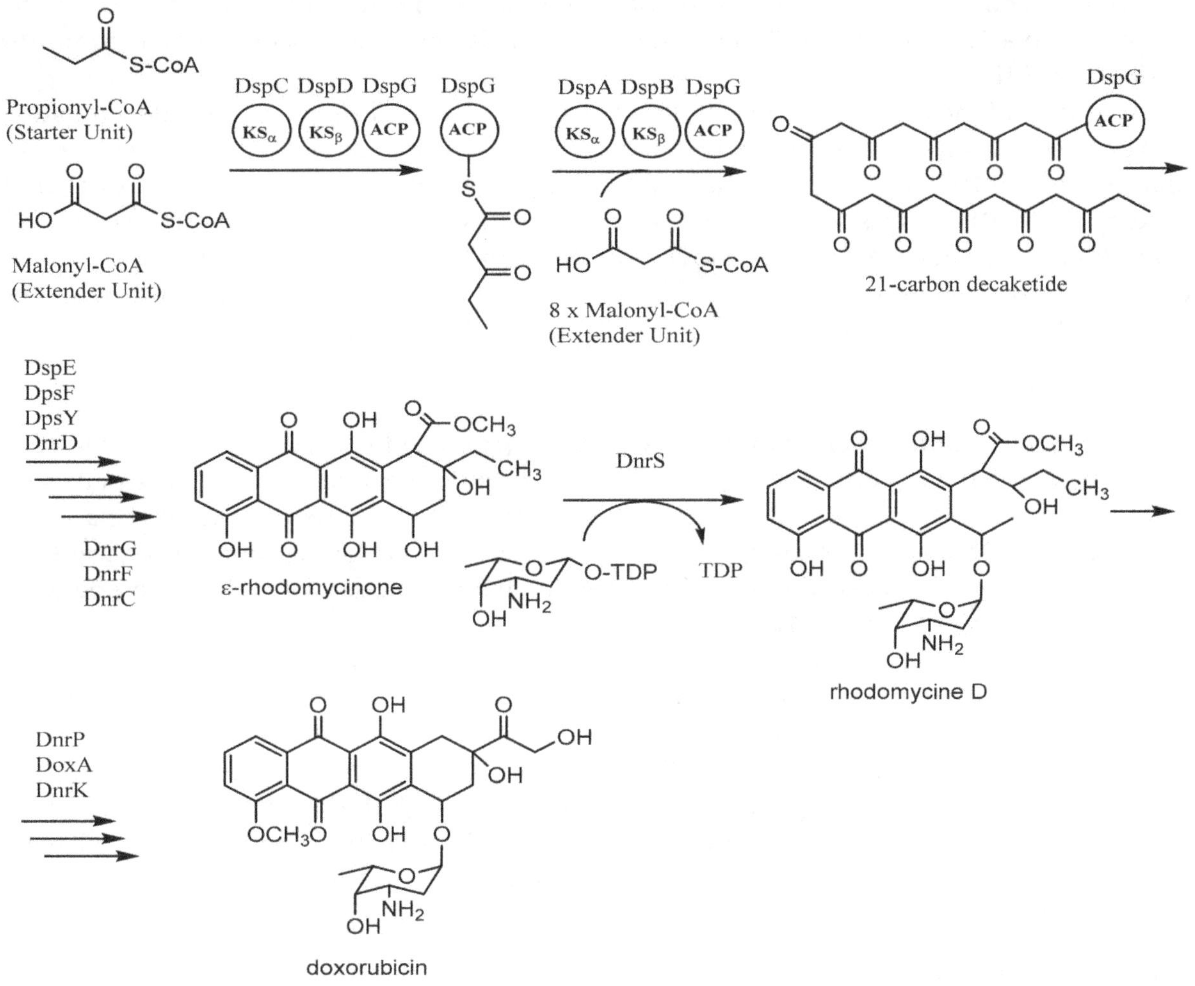

**Figure 6.** Biosynthesis of doxorubicin. Adapted with permission of The Royal Society of Chemistry 2009, from Chan, Y.A.; Podevels, A.M.; Kevany, B.M.; Thomas, M.G. Biosynthesis of polyketide synthase extender units. *Nat. Prod. Rep.* **2009**, *26*, 90–114 [60]; permission conveyed through Copyright Clearance Center, Inc.

**Table 2.** Some polyketides produced by *Streptomyces* and their minimal type II PKSs.

| Polyketide | Intermediate Backbone Structure | Producer | Minimal Type II PKSs | Ref. |
|---|---|---|---|---|
| Medermycin | octaketide | *Streptomyces* sp. *K73* | Med-1,2,23 | [69,70] |
| Doxorubicin | decaketide | *Streptomyces peucetius* | DpsABCDG | [60,65] |
| Oxytetracycline | decaketide | *Streptomyces rimosus* | OxyABCD | [71] |
| Gilvocarcin | decaketide | *Streptomyces griseoflavus* Gö 3592 | GilABC | [72] |
| Oviedomycin | decaketide | *Streptomyces antibioticus* | OvmPKS | [73] |
| Chartreusin | decaketide | *Streptomyces chartreusis* | ChaABC | [74] |
| Cervimycin | decaketide | *Streptomyces tendae* HKI-179 | CerABC | [75,76] |
| Resistomycin | decaketide | *Streptomyces resistomycificus* | RemABC | [77] |
| Chromomycin | decaketide | *Streptomyces griseus* subsp. *griseus* | CmmPKS | [78,79] |
| Hedamycin | dodecaketide | *Streptomyces griseoruber* | HedCDE | [80,81] |
| Fredericamycin | pentadecaketide | *Streptomyces griseus* ATCC 49344 | FdmFSGH | [82,83] |

### *3.2. Biosynthesis of Medermycin*

Medermycin is a benzoisochromanequinone (BIQ) antibiotic, isolated from *Streptomyces* sp. K73. It has high activity against some Gram-positive bacteria such as *Staphylococcus aureus*, *Staphylococcus epidermidis*, *Sarcina lutea*, *Bacillus subtilis*, and *Bacillus cereus*. Besides antibiotic activity, medermycin also has potent activity as a platelet aggregation inhibitor. Because of its unique ability to give different colors in acidic and alkaline aqueous solution, medermycin is considered as an indicator type antibiotic [69,70,84].

Biosynthesis of medermycin requires eight malonyl-CoAs and a sugar molecule, angolosamine, which is derived from the deoxyhexose (DOH) pathway. In the first step of biosynthesis, the minimal PKS that consists of ACP (encoded by the *med*-ORF23 gene), $KS_{\alpha}$ (encoded by *med*-ORF1), and $KS_{\beta}$ (encoded by *med*-ORF2) forms an octaketide moiety. The next process employs several enzymes such as keto reductase, aromatase, cyclase, enoyl reductase, and oxygenase/hydroxylase to produce the aglycone compound dihydrokalafungin. The aglycone structure then is combined by C-glycosyl transferase with an angolosamine structure to yield the final structure medermycin [69].

### 3.3. Biosynthesis of Hedamycin

Hedamycin is a pluramycin antitumor antibiotic, produced by *Streptomyces griseoruber*. This aromatic polyketide has a planar anthrapyrantrione chromophore, two amino sugars in its structure (α-L-*N*,*N*-dimethylvancosamine and β-D-angolosamine), and a bisepoxide-containing a side chain. The compound could inhibit 50% of human cancer cell growth at a subnanomolar concentration in three days. It is a monofunctional DNA alkylating agent, and because of its low therapeutic index, hedamycin is not clinically used [81,85].

Biosynthesis of hedamycin uses twelve malonyl-CoAs and two amino sugars, vancosamine and an angolosamine moiety. The minimal type II PKSs of hedamycin biosynthesis consist of HedC ($KS_{\alpha}$), HedD (CLF), and HedE (ACP). Uniquely, the initial process involves type I PKSs (HedT and HedU proteins) that produce the 2,4-hexadienyl primer unit from three malonyl-CoAs, and then, it is transferred to the minimal type II PKSs of hedamycin biosynthesis. After that, a dodecaketide structure is formed by processing nine malonyl-CoAs. The structure then is modified with keto reductase, aromatase/cyclase, and oxygenase into the aglycone compound. In the last step, two glycosyltransferases are used for incorporating two amino sugars to produce hedamycin [80,81].

### 3.4. Biosynthesis of Fredericamycin

Fredericamycin, isolated from *Streptomyces griseus* ATCC 49344, is an aromatic polyketide that contains a spirocyclic structure. It has moderate antitumor and cytotoxic activity in various cell lines. These bioactivities are suggested because of the blockage of topoisomerases I and II or the peptidyl-prolyl cis-trans isomerase Pin1 [83].

The biosynthesis of fredericamycin employs the minimal type II PKSs that contains $KS_{\alpha}$ (FdmF and FdmS), $KS_{\beta}$ (FdmG), and ACP (FdmH). There are two alternative mechanisms for chain initiation in the biosynthesis of fredericamycin. The first one requires acetyl-CoA and two malonyl-CoAs to produce the hexadienyl-priming unit. The second mechanism is by utilizing butyryl- or crotonyl-CoA and one malonyl-CoA to yield the hexadienyl-priming unit. The next step is carried out by processing twelve malonyl-CoAs as extender units to give the pentadecaketide intermediate, and then, the cyclases and oxygenases modify the intermediate compound into the final product [82,83,86].

## 4. Polyketide Synthases Type III

Unlike the type I and type II PKSs, the type III PKSs do not utilize ACP as an anchor for the production of polyketide metabolite. In this case, acyl-CoAs are used directly as substrates for generating polyketide compounds. In order to create polyketides, this system contains enzymes that construct homodimers and catalyzes many reactions such as priming, extension, and cyclization in an iterative way. With this fact, the type III PKSs are the simplest structures among the other types of PKSs. The type III PKSs found in bacteria were reported the first time in1999, and before that time, the type III PKSs were known only to be detected in plants [87–89].

Some studies previously revealed that type III PKSs could also be identified in the *Streptomyces* such as RppA, found in *Streptomyces griseus*, which is responsible in the synthesis of

1,3,6,8-tetrahydroxynaphthalene (THN) [90]. Gcs, identified in *Streptomyces coelicolor* A3(2), is reported to have an important role in the biosynthesis of germicidin [91]. SrsA, encoded by the *srsA* gene and isolated from *Streptomyces griseus*, is known to have an important role in the biosynthesis of phenolic lipids, i.e., alkylresorcinols and alkylpyrones [29].

The type III PKS Ken2, isolated from *Streptomyces violaceoruber*, was suggested to be involved in the production of 3,5-dihydroxyphenylglycine (3,5-DHPG). This compound is a nonproteinogenic amino acid needed for the formation of kendomycin and several other glycopeptide antibiotics such as balhimycin, chloroeremomycin, and also vancomycin [92]. Cpz6, encoded by the *cpz6* gene and isolated from *Streptomyces* sp. MK730–62F2, was reported to be engaged in the biosynthesis of caprazamycins by producing a group of new triketidepyrenes (presulficidins) [93]. Moreover, another finding also suggested that DpyA catalyzes the formation of alkyldihydropyrones in *Streptomyces reveromyceticus* (Figure 7) [94].

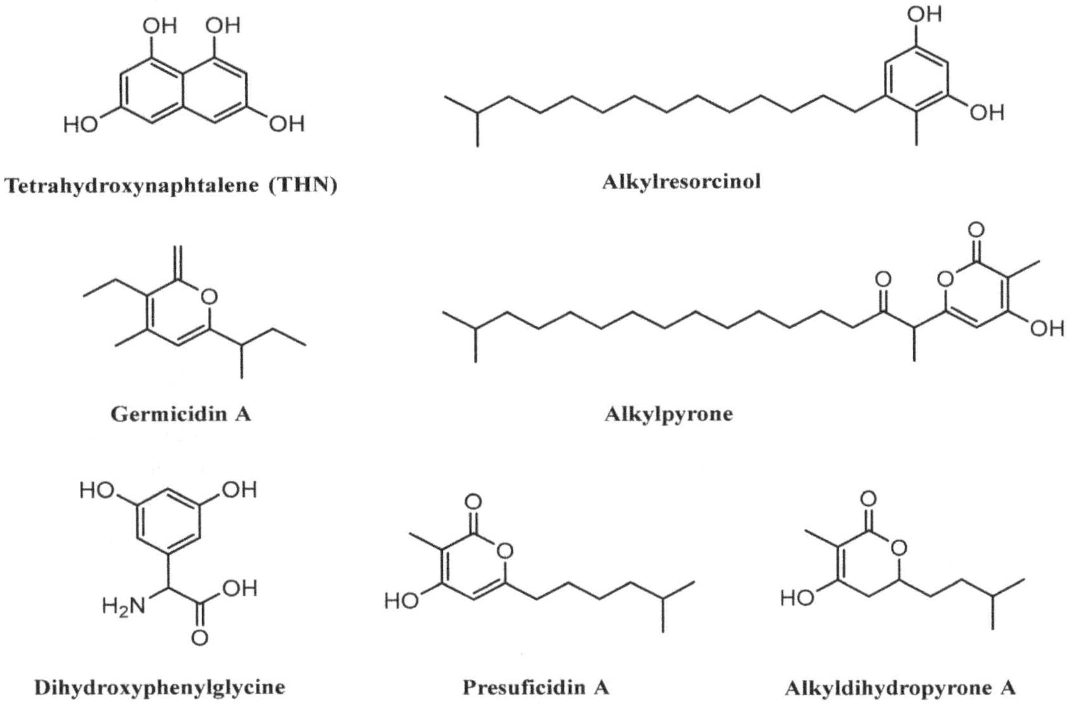

OH OH
HO OH
**Tetrahydroxynaphtalene (THN)**

OH
OH
**Alkylresorcinol**

O
**Germicidin A**

O
O O
OH
**Alkylpyrone**

HO OH
$H_2N$ OH
O
**Dihydroxyphenylglycine**

O
O
HO
**Presuficidin A**

O
O
HO
**Alkyldihydropyrone A**

**Figure 7.** Some compounds produced by type III PKSs.

*4.1. Biosynthesis of Germicidin*

Germicidin, a pyrone-derived polyketide, is produced by a type III PKS germicidin synthase (Gcs) and is known to inhibit spore germination. Germicidin A, produced by *Streptomyces viridochromogenes* and *Streptomyces coelicolor*, prevents the spore germination reversibly at a very low concentration (40 pg/mL). The mechanism of inhibition is suggested by affecting the sporal respiratory chain and blocking $Ca^{2+}$-activated ATPase, thus resulting in inadequate energy for spore germination. Furthermore, germicidin A also has antibacterial properties against various Gram-positive bacteria [95,96].

Although many bacterial type III PKSs use only malonyl-CoA as both starter and extender units, the type III PKS Gcs, which is responsible for germicidin biosynthesis, is suggested to have the ability to utilize either acyl-ACP or acyl-CoA such as medium-chain acyl-CoAs (C4–C8) as starter units and malonyl-CoA, methylmalonyl-CoA, and ethylmalonyl-CoA as extender units [97,98]. In the first step, the starter unit is transacylated onto the cysteine residue of Gcs, and then, Gcs catalyzes the condensation reaction between the starter unit and extender unit concomitantly with the decarboxylation process, resulting in β-ketoacyl-CoA. The process continues with β-ketoacyl-CoA, which transacylates back onto the cysteine residue of Gcs (repetitive process) and subsequently undergoes a condensation reaction with either methylmalonyl-CoA or ethylmalonyl-CoA simultaneously with decarboxylation to formulate β,δ-diketothioester of CoA or a triketide intermediate. In the end of the reaction, cyclization of the β,δ-diketothioester of CoA is carried out to produce various types of germicidins (Figure 8) [91].

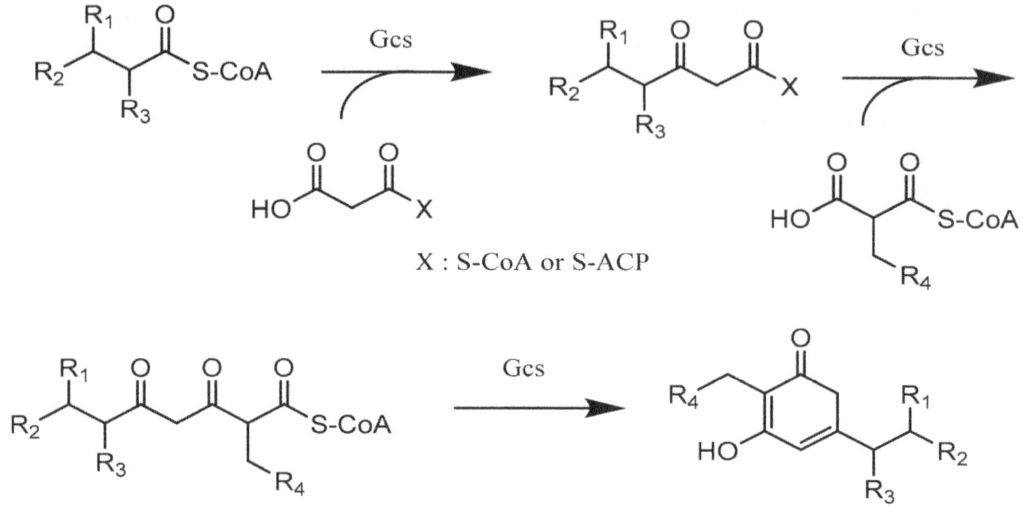

X : S-CoA or S-ACP

Germicidins

Germicidin A : $R_1$= H, $R_2 = R_3 = R_4 = CH_3$
Isogermicidin A : $R_1 = R_2 = R_4 = CH_3$, $R_3$ = H
Germicidin B : $R_1 = R_2$ = H, $R_3 = R_4 = CH_3$
Isogermicidin B : $R_1 = R_3$ = H, $R_2 = R_4 = CH_3$
Germicidin C : $R_1 = R_4$ = H, $R_2 = R_3 = CH_3$

**Figure 8.** Biosynthesis of germicidins. Gcs: germicidin synthase. Adapted with permission from Song, L.; Barona-Gomez, F.; Corre, C.; Xiang, L.; Udwary, D.W.; Austin, M.B.; Noel, J.P.; Moore, B.S.; Challis, G.L. Type III polyketide synthase β-ketoacyl-ACP starter unit and ethylmalonyl-coA extender unit selectivity discovered by Streptomyces coelicolor genome mining. *J. Am. Chem. Soc.* **2006**, *128*, 14754–14755 [91].

*4.2. Biosynthesis of Tetrahydroxynaphthalene*

Tetrahydroxynaphthalene or THN is a small aromatic compound that is produced by utilizing type III PKSs (RppA). The biosynthesis process of THN requires five molecules of malonyl-CoA to form a pentaketide intermediate structure, and then, it is cyclized and aromatized to yield THN product. Spontaneous oxidation of THN may result flaviolin (red pigment) [90,99,100].

*4.3. Biosynthesis of Dihydroxyphenylglycine*

In order to synthesize 3,5-dihydroxyphenylglycine (3,5-DHPG), four malonyl-CoAs are needed, and the process is catalyzed by type III PKS (Ken2 or DpgA), which leads to the formation of the intermediate tetraketide compound. The tetraketide further is modified by hydratase/dehydratase, and oxidase/thioesterase to form 3,5-dihydroxyphenylacetic acid. The final step involves transaminase and tyrosine, as the amino group donor, to yield 3,5-DHPG, which is known as a nonproteinogenic amino acid [92,100,101].

*4.4. Biosynthesis of Alkylresorcinol*

The alkylresorcinol biosynthesis in *Streptomyces griseus* is catalyzed by SrsA. The reaction needs fatty acid (starter unit), one methylmalonyl-CoA, and two malonyl-CoAs (extender unit), and the intermediate structure is tetraketide. The tetraketide structure then transforms into the aromatic compound nonenzymatically (alkylresorcinol). This reaction may occur because of the nucleophilic attack on the thioester group by the methine carbon of the intermediate tetraketide compound [29].

## 5. Conclusions

*Streptomyces* has various systems in order to produce polyketides with different structures and functions. Knowing the polyketide structures, activities, producing enzymes, starter units, extender units, and the structural genes are very important in the development of new drugs. Some mechanisms of polyketide biosynthesis in *Streptomyces* that have been reported previously could provide strong basic knowledge not only for the biosynthesis investigation of the new polyketides, but also engineering the producing system in the future.

**Acknowledgments:** The authors gratefully acknowledge support from German Federal Ministry of Education and Research (BMBF) under the German-Indonesian anti-infective cooperation (GINAICO) project, a fellowship awarded by the German Academic Exchange Service (German: Deutscher Akademischer Austauschdienst or DAAD), and The President's Initiative and Networking Funds of the Helmholtz Association of German Research Centres (German: Helmholtz Gemeinschaft Deutscher Forschungszentren or HGF) under Contract Number VH-GS-202.

## References

1. Moore, B.S.; Hopke, J.N. Discovery of a new bacterial polyketide biosynthetic pathway. *ChemBioChem* **2001**, *2*, 35–38. [CrossRef]
2. Rokem, J.S.; Lantz, A.E.; Nielsen, J. Systems biology of antibiotic production by microorganisms. *Nat. Prod. Rep.* **2007**, *24*, 1262. [CrossRef]
3. Katsuyama, Y.; Funa, N.; Miyahisa, I.; Horinouchi, S. Synthesis of unnatural flavonoids and stilbenes by exploiting the plant biosynthetic pathway in *Escherichia coli*. *Chem. Biol.* **2007**, *14*, 613–621. [CrossRef]
4. Chopra, I.; Roberts, M. Tetracycline antibiotics: Mode of action, applications, molecular biology, and epidemiology of bacterial resistance tetracycline antibiotics: Mode of action, applications, molecular biology, and epidemiology of bacterial resistance. *Microbiol. Mol. Biol. Rev.* **2001**, *65*, 232–260. [CrossRef] [PubMed]
5. Ghannoum, M.A.; Rice, L.B. Antifungal agents: Mode of action, mechanisms of resistance, and correlation of these mechanisms with bacterial resistance. *Clin. Microbiol. Rev.* **1999**, *12*, 501–517. [CrossRef]
6. Tacar, O.; Sriamornsak, P.; Dass, C.R. Doxorubicin: An update on anticancer molecular action, toxicity and novel drug delivery systems. *J. Pharm. Pharmacol.* **2013**, *65*, 157–170. [CrossRef]
7. Shushni, M.A.M.; Singh, R.; Mentel, R.; Lindequist, U. Balticolid: A new 12-membered macrolide with antiviral activity from an *Ascomycetous* fungus of marine origin. *Mar. Drugs* **2011**, *9*, 844–851. [CrossRef]
8. Li, J.; Kim, S.G.; Blenis, J. Rapamycin: One drug, many effects. *Cell Metab.* **2014**, *19*, 373–379. [CrossRef] [PubMed]
9. Van de Donk, N.W.C.J.; Kamphuis, M.M.J.; Lokhorst, H.M.; Bloem, A.C. The cholesterol lowering drug lovastatin induces cell death in myeloma plasma cells. *Leukemia* **2002**, *16*, 1362–1371. [CrossRef]
10. Hleba, L.; Charousova, I.; Cisarova, M.; Kovacik, A.; Kormanec, J.; Medo, J.; Bozik, M.; Javorekova, S. Rapid identification of *Streptomyces* tetracycline producers by MALDI-TOF mass spectrometry. *J. Environ. Sci. Heal. Part A* **2018**. [CrossRef] [PubMed]
11. Parthasarathy, R.; Sathiyabama, M. Lovastatin-producing endophytic fungus isolated from a medicinal plant *Solanum xanthocarpum*. *Nat. Prod. Res.* **2015**, *29*, 2282–2286. [CrossRef]
12. Huang, Q.; Lu, G.; Shen, H.-M.; Chung, M.C.M.; Ong, C.N. Anti-cancer properties of anthraquinones from rhubarb. *Med. Res. Rev.* **2006**, *27*, 609–630. [CrossRef]
13. Kohli, G.S.; John, U.; Figueroa, R.I.; Rhodes, L.L.; Harwood, D.T.; Groth, M.; Bolch, C.J.S.; Murray, S.A. Polyketide synthesis genes associated with toxin production in two species of *Gambierdiscus* (Dinophyceae). *BMC Genomics* **2015**, *16*, 410. [CrossRef]
14. Florian, P.; Monika, H. Polyketides in insects: Ecological role of these widespread chemicals and evolutionary aspects of their biogenesis. *Biol. Rev.* **2008**, *83*, 209–226.
15. Adele, C.; Guido, C.; Guido, V.; Angelo, F. Shaping the polypropionate biosynthesis in the solar-powered mollusc *Elysia viridis*. *ChemBioChem* **2008**, *10*, 315–322.
16. Müller, R.; Wink, J. Future potential for anti-infectives from bacteria—How to exploit biodiversity and genomic potential. *Int. J. Med. Microbiol.* **2014**, *304*, 3–13. [CrossRef]

17. Widyastuti, Y.; Lisdiyanti, P.; Ratnakomala, S.; Kartina, G.; Ridwan, R.; Rohmatussolihat, R.; Rosalinda Prayitno, N.; Triana, E.; Widhyastuti, N.; et al. Genus diversity of Actinomycetes in Cibinong Science Center, West Java, Indonesia. *Microbiol. Indones.* **2012**, *6*, 165–172. [CrossRef]
18. Lucas, X.; Senger, C.; Erxleben, A.; Grüning, B.A.; Döring, K.; Mosch, J.; Flemming, S.; Günther, S. StreptomeDB: A resource for natural compounds isolated from *Streptomyces* species. *Nucleic Acids Res.* **2013**, *41*, D1130–D1136. [CrossRef]
19. Igarashi, M.; Takahashi, Y.; Shitara, T.; Nakamura, H.; Naganawa, H.; Miyake, T.; Akamatsu, Y. Caprazamycins, novel lipo-nucleoside antibiotics, from *Streptomyces* sp. *J. Antibiot.* **2005**, *58*, 327–337. [CrossRef]
20. Dutta, S.; Basak, B.; Bhunia, B.; Chakraborty, S.; Dey, A. Kinetics of rapamycin production by *Streptomyces hygroscopicus* MTCC 4003. *3 Biotech.* **2014**, *4*, 523–531. [CrossRef]
21. Rodríguez, L.; Rodríguez, D.; Olano, C.; Braña, A.F.; Méndez, C.; Salas, J.A. Functional analysis of OleY L-oleandrosyl 3-O-methyltransferase of the oleandomycin biosynthetic pathway in *Streptomyces antibioticus*. *J. Bacteriol.* **2001**, *183*, 5358–5363. [CrossRef] [PubMed]
22. Elibol, M. Optimization of medium composition for actinorhodin production by *Streptomyces coelicolor* A3(2) with response surface methodology. *Process Biochem.* **2004**, *39*, 1057–1062. [CrossRef]
23. Pokhrel, A.R.; Chaudhary, A.K.; Nguyen, H.T.; Dhakal, D.; Le, T.T.; Shrestha, A.; Liou, K.; Sohng, J.K. Overexpression of a pathway specific negative regulator enhances production of daunorubicin in *bldA* deficient *Streptomyces peucetius* ATCC 27952. *Microbiol. Res.* **2016**, *192*, 96–102. [CrossRef]
24. Shelest, E.; Heimerl, N.; Fichtner, M.; Sasso, S. Multimodular type I polyketide synthases in algae evolve by module duplications and displacement of AT domains *in trans*. *BMC Genomics* **2015**, *16*, 1–15. [CrossRef]
25. Hopwood, D.A. Cracking the polyketide code. *PLoS Biol.* **2004**, *2*, 166–169. [CrossRef]
26. Staunton, J.; Weissman, K.J. Polyketide biosynthesis: A millennium review. *Nat. Prod. Rep.* **2001**, *18*, 380–416. [CrossRef]
27. Lal, R.; Kumari, R.; Kaur, H.; Khanna, R.; Dhingra, N.; Tuteja, D. Regulation and manipulation of the gene clusters encoding type I PKSs. *Trends Biotechnol.* **2000**, *18*, 264–274. [CrossRef]
28. Okamoto, S.; Taguchi, T.; Ochi, K.; Ichinose, K. Biosynthesis of actinorhodin and related antibiotics: Discovery of alternative routes for quinone formation encoded in the act gene cluster. *Chem. Biol.* **2009**, *16*, 226–236. [CrossRef] [PubMed]
29. Funabashi, M.; Funa, N.; Horinouchi, S. Phenolic lipids synthesized by type III polyketide synthase confer penicillin resistance on *Streptomyces griseus*. *J. Biol. Chem.* **2008**, *283*, 13983–13991. [CrossRef]
30. Vance, S.; Tkachenko, O.; Thomas, B.; Bassuni, M.; Hong, H.; Nietlispach, D.; Broadhurst, W. Sticky swinging arm dynamics: Studies of an acyl carrier protein domain from the mycolactone polyketide synthase. *Biochem. J.* **2016**, *473*, 1097–1110. [CrossRef] [PubMed]
31. Ruan, X.; Stassi, D.; Lax, S.A.; Katz, L. A second type I PKS gene cluster isolated from *Streptomyces hygroscopicus* ATCC 29253, a rapamycin-producing strain. *Gene* **1997**, *203*, 1–9. [CrossRef]
32. Choi, S.S.; Hur, Y.A.; Sherman, D.H.; Kim, E.S. Isolation of the biosynthetic gene cluster for tautomycetin, a linear polyketide T cell-specific immunomodulator from *Streptomyces* sp. CK4412. *Microbiology* **2007**, *153*, 1095–1102. [CrossRef] [PubMed]
33. Mazzei, T.; Mini, E.; Noveffi, A.; Periti, P. Chemistry and mode of action of macrolides. *J. Antimicrob. Chemother.* **1993**, *31*, 1–9. [CrossRef]
34. Schwecke, T.; Aparicio, J.F.; Molnár, I.; König, A.; Khaw, L.E.; Haydock, S.F.; Oliynyk, M.; Caffrey, P.; Cortés, J.; Lester, J.B. The biosynthetic gene cluster for the polyketide immunosuppressant rapamycin. *Proc. Natl. Acad. Sci. USA* **1995**, *92*, 7839–7843. [CrossRef] [PubMed]
35. Motamedi, H.; Shafiee, A. The biosynthetic gene cluster for the macrolactone ring of the immunosuppressant FK506. *Eur. J. Biochem.* **1998**, *256*, 528–534. [CrossRef]
36. Karray, F.; Darbon, E.; Oestreicher, N.; Dominguez, H.; Tuphile, K.; Gagnat, J.; Blondelet-Rouault, M.H.; Gerbaud, C.; Pernodet, J.L. Organization of the biosynthetic gene cluster for the macrolide antibiotic spiramycin in *Streptomyces ambofaciens*. *Microbiology* **2007**, *153*, 4111–4122. [CrossRef] [PubMed]
37. Xue, Y.; Zhao, L.; Liu, H.W.; Sherman, D.H. A gene cluster for macrolide antibiotic biosynthesis in *Streptomyces venezuelae*: Architecture of metabolic diversity. *Proc. Natl. Acad. Sci. USA* **1998**, *95*, 12111–12116. [CrossRef]
38. Cruz, M.C.; Cavallo, L.M.; Görlach, J.M.; Cox, G.; Perfect, J.R.; Cardenas, M.E.; Heitman, J. Rapamycin antifungal action is mediated via conserved complexes with FKBP12 and TOR kinase homologs in *Cryptococcus neoformans*. *Mol. Cell Biol.* **1999**, *19*, 4101–4112. [CrossRef]

39. Bastidas, R.J.; Shertz, C.A.; Lee, S.C.; Heitman, J.; Cardenas, M.E. Rapamycin exerts antifungal *activity in vitro* and *in vivo* against *Mucor circinelloides* via FKBP12-dependent inhibition of tor. *Eukaryot. Cell* **2012**, *11*, 270–281. [CrossRef]
40. Kwan, D.H.; Schulz, F. The stereochemistry of complex polyketide biosynthesis by modular polyketide synthases. *Molecules* **2011**, *16*, 6092–6115. [CrossRef]
41. Ikeda, H.; Nonomiya, T.; Usami, M.; Ohta, T.; Omura, S. Organization of the biosynthetic gene cluster for the polyketide anthelmintic macrolide avermectin in *Streptomyces avermitilis*. *Proc. Natl. Acad. Sci. USA* **1999**, *96*, 9509–9514. [CrossRef] [PubMed]
42. Ward, S.L.; Hu, Z.; Schirmer, A.; Reid, R.; Revill, W.P.; Reeves, C.D.; Petrakovsky, O.V.; Dong, S.D.; Katz, L. Chalcomycin biosynthesis gene cluster from *Streptomyces bikiniensis*: Novel features of an unusual ketolide produced through expression of the *chm* polyketide synthase in *Streptomyces fradiae*. *Antimicrob. Agents Chemother.* **2004**, *48*, 4703–4712. [CrossRef]
43. Gil, J.A.; Campelo-Diez, A.B. Candicidin biosynthesis in *Streptomyces griseus*. *Appl. Microbiol. Biotechnol.* **2003**, *60*, 633–642. [CrossRef]
44. Campelo, A.B.; Gil, J.A. The candicidin gene cluster from *Streptomyces griseus* IMRU 3570. *Microbiology* **2002**, *148*, 51–59. [CrossRef]
45. Kino, T.; Hatanaka, H.; Hashimoto, M.; Nishiyama, M.; Goto, T.; Okuhara, M.; Kohsaka, M.; Aoki, H.; Imanaka, H. FK-506, a novel immunosuppressant isolated from a *Streptomyces*. I. Fermentation, isolation, and physico-chemical and biological characteristics. *J. Antibiot. (Tokyo)* **1987**, *40*, 1249–1255. [CrossRef]
46. Wu, K.; Chung, L.; Revill, W.P.; Katz, L.; Reeves, C.D. The FK520 gene cluster of *Streptomyces hygroscopicus* var. *ascomyceticus* (ATCC 14891) contains genes for biosynthesis of unusual polyketide extender units. *Gene* **2000**, *251*, 81–90. [CrossRef]
47. Aparicio, J.; Colina, A.; Ceballos, E.; Martin, J. The biosynthetic gene cluster for the 26-membered ring polyene macrolide pimaricin. A new polyketide synthase organization encoded by two subclusters separated by functionalization genes. *J. Biol. Chem.* **1999**, *274*, 10133–10139. [CrossRef]
48. Fiers, W.D.; Dodge, G.J.; Li, Y.; Smith, J.L.; Fecik, R.A.; Aldrich, C.C. Tylosin polyketide synthase module 3: Stereospecificity, stereoselectivity and steady-state kinetic analysis of β-processing domains *via* diffusible, synthetic substrates. *Chem. Sci.* **2015**, *6*, 5027–5033. [CrossRef]
49. Zhang, X.; Chen, Z.; Li, M.; Wen, Y.; Song, Y.; Li, J. Construction of ivermectin producer by domain swaps of avermectin polyketide synthase in *Streptomyces avermitilis*. *Appl. Microbiol. Biotechnol.* **2006**, *72*, 986–994. [CrossRef]
50. Hammond, S.M.; Kliger, B.N. Mode of action of the polyene antibiotic candicidin: Binding factors in the wall of *Candida albicans*. *Antimicrob. Agents Chemother.* **1976**, *9*, 561–568. [CrossRef]
51. Cheng, X.-C.; Kihara, T.; Ying, X.; Uramoto, M.; Osada, H.; Kusakabe, H.; Wang, B.-N.; Kobayashi, Y.; Ko, K.; Yamaguchi, I.; Shen, Y.-C.; Isono, K. A new antibiotic, tautomycetin. *J. Antibiot.* **1989**, *42*, 141–144.
52. Hertweck, C.; Luzhetskyy, A.; Rebets, Y.; Bechthold, A. Type II polyketide synthases: Gaining a deeper insight into enzymatic teamwork. *Nat. Prod. Rep.* **2007**, *24*, 162–190. [CrossRef]
53. Malla, S.; Prasad Niraula, N.; Singh, B.; Liou, K.; Kyung Sohng, J. Limitations in doxorubicin production from *Streptomyces peucetius*. *Microbiol. Res.* **2010**, *165*, 427–435. [CrossRef]
54. Jakeman, D.L.; Bandi, S.; Graham, C.L.; Reid, T.R.; Wentzell, J.R.; Douglas, S.E. Antimicrobial activities of jadomycin B and structurally related analogues. *Antimicrob. Agents Chemother.* **2009**, *53*, 1245–1247. [CrossRef]
55. Lombó, F.; Menéndez, N.; Salas, J.A.; Méndez, C. The aureolic acid family of antitumor compounds: Structure, mode of action, biosynthesis, and novel derivatives. *Appl. Microbiol. Biotechnol.* **2006**, *73*, 1–14. [CrossRef]
56. Petkovic, H.; Cullum, J.; Hranueli, D.; Hunter, I.S.; Peric-Concha, N.; Pigac, J.; Thamchaipenet, A.; Vujaklija, D.; Long, P.F. Genetics of *Streptomyces rimosus*, the oxytetracycline producer. *Microbiol. Mol. Biol. Rev.* **2006**, *70*, 704–728. [CrossRef] [PubMed]
57. Shen, B.; Hutchinson, C.R. Triple hydroxylation of tetracenomycin A2 to tetracenomycin C in *Streptomyces glaucescens*. Overexpression of the *tcmG* gene in *Streptomyces lividans* and characterization of the tetracenomycin A2 oxygenase. *J. Biol. Chem.* **1994**, *1326*, 30726–30733.
58. Zhang, Z.; Pan, H.-X.; Tang, G.-L. New insights into bacterial type II polyketide biosynthesis. *F1000Research* **2017**, *6*, 172. [CrossRef] [PubMed]

59. Komaki, H.; Harayama, S. Sequence diversity of type II polyketide synthase genes in *Streptomyces*. *Actinomycetologica* **2006**, *20*, 42–48. [CrossRef]
60. Chan, Y.A.; Podevels, A.M.; Kevany, B.M.; Thomas, M.G. Biosynthesis of polyketide synthase extender units. *Nat. Prod. Rep.* **2009**, *26*, 90–114. [CrossRef] [PubMed]
61. Minotti, G. Anthracyclines: Molecular advances and pharmacologic developments in antitumor activity and cardiotoxicity. *Pharmacol. Rev.* **2004**, *56*, 185–229. [CrossRef]
62. Yang, F.; Teves, S.S.; Kemp, C.J.; Henikoff, S. Doxorubicin, DNA torsion, and chromatin dynamics. *Biochim. Biophys. Acta* **2014**, *1845*, 84–89. [CrossRef] [PubMed]
63. Thorn, C.F.; Oshiro, C.; Marsh, S.; Hernandez-Boussard, T.; McLeod, H.; Klein, T.E.; Altman, R.B. Doxorubicin pathways: Pharmacodynamics and adverse effects. *Pharmacogenet. Genomics* **2011**, *21*, 440–446. [CrossRef] [PubMed]
64. Bao, W.; Sheldon, P.J.; Wendt-Pienkowski, E.; Richard Hutchinson, C. The *Streptomyces peucetius dpsC* gene determines the choice of starter unit in biosynthesis of the daunorubicin polyketide. *J. Bacteriol.* **1999**, *181*, 4690–4695.
65. Grimm, A.; Madduri, K.; Ali, A.; Hutchinson, C.R. Characterization of the *Streptomyces peucetius* ATCC 29050 genes encoding doxorubicin polyketide synthase. *Gene* **1994**, *151*, 1–10. [CrossRef]
66. Rajgarhia, V.B.; Strohl, W.R. Minimal *Streptomyces* sp. strain C5 daunorubicin polyketide biosynthesis genes required for aklanonic acid biosynthesis. *J. Bacteriol.* **1997**, *179*, 2690–2696. [CrossRef]
67. Hutchinson, C.R. Biosynthetic studies of daunorubicin and tetracenomycin C. *Chem. Rev.* **1997**, *97*, 2525–2536. [CrossRef]
68. Lomovskaya, N.; Doi-Katayama, Y.; Filippini, S.; Nastro, C.; Fonstein, L.; Gallo, M.; Colombo, A.L.; Hutchinson, C.R. The *Streptomyces peucetius dpsY* and *dnrX* genes govern early and late steps of daunorubicin and doxorubicin biosynthesis. *J. Bacteriol.* **1998**, *180*, 2379–2386.
69. Ichinose, K.; Ozawa, M.; Itou, K.; Kunieda, K.; Ebizuka, Y. Cloning, sequencing and heterologous expression of the medermycin biosynthetic gene cluster of *Streptomyces* sp. AM-7161: Towards comparative analysis of the benzoisochromanequinone gene clusters. *Microbiology* **2003**, *149*, 1633–1645. [CrossRef] [PubMed]
70. Takano, S.; Hasuda, K.; Ito, A.; Koide, Y.; Ishii, F. A new antibiotic, medermycin. *J. Antibiot. (Tokyo)* **1976**, *29*, 765–768. [CrossRef]
71. Pickens, L.B.; Tang, Y. Oxytetracycline biosynthesis. *J. Biol. Chem.* **2010**, *285*, 27509–27515. [CrossRef]
72. Fischer, C.; Lipata, F.; Rohr, J. The complete gene cluster of the antitumor agent gilvocarcin V and its implication for the biosynthesis of the gilvocarcins. *J. Am. Chem. Soc.* **2003**, *125*, 7818–7819. [CrossRef] [PubMed]
73. Lombó, F.; Abdelfattah, M.S.; Braça, A.F.; Salas, J.A. Elucidation of oxygenation steps during oviedomycin biosynthesis and generation of derivatives with increased antitumor activity. *ChemBioChem* **2009**, *10*, 296–303. [CrossRef]
74. Xu, Z.; Jakobi, K.; Welzel, K.; Hertweck, C. Biosynthesis of the antitumor agent chartreusin involves the oxidative rearrangement of an anthracyclic polyketide. *Chem. Biol.* **2005**, *12*, 579–588. [CrossRef] [PubMed]
75. Bretschneider, T.; Zocher, G.; Unger, M.; Scherlach, K.; Stehle, T.; Hertweck, C. A ketosynthase homolog uses malonyl units to form esters in cervimycin biosynthesis. *Nat. Chem. Biol.* **2011**, *8*, 154–161. [CrossRef] [PubMed]
76. Herold, K.; Xu, Z.; Gollmick, F.A.; Gräfe, U.; Hertweck, C. Biosynthesis of cervimycin C, an aromatic polyketide antibiotic bearing an unusual dimethylmalonyl moiety. *Organ. Biomol. Chem.* **2004**, *2*, 2411–2414. [CrossRef]
77. Jakobi, K.; Hertweck, C. A gene cluster encoding resistomycin biosynthesis in *Streptomyces resistomycificus*; exploring polyketide cyclization beyond linear and angucyclic patterns. *J. Am. Chem. Soc.* **2004**, *126*, 2298–2299. [CrossRef] [PubMed]
78. Menendez, N.; Nur-e-Alam, M.; Bran, A.F.; Rohr, J.; Salas, J.A.; Mendez, C. Biosynthesis of the antitumor chromomycin A3 in *Streptomyces griseus*: Analysis of the gene cluster and rational design of novel chromomycin analogs. *Chem. Biol.* **2004**, *11*, 21–32.
79. Sun, L.; Zeng, J.; Cui, P.; Wang, W.; Yu, D.; Zhan, J. Manipulation of two regulatory genes for efficient production of chromomycins in *Streptomyces reseiscleroticus*. *J. Biol. Eng.* **2018**, *12*, 1–11. [CrossRef]
80. Das, A.; Khosla, C. *In vivo* and *in vitro* analysis of the hedamycin polyketide synthase. *Chem. Biol.* **2009**, *16*, 1197–1207. [CrossRef]

81. Bililign, T.; Hyun, C.-G.; Williams, J.S.; Czisny, A.M.; Thorson, J.S. The hedamycin locus implicates a novel aromatic PKS priming mechanism. *Chem. Biol.* **2004**, *11*, 959–969. [CrossRef]
82. Das, A.; Szu, P.; Fitzgerald, J.T.; Khosla, C. Mechanism and engineering of polyketide chain initiation in fredericamycin biosynthesis. *J. Am. Chem. Soc.* **2010**, *132*, 8831–8833. [CrossRef]
83. Chen, Y.; Wendt-Pienkowski, E.; Shen, B. Identification and utility of FdmR1 as a *Streptomyces* antibiotic regulatory protein activator for fredericamycin production in *Streptomyces griseus* ATCC 49344 and heterologous hosts. *J. Bacteriol.* **2008**, *190*, 5587–5596. [CrossRef] [PubMed]
84. Nakagawa, A.; Fukamachi, N.; Yamaki, K.; Hayashi, M.; Oh-ishi, S.; Kobayashi, B.; Omura, S. Inhibition of platelet aggregation by medermycin and it's related isochromanequinone antibiotics. *J. Antibiot.* **1987**, *40*, 1075–1076. [CrossRef]
85. Tu, L.C.; Melendy, T.; Beerman, T.A. DNA damage responses triggered by a highly cytotoxic monofunctional DNA alkylator, hedamycin, a pluramycin antitumor antibiotic. *Mol. Cancer Ther.* **2004**, *3*, 577–585. [PubMed]
86. Szu, P.-H.; Govindarajan, S.; Meehan, M.J.; Das, A.; Nguyen, D.D.; Dorrestein, P.C.; Minshull, J.; Khosla, C. Analysis of the ketosynthase-chain length factor heterodimer from the fredericamycin polyketide synthase. *Acta Neurobiol. Exp. (Wars)* **2011**, *18*, 1021–1031. [CrossRef]
87. Shen, B. Polyketide biosynthesis beyond the type I, II and III polyketide synthase paradigms. *Curr. Opin. Chem. Biol.* **2003**, *7*, 285–295. [CrossRef]
88. Yu, D.; Xu, F.; Zeng, J.; Zhan, J. Type III polyketide synthases in natural product biosynthesis. *IUBMB Life* **2012**, *64*, 285–295. [CrossRef] [PubMed]
89. Nakano, C.; Ozawa, H.; Akanuma, G.; Funa, N.; Horinouchi, S. Biosynthesis of aliphatic polyketides by type III polyketide synthase and methyltransferase in *Bacillus subtilis*. *J. Bacteriol.* **2009**, *191*, 4916–4923. [CrossRef] [PubMed]
90. Funa, N.; Funabashi, M.; Ohnishi, Y.; Horinouchi, S. Biosynthesis of hexahydroxyperylenequinone melanin via oxidative aryl coupling by cytochrome P-450 in *Streptomyces griseus*. *J. Bacteriol.* **2005**, *187*, 8149–8155. [CrossRef]
91. Song, L.; Barona-Gomez, F.; Corre, C.; Xiang, L.; Udwary, D.W.; Austin, M.B.; Noel, J.P.; Moore, B.S.; Challis, G.L. Type III polyketide synthase β-ketoacyl-ACP starter unit and ethylmalonyl-coA extender unit selectivity discovered by *Streptomyces coelicolor* genome mining. *J. Am. Chem. Soc.* **2006**, *128*, 14754–14755. [CrossRef]
92. Wenzel, S.C.; Bode, H.B.; Kochems, I.; Müller, R. A type I/type III polyketide synthase hybrid biosynthetic pathway for the structurally unique *ansa* compound kendomycin. *ChemBioChem* **2008**, *9*, 2711–2721. [CrossRef]
93. Tang, X.; Eitel, K.; Kaysser, L.; Kulik, A.; Grond, S.; Gust, B. A two-step sulfation in antibiotic biosynthesis requires a type III polyketide synthase. *Nat. Chem. Biol.* **2013**, *9*, 610–615. [CrossRef]
94. Aizawa, T.; Kim, S.Y.; Takahashi, S.; Koshita, M.; Tani, M.; Futamura, Y.; Osada, H.; Funa, N. Alkyldihydropyrones, new polyketides synthesized by a type III polyketide synthase from *Streptomyces reveromyceticus*. *J. Antibiot. (Tokyo)* **2014**, *67*, 819–823. [CrossRef] [PubMed]
95. Čihák, M.; Kameník, Z.; Šmídová, K.; Bergman, N.; Benada, O.; Kofroňová, O.; Petříčková, K.; Bobek, J. Secondary metabolites produced during the germination of *Streptomyces coelicolor*. *Front. Microbiol.* **2017**, *8*, 1–13. [CrossRef]
96. Petersen, F.; Zähner, H.; Metzger, J.W.; Freund, S.; Hummel, R.P. Germicidin, an autoregulative germination inhibitor of *Streptomyces viridochromogenes* NRRL B-1551. *J. Antibiot. (Tokyo)* **1993**, *46*, 1126–1138. [CrossRef] [PubMed]
97. Lim, Y.P.; Go, M.K.; Yew, W.S. Exploiting the biosynthetic potential of type III polyketide synthases. *Molecules* **2016**, *21*, 1–37. [CrossRef] [PubMed]
98. Chemler, J.A.; Buchholz, T.J.; Geders, T.W.; Akey, D.L.; Rath, C.M.; Chlipala, G.E.; Smith, J.L.; Sherman, D.H. Biochemical and structural characterization of germicidin synthase: Analysis of a type III polyketide synthase that employs acyl-ACP as a starter unit donor. *J. Am. Chem. Soc.* **2012**, *134*, 7359–7366. [CrossRef]
99. Izumikawa, M.; Shipley, P.R.; Hopke, J.N.; O'Hare, T.; Xiang, L.; Noel, J.P.; Moore, B.S. Expression and characterization of the type III polyketide synthase 1,3,6,8-tetrahydroxynaphthalene synthase from *Streptomyces coelicolor* A3(2). *J. Ind. Microbiol. Biotechnol.* **2003**, *30*, 510–515. [CrossRef]
100. Chen, H.; Tseng, C.C.; Hubbard, B.K.; Walsh, C.T. Glycopeptide antibiotic biosynthesis: Enzymatic assembly

of the dedicated amino acid monomer (S)-3,5-dihydroxyphenylglycine. *Proc. Natl. Acad. Sci. USA* **2001**, *98*, 14901–14906. [CrossRef] [PubMed]

101. Pfeifer, V.; Nicholson, G.J.; Ries, J.; Recktenwald, J.; Schefer, A.B.; Shawky, R.M.; Schröder, J.; Wohlleben, W.; Pelzer, S. A polyketide synthase in glycopeptide biosynthesis: The biosynthesis of the non-proteinogenic amino acid (S)-3,5-dihydroxyphenylglycine. *J. Biol. Chem.* **2001**, *276*, 38370–38377. [CrossRef] [PubMed]

# 5

# Taxonomic Characterization, and Secondary Metabolite Analysis of *Streptomyces triticiradicis* sp. nov.: A Novel Actinomycete with Antifungal Activity

**Zhiyin Yu [1,2,†], Chuanyu Han [1,†], Bing Yu [1], Junwei Zhao [1], Yijun Yan [2], Shengxiong Huang [2], Chongxi Liu [1,2,*] and Wensheng Xiang [1,3,*]**

[1] Key Laboratory of Agricultural Microbiology of Heilongjiang Province, Northeast Agricultural University, Harbin 150030, China; zhiyinyu@yeah.net (Z.Y.); chuanyuhan@yeah.net (C.H.); yubing95yb@163.com (B.Y.); guyan2080@126.com (J.Z.)

[2] State Key Laboratory of Phytochemistry and Plant Resources in West China, Kunming Institute of Botany, Chinese Academy of Sciences, Kunming 650201, China; yanyijun@mail.kib.ac.cn (Y.Y.); sxhuang@mail.kib.ac.cn (S.H.)

[3] State Key Laboratory for Biology of Plant Diseases and Insect Pests, Institute of Plant Protection, Chinese Academy of Agricultural Sciences, Beijing 100193, China

* Correspondence: xizi-ok@163.com (C.L.); xiangwensheng@neau.edu.cn (W.X.)

† These authors have contributed equally to this work.

**Abstract:** The rhizosphere, an important battleground between beneficial microbes and pathogens, is usually considered to be a good source for isolation of antagonistic microorganisms. In this study, a novel actinobacteria with broad-spectrum antifungal activity, designated strain NEAU-H2$^T$, was isolated from the rhizosphere soil of wheat (*Triticum aestivum* L.). 16S rRNA gene sequence similarity studies showed that strain NEAU-H2$^T$ belonged to the genus *Streptomyces*, with high sequence similarities to *Streptomyces rhizosphaerihabitans* NBRC 109807$^T$ (98.8%), *Streptomyces populi* A249$^T$ (98.6%), and *Streptomyces siamensis* NBRC 108799$^T$ (98.6%). Phylogenetic analysis based on 16S rRNA, *atpD*, *gyrB*, *recA*, *rpoB*, and *trpB* gene sequences showed that the strain formed a stable clade with *S. populi* A249$^T$. Morphological and chemotaxonomic characteristics of the strain coincided with members of the genus *Streptomyces*. A combination of DNA–DNA hybridization results and phenotypic properties indicated that the strain could be distinguished from the abovementioned strains. Thus, strain NEAU-H2$^T$ belongs to a novel species in the genus *Streptomyces*, for which the name *Streptomyces triticiradicis* sp. nov. is proposed. In addition, the metabolites isolated from cultures of strain NEAU-H2$^T$ were characterized by nuclear magnetic resonance (NMR) and mass spectrometry (MS) analyses. One new compound and three known congeners were isolated. Further, genome analysis revealed that the strain harbored diverse biosynthetic potential, and one cluster showing 63% similarity to natamycin biosynthetic gene cluster may contribute to the antifungal activity. The type strain is NEAU-H2$^T$ (= CCTCC AA 2018031$^T$ = DSM 109825$^T$).

**Keywords:** *Streptomyces triticiradicis* sp. nov.; antifungal activity; rhizosphere soil; new compound; genome analysis

---

## 1. Introduction

It is well known that plant pathogenic fungi can cause a tremendous loss of global agricultural production [1]. Despite synthetic fungicides being effective and playing an indispensable role against pathogenic fungi, the available antifungal agents are far from satisfactory as a result of several drawbacks, such as severe drug resistance, drug-related toxicity, and many other problems [2]. Therefore, novel antifungal agents and antagonistic microorganisms are needed to effectively control the fungal diseases of agricultural crops. Natural products and their derivatives, in particular secondary metabolites derived from *Streptomyces*, have always been valuable sources for lead discovery in medicinal and agricultural chemistry because their novel scaffolds can provide new modes of action [3,4]. Members of the genus *Streptomyces* species are gram-positive, filamentous, and sporulating actinobacteria containing a number of biosynthetic gene clusters, indicating their potential ability to produce large numbers of secondary metabolites with diverse biological activities [5,6], and they represent the source of 75% of clinically useful antibiotics presently available [7]. Many *Streptomyces* species have been successfully developed as commercial biofungicides based on *Streptomyces griseoviridis* [8]. Thus, *Streptomyces* are still an attractive and indispensable resource for drug discovery.

The rhizosphere is an environment where pathogenic and beneficial microorganisms constitute a major influential force on plant growth and health, which differs from the bulk soil [9]. Plants not only provide nutrients for microbial growth but also change the microbial diversity and increase the numbers of bioactive microorganisms in the rhizosphere. The rhizosphere provides an excellent place for pursuing actinobacteria producing novel antibiotics [10,11]. During our search for antagonistic actinobacteria from the rhizosphere soil of wheat (*Triticum aestivum* L.), the strain NEAU-H2$^T$ was isolated, which showed broad inhibitory activities against phytopathogenic fungi. Based on the polyphasic taxonomy analysis, this strain was classified as representative of a novel species in the genus *Streptomyces*. In addition, the secondary metabolites of this strain were investigated by spectroscopic and genomic analyses.

## 2. Materials and Methods

### 2.1. Isolation of Actinobacterial Strain

Strain NEAU-H2$^T$ was isolated from the rhizosphere soil of wheat (*Triticum aestivum* L.) collected from Zhumadian, Henan Province, Central China (32°98′ N, 114°02′ E). The root sample was air-dried for 24 h at room temperature, and then the surface soil was shaken off gently. After, the sample was shaken at 250 rpm in 100 mL of sterile water with glass beads for 30 min at 20 °C and then filtered with a single layer of gauze to obtain the rhizosphere soil suspension. The suspension was serially diluted and spread on cellulose-proline agar (CPA) [12] supplemented with cycloheximide (50 mg·L$^{-1}$) and nalidixic acid (20 mg·L$^{-1}$), and cultured at 28 °C for 3 weeks. Strain NEAU-H2$^T$ was isolated and purified on the International *Streptomyces* Project (ISP) medium 3 [13], and maintained as glycerol suspensions (20%, *v/v*) at −80 °C.

### 2.2. Morphological and Biochemical Characteristics

Gram staining was performed by the Hucker method [14]. Morphological characteristics were observed by light microscopy (Nikon ECLIPSE E200, Nikon Corporation, Tokyo, Japan) and scanning electron microscopy (Hitachi SU8010, Hitachi Co., Tokyo, Japan) using cultures grown on ISP 3 agar at 28 °C for 2 weeks. Samples for scanning electron microscopy were prepared as described by Jin et al. [15]. Cultural characteristics were determined on the ISP media 1–7 [13], Bennett's agar [16], Czapek's agar [17], and Nutrient agar [18] after 2 weeks at 28 °C. The color of substrate mycelium, aerial mycelium, and diffusible pigment on the different tested media were determined using color chips from the ISCC-NBS color charts [19]. Temperature tolerance for growth was evaluated at 4, 10, 15, 20, 25, 28, 35, 37, 40, and 45 °C on ISP 3 agar after incubation for 2 weeks. The pH range for growth (pH 3.0–12.0, at intervals of 1.0 pH unit) was tested in GY broth [20] using the buffer system described by Zhao et al. [21] and NaCl tolerance (0%–15% (*w/v*) in 1% intervals) for growth was determined after 2 weeks growth in GY broth at 28 °C with shaking at 250 rpm. Hydrolysis of Tweens (20, 40, and 80) and production of urease were tested according to the method of Smibert and Krieg [14]. The utilization of sole carbon and nitrogen sources were determined following the methods of Gordon et al. [22]. The decomposition of cellulose, hydrolysis of starch, coagulation of milk, aesculin, reduction of nitrate, liquefaction of gelatin, and production of $H_2S$ were examined as described previously [23].

### 2.3. Chemotaxonomic Analysis

The freeze-dried cells used for chemotaxonomic analysis were obtained from cultures grown in GY medium on a rotary shaker for seven days at 28 °C. Cells were acquired and washed twice with sterile distilled water and freeze-dried. The isomer of diaminopimelic acid (DAP) in the cell wall hydrolysates was derivatized and analyzed by HPLC (Agilent TC-$C_{18}$ Column, 250 × 4.6 mm, i.d. 5 μm) with a mobile phase consisting of acetonitrile/phosphate buffer (0.05 mol·$L^{-1}$, pH 7.2, 15:85, *v*/*v*), and a flow rate of 0.5 mL·$min^{-1}$ at a column temperature of 28 °C [24]. An Agilent G1321A fluorescence detector was used to detect the peak with a 365 nm excitation and 455 nm longpass emission filters. The whole-cell sugars were analyzed according to Lechevalier [25]. The polar lipids were extracted and examined by two-dimensional TLC (thin-layer chromatography, Qingdao Marine Chemical Inc., Qingdao, China) and identified according to the method of Minnikin et al. [26]. Menaquinones were extracted and purified from freeze-dried biomass following the methods of Collins [27]. The extracts were analyzed by HPLC-UV (Agilent Extend-$C_{18}$ Column, 150 × 4.6 mm, i.d. 5 μm, 1.0 mL·$min^{-1}$ acetonitrile: iso-propyl alcohol = 60:40) at 270 nm [28]. Fatty acid methyl esters were performed by GC-MS according to the method of Xiang et al. [29] and identified with the NIST 14 database.

### 2.4. Phylogenetic Analysis

Strain NEAU-H2$^T$ was grown on ISP 3 agar plates for one week at 28 °C. Then, it was inoculated into 250-mL baffle Erlenmeyer flasks containing 50 mL of GY broth and cultivated for two days at 28 °C with shaking at 250 rpm. After that, the total DNA was extracted according to the lysozyme-sodium dodecyl sulfate-phenol/chloroform method [30]. The primers and procedure for PCR amplification were carried out as described by Yi et al. [31]. The PCR product was purified and cloned into the vector pMD19-T (Takara, Shiga, Japan) and sequenced using an Applied Biosystems DNA sequencer (model 3730XL, Applied Biosystems Inc., Foster City, CA, USA). Almost full-length 16S rRNA gene sequence (1519 bp) was multiply aligned in MEGA (Molecular Evolutionary Genetics Analysis) using the Clustal W algorithm and trimmed manually if necessary. Phylogenetic trees were constructed with neighbor-joining [32] and maximum likelihood [33] algorithms using MEGA software version 7.0 (Kumar S, Philly, PA, USA) [34]. The stability of the topology of the phylogenetic tree was assessed using the bootstrap method with 1000 repetitions [35]. A distance matrix was calculated using Kimura's two-parameter model [36]. All positions containing gaps and missing data were eliminated from the dataset (complete deletion option). The calculation of 16S rRNA gene sequence similarities between strains was carried out on the basis of pairwise alignment using the EzBioCloud server (https://www.ezbiocloud.net/) [37]. Phylogenetic relationships of strain NEAU-H2$^T$ were also confirmed using sequences for five individual housekeeping genes (*atpD*, *gyrB*, *recA*, *rpoB*, and *trpB*). These sequences of housekeeping genes of strain NEAU-H2$^T$ were obtained from the Whole Genome sequences. The sequences of each locus were aligned using the software package MEGA version 7.0 and trimmed manually at the same position before being used for further analysis. Phylogenetic analysis was performed as described above.

### 2.5. DNA–DNA Relatedness Tests

The total DNA was extracted according to the method in the Section 2.4. The harvested DNA was detected by agarose gel electrophoresis and quantified by a Qubit 2.0 Fluorometer (Thermo Scientific, Ashville, NC, USA). The Illumina Novaseq PE150 (Illumina, San Diego, CA, USA) platform was used to perform whole-genome sequencing. A-tailed, ligated to paired-end adaptors, and PCR-amplified samples with a 350-bp insert were used for the library construction at the Beijing Novogene Bioinformatics Technology Co., Ltd. Illumina PCR adapter reads and low-quality reads from the paired end were filtered with a quality control step by our own compling pipeline. All good-quality paired reads were assembled by the SOAP (Short Oligonucleotide Alignment Program) denovo [38,39] (https://github.com/aquaskyline) into a number of contigs. After that, the filter reads were handled by the next step of the gap closing.

Because of a lack of the whole genome sequence of strains *Streptomyces rhizosphaerihabitans* NBRC 109807$^T$ and *Streptomyces siamensis* NBRC 108799$^T$, a DNA–DNA relatedness test was carried out as described by De Ley et al. [40] under consideration of modifications [41] with a model Cary 100 Bio UV/VIS-spectrophotometer (Hitachi U-3900, Hitachi Co., Tokyo, Japan) and a temperature controller. The DNA hybridization samples were diluted to $OD_{260}$ around 1.0 using 0.1 × SSC (saline sodium citrate buffer), then sheared using a JY92-II ultrasonic cell disruptor (ultrasonic time 3s, interval time 4 s, 90 times). The DNA renaturation rates were measured in 2 × SSC at 70 °C. This experiment was repeated three times to calculate the average value. The DNA–DNA relatedness value was determined between the genomes of strain NEAU-H2$^T$ and *Streptomyces populi* A249$^T$ (PJOS01000000) using the genome-to-genome distance calculator (GGDC 2.0) at http://ggdc.dsmz.de [42]. Genome mining analysis was performed with antiSMASH (version 4.0, Blin K, Oxford, UK) [43].

### 2.6. *In Vitro Antifungal Activity Test*

Antifungal screening was performed against 10 different phytopathogenic fungi: *Sclerotinia sclerotiorum*, *Exserohilum turcicum*, *Colletotrichum orbiculare*, *Corynespora cassiicola*, *Rhizoctonia solani*, *Fusarium graminearum*, *Fusarium oxysporum*, *Sphacelotheca reiliana*, *Curvularia lunata*, and *Helminthosporium maydis*. Ten phytopathogenic fungi were preserved in the Key Laboratory of Agricultural Microbiology within the Heilongjiang province, China. Antifungal activity of strain NEAU-H2$^T$ was assessed using the dual culture plate assay [44]. The strain was point-inoculated at the margin of potato dextrose agar (PDA) [45] plates and cultivated for three days at 28 °C, after which a fresh mycelial PDA agar plug of the fungus was transferred into the opposite margin of the corresponding plate. Inhibition of hyphal growth of the fungus was recorded after incubated for seven days at 28 °C. The percentage inhibition rates were calculated using the formula: Inhibition rate (%) = Wi/W × 100%, where Wi is the width of inhibition and W is the width between the pathogen and actinobacteria. The assay was repeated three times and the average was calculated.

### 2.7. *Isolation and Characterization of Secondary Metabolites*

Strain NEAU-H2$^T$ was grown on ISP 3 agar plates for five days at 28 °C. Then, it was inoculated into 250-mL baffle Erlenmeyer flasks containing 50 mL of tryptone soy broth (TSB) and cultivated for one day at 28 °C with shaking at 250 rpm. After that, aliquots (15 mL) of the culture were transferred into 1-L baffled Erlenmeyer flasks filled with 250 mL of the production medium (tryptone 0.1%, glucose 3%, beef extract 0.5%, 0.25% $CaCO_3$, 0.5% NaCl, 0.1% minor elements concentrate ($FeSO_4{\cdot}7H_2O$ 1.0 g, $CuSO_4{\cdot}5H_2O$ 0.45 g, $ZnSO_4{\cdot}7H_2O$ 1.0 g, $MnSO_4{\cdot}4H_2O$ 0.1 g, $K_2MoO_4$ 0.1 g, distilled water 1 L), pH 7.2–7.4), and cultured at 30 °C for six days with shaking at 250 rpm.

The fermentation broth (25 L) was centrifuged (4000 rev/min, 20 min), and the supernatant was extracted with ethylacetate three times. The ethylacetate extract was evaporated under reduced pressure at temperatures within 40 °C to yield an oily crude extract (5.0 g). The mycelia were extracted with methanol (1 L) and then concentrated in vacuo to remove the methanol to yield the aqueous concentrate. The mycelia concentrate was extracted with ethylacetate (1 L × 3) to afford 1.0 g of crude extract after removing the ethylacetate. Both extracts displayed most of the similar secondary metabolites based on HPLC analyses. Thus, they were combined for further purification. The samples were applied to reverse-phase HPLC analysis eluted with a flow rate of 1 $mL{\cdot}min^{-1}$ over a 28 min gradient with water and methanol (T = 0 min, 10% methanol; T = 20.0 min, 100% methanol; T = 24.0 min, 100% methanol; T = 24.1 min, 10% methanol; T = 28.0 min, 10% methanol) and at 25 °C.

The crude extract (6.1 g) was subjected to silica gel CC using a successive elution of petroleum ether/ethylacetate (1:0, 10:1, 5:1, 1:1 and 0:1, *v*/*v*) to yield A–F fractions. Fraction D (petroleum ether/ ethylacetate = 1:1, *v*/*v*) was subjected to semipreparative HPLC (YMC- Hydrosphere $C_{18}$ column, 250 mm × 10 mm i.d., 5 µm, 0–20.0 min $CH_3OH{:}H_2O$ = 35:55, *v*/*v*, 3 mL/min) to afford **1** ($t_R$ = 25.4 min,

2.2 mg) and **2** ($t_R$ = 30.4 min, 2.6 mg). Compound **4** ($t_R$ = 17.2 min, 5.5 mg) was obtained from the fraction C (petroleum ether/ethylacetate = 5:1, *v*/*v*) by semipreparative HPLC (YMC-Triart $C_{18}$ column, 250 mm × 10 mm i.d., 5 μm, $CH_3OH$:$H_2O$ = 25:75, *v*/*v*, 3 mL/min). Fraction E (petroleum ether/ethylacetate = 0:1, *v*/*v*) was further purified by semipreparative HPLC (YMC-Hydrosphere $C_{18}$ column, 250 mm × 10 mm i.d., 5 μm, 0–15.0 min, $CH_3OH$:$H_2O$ = 25:75; 15.1–30 min, $CH_3OH$:$H_2O$ = 35:65, *v*/*v*, 3 mL/min) to give compound **3** ($t_R$ = 21.4 min, 2.0 mg).

NMR spectra were recorded in methanol-d4 or DMSO-d6 using a Bruker AVANCE III-600 or a Bruker AVANCE III-400 spectrometer (Bruker Corp., Karlsruhe, Germany) and TMS was used as the internal standard. HR-ESI-MS data were obtained using an Agilent G6230 Q-TOF mass instrument (Agilent Technologies Inc. Santa Clara, CA, USA). Thin-layer chromatography (TLC) was performed using precoated silica gel GF254 plates (Qingdao Marine Chemical Inc., Qingdao, China), and spots were visualized by UV light (254 nm) and colored by iodine, or by spraying heated silica gel plates with 10% $H_2SO_4$ in ethanol. Semipreparative HPLC was conducted on a HITACHI Chromaster system (Hitachi-DAD, Tokyo, Japan).

### *2.8. In Vitro Antifungal Activity Test of Compounds*

The fungi were retrieved from the storage tube and cultured for seven days at 28 °C on PDA. All fungi were further cultured for one week to get new mycelium for the antifungal assays in PDA at 28 °C. The medium was mixed with the pathogenic fungi suspension at about 45 °C, ensuring the abundance of the strains was about $10^8$ cfu/mL. Next, the mixture was poured on 9-cm Petri dishes. The tested compounds were dissolved in DMSO at a concentration of 2 mg/mL. Each filter paper (5 mm in diameter) was impregnated with 10 μL of the tested compounds. The inoculated Petri dishes were cultured for 3 to 4 d at 28 °C. DMSO served as a blank control. The assay was measured three times. The inhibition diameter was measured by the cross bracketing method [46].

## **3. Results and Discussion**

### *3.1. Polyphasic Taxonomic Characterization of NEAU-H2$^T$*

Morphological observation of the two-week culture of strain NEAU-H2$^T$ grown on ISP 3 medium revealed that it had characteristics typical of the genus *Streptomyces* [47]. Aerial and substrate mycelium were well developed without fragmentation. Spiral spore chains with spiny surfaced spores (0.8–1.0 × 1.0–1.3 μm) were borne on the aerial mycelium (Figure 1). Strain NEAU-H2$^T$ exhibited good growth on ISP 1–4, ISP 7, and Nutrition agar media; moderate growth on Bennett's and Czapek's agar media; and poor growth on ISP 5 and ISP 6 media. The colony colors varied from white to moderate yellow. A dark grayish olive pigment was produced on ISP 6 medium. The detailed cultural characteristics of strain NEAU-H2$^T$ are shown in Table S1. Strain NEAU-H2$^T$ was found to grow at a temperature range of 4 to 40 °C (optimum temperature 28 °C), pH 5 to 10 (optimum pH 7), and NaCl tolerance of 0% to 9% (optimum NaCl of 0% to 1%). The physiological and biochemical properties of strain NEAU-H2$^T$ are given in Table 1 and the species description.

**Table 1.** Differential characteristics of strain NEAU-H2$^{T}$ and its most closely related *Streptomyces* species. Strains: 1, NEAU-H2$^{T}$; 2, *S. rhizosphaerihabitans* NBRC 109807$^{T}$; 3, *S. populi* A249$^{T}$; 4, *S. siamensis* NBRC 108799$^{T}$. All data are from this study except where marked. +, positive; –, negative. ‡ Data from Lee et al. [48]; † Data from Wang et al. [49]; § Data from Sripreechasak et al. [50].

| Characteristic | 1 | 2 | 3 | 4 |
| --- | --- | --- | --- | --- |
| Spore chain | Spiral | Straight ‡ | Straight † | Spiral § |
| Spore surface | Spiny | Hairy ‡ | Rough † | Smooth § |
| Growth temperature range (°C) | 4–40 | 10–40 | 10–37 | 10–40 |
| Growth pH range | 5–10 | 5–11 | 6–12 | 5–11 |
| NaCl tolerance range (*w*/*v*, %) | 0–9.0 | 0–10.0 | 0–4.0 | 0–10.0 |
| Cellulose decomposition | + | + | – | – |
| Gelatin liquefaction | – | + | + | – |
| Catalase production | – | + | + | – |
| $H_2S$ production | + | – | + | + |
| Milk coagulation | + | + | + | – |
| Nitrate reduction | – | – | + | – |
| Starch hydrolysis | + | – | – | – |
| Tween 20 hydrolysis | + | + | + | – |
| Tween 80 hydrolysis | + | + | – | + |
| Nitrogen source utilization | | | | |
| L-serine | – | + | + | + |
| L-threonine | + | + | – | + |
| L-tyrosine | – | + | + | + |
| Carbon source utilization | | | | |
| L-arabinose | – | + | + | + |
| Dulcitol | – | – | – | + |
| *meso*-inositol | – | + | + | + |
| D-mannitol | – | + | – | + |
| L-rhamnose | + | – | + | – |
| D-ribose | – | + | – | + |
| D-sorbitol | – | + | – | + |
| D-xylose | – | + | + | – |
| Phospholipids * | DPG, PE, PL, PI, PIM | AL, DPG, GL, PE, PG, PI, PIM, 2PLs ‡ | AL, APL, DPG, 2Ls, PE, PL, PIM † | DPG, PE, PG, PI, PL § |
| Menaquinones | MK-9($H_6$), MK-9($H_8$), MK-9($H_4$) | MK-9($H_6$), MK-9($H_8$) ‡ | MK-9($H_6$), MK-9($H_8$), MK-9($H_2$), MK-9($H_4$) † | MK-9($H_6$), MK-9($H_4$), MK-9($H_8$) § |
| Whole cell-wall sugars | Glucose | Glucose, ribose ‡ | Xylose, galactose † | ND |

* APL, aminophopholipid; DPG, diphosphatidylglycerol; PE, phosphatidylethanolamine; PI, phosphatidylinositol; PIM, phosphatidylinositol mannoside; AL, unidentified aminolipid; GL, unknown glycolipid; L, unidentified lipid; PL, unidentified phospholipid; MK, menaquinone; ND, no detection.

**Figure 1.** Scanning electron micrograph of strain NEAU-H2$^T$ grown on International *Streptomyces* Project (ISP) medium 3 (ISP 3 ) for 2 weeks at 28 °C; Bar 1 µm.

Chemotaxonomic analyses revealed that strain NEAU-H2$^T$ also exhibited the typical characteristics of the genus *Streptomyces* [47]. It contained LL-diaminopimelic acid as cell wall diamino acid, indicating that the strain is of cell wall chemotype I [51]. The whole-cell sugar was found to contain glucose. The phospholipid profile consisted of diphosphatidylglycerol (DPG), phosphatidylethanolamine (PE), phosphatidylinositol (PI), phosphatidylinositolmannosides (PIM), and an unidentified phospholipid (PL), corresponding to phospholipid type II [52] (Figure S1). The major cellular fatty acids (>10%) were iso-$C_{16:0}$ (21.6%), anteiso-$C_{15:0}$ (19.4%), iso-$C_{15:0}$ (16.9%), and anteiso-$C_{17:0}$ (13.0 %), which is fatty acid type IIc [53]; minor amounts of $C_{16:1}\omega7c$ (8.5 %), $C_{16:0}$ (7.5 %), iso-$C_{14:0}$ (7.2%), $C_{17:0}$ cyclo (2.1%), $C_{17:1}\omega8c$ (2.1%), $C_{18:0}$ (1.2%), and $C_{15:0}$ (0.5%) were also present. The menaquinones detected were MK-9($H_8$) (57.5%), MK-9($H_6$) (32.3%), and MK-9($H_4$) (10.2%), which have been reported for most species of the genus *Streptomyces* [47].

EzBioCloud analysis suggests that strain NEAU-H2$^T$ belongs to the genus *Streptomyces*. The novel strain shared the highest 16S rRNA gene sequence similarities with *S. rhizosphaerihabitans* NBRC 109807$^T$ (98.8%), *S. populi* A249$^T$ (98.6%), and *S. siamensis* NBRC 108799$^T$ (98.6%). In the neighbor-joining phylogenetic tree based on 16S rRNA gene sequences, strain NEAU-H2$^T$ formed a separate clade with *S. populi* A249$^T$ (Figure 2), a relationship also recovered by the maximum likelihood algorithm (Figure S2). To further clarify the affiliation of strain NEAU-H2$^T$ to closely related strains, phylogenetic trees were constructed from the concatenated sequence alignment of the five housekeeping genes based on the neighbor-joining and maximum likelihood algorithms (Figure 3 and Figure S3), which showed the same topology as the 16S rRNA gene tree. Furthermore, the concatenated sequences of *atp*D-*gyr*B-*rec*A-*rpo*B-*trp*B were used to calculate pairwise distances well above 0.007 (Table S2) for the related species, which was considered to be the threshold for species determination [54]. Based on the 16S rRNA gene sequence similarities and phylogenetic trees, *S. rhizosphaerihabitans* NBRC 109807$^T$, *S. populi* A249$^T$, and *S. siamensis* NBRC 108799$^T$ were selected as the closely related strains for subsequent comparative analysis.

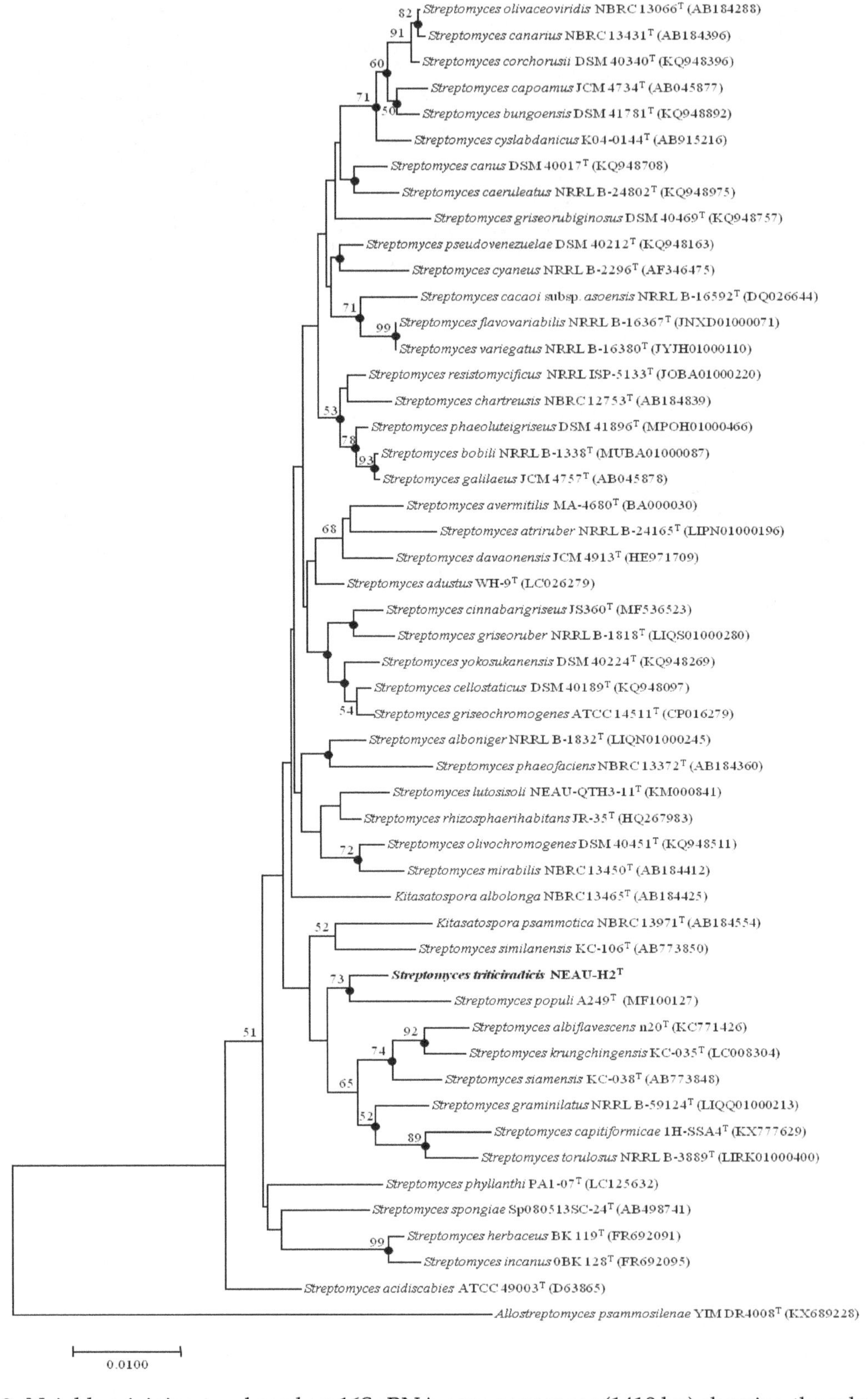


**Figure 2.** Neighbor-joining tree based on 16S rRNA gene sequences (1418 bp) showing the relationship of strain NEAU-H2$^{T}$ (in bold) with related taxa, which are the top 50 type strains of *Streptomyces* species of gene sequence similarities based on analysis using EzTaxon-e. Filled circles indicate branches that were also recovered using the maximum likelihood methods. Only bootstrap values above 50% (percentages of 1000 replications) are indicated. *Allostreptomyces psammosilenae* YIM DR4008$^{T}$ (KX689228) was used as an outgroup. Bar, 0.01 nucleotide substitutions per site.

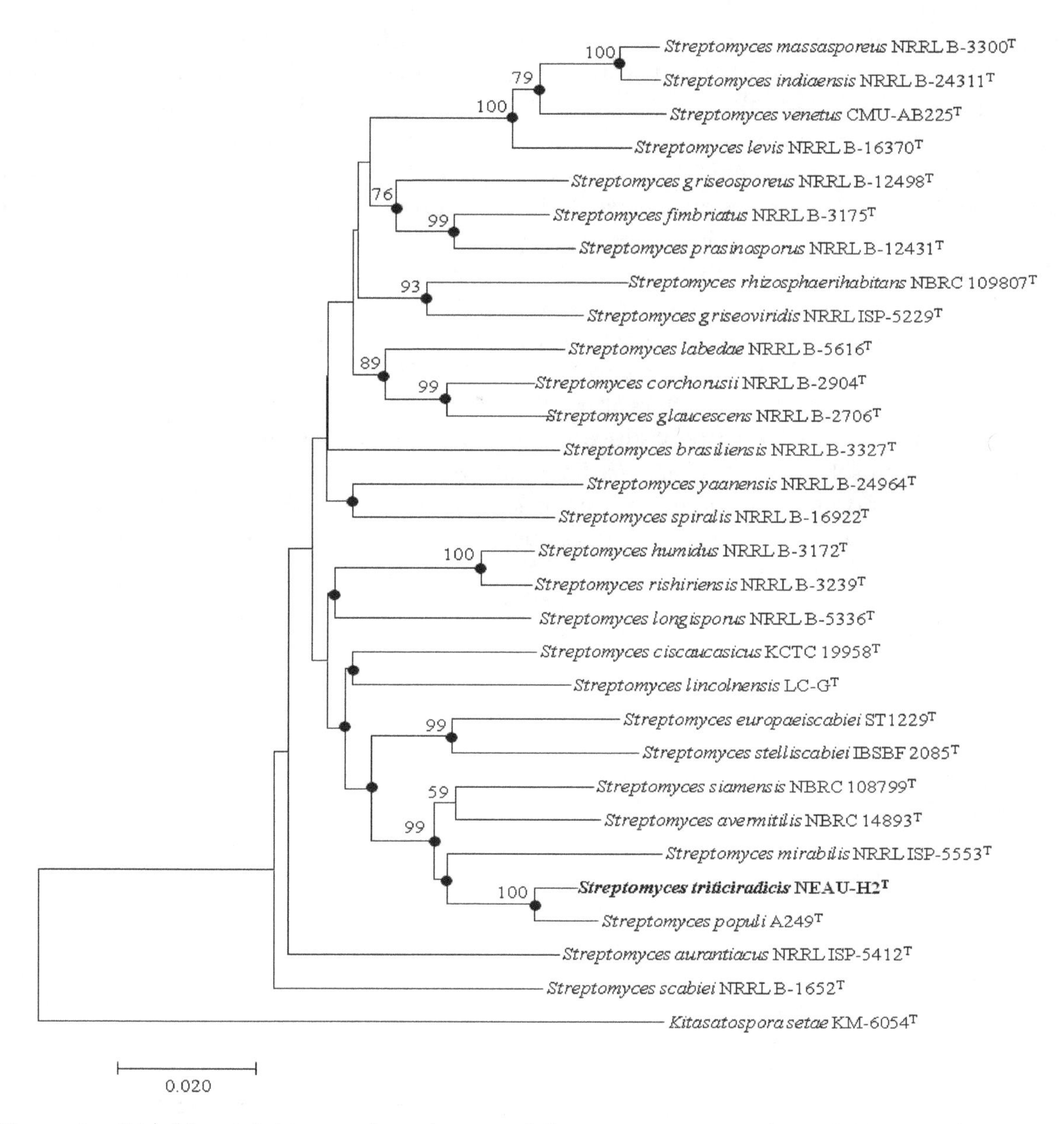


**Figure 3.** Neighbor-joining tree based on multilocus sequence analysis (MLSA)analysis of the concatenated partial sequences (2060 bp) from five housekeeping genes (*atpD, gyrB, recA, rpoB,* and *trpB*) of strain NEAU-H2$^T$ (in bold) with related taxa. Filled circles indicate branches that were also recovered using the maximum likelihood methods. Only bootstrap values above 50% (percentages of 1000 replications) are indicated. *Kitasatospora setae* KM-6054$^T$ was used as an outgroup. Bar, 0.02 nucleotide substitutions per site.

The assembled genome sequence of strain NEAU-H2$^T$ was found to be 9,921,301 bp long and composed of 135 contigs with an N50 of 167,996 bp, a DNA G+C content of 71.5 mol%, and a coverage of 152.0×. It was deposited into GenBank under the accession number WBKG00000000. Detailed genomic information is presented in the Table S3. DNA–DNA hybridization was employed to further clarify the relatedness between strain NEAU-H2$^T$ and *S. rhizosphaerihabitans* NBRC 109807$^T$ and *S. siamensis* NBRC 108799$^T$. The DNA–DNA relatedness values were 33.3 ± 2.5% and 44.5 ± 3.5%, respectively. Digital DNA–DNA hybridization was employed to clarify the relatedness between strain NEAU-H2$^T$ and *S. populi* A249$^T$. The level of DNA–DNA relatedness between them was 56.5 to 62.1%. According to the description proposed by Wayne et al. [55], the relatedness values are below the threshold value of 70% for assigning bacterial strains to the same genomic species.

Besides the genotypic evidence above, some obvious differences can also be found between strain NEAU-H2$^T$ with its closely related strains regarding several phenotypic and chemotaxonomic characteristics. Strain NEAU-H2$^T$ could be easily distinguished from its most closely related species by cultural characteristics, such as colony colors and diffusible pigment production (Table S1 and Figure S4). Morphological characteristics, including spore chain and surface ornamentation, could also distinguish the isolate from its closely related strains (Table 1). In addition, the isolate was able to grow at 4 °C, in contrast to its closely related strains, which could not. The novel strain could not utilize L-serine, L-tyrosine, L-arabinose, and meso-inositol while the closely related species could. Strain NEAU-H2$^T$ was found to contain both PI and PIM in its phospholipid profile, which could distinguish it from *S. siamensis* NBRC 108799$^T$ and *S. populi* A249$^T$. The presence of MK-9($H_4$) could differentiate the isolate from *S. rhizosphaerihabitans* NBRC 109807$^T$. Most notably, the whole-cell sugar of strain NEAU-H2$^T$ was evidently different from that of *S. rhizosphaerihabitans* NBRC 109807$^T$ and *S. populi* A249$^T$, with the only presence of glucose. The detailed characteristics of strain NEAU-H2$^T$ in comparison with its closely related strains are listed in Table 1.

Therefore, it is evident from the genotypic, phenotypic, and chemotaxonomic data that strain NEAU-H2$^T$ represents a novel species of the genus *Streptomyces*, for which the name *Streptomyces triticiradicis* sp. nov. is proposed.

### 3.2. *Description of Streptomyces triticiradicis sp. nov.*

*Streptomyces triticiradicis* (tri.ti.ci.ra'di.cis. L. neut. n. *triticum* wheat; L. fem. n. *radix* a root; N.L. gen. n. *triticiradicis* of a wheat root).

This is an aerobic gram-staining-positive actinomycete that forms well-developed, branched substrate hyphae and aerial mycelium that differentiate into spiral spore chains consisting of spiny surfaced spores. It has good growth on ISP 1–4, ISP 7 and Nutrient agar media, moderate growth on Bennett's and Czapek's agar media, and poor growth on ISP 5 and ISP 6 media. A dark grayish olive pigment is produced on ISP 6 medium. Growth is observed at temperatures between 4 and 40 °C, with an optimum temperature of 28 °C. Growth occurs in the pH range from 5.0 to 10.0 (optimum pH 7.0) with 0% to 9.0% (*w/v*) NaCl tolerance (optimum 0%–1%). It is positive for coagulation of milk; decomposition of cellulose; hydrolysis of aesculin, starch, and Tweens (20, 40, and 80); and production of $H_2S$ and urease; but negative for liquefaction of gelatin, production of catalase, and reduction of nitrate. L-alanine, L-arginine, L-asparagine, L-aspartic acid, creatine, L-glutamic acid, L-glutamine glycine, L-proline, and L-threonine are utilized as sole nitrogen sources but not L-serine or L-tyrosine. D-Fructose, D-galactose, D-glucose, lactose, D-maltose, D-mannose, D-raffinose, L-rhamnose, and D-sucrose are utilized as sole carbon sources but not L-arabinose, dulcitol, meso-inositol, D-mannitol, D-ribose, D-sorbitol, or D-xylose. The cell wall contains LL-diaminopimelic acid as diagnostic diamino acid and the whole cell hydrolysate contains glucose. The major menaquinones are MK-9($H_8$), MK-9($H_6$), and MK-9($H_4$). The polar lipids profile contains DPG, PE, PI, PIM, and PL. Major fatty acids (>10%) are iso-$C_{16:0}$, anteiso-$C_{15:0}$, iso-$C_{15:0}$, and anteiso-$C_{17:0}$. The DNA G + C content of the type strain is 71.5 mol%.

The type strain is NEAU-H2$^T$ (= CCTCC AA 2018031$^T$ = DSM 109825$^T$), isolated from the rhizosphere soil of wheat (*Triticum aestivum* L.) collected from Zhumadian, Henan Province, Central China. The GenBank/EMBL/DDBJ accession number for the 16S rRNA gene sequence of strain NEAU-H2$^T$ is MN512450. This Whole Genome Shotgun project has been deposited at DDBJ/ENA/GenBank under the accession WBKG00000000. The version described in this paper is version WBKG01000000.1.

### 3.3. *Antifungal Activity Evaluation*

Strain NEAU-H2$^T$ showed a wide range of inhibitory effects on the mycelial growth of the 10 tested phytopathogenic fungi (Figure 4A). It displayed significant inhibitory effects against four phytopathogenic fungi, including *C. orbiculare*, *C. cassiicola*, *S. sclerotiorum*, and *E. turcicum*, with the inhibition rate ranging from 48.4% to 69.0% (Figure 4B).

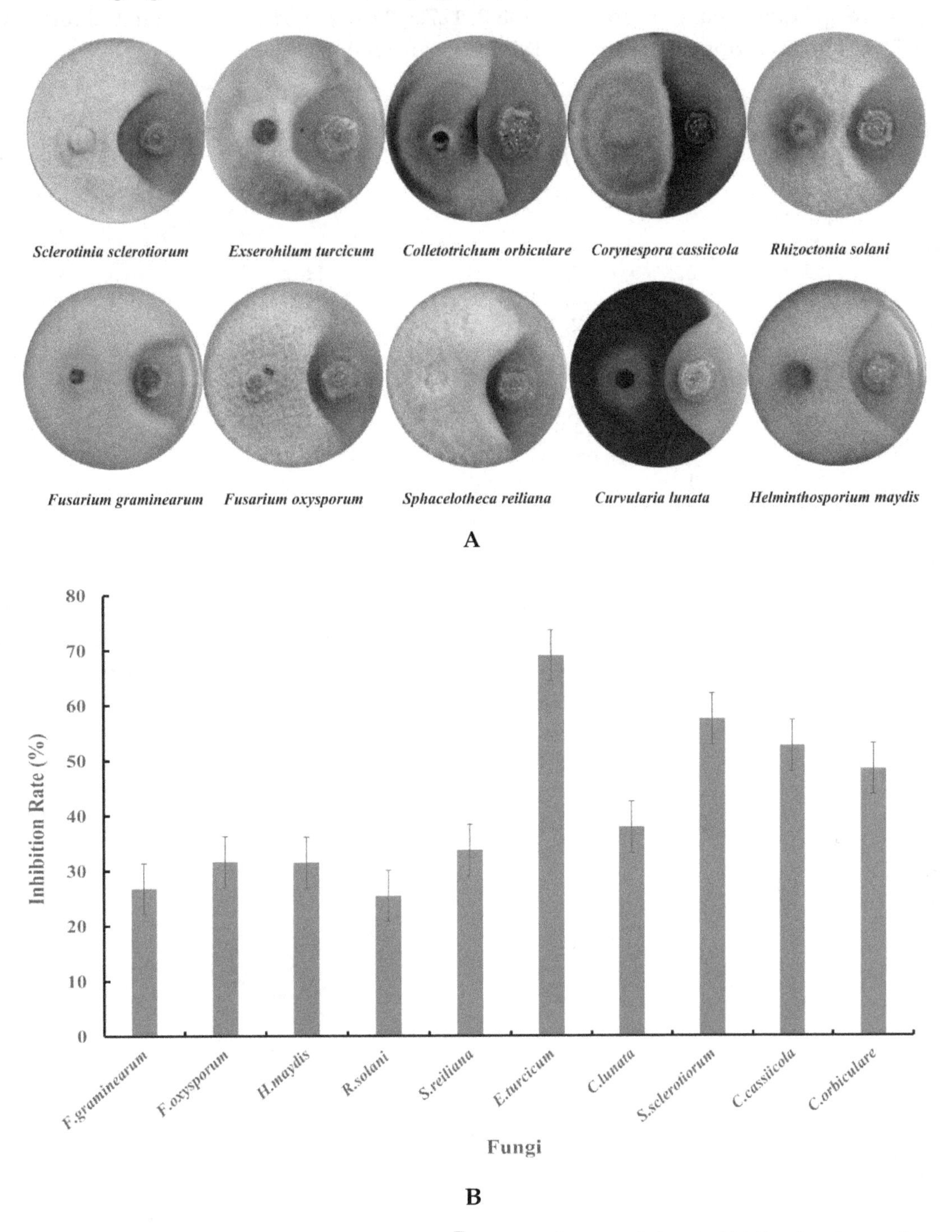


**Figure 4.** Antifungal activity of strain NEAU-H2$^T$ against the tested fungi. (**A**) Dual culture plate assay against tested fungi; (**B**) Inhibition rate against the tested fungi.

### *3.4. Identified of Secondary Metabolites from Strain NEAU-H2$^T$*

Only major components were identified from the liquid fermentation extract. Compound **1** was obtained as white amorphous powder, and its molecular formula, $C_{12}H_{13}NO_3$, was determined by high resolution electrospray ionization mass spectrometry (HRESIMS) data (m/z 242.0792 $[M + Na]^+$, calculatedd for 242.0788), corresponding to 7 degrees of unsaturation (Figure S5). The $^1$H NMR showed the presence of five aromatic protons with signals at $\delta_H$ 8.29 (s, 1H), 8.27 (d, $J$ = 7.3 Hz, 1H), 7.46 (d, $J$ = 7.4 Hz, 1H), 7.24 (td, $J$ = 7.2, 1.2 Hz, 1H), and 7.21 (td, $J$ = 7.2, 1.2 Hz, 1H), which indicated a three-substituted indole moiety (Table 2, Figure S5). The $^{13}C$ NMR and HSQC spectra revealed 12 carbons, which were classified into one methyl ($\delta_C$ 17.9), five $sp^2$ methines ($\delta_C$ 135.8, 124.5, 123.4, 122.9, 112.9), three $sp^2$ quaternary carbons ($\delta_C$ 138.2, 127.3, 116.3), and two oxygenated tertiary carbons ($\delta_C$ 79.7 and 71.1) and a carbonyl carbon ($\delta_C$ 197.2) (Figure S5).

**Table 2.** $^1H$ (600 MHz) and $^{13}C$ (150 MHz) NMR Data of **1** in $CD_3OD$.

| No. | $\delta_C$ | $\delta_H$ (mult, $J$ in Hz) | $^1H$-$^1H$ COSY | HMBC (H→C) |
|---|---|---|---|---|
| 2 | 135.8 | 8.29 (s, 1H) | | C-3, 3a, 7a, 8 |
| 3 | 116.3 | | | |
| 3a | 127.3 | | | |
| 4 | 122.9 | 8.27 (d, $J$ = 7.3 Hz, 1H) | H-5 | C-6 |
| 5 | 123.4 | 7.21 (td, $J$ = 7.2, 1.2 Hz, 1H) | H-5, H-6 | C-4, 6, 3a, 7 |
| 6 | 124.5 | 7.24 (td, $J$ = 7.2, 1.2 Hz, 1H) | H-5, H-7 | C-4, 7a |
| 7 | 112.9 | 7.46 (d, $J$ = 7.4 Hz, 1H) | H-6 | C-5, 3a |
| 7a | 138.2 | | | |
| 8 | 197.2 | | | |
| 9 | 79.7 | 4.74 (d, $J$ = 4.8 Hz, 1H) | H-10 | C-11, 10, 8 |
| 10 | 71.1 | 4.10 (m, 1H) | H-11, H-9 | C-11, 8 |
| 11 | 17.9 | 1.16 (d, $J$ = 6.4 Hz, 3H) | H-10 | C-10, 9 |

* $\delta_{C\ or\ H}$: chemical shift; $J$: coupling constant; COSY: correlated spectroscopy; HMBC: $^1H$ detected heteronuclear multiple bond correlation.

The $^1H$-$^1H$ COSY and HSQC spectra of **1** showed two spin-coupling systems, H-9/H-10/H-11 and H-4/H-5/H-6/H-7 (Figure 5B). The HMBC cross-peaks from H-5 to C-3a, from H-6 to C-7a, and from H-2 to C-3/3a/C-7a further revealed the presence of an indole moiety. Cross-peaks from H-9 to C-8 and from H-10 to C-8 were observed in the HMBC spectrum, which suggested a 2,3-dihydroxybutanone connected with indole moiety at C-3 (Figure 5B). Therefore, the planar structure **1** was elucidated as depicted in Figure 5A.

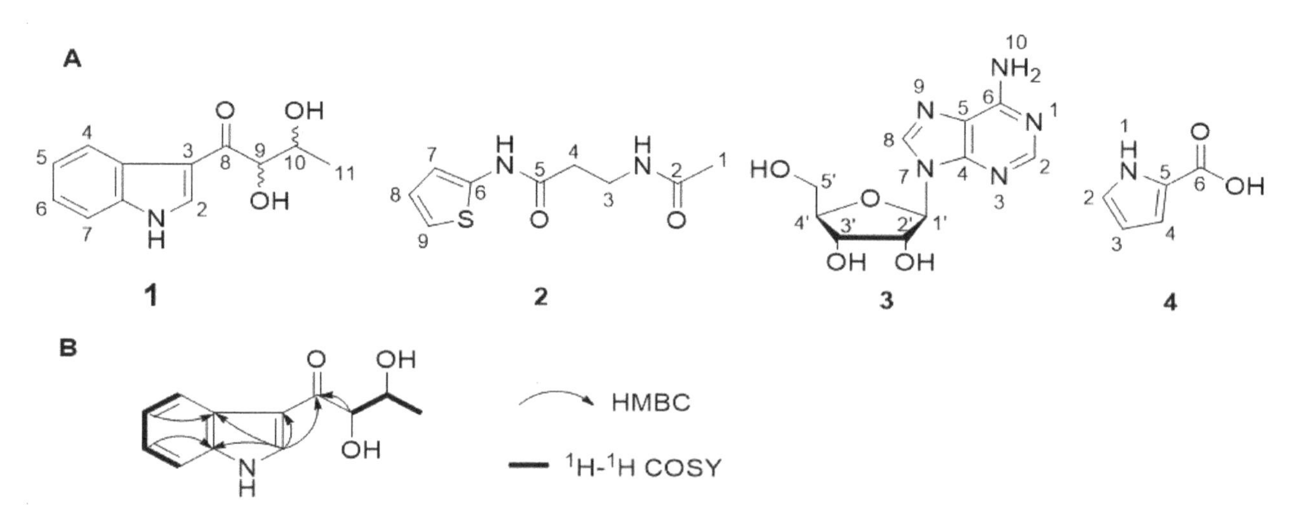


**Figure 5.** (**A**)The structure of compounds **1–4**; (**B**) 2D NMR correlations of **1**.

Compound **2** was isolated as a colorless powder, HRESIMS m/z 235.0532 $[M + Na]^+$(calculated for $C_9H_{12}N_2O_2S$, 235.0512); $^1H$ NMR data (600 MHz) $\delta_H$ 7.03 (1H, dd, $J$ = 2.4, 1.4 Hz, H-7), 6.96 (1H, dd, $J$ = 3.9, 1.3 Hz, H-9), 6.20 (1H, dd, $J$ = 3.8, 2.5 Hz, H-8), 3.13 (2H, t, $J$ = 6.7 Hz, H-4), 3.38 (2H, t, $J$ = 6.7 Hz, H-3), 1.92 (3H, s, H-1); $^{13}C$ NMR data (150 MHz, $CD_3OD$) $\delta_C$ 181.9 (C-5), 173.5 (C-2), 131.2 (C-6), 125.7 (C-7), 116.4 (C-9), 111.1 (C-8), 40.7 (C-3), 28.2 (C-4), and 22.5 (C-1) (Figure S6). Compound **2** was proven to be 3-Acetylamino-N-2-thienyl-propanamide by direct comparison of these data with those from the literature [56].

Compound **3**: $^{1}$H NMR data (600 MHz, DMSO-$d_6$) $\delta_H$ 8.37 (1H, s, H-8), 8.21 (1H, s, H-2), 5.90 (1H, d, $J$ = 5.9 Hz, H-1′), 4.55 (2H, t, $J$ = 5.4 Hz, H-2′), 4.14 (1H, m, H-3′), 3.95 (1H, q, $J$ = 3.3 Hz, H-4′), 3.67 (1H, dd, $J$ = 12.1, 3.5 Hz, H-5′), 3.55 (1H, dd, $J$ = 12.1, 3.5 Hz, H-5′); $^{13}$C NMR data (150 MHz, DMSO-$d_6$) $\delta_C$ 154.3 (C-6), 151.7 (C-2), 149.9 (C-4), 138.6 (C-8), 119.8 (C-5), 87.8 (C-1′), 85.8 (C-4′), 73.6 (C-2′), 70.5 (C-3′), and 61.6 (C-5′) (Figure S7). Compound **3** was proven to be *ß*-adenosine by direct comparison of these data with those from the literature [57].

Compound **4**: $^{1}$H NMR data (400 MHz, $CD_3OD$) $\delta_H$ 6.94 (1H, s, H-4), 6.85 (1H, d, $J$ = 2.7 Hz, H-2), 6.18 (1H, m, H-3); $^{13}$C NMR (100 MHz, $CD_3OD$) $\delta_C$164.6 (C-6), 124.4 (C-2), 124.1 (C-5), 116.6 (C-4), and 110.6 (C-3) (Figure S8). Compound **4** was proven to be 2-minaline by direct comparison of these data with those from the literature [58].

### 3.5. *Mining the Biosynthetic Potential of the Strain*

All the compounds were evaluated for their antifungal activity, which showed no significant inhibitory activity. In order to further discover the biosynthetic potential of the strain, we performed draft genome sequencing analysis. AntiSMASH analysis led to the identification of 38 putative gene clusters in the genome of strain NEAU-H2$^{T}$. Eleven clusters were identified belonging to a family of polyketide synthases (PKSs), including four type I PKSs, one type II PKS, and three type III PKSs. Likewise, further genome sequence analysis revealed eight additional gene clusters comprising modular enzyme coding genes, such as non-ribosomal peptide synthetase (NRPS, four clusters) and hybrid PKS-NRPS genes (four clusters). Other gene clusters included seven terpene gene clusters, three bacteriocin gene clusters, two siderophore gene clusters, one lanthipeptide gene cluster, one lassopeptide gene cluster, one melanin gene cluster, one ectoine gene cluster, and three butyrolactone gene clusters.

The important feature of NRPs is their ability to use nonproteinogenic amino acids as building blocks. By using such building blocks, NRPSs are able to produce peptides with diverse structures and bioactivities. As such, many NRPSs have been developed into pharmaceuticals, such as vancomycin, daptomycin, and β-lactam [59].

However, only a few metabolites were isolated and identified in culture broth under laboratory conditions from strain NEAU-H2$^{T}$. One answer is that most biosynthetic gene clusters of secondary metabolites are cryptic in culture broth under conventional laboratory culture conditions [60]. In addition, an active ingredient has not been isolated, possibly due to the low production of these metabolites under our culture conditions.

One of the secondary metabolite biosynthetic gene clusters of strain NEAU-H2$^{T}$ shows a 63% similarity to the biosynthetic gene cluster of natamycin, which is a 26-membered polyene macrolide antifungal agent produced by *Streptomyces chattanoogensis* L10, and the macrolide core was synthesized by five PKSs (ScnS0, ScnS1, ScnS2, ScnS3, and ScnS4) in turn [61]. Natamycin is currently widely used as an antifungal agent in human therapy and the food industry [62]. However, considering the poor quality of the genome sequence, with a large number of contigs, this may not be related to the antifungal active components identified with antibiotics and secondary metabolite analysis shell–antiSMASH. In the following research, we will focus on the study of secondary metabolites using activity tracking, amplification fermentation, and other approaches involving modification of the nutrient conditions in the medium and the genetic recombination of biosynthetic gene clusters.

## 4. Conclusions

A novel strain, NEAU-H2$^{T}$, with antifungal activity was isolated from the rhizosphere soil of wheat (*Triticum aestivum L.*). Four compounds, including one new compound, along with three known congeners (3-Acetylamino-N-2-thienyl-propanamide, β-adenosine, 2-minaline), were isolated. Morphological and chemotaxonomic features together with phylogenetic analysis and genomes suggested that strain NEAU-H2$^{T}$ belonged to the genus *Streptomyces*. Cultural and biochemical characteristics combined with DNA–DNA relatedness values clearly revealed that strain NEAU-H2$^{T}$

was differentiated from its closely related strains. Based on the polyphasic taxonomic analysis, it is suggested that strain NEAU-H2$^T$ represents a novel species of the genus *Streptomyces*, for which the name *Streptomyces triticiradicis* sp. nov. is proposed. The type strain is NEAU-H2$^T$ (=CCTCC AA 2018031$^T$ = DSM 109825$^T$).

**Author Contributions:** Z.Y., C.H., B.Y. and J.Z. performed the experiments. S.H. and Y.Y. analyzed the data. Z.Y. and C.H. wrote the paper. C.L. and W.X. designed the experiments and reviewed the manuscript. All authors have read and agreed to the published version of the manuscript.

**Acknowledgments:** We are grateful to Aharon Oren for helpful advice on the species epithet.

## References

1. Volova, T.; Prudnikova, S.; Boyandin, A.; Zhila, N.; Kiselev, E.; Shumilova, A.; Baranovskiy, S.; Demidenko, A.; Shishatskaya, A.; Thomas, S. Constructing slow-release fungicide formulations based on poly (3-hydroxybutyrate) and natural materials as a degradable matrix. *J. Agric. Food Chem.* **2019**, *67*, 9220–9231. [CrossRef] [PubMed]
2. Bai, Y.B.; Gao, Y.Q.; Nie, X.D.; Tuong, T.M.; Li, D.; Gao, J.M. Antifungal activity of griseofulvin derivatives against phytopathogenic fungi in vitro and in vivo and three-dimensional quantitative structure-activity relationship analysis. *J. Agric. Food Chem.* **2019**, *67*, 6125–6132. [CrossRef] [PubMed]
3. Onaka, H. Novel antibiotic screening methods to awaken silent or cryptic secondary metabolic pathways in actinomycetes. *J. Antibiot.* **2017**, *70*, 865–870. [CrossRef] [PubMed]
4. Liu, B.; Li, R.; Li, Y.A.; Li, S.Y.; Yu, J.; Zhao, B.F.; Liao, A.C.; Wang, Y.; Wang, Z.W.; Lu, A.D.; et al. Discovery of pimprinine alkaloids as novel agents against a plant virus. *J. Agric. Food Chem.* **2019**, *67*, 1795–1806. [CrossRef] [PubMed]
5. Chen, Y.; Zhou, D.; Qi, D.; Gao, Z.; Xie, J.; Luo, Y. Growth promotion and disease suppression ability of a *Streptomyces* sp. CB-75 from Banana rhizosphere soil. *Front. Microbiol.* **2018**, *8*, 2704. [CrossRef] [PubMed]
6. Kemung, H.M.; Tan, L.T.; Khan, T.M.; Chan, K.G.; Pusparajah, P.; Goh, B.H.; Lee, L.H. *Streptomyces* as a prominent resource of future anti-MRSA drugs. *Front. Microbiol.* **2018**, *9*, 2221. [CrossRef] [PubMed]
7. Janardhan, A.; Kumar, A.P.; Viswanath, B.; Saigopal, D.V.; Narasimha, G. Production of bioactive compounds by Actinomycetes and their antioxidant properties. *Biotechnol. Res. Int.* **2014**, *2014*, 217030. [CrossRef]
8. Minuto, A.; Spadaro, D.; Garibaldi, A.; Gullino, M.L. Control of soilborne pathogens of tomato using a commercial formulation of *Streptomyces griseoviridis* and *solarization*. *Crop Protect.* **2006**, *25*, 468–475. [CrossRef]
9. Raaijmakers, J.M.; Paulitz, T.C.; Steinberg, C.; Alabouvette, C.; Moënne-Loccoz, Y. The rhizosphere: A playground and battlefield for soilborne pathogens and beneficial microorganisms. *Plant. Soil.* **2009**, *321*, 341–361. [CrossRef]
10. Che, Q.; Zhu, T.J.; Keyzers, R.A.; Liu, X.F.; Li, J.; Gu, Q.Q.; Li, D.H. Polycyclic hybrid isoprenoids from a reed rhizosphere soil derived *Streptomyces* sp. CHQ-64. *J. Nat. Prod.* **2013**, *76*, 759–763. [CrossRef]
11. Zhao, K.; Penttinen, P.; Chen, Q.; Guan, T.W.; Lindström, K.; Ao, X.L.; Zhang, L.L.; Zhang, X.P. The rhizospheres of traditional medicinal plants in Panxi, China, host a diverse selection of actinobacteria with antimicrobial properties. *Appl. Microbiol. Biotechnol.* **2012**, *94*, 1321–1335. [CrossRef]
12. Zhao, J.W.; Shi, L.L.; Li, W.C.; Wang, J.B.; Wang, H.; Tian, Y.Y.; Xiang, W.S.; Wang, X.J. *Streptomyces tritici* sp. nov. a novel actinomycete isolated from rhizosphere soil of wheat (*Triticum aestivum* L.). *Int. J. Syst. Evol. Microbiol.* **2018**, *68*, 492–497. [CrossRef]
13. Shirling, E.B.; Gottlieb, D. Methods for characterization of *Streptomyces* species. *Int. J. Syst. Bacteriol.* **1966**, *16*, 313–340. [CrossRef]
14. Smibert, R.M.; Krieg, N.R. Phenotypic Characterization. In *Methods for General and Molecular Bacteriology*; Gerhardt, P., Murray, R.G.E., Wood, W.A., Krieg, N.R., Eds.; American Society for Microbiology: Washington, DC, USA, 1994; pp. 607–654.
15. Jin, L.Y.; Zhao, Y.; Song, W.; Duan, L.P.; Jiang, S.W.; Wang, X.J.; Zhao, J.W.; Xiang, W.S. *Streptomyces inhibens* sp. nov., a novel actinomycete isolated from rhizosphere soil of wheat (*Triticum aestivum* L.). *Int. J. Syst. Evol. Microbiol.* **2019**, *69*, 688–695. [CrossRef]
16. Jones, K.L. Fresh isolates of actinomycetes in which the presence of sporogenous aerial mycelia is a fluctuating characteristic. *J. Bacteriol.* **1949**, *57*, 141–145. [CrossRef]

17. Waksman, S.A. *The Actinomycetes. A Summary of Current Knowledge*; The Ronald Press Co.: New York, NY, USA, 1967; p. 286.
18. Waksman, S.A. *The Actinomycetes, Volume 2, Classification, Identification and Descriptions of Genera and Species*; Williams and Wilkins Company: Philadelphia, PA, USA, 1961; p. 363.
19. Kelly, K.L. Color-Name Charts Illustrated with Centroid Colors. In *Inter-Society Color Council-National Bureau of Standards*; U.S. National Bureau of Standards: Washington, DC, USA, 1965.
20. Jia, F.Y.; Liu, C.X.; Wang, X.J.; Zhao, J.W.; Liu, Q.F.; Zhang, J.; Gao, R.X.; Xiang, W.S. *Wangella harbinensis* gen. nov., sp. nov., a new member of the family *Micromonosporaceae*. *Antonie Leeuwenhoek* **2013**, *103*, 399–408. [CrossRef]
21. Zhao, J.W.; Han, L.Y.; Yu, M.Y.; Cao, P.; Li, D.; Guo, X.W.; Liu, Y.Q.; Wang, X.J.; Xiang, W.S. Characterization of *Streptomyces sporangiiformans* sp. nov., a novel soil actinomycete with antibacterial activity against Ralstonia solanacearum. *Microorganisms* **2019**, *7*, 360–376.
22. Gordon, R.E.; Barnett, D.A.; Handerhan, J.E.; Pang, C. *Nocardia coeliaca*, *Nocardia autotrophica*, and the nocardin strain. *Int. J. Syst. Bacteriol.* **1974**, *24*, 54–63. [CrossRef]
23. Yokota, A.; Tamura, T.; Hasegawa, T.; Huang, L.H. *Catenuloplanes japonicas* gen. nov., sp. nov., nom. rev., a new genus of the order actinomycetales. *Int. J. Syst. Bacteriol.* **1993**, *43*, 805–812. [CrossRef]
24. McKerrow, J.; Vagg, S.; McKinney, T.; Seviour, E.M.; Maszenan, A.M.; Brooks, P.; Sevious, R.J. A simple HPLC method for analysing diaminopimelic acid diastereomers in cell walls of Gram-positive bacteria. *Lett. Appl. Microbiol.* **2000**, *30*, 178–182. [CrossRef]
25. Lechevalier, M.P.; Lechevalier, H.A. The Chemotaxonomy of Actinomycetes. In *Actinomycete Taxonomy*; Dietz, A., Thayer, D.W., Eds.; Special Publication for Society of Industrial Microbiology: Arlington, TX, USA, 1980; pp. 227–291.
26. Minnikin, D.E.; O'Donnell, A.G.; Goodfellow, M.; Alderson, G.; Athalye, M.; Schaal, A.; Parlett, J.H. An integrated procedure for the extraction of bacterial isoprenoid quinones and polar lipids. *J. Microbiol. Methods.* **1984**, *2*, 233–241. [CrossRef]
27. Collins, M.D. Isoprenoid Quinone Analyses in Bacterial Classification and Identification. In *Chemical Methods in Bacterial Systematics*; Goodfellow, M., Minnikin, D.E., Eds.; Academic Press: Cambridge, MA, USA, 1985; pp. 267–284.
28. Wu, C.; Lu, X.; Qin, M.; Wang, Y.; Ruan, J. Analysis of menaquinone compound in microbial cells by HPLC. *Microbiology* **1989**, *16*, 76–178.
29. Xiang, W.S.; Liu, C.X.; Wang, X.J.; Du, J.; Xi, L.J.; Huang, Y. *Actinoalloteichus nanshanensis* sp. nov., isolated from the rhizosphere of a fig tree (*Ficus religiosa*). *Int. J. Syst. Evol. Microbiol.* **2011**, *61*, 1165–1169. [CrossRef]
30. Zhou, J.Z.; Bruns, M.A.; Tiedje, J.M. DNA recovery from soils of diverse composition. *Appl. Environ. Microbiol.* **1996**, *62*, 316–322.
31. Yi, R.K.; Tan, F.; Liao, W.; Wang, Q.; Mu, J.F.; Zhou, X.R.; Yang, Z.N.; Zhao, X. Isolation and identification of *Lactobacillus plantarum* HFY05 from natural fermented Yak Yogurt and Its effect on alcoholic liver injury in mice. *Microorganisms* **2019**, *7*, 530. [CrossRef]
32. Saitou, N.; Nei, M. The neighbor-joining method: A new method for reconstructing phylogenetic trees. *Mol. Biol. Evol.* **1987**, *4*, 406–425.
33. Felsenstein, J. Evolutionary trees from DNA sequences: A maximum likelihood approach. *J. Mol. Evol.* **1981**, *17*, 368–376. [CrossRef]
34. Kumar, S.; Stecher, G.; Tamura, K. Mega7: Molecular evolutionary genetics analysis version 7.0 for bigger datasets. *Mol. Biol. Evol.* **2016**, *33*, 1870–1874. [CrossRef]
35. Felsenstein, J. Confidence limits on phylogenies: An approach using the bootstrap. *Evolution* **1985**, *39*, 83–791. [CrossRef]
36. Kimura, M. A simple method for estimating evolutionary rates of base substitutions through comparative studies of nucleotide sequences. *J. Mol. Evol.* **1980**, *16*, 111–120. [CrossRef]
37. Yoon, S.H.; Ha, S.M.; Kwon, S.; Lim, J.; Kim, Y.; Seo, H.; Chun, J. Introducing EzBioCloud: A taxonomically united database of 16S rRNA and whole genome assemblies. *Int. J. Syst. Evol. Microbiol.* **2017**, *67*, 1613–1617. [CrossRef] [PubMed]
38. Li, R.Q.; Zhu, H.M.; Ruan, J.; Qian, W.B.; Fang, X.D.; Shi, Z.B.; Li, Y.R.; Li, S.T.; Shan, G.; Kristiansen, K.; et al. De novo assembly of human genomes with massively parallel short read sequencing. *Genome Res.* **2010**, *20*, 265–272. [CrossRef] [PubMed]

39. Li, R.; Li, Y.; Kristiansen, K.; Wang, J. SOAP: Short oligonucleotide alignment program. *Bioinformatics* **2008**, *24*, 713–714. [CrossRef]
40. Ley, J.D.; Cattoir, H.; Reynaerts, A. The quantitative measurement of DNA hybridization from renaturation rates. *Eur. J. Biochem.* **1970**, *12*, 133–142. [CrossRef]
41. Huss, V.A.R.; Festl, H.; Schleifer, K.H. Studies on the spectrometric determination of DNA hybridisation from renaturation rates. *Syst. Appl. Microbiol.* **1983**, *4*, 184–192. [CrossRef]
42. Meier-Kolthoff, J.P.; Auch, A.F.; Klenk, H.P.; Goker, M. Genome sequence-based species delimitation with confidence intervals and improved distance functions. *BMC Bioinform.* **2013**, *14*, 60. [CrossRef]
43. Blin, K.; Wolf, T.; Chevrette, M.G.; Lu, X.; Schwalen, C.J.; Kautsar, S.A.; Duran, H.G.S.; Santos, E.L.C.D.L.; Kim, H.U.; Nave, M.; et al. Antismash 4.0-improvements in chemistry prediction and gene cluster boundary identification. *Nucleic Acids Res.* **2017**, *45*, W36–W41. [CrossRef]
44. Liu, C.X.; Zhuang, X.X.; Yu, Z.Y.; Wang, Z.Y.; Wang, Y.J.; Guo, X.W.; Xiang, W.S.; Huang, S.X. Community structures and antifungal activity of root-associated endophytic actinobacteria of healthy and diseased soybean. *Microorganisms* **2019**, *7*, 243. [CrossRef]
45. Zhang, J.; Wang, X.J.; Yan, Y.J.; Jiang, L.; Wang, J.D.; Li, B.J.; Xiang, W.S. Isolation and identification of 5-hydroxyl-5-methyl-2-hexenoic acid from *Actinoplanes* sp. HBDN08 with antifungal activity. *Bioresour. Technol.* **2010**, *101*, 8383–8388. [CrossRef]
46. Yu, Z.Y.; Wang, L.; Yang, J.; Zhang, F.; Sun, Y.; Yu, M.M.; Yan, Y.J.; Ma, Y.T.; Huang, S.X. A new antifungal macrolide from *Streptomyces* sp. KIB-H869 and structure revision of halichomycin. *Tetrahedron Lett.* **2016**, *57*, 1375–1378. [CrossRef]
47. Kämpfer, P.; Genus, I. *Streptomyces* Waksman and Henrici 1943, 339 [AL]. In *Bergey's Manual of Systematic Bacteriology*, 2nd ed.; Springer: New York, NY, USA, 2012; pp. 1679–1680.
48. Lee, H.J.; Whang, K.S. Streptomyces rhizosphaerihabitans sp. nov. and Streptomyces adustus sp. nov. isolated from bamboo forest soil. *Int. J. Syst. Microbiol.* **2016**, *66*, 3573–3578.
49. Wang, Z.K.; Jiang, B.J.; Li, X.G.; Gan, L.Z.; Long, X.F.; Zhang, Y.Q.; Tian, Y.Q. Streptomyces populi sp. nov. a novel endophytic actinobacterium isolated from stem of Populus adenopoda Maxim. *Int. J. Syst. Microbiol.* **2018**, *68*, 2568–2573.
50. Sripreechasak, P.; Matsumoto, A.; Suwanborirux, K.; Inahashi, Y.; Shiomi, K.; Tanasupawat, S.; Takahashi, Y. Streptomyces siamensis sp. nov., and Streptomyces similanensis sp. nov., isolated from Thai soils. *J. Antibiot.* **2013**, *66*, 633–640.
51. Kim, S.B.; Lonsdale, J.; Seong, C.N.; Goodfellow, M. *Streptacidiphilus* gen. nov., acidophilic actinomycetes with wall chemotype I and emendation of the family *Streptomycetaceae* (Waksman and Henrici (1943) AL) emend. Rainey et al. 1997. *Antonie. Leeuwenhoek.* **2003**, *83*, 107–116. [CrossRef]
52. Kroppenstedt, R.M. Fatty Acid and Menaquinone Analysis of Actinomycetes and Related Organisms. In *Chemical Methods in Bacterial Systematics*; Goodfellow, M., Minnikin, D.E., Eds.; Academic Press: London, UK, 1985; pp. 173–199.
53. Lechevalier, M.P.; Lechevalier, H. Chemical composition as a criterion in the classification of aerobic actinomycetes. *Int. J. Syst. Bacteriol.* **1970**, *20*, 435–443. [CrossRef]
54. Rong, X.Y.; Huang, Y. Taxonomic evaluation of the *Streptomyces hygroscopicus* clade using multilocus sequence analysis and DNA-DNA hybridization, validating the MLSA scheme for systematics of the whole genus. *Syst. Appl. Microbiol.* **2012**, *35*, 7–18. [CrossRef]
55. Krichevsky, M.I.; Moore, L.H.; Moore, W.E.C.; Murray, R.G.E.; Stackebrandt, E.; Starr, M.P.; Trper, H.G. International Committee on Systematic Bacteriology. Report of the ad hoc committee on reconciliation of approaches to bacterial systematics. *Int. J. Syst. Bacteriol.* **1987**, *37*, 463–464.
56. Ye, X.W.; Chai, W.Y.; Lian, X.Y.; Zhang, Z.Z. Novel propanamide analogue and antiproliferative diketopiperazines from mangrove *Streptomyces* sp. Q 24. *Nat. Prod. Res.* **2017**, *31*, 1390–1396. [CrossRef]
57. Domondon, D.L.; He, W.; De Kimpe, N.; Höfte, M.; Poppe, J. $\beta$-Adenosine, a bioactive compound in grass chaff stimulating mushroom production. *Phytochemistry* **2004**, *65*, 181–187. [CrossRef]
58. Cheng, Y.X.; Zhou, J.; Teng, R.W.; Tan, N.H. Nitrogen-containing compounds from *Brachystemma calycinum*. *Acta Bot. Yunnanica* **2001**, *23*, 527–530.
59. Katsuyama, Y. Mining novel biosynthetic machineries of secondary metabolites from actinobacteria. *Biosci. Biotechnol. Biochem.* **2019**, *83*, 1606–1615. [CrossRef] [PubMed]

60. Rutledge, P.J.; Challis, G.L. Discovery of microbial natural products by activation of silent biosynthetic gene clusters. *Nat. Rev. Microbiol.* **2015**, *13*, 509–523. [CrossRef] [PubMed]
61. Liu, S.P.; Yuan, P.H.; Wang, Y.Y.; Liu, X.F.; Zhou, Z.X.; Bu, Q.T.; Yu, P.; Jiang, H.; Li, Y.Q. Generation of the natamycin analogs by gene engineering of natamycin biosynthetic genes in *Streptomyces chattanoogensis* L10. *Microbiol. Res.* **2015**, *173*, 25–33. [CrossRef] [PubMed]
62. Du, Y.L.; Li, S.Z.; Zhou, Z.; Chen, S.F.; Fan, W.M.; Li, Y.Q. The pleitropic regulator AdpAch is required for natamycin biosynthesis and morphological differentiation in *Streptomyces chattanoogensis*. *Microbiology* **2011**, *157*, 1300–1311. [CrossRef] [PubMed]

# 6

# Isolation and Characterization of a New Endophytic Actinobacterium *Streptomyces californicus* Strain ADR1 as a Promising Source of Anti-Bacterial, Anti-Biofilm and Antioxidant Metabolites

**Radha Singh and Ashok K. Dubey ***

Department of Biological Sciences & Engineering, Netaji Subhas Institute of Technology, New Delhi 110078, India; dharana.radha@gmail.com
* Correspondence: adubey.nsit@gmail.com or akdubey@nsut.ac.in

**Abstract:** In view of the fast depleting armamentarium of drugs against significant pathogens, like methicillin-resistant *Staphylococcus aureus* (MRSA) and others due to rapidly emerging drug-resistance, the discovery and development of new drugs need urgent action. In this endeavor, a new strain of endophytic actinobacterium was isolated from the plant *Datura metel*, which produced secondary metabolites with potent anti-infective activities. The isolate was identified as *Streptomyces californicus* strain ADR1 based on 16S rRNA gene sequence analysis. Metabolites produced by the isolate had been investigated for their antibacterial attributes against important pathogens: *S. aureus*, MRSA, *S. epidermis*, *Enterococcus faecium* and *E. faecalis*. Minimum inhibitory concentration ($MIC_{90}$) values against these pathogens varied from 0.23 ± 0.01 to 5.68 ± 0.20 μg/mL. The metabolites inhibited biofilm formation by the strains of *S. aureus* and MRSA (Biofilm inhibitory concentration [$BIC_{90}$] values: 0.74 ± 0.08–4.92 ± 0.49 μg/mL). The $BIC_{90}$ values increased in the case of pre-formed biofilms. Additionally, the metabolites possessed good antioxidant properties, with an inhibitory concentration ($IC_{90}$) value of 217.24 ± 6.77 μg/mL for 1, 1-diphenyl-2-picrylhydrazyl (DPPH) free radical scavenging. An insight into different classes of compounds produced by the strain ADR1 was obtained by chemical profiling and GC-MS analysis, wherein several therapeutic classes, for example, alkaloids, phenolics, terpenes, terpenoids and glycosides, were discovered.

**Keywords:** endophytic actinobacteria; *Streptomyces* sp., anti-*S. aureus*; anti-MRSA; anti-biofilm

## 1. Introduction

The emergence of drug resistance among pathogens has assumed alarming proportions in recent times, causing rapid depletion in the current armamentarium of drugs to fight such infections [1]. This has posed a serious threat to human health globally and required the rapid development of new and effective drugs at a pace faster than resistance to achieve desirable outcomes in the treatment of infectious diseases. Some of these pathogens have gained exceptional notoriety due to their tremendous ability to adapt, to evade the host immune response and to develop drug resistance. This has led World Health Organization to enlist significant human pathogens under critical, high and medium priority categories [2]. *Enterococcus faecium* (vancomycin resistant) and *Staphylococcus aureus* (methicillin resistant) are considered as high priority pathogens for whom new antibiotics are required the most urgently. Infections involving such pathogens are often associated with biofilms, which are responsible for multi-fold increase in drug-resistance of the pathogens [3,4]. Therefore, it is highly desirable that the new antibiotics possess anti-biofilm activities: disruption of pre-formed biofilms and inhibition of biofilms formation.

Some of the recent studies have reported that the reactive oxygen species were causing antibiotic tolerance in *S. aureus* during systemic infections [5]. Further, oxidative immune response of the host appeared to be switched on during bacterial infections, resulting in increased oxidative stress to the host [6]. Additionally, the antibiotics used to treat infections might also cause an increase in the level of oxidative stress [7,8]. It was reported that antioxidants might prevent oxidative stress-induced pathology [9]. Biofilm formation in *S. aureus* is also enhanced in the presence of oxidative stress [10]. Therefore, providing antioxidants may help in the inhibition of biofilms formation and thus in the prevention of concomitant resistance to antibiotics among the pathogens.

In view of the facts mentioned above, our research endeavors have focused on the discovery of novel anti-infective therapeutics for the treatment and cure of drug resistant infections. Characterization of antioxidant properties of metabolites was also part of our study design due to its foreseeable application in therapy of infectious diseases. There are several approaches to develop drugs, for example, rational drug design (structure-based design of inhibitors against target), synthetic and combinatorial chemistry, high throughput screen of chemical libraries and mining of natural products [11,12]. However, from the stand point of discovery of novel pharmacophore or new classes of drug working on as yet unknown targets, mining of natural products is the obvious choice. Therefore, we chose natural products in the search for new antibiotics. Furthermore, we have considered actinobacteria from vast pool of natural product resources due to their versatility, ubiquity and ability to produce therapeutic compounds with extensive chemical diversity [13,14]. The genus *Streptomyces* of actinobacteria has been regarded as containing the most prolific producers of therapeutic compounds [15,16]. However, repeat discovery of known molecules remains a challenge while hunting the actinobacteria for drugs [17,18]. One of the possible approaches to avoid the repeat discovery of the drugs could be the sourcing of actinobacteria from niche habitats instead of common sources like soil. Accordingly, we have explored endophytic actinobacteria in an effort to enhance the chances of finding new compounds as potential drug candidates. Endophytic actinobacteria are the microorganisms that reside within the plant tissues without causing any adverse effect to plants [19]. Further, for increasing the prospect of strain novelty, we have selected a medicinally important plant, *Datura metel*, which had largely remained unexplored for endophytic population of actinobacteria.

In the present communication, identification of a novel strain, *Streptomyces californicus* strain ADR1 is reported from the plant *D. metel*. Secondary metabolites produced by the isolate ADR1 were characterized for their antibacterial, antibiofilm and antioxidant properties. Further, the metabolite preparations were analyzed for different class of therapeutically significant compounds produced by the isolate.

## 2. Materials and Methods

### 2.1. Isolation of Endophytic Actinobacteria

The endophytic actinobacteria were isolated from the plant, *Datura metel*. The ex-plants were surface sterilized by following the method reported elsewhere [20]. The sterilized plant parts were aseptically grinded by using autoclaved mortar pestle in phosphate-buffered saline (PBS) pH 7.0. The ground paste was spread over the following isolation media: nutrient agar, asparagine glycerol (AGS) agar) [21], humic acid—vitamin agar [22] and starch casein nitrate (SCN) agar [23]. The media were supplemented with cycloheximide (50 μg/mL). The plates were incubated at 28 °C for up to four weeks with regular observations for potential actinobacterial colonies.

The putative actinobacterial colonies were transferred and maintained on AGS medium. Purification of the isolates was achieved by repeated cycles of streaking on fresh plate. The purified cultures were screened for anti-bacterial against *S. aureus* ATCC 29213 by well diffusion method [24].

*2.2. Molecular Identification and Characterization of the Isolate ADR1*

Molecular identification of the strain ADR1 was based on 16S rRNA gene sequence analysis. The genomic DNA of the strain ADR1 was isolated using the method developed for Gram-positive bacteria [20] with a few modifications. Briefly, amplification of 16S rRNA gene was carried out using universal primers: V1f (50-AGAGTTTGATCMTGGCTCAG-30), V9r (50-AAGGAGGTGATCCANCCRCA-30), V3f (50-CCAGACTCCTACGGGAGGCAG-30) and V6r (50-ACGAGCTGACGACARCCATG-30) in a PCR machine (Mastercycler® nexus, Eppendorf International, Germany) by using the programme described elsewhere [25]. The amplified product was sequenced by Sanger's method using a 3130XL sequencer (Applied Biosystems, California, USA) for the 16S rRNA gene using universal primers as described above. The sequences were aligned in MEGA 6.0 to generate single consensus sequence. Homology search was performed using the standard Basic Local Alignment Search Tool (BLAST) sequence similarity search tool of the NCBI database to establish the identity of the isolate ADR1. Nucleotide sequences producing significant alignments after BLAST analysis with the 16S rRNA gene sequence of ADR1 were retrieved in FASTA format. These sequences were used to generate phylogenetic relationship of ADR1 with them by using the software, Phylogeny.fr [26,27]. The analysis was done by using advanced mode of this tool, which is an automated programme that performs step-by-step analysis starting from the multiple alignment of the sequences (MUSCLE 3.8.31) [28], alignment curation (Gblocks 0.91b), construction of phylogenetic tree (PhyML 3.1/3.0 aLRT) [29,30] to the visualization of phylogenetic tree (TreeDyn 198.3) [31]. The culture was characterized for its morphological features on different international *Streptomyces* protocol (ISP) media [32]. Isolate ADR1 was streaked on ISP1, ISP2, ISP3, ISP4, ISP5, ISP6 and ISP7 media plates and was incubated for 7 days at 28 °C for phenotypic and morphological observations. Single colony morphology of the culture was observed under Nikon stereo zoom microscope SMZ1270 at zooming ratio of 12.7:1 and resolution of 8×. Mycelial structure was observed under Nikon E600 microscope (Nikon, Tokyo, Japan) at a resolution of 100×.

*2.3. Production of Secondary Metabolites*

A single colony from freshly grown culture plate (72 h) was inoculated in 50 mL SCN broth (pH 7.4), which was incubated at 28 °C for 72 h to develop the pre-seed culture. The production medium (SCN broth, pH 7.2) was inoculated with the pre-seed culture (1%; *v*/*v*) to commence production of the secondary metabolites, which was carried out for 7 days at 28 °C in an incubator shaker (Adolf Kuhner AG, Birsfelden, Basel, Switzerland) run at 200 rpm. The cell-free broth was recovered by centrifugation at 5000× *g* for 20 min in Sorvall RC 5C plus centrifuge (Kendro Laboratory Products, Newtown, Connecticut, USA). The metabolites were recovered from the supernatant by using liquid-liquid extraction with equal volume of ethyl acetate. The extracted metabolites were dried by using rotary evaporator (50 °C) and vacuum oven (35 °C). The dried metabolite preparations were stored at 25 °C ± 2 °C till further use.

*2.4. Antimicrobial Susceptibility Testing*

The reference strains of bacterial pathogens used in this study were: *S. aureus* ATCC 29213, *S. aureus* ATCC 25923, *S. aureus* ATCC 13709, MRSA ATCC 43300, MRSA 562, *S. epidermis* ATCC 12228, *Enterococcus faecalis* ATCC 29212, *E. faecium* ATCC 49224 and *E. faecium* AIIMS. In-vitro antibacterial activity of the metabolite extract was determined on cation adjusted Muller Hinton agar (MHA) (Himedia, Mumbai, India) plates using well diffusion method [24]. The minimum inhibitory concentration ($MIC_{90}$) values were measured in a 96-well microtiter plates by the broth microdilution method as per the guidelines of Clinical and Laboratory Standards Institute (CLSI) [33]. Briefly, a stock solution of the metabolite extract (1mg/mL) was prepared in 0.2% DMSO and cation adjusted Muller Hinton broth. Bacterial pathogens (100 μL; $2 \times 10^8$ CFU/mL) and metabolite extract (100 μL) at concentrations varying from 125 to 0.122 μg/mL were added to the individual well in the microtitre plate. A sample control (ADR1 extract alone) and blank (media only) were included in each assay. After incubation for 24 h at 37 °C, iodonitrotetrazolium chloride (INT) (Sisco Research Laboratories, Mumbai, India) was added to the wells and the plates were incubated further for 30 min. The absorbance was measured on a multimode reader (Biotek Instruments, Winooski, Vermont, USA) at 490 nm. The value of $MIC_{90}$ was considered to be the minimum concentration at which no visible growth could be observed. The following equation was used to compute the percent inhibition [34].

$$\text{Growth inhibition of pathogen (\%)} = [(\text{control } OD_{490\text{ nm}} - \text{test } OD_{490\text{ nm}})/\text{control } OD_{490\text{ nm}}] \times 100 \quad (1)$$

### 2.5. *Antibiofilm Assay*

Biofilms of *S. aureus* ATCC 25923, *S. aureus* ATCC 29213, MRSA ATCC 43300 and MRSA 562 were produced by using the method published elsewhere [35] in accordance with the CLSI guidelines [33]. Briefly, overnight grown reference cultures were suspended in tryptic soy broth (Himedia, Mumbai, India) supplemented with 2% glucose to attain turbidity equivalent to 0.5 McFarland standard ($2 \times 10^8$ CFU/mL). A total of 100 µL of the cell suspension was transferred to the wells on the microtiter plate and was incubated at 37 °C for 24 h under static condition. Non-adherent cells were aspired out along with the medium. The wells were rinsed with 100 µL of phosphate-buffered-saline (PBS). Fresh medium containing desired concentrations of ADR1 metabolites (from 250 to 0.49 µg/mL) were added to the wells on the microtiter plate, which was then incubated for the next 24 h at 37 °C under static condition. Viability of the biofilms was quantified by INT-calorimetric assay as described above. The following equation was used to find % inhibition of biofilm [35].

$$\text{Biofilm inhibition (\%)} = [(\text{control OD }_{490\text{ nm}} - \text{test OD }_{490\text{ nm}})/\text{control OD }_{490\text{ nm}}] \times 100 \quad (2)$$

### 2.6. *Antioxidant Activity*

The antioxidant potential of ADR1 metabolites was assessed by measuring reduction of DPPH (1, 1-diphenyl-2-picrylhydrazyl) free radicals as reported earlier [36]. Briefly, DPPH solution (0.1 mM) was prepared in methanol; 100 µL of this solution was added to 100 µL of the ADR1 metabolite preparations at different concentrations varying from 1000 to 7.81 µg/mL in 96-well microtitre plate. The plate was then incubated at 25 °C for 20 min in dark and the absorbance was measured at 517 nm. The scavenging strength was calculated using the following formula [36].

$$\text{\% scavenging activity} = [(\text{absorbance of DPPH control} - \text{absorbance of DPPH in the presence of metabolite})/\text{absorbance of DPPH control}] \times 100 \quad (3)$$

### 2.7. *Hemolytic Activities*

Hemolytic activity was determined by disc diffusion assay using sheep blood agar (SBA) plates (Himedia, Mumbai, India). A total of 10 µL solution containing varying concentrations of the ADR1 metabolites (1000 to 7.8125 µg/mL) were dispensed on discs placed aseptically on the SBA plates and were incubated for 24 h at 37 °C. The type of hemolysis was observed as alpha, beta and gamma [37].

### 2.8. *Secondary Metabolite Profiling and GC-MS Analysis*

The ADR1 metabolite extract was used at a concentration of 5 mg/mL for chemical profiling of the classes of metabolites present in the extract. The tests for the different class of metabolites, for example, anthraquinones, glycosides, terpenoids, flavonoids, tannins, alkaloids, saponins, sterols, anthocyanins, coumarins, tannins, lactones, terpenes, fatty acids, proteins/amino acids and carbohydrates were carried out by using standard methods reported earlier [38–40].

Further analysis of the metabolites was carried out by employing GC-MS (GC-MS-QP2010 plus; Shimadzu, Kyoto, Japan) as outlined below. A constant column flow rate of 1.21 mL/min with helium gas was maintained in RESTEK capillary column (30 m × 0.25 mm I.D. × 0.25 µm film thickness).

Initial oven temperature was 100 °C for 3 min, which was increased to 250 °C for a hold time of 5 min, was further increased gradually to 280 °C where it was kept constant for 15 min. A total of 3 μL of sample (3 mg/mL) was injected in split mode (split ratio of 10.0) and linear velocity of the column was maintained at 40.9 cm/s. The mass fragmentation patterns (spectra) of the metabolites were obtained at electron ionization (EI) of 70 eV scanned over a m/z range of 40–650. The compounds detected were identified on the basis of comparison of the mass spectra with those available in the NIST14 and Wiley8 spectral library. The spectra having a match limit value lower than 700 were not considered.

### 2.9. Statistical Analysis

All the experiments were performed in triplicates. The data were expressed as mean ± standard deviation. The statistical analysis and significance of the test was performed by analysis of variance (ANOVA), using the software, graph pad prism 5.01. The graphs were also generated using grouped analysis in the graph pad prism and represented with SEM in the form of error bar.

## 3. Results and Discussion

Eight putative actinobacterial endophytes, designated as ADR1 to ADR8, were isolated from the plant, *Datura metel*. While no isolate could be found from the stem part of the plant, six were obtained from the root tissues and two were from the leaves. The antibacterial potency of these isolates was examined against the reference strain, *S. aureus* ATCC 25923. Based on the size of the zone of inhibition, the isolate ADR1 was chosen for further studies as it produced largest zone (22.5 ± 0.58 mm). Production of the metabolites by ADR1 was carried out under the conditions as described under Section 2.3. The metabolite extract was recovered as red colour hygroscopic sticky mass (approximately 120 mg/L).

### 3.1. Identification and Characterization of the Isolate ADR1

Amplification of the 16S rRNA gene from the genome of ADR1 produced a sequence of 1452 nucleotides. Blast analysis revealed 99.17% sequence identity of the ADR1 sequence with *S. californicus* strains with a query coverage of 99%. The phylogenetic relationship of the strain ADR1 can be seen in Figure 1, where it showed closest relationship with *S. californicus* strains. The 16S rRNA gene sequence obtained in this study was submitted as '*Streptomyces californicus* strain ADR1' to NCBI GenBank with accession no. KU299789.1.

A few strains of *S. californicus* had been reported earlier from the soil [41–44]. However, there are no reports of any endophytic strain of *S. californicus* till date to the best of our knowledge, making the present isolate as a new endophytic actinobacterial strain, designated as *S. californicus* strain ADR1.

The strain ADR1 was characterized for its cultural attributes on ISP media 1 to 7. The results (Table 1) suggested that the extent of growth of the culture varied from scanty to abundant on different ISP media. Further, differences with respect to the colour of substrate and aerial mycelia, and production of diffusible pigments were also noted as described in the Table 1. When compared with the cultural characteristics of non-endophytic *S. californicus* strains JCM 6910, MNM-1400 and G16, it was observed that ADR1 shared a few similarities, for example, colour of aerial mycelium on ISP 2, 3 and diffusible pigments on 4, with *S. californicus* strain JCM 6910, a soil isolate from Japan [41,44,45]. However, the growth of ADR1 was abundant on ISP3, while that of JCM 6910 was poor. No diffusible pigment was produced by ADR1 in ISP5, while violet pigment was produced by JCM 6910. Other soil isolates of *S. californicus*, strain MNM-1400 and strain G16, were morphologically very different from the strain ADR1 [44,45] Thus, the endophytic *S. californicus* strain ADR1 was evidently distinct from the soil isolates.

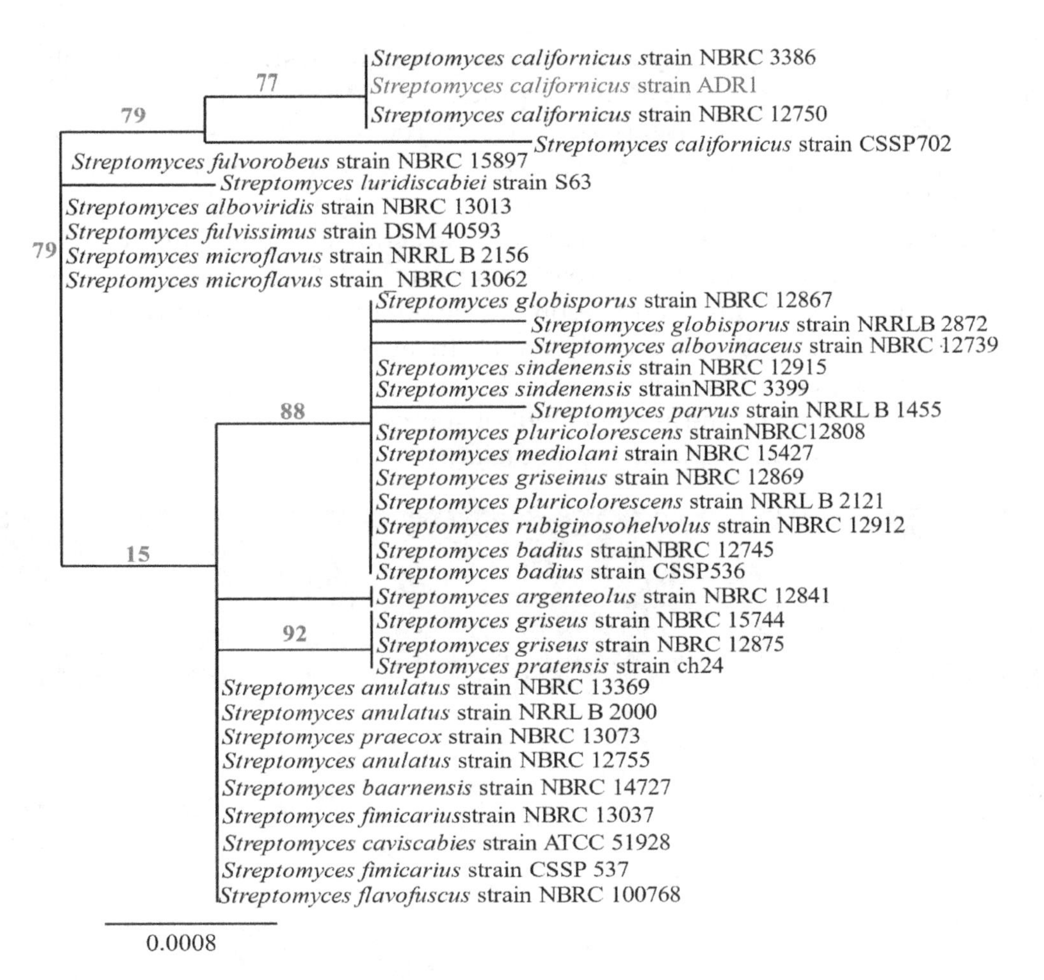


**Figure 1.** Phylogenetic analysis of isolate ADR1. Neighbour-joining phylogenetic tree showed maximum likelihood model showing the phylogenetic relationship of selected isolate (highlighted in blue) based on 16S rRNA gene sequence alignments. The numbers at the branching points are the percentages of occurrence in 500 bootstrapped trees. Bar indicated 0.0008 substitutions per nucleotide position.

**Table 1.** Cultural characteristics of the isolate ADR1 on international *Streptomyces* project (ISP) media.

| ISP Media | Growth | Substrate Mycelium | Aerial Mycelium | Diffusible Pigments | Appearance |
|---|---|---|---|---|---|
| ISP-1 (Tryptone-Yeast Extract Broth) | Abundant | Crimson red | White | No | Shrinked and depressed with irregular edges |
| ISP-2 (Yeast extract- Malt extract Agar) | moderate | Wine Red | Pale green | Yellow | Elevated, smooth, regular edges |
| ISP-3 (Oatmeal agar) | Abundant | Wine Red | Dusty green | Light violet | Shrinked, pits formation, regular edges |
| ISP-4 (Inorganic salt starch agar) | Moderate | Dark pink | Light pink | Light pink | Flat, wavy edges, pointed centre |
| ISP-5 (Glycerol asparagine agar base) | Abundant | Pink | Dusty green | No | Elevated, Round, smooth edges |
| ISP-6 (Peptone yeast extract iron agar) | Moderate | Rusty red | White | Light pink | Elevated at centre, Round, smooth edges |
| ISP-7 (Tyrosine agar) | Scanty | Light pink | Whitish Pink | No | Pin pointed at centre, flat, round and smooth edges |

A detailed view of the morphology was obtained through the study of single colonies (Figure 2; Panel A and B). Growth on different ISP media produced differences in colour and appearance of the colonies, which appeared as dense, depressed and rocky on ISP1, while on ISP2, 3 and 5 the aerial mycelia appeared to be fluffy and dusty. Clear exudates can be seen over the colony on ISP6. The colonies on ISP7 appeared scantly grown lacking distinct structures that were observed on other media. Prominent differences in the extent of growth, structure and pigmentation of the colonies on different ISP media were consistent with the earlier reports [46]. Such media-dependent phenotypic variations suggested that the primary and secondary metabolism of the culture varied significantly with changes in composition of the medium, which is in agreement with the current understanding of the physiology of the genus *Streptomyces* [47,48]. The microscopic observations showed highly branched flexuous mycelium and the arrangement of spores in a chain inside mycelium as shown in (Figure 2, panel C and D). These are some explicit features of *Streptomyces* species [32].

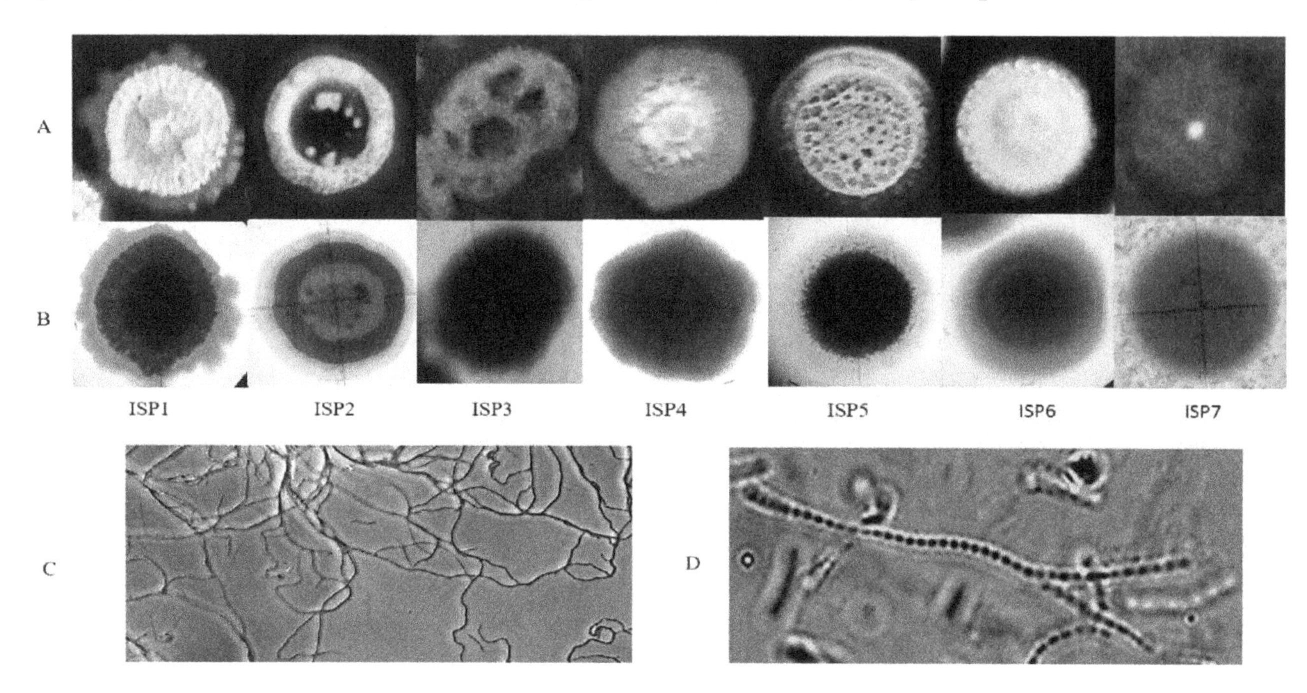


**Figure 2.** Morphological characterization of *S. californicus* ADR1. Colony observations were made on different ISP media. Panel (**A**) represented the magnified view showing exterior structures of the single colony while panel (**B**) showed interior view of the colony as observed under a Nikon Stereo Zoom Microscope SMZ1270; the mycelial network of branched hyphae could be observed in panel (**C**) and arrangement of spores in chains were viewed in panel (**D**).

### *3.2. Antibacterial Spectrum of ADR1 Metabolites Against Significant Gram-Positive Pathogens*

The spectrum of activity of the ADR1 metabolites against different Gram-positive pathogens was determined as described under Section 2. The results presented in Table 2 showed that the metabolite extract possessed broad spectrum activity against Gram-positive pathogens as it effectively inhibited growth of all the reference strains. It may be noted that the activities of metabolite extract against MRSA strains were similar to those against *S. aureus* strains.

However, the above data provided more of a qualitative rather than quantitative estimate of the antibacterial activity. An accurate view of the potency of the metabolites was achieved by assessment of $MIC_{90}$ values against the pathogenic strains used in this study. As presented in Figure 3a, the $MIC_{90}$ values of the ADR1 metabolites against different strains of *S. aureus* were between $0.44 \pm 0.07 - 0.84 \pm 0.03$ µg/mL, while against *S. epidermis* ATCC1222 it was even lower ($0.23 \pm 0.01$ µg/mL). It may be noted that the potency against MRSA strains (Figure 3b) was as good as it was against *S. aureus* strains. These results can be considered as significant since the $MIC_{90}$ values of the ADR1 metabolites stood better than the standard drug vancomycin (0.5 to 2 µg/mL) for MRSA strains [49]. Other than *S. aureus* strains, the $MIC_{90}$ values of 1.92 ± 0.03 and 3.35 ± 0.18 μg/mL were observed against *E. faecium* strains AIIMS and ATCC 49214, respectively, which demonstrated good sensitivity of these strains to ADR1 metabolites. *E. faecalis* ATCC 29212 was the least sensitive to ADR1 metabolites among the reference pathogens since its $MIC_{90}$ value (5.68 ± 0.20 μg/mL) stood at the maximum, but it was still better than some of the previously reported activities [50,51]. The $MIC_{90}$ values of the metabolites from some of the recently reported *Streptomyces* spp. against the strains of *S. aureus* and MRSA were in the ranges of 2–125 μg/mL [52–54], which indicated wide variation in the anti-bacterial potency of the metabolites. The results of the present study clearly demonstrated that the anti-bacterial potency of ADR1 metabolites appeared at the bottom of the range or even lower, validating its promising potential for discovery of drugs against priority pathogens.

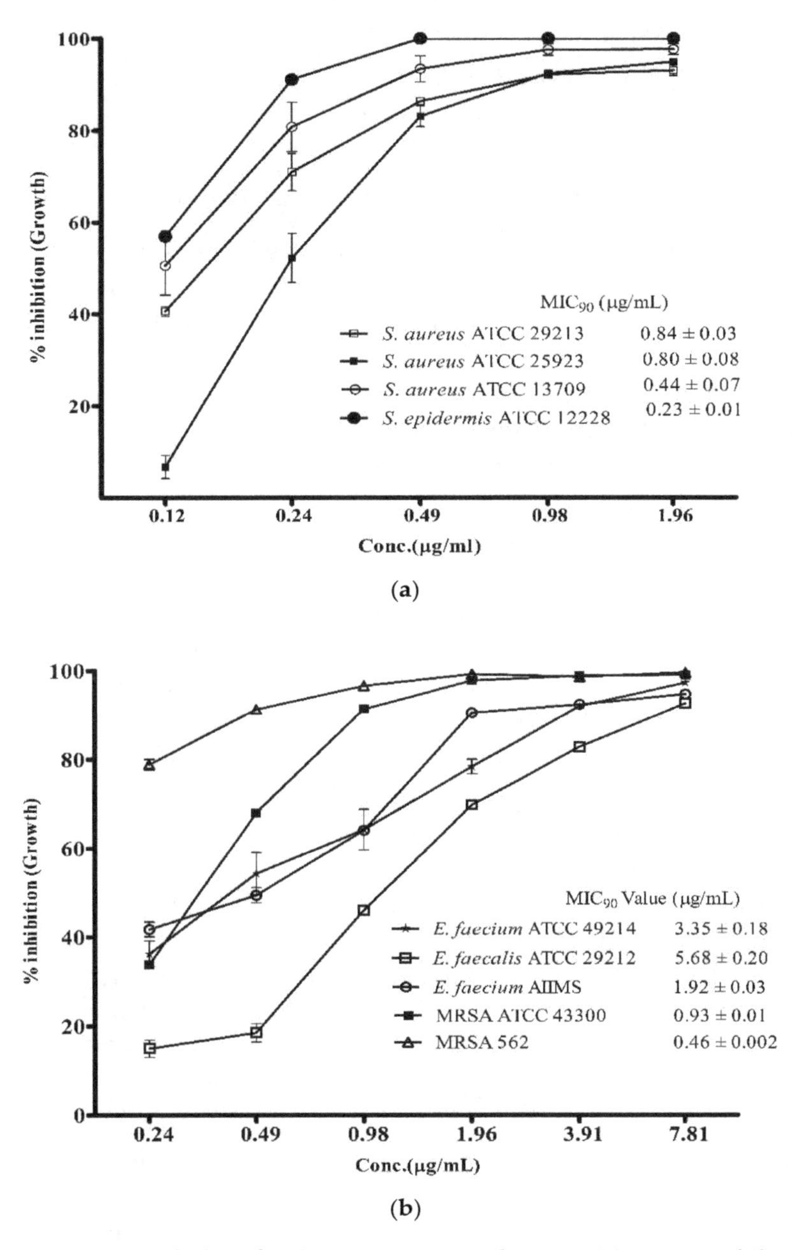


**Figure 3.** Potency of ADR1 metabolites for Gram-positive pathogens. Minimum inhibitory concentration ($MIC_{90}$) values of the metabolites against the target pathogens were corelated with its antibacterial potency. Percent inhibition in growth at different concentration of the metabolites was measured against the reference strains as shown in (**a**) and in (**b**). The data represented mean ± SEM values of experiments done in triplicate (*p* value < 0.0001).

**Table 2.** Spectrum of antibacterial activity of the metabolites produced by *S. californicus* ADR1.

| S. No. | Reference Strains of Gram-Positive Pathogens | Zone of Inhibition (mm) |
|---|---|---|
| 1 | *Staphylococcus aureus* ATCC 29213 | 22.5 ± 0.58 |
| 2 | *S. aureus* ATCC 25923 | 19 ± 0.42 |
| 3 | *S. aureus* ATCC 13709 | 20 ± 0.5 |
| 4 | *S. epidermis* ATCC 12228 | 18 ± 0.45 |
| 5 | Methicillin-resistant *S. aureus* (MRSA) ATCC 43300 | 21.3 ± 0.27 |
| 6 | MRSA 562 (clinical strain) | 19 ± 0.25 |
| 7 | *Enterococcus faecium* ATCC 49224 | 16.5 ± 0.4 |
| 8 | *E. faecium* AIIMS | 19.4 ± 0.47 |
| 9 | *E. faecalis* ATCC 29212 | 17 ± 0.52 |

### *3.3. Anti-Biofilm Activity of ADR1 Metabolites Against S. aureus and MRSA*

Biofilm-associated infections posed a greater challenge in treatment of infectious diseases as it is one of the major contributing factors in enhancing antibiotic-resistance among *S. aureus* and its methicillin-resistant strains [3]. Therefore, the discovery of drugs with potent anti-biofilm activity is needed more than ever before to combat the ever-growing global challenge of antibiotic resistance against the notorious pathogens like *S. aureus* and MRSA. In view of this, the biofilm inhibitory potential of the ADR1 metabolites was investigated. The results (Figure 4a) suggested that the ADR1 metabolites were able to effectively inhibit formation of biofilm by the *S. aureus* and the MRSA strains. Up to 90% reduction in the formation of biofilm could be achieved at significantly lower concentration of the metabolites; the $BIC_{90}$ values were noted to be in the range of 0.74 ± 0.08 to 4.59 ± 0.71 μg/mL. The effectiveness of the ADR1 metabolites in the inhibition of the biofilm was found to be better than the previously reported activity of biofilm inhibition by actinobacterial metabolites where maximum inhibition of 83% was recorded at 265 μg/mL [35]. However, when the activity against pre-formed biofilms (24 h) was examined, the $BIC_{90}$ values for *S. aureus* strains increased by many folds, for example, up to 45.69 ± 3.32 and 89.54 ± 0.40 μg/mL for *S. aureus* ATCC 25923 and *S. aureus* ATCC 29213, respectively. The biofilms produced by MRSA proved even more resistant (Figure 4b). It was reported that some of the well-known antibiotics like pyrrolomycin and related compounds could inhibit the biofilm only up to 67%–87% [55]. Similarly, in a study on the effect of antibiotics like rifampicin, polymyxin B, kanamycin and doxycyclin on reduction of *S. aureus* biofilm formation, only rifampicin was found to inhibit the biofilm by about 50% [56]. The effect of ADR1 metabolites on inhibition of biofilm formation as well as on the preformed biofilms was better than some previously reported metabolite extracts [57–59]. Inhibition of biofilm formation strongly suggested that the metabolites prevented adherence of *S. aureus* and MRSA cell to the polystyrene surface. Further, their ability to disrupt pre-formed biofilms might limit the biofilm-associated drug resistance among the pathogens.

### *3.4. Antioxidant Activity of the ADR1 Metabolites*

Antioxidants assume significance for therapeutic applications in view of their role in neutralizing reactive oxygen species in patients fighting diseases involving infectious agents or metabolic disorders [60–62]. Hence, it was prudent to examine if the ADR1 metabolites possessed any such activity. The standard assay, which measured the reduction of DPPH free radical, revealed the antioxidant properties of ADR1 metabolites (Figure 5). The free radical scavenging activity of the metabolite extract and of ascorbic acid (a well-known antioxidant agent), followed a different pattern, where a sharp increase in DPPH scavenging activity (from 40% to 80%) was observed when concentration of the metabolite was increased from 62.5 to 125 μg/mL. However, in the case of ascorbic acid, there was no significant change in the activity over the above concentration range. The $IC_{90}$ value for DPPH scavenging by ADR1 metabolite was achieved at the concentration of 217.24 ± 6.77 µg/mL, while that for the ascorbic acid it was 904.32 ± 12.93 µg/mL (Figure 5), which was approximately 4-fold higher compared to ADR1 metabolites. However, interestingly $IC_{50}$ of the ascorbic acid was 4.617 ± 0.89 µg/mL while that of ADR1 extract was 77.41 ± 1.02 µg/mL (not plotted). It was recently reported that the endophytic actinobacterial strains BPSAC77, 101, 121 and 147 showed DPPH scavenging, with an $IC_{50}$ value of 43.2 µg/mL, but the $IC_{90}$ value was not reported in this study [63]. In another study, *Streptomyces* sp. had been found to scavenge DPPH free radicals at a much higher concentration, with an $IC_{50}$ value at 435.31 ± 1.79 µg/mL [64]. In one of the recent studies, it was reported that $IC_{50}$ values for DPPH scavenging activities of several isolates of actinobacteria varied from 12 ± 1.8 to 65 ± 3.2 µg/mL [65]. Such variations were also noted by others [54]. These reports indicated that there was a wide variation in the antioxidant activities of the metabolites produced by different actinobacterial strains. An in-depth characterization of relevant compounds in pure form may offer further insight into such differences in the activities.

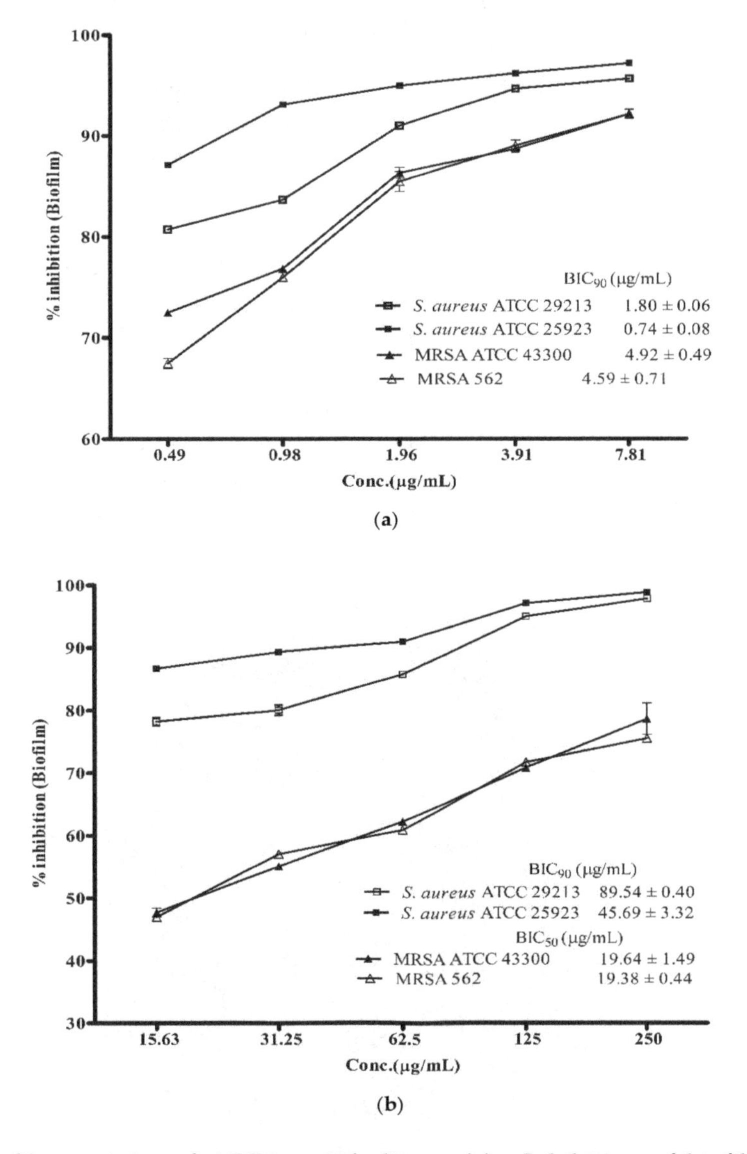


**Figure 4.** Anti-biofilm activity of ADR1 metabolites. (**a**). Inhibition of biofilm formation and (**b**) inhibition of pre-formed biofilm of *S. aureus* ATCC 29213, 25923, MRSA 562 and ATCC 43300 are shown at various concentration of the ADR1 metabolites. The data represented mean ± SEM values of experiments done in triplicate ($p$ value < 0.0001).

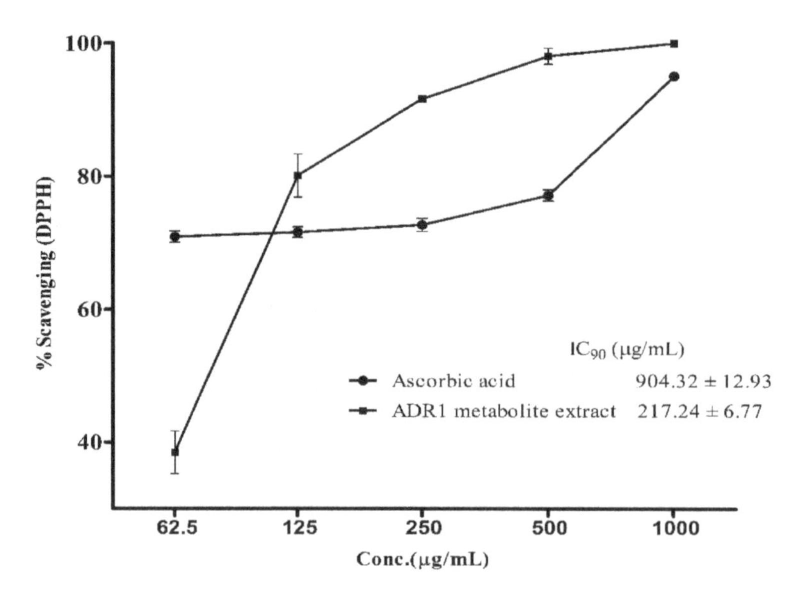


**Figure 5.** DPPH (1, 1-diphenyl-2-picrylhydrazyl) radical scavenging by ADR1 metabolites. Oxidation of DPPH free radicals were measured at the different concentrations of the metabolite. The data represented mean ± SEM values of the experiments done in triplicate ($p$ value < 0.0001).

### 3.5. *Haemolytic Activity*

Drug-induced haemolytic anaemia (DIHA) and thrombocytopenia (DIT) are common adverse effects associated with antibiotics [66]. Reports suggested that the haemolytic effects of antimicrobial peptides tyrocidine A and gramicidin S, limited their use as topical agents [67,68]. Ceftriaxone, a third-generation cephalosporin was also reported to cause haemolysis, attracting advise for restricted administration of the drug [69]. So, it is very important for a potential drug to be tested for its haemolytic activity to secure critical information on its clinical suitability. In this context, the haemolytic effect of the ADR1 metabolites was tested. It was noted that the ADR1 metabolites did not show any sign of haemolysis in the concentration range from 7.8125 to 1000 μg/mL (Figure 6), which was far greater than the $MIC_{90}$ values against test pathogens.

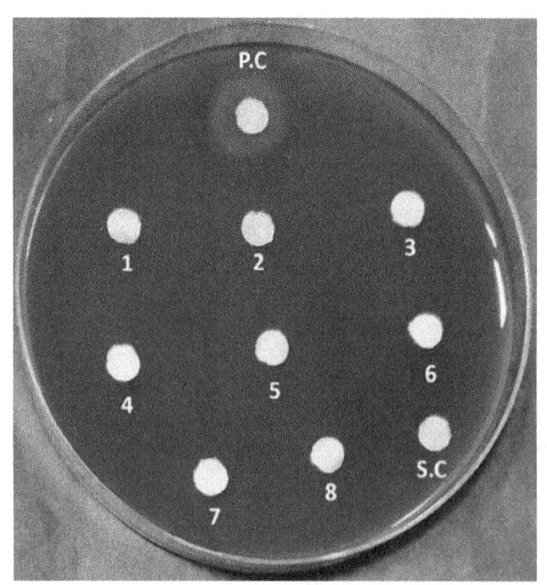


**Figure 6.** Haemolytic activity of the ADR1 metabolites. Spot No. 1 to 8 showed metabolite dilutions from higher (1000 μg/mL) to lower concentration (7.8125 μg/mL). 'PC' was positive control with 0. 1% SDS. SC represented solvent control.

In some previous studies, the extracts with low or no haemolytic activity were considered suitable for further characterization of their antimicrobial properties [70]. This implied that the metabolite extract of strain ADR1 could be safe for further investigation into specific metabolites as probable drug candidates.

### *3.6. Secondary Metabolite Profiling and GC-MS Analysis of Metabolite Extract*

Characterization of bioactivities of the metabolites had demonstrated that the present endophytic strain, *S. californicus* ADR1 was an unexplored source of antimicrobial, antibiofilm and antioxidant agents, which might have potential clinical applications. It was therefore prudent to analyze the composition of the metabolite extract to unravel the class of compounds produced. Therefore, the ADR1 metabolite extract was subjected to chemical profiling using specific reagents. The tests revealed the classes of compounds present in the extract of metabolites produced by the strain ADR1, which included anthraquinones, anthocyanins, terpenoids, terpenes, flavonoids, phenols, alkaloids and glycosides (Table 3). Various compounds, which belonged to these classes have therapeutic significance and are routinely reported from phytochemical screening of plant extracts. But recent literature suggested that actinobacteria too are proving to be excellent sources of such class of compounds [14,71].

**Table 3.** Chemical profiling of ethyl acetate extract of secondary metabolites produced by *S. californicus* strain ADP4.

| Chemical Class of Metabolites | Testes/Reagents Used | Observations | Results |
|---|---|---|---|
| Terpenoids | Salkowski Test | Reddish brown coloration at the interface | + |
| Phenols | Folin–Ciocalteu Test | Blue coloration was appeared | + |
| Flavonoids | Ferric chloride Test | Formation of greenish colour | + |
| | NaOH, HCl | Intense yellow coloration after adding HCl | |
| Terpenes | Salkowski Regent | Appearance of golden colour in the chloroform layer | + |
| Alkaloids | Wagner's Test | Formation of reddish-brown precipitate | + |
| Anthocyanins | HCl, Ammonia | Appearance of pink-red, turns blue | + |
| Anthraquinones | $H_2SO_4$, Chloroform, Ammonia | Light pink coloured layer of ammonia | + |
| Glycosides | Keller–Killiani Test | A reddish-brown colour ring at the junction of the two layers | + |
| Tannins | Lead Acetate Test | No precipitation was observed | – |
| Saponins | Foam Test | No frothing was observed | – |
| Lactones | Pyridine, sodium nitroprusside, NaOH | No change in coloration was observed | – |
| Coumarins | Alcoholic NaOH | Yellow fluorescence was not appeared on the paper soaked in NaOH | – |
| Sterols | Salkowski Test | No red colour was appeared in the lower layer (two layers formed) | – |
| Lignins | Gallic acid | No appearance of Olive-green coloration | – |
| Carbohydrates | Fehling's Test | No reddish violet ring appeared | – |
| Fatty acids | Ether | No appearance of transparence on filter paper | – |
| Proteins | Biuret Test | No violet coloration was observed | – |

Note: '+': Present; '–': Absent.

While the results of chemical profiling were primarily qualitative in nature suggesting the presence or the absence of a specific chemical class, it did not offer any clue about the identity of the compounds. But it is important to have a deeper insight into the chemical class and identity of the specific compounds produced by the culture if these are to be considered for drug development. In pursuit of this goal,

GC-MS analysis of the ADR1 metabolites was undertaken to probe the identity of metabolites based on their mass spectrum.

The GC-MS analysis revealed the presence of alkaloids, terpene, terpenoids and glycosides in the ADR1 metabolite extract, thus confirming the findings of chemical profiling. The compounds were assigned probable identities based on similarity index (SI) value, generated from the mass spectral analyses of the test compounds and their comparison with the reference compounds listed in the NIST and Wiley library. An SI of more than 85 was considered for assignment of probable identity to the compounds [72]. Out of the total 30 peaks detected in the chromatogram, 20 peaks were specific to the culture *S. californicus* ADR1 while rest of the peaks were related to the production media and the solvent control (Figure S1). Only six of ADR1 metabolite peaks were found to have SI value greater than 85. Thus, identities could be assigned to only these compounds (Table 4). Majority of the compounds detected in the GC-MS chromatogram showed lower degree of similarity with SI values between 60–80 (Table S1), suggesting a high probability of occurrence of novel compounds among the secondary metabolites produced by *S. californicus* strain ADR1.

The first compound, detected at RT 11.945 min, showed SI of 88 with Methanoazulen-9-ol, decahydro-2, 2, 4, 8-tetramethyl-stereoisomer, which belonged to the class sesquiterpene and possessed antibacterial as well as antioxidant properties [73]. This compound was earlier reported as a major component of plant essential oils [74,75] but there are no reports on this from microbial sources as yet. Another compound in ADR1 metabolite extract was detected at RT 14.519 min, which showed SI of 90 with 5-z-methyl-2-z-hydroxycarbonyl-5-e-ethenyl-4-z-propen-2-ylcyclohexanone, was a terpenoid having structure similar to Asperaculane B, a known GABA-transaminase inhibitor having significance in epilepsy treatment [76]. Terpenes and terpenoids have wide therapeutic potential like antimalarial, antibacterial, antiinflammatory, antioxidant as well as wound healing properties [77]. A flavonoid that was found most abundant among the ADR1 metabolites, showed SI of 85 with 4′-Methoxy-2′-(trimethylsiloxy) acetophenone. Flavonoids are well regarded molecules with broad-spectrum biological activities as well as applications in nutraceuticals and cosmetic industry [78]. However, there are no reports on the activity of this molecule till date. An alkaloid with SI of 88 in comparison to pyridineethanamine, n-methyl-n-[2-(4-pyridinyl) ethyl] was detected among the ADR1 metabolites at RT 16.068 min. The compound is known for its importance in the treatment of vertigo [79]. Alkaloids are well regarded for their wide spectrum of therapeutic activities including antibacterial, antioxidant, anti-inflammatory, anti-diabetic, antimalarial and anticancer properties [80]. The presence of therapeutically important chemical classes of compounds in ADR1 metabolite extract underscored its importance in the discovery of novel compounds with prominent therapeutic potential.

**Table 4.** Similarity Index-based analysis of the compounds in the secondary metabolite extract of *S. californicus* ADR1. The data were obtained from GC-MS chromatogram. The reference compounds were from NIST and Wiley Library.

| S. No. | RT (min) | Similarity Index (SI) | Reference Compounds | Chemical Class | Therapeutic Properties |
|---|---|---|---|---|---|
| 1 | 11.945 | 88 | Methanoazulen-9-ol, decahydro-2,2,4,8-tetramethyl-stereoisomer | Sesquiterpene (alpha-Caryophyllene alcohol) | Antibacterial, antioxidant, antiinflammatory [73] |
| 2 | 13.089 | 96 | Naphtho[2,3-g]-1,6,2,5-dioxasilaborocin | Organoboranic acid | No activity reported |
| 3 | 13.231 | 85 | 2-[(trimethylsilyl)oxy]-4-methoxyacetophenone | Flavonoid | No activity reported |
| 4 | 14.519 | 90 | 5-z-methyl-2-z-hydroxycarbonyl-5-e-ethenyl-4-z-propen- 2-ylcyclohexanone | Terpenoids (Asperaculane B type) | Anticonvulsant drug design pharmacophore, sodium channel blocker, GABA-transaminase inhibitors [76] |
| 5 | 15.102 | 92 | 1,2-benzenedicarboxylic acid | Diisobutyl phthalate | No activity reported |
| 6 | 16.068 | 88 | 2-pyridineethanamine, n-methyl-n-[2-(4-pyridinyl)ethyl] | Alkaloid (betahistine types) | Vasodilation and reduction of endolymphatic pressure [79] |

## 4. Conclusions

The metabolites produced by *S. californicus* ADR1 had shown very low $MIC_{90}$ values against a range of Gram-positive pathogens including those notified by the WHO as high priority pathogens. This suggested that it could be an excellent source for effective anti-infective compounds against some of the most challenging pathogens like *S. aureus*, MRSA, *S. epidermis*, *E. faecalis* and *E. faecium*. The ADR1 metabolites came across as potent inhibitors of biofilms of both methicillin-sensitive and methicillin-resistant strains of *S. aureus*. Along with the antimicrobial activity, the metabolites exhibited good antioxidant activity. Antioxidant property with no haemolytic activity indicated the potential of ADR1 metabolites for use as antioxidants. The above bioactivities could be attributed to the presence of several therapeutically significant class of compounds as revealed by the biochemical profiling and the GC-MS analysis. These constituents, individually or in combination, may account for the pharmacological actions of the extract. Low SI for several compounds in the ADR1 metabolite preparations was noteworthy as it indicated greater possibility of finding novel molecules.

**Author Contributions:** A.K.D. has conceptualized the research project and guided the experimental work. R.S. has carried out the experiments and prepared the initial draft of the manuscript, which was corrected and finalized by A.K.D. All authors have read and agreed to the published version of the manuscript.

**Acknowledgments:** The authors acknowledge the administrative and material support given by the Netaji Subhas University of Technology, New Delhi to carry out this work. Research fellowship to RS by NSUT is duly acknowledged. Services of Advanced Instrumentation Centre, Jawahar Lal Nehru University, New Delhi for GC-MS analysis, is duly acknowledged.

## References

1. Chernov, V.M.; Chernova, O.A.; Mouzykantov, A.A.; Lopukhov, L.L.; Aminov, R.I. Omics of antimicrobials and antimicrobial resistance. *Expert Opin. Drug Discov.* **2019**, *14*, 455–468. [CrossRef] [PubMed]
2. *Prioritization of Pathogens to Guide Discovery, Research and Development of New Antibiotics for Drug-Resistant Bacterial Infections, Including Tuberculosis*; World Health Organization: Geneva, Switzerland, 2017.
3. Ch'ng, J.H.; Chong, K.; Lam, L.N.; Wong, J.J.; Kline, K.A. Biofilm-associated infection by enterococci. *Nat. Rev. Microbiol.* **2019**, *17*, 82–94. [CrossRef] [PubMed]
4. Xu, K.D.; McFeters, G.A.; Stewart, P.S. Biofilm resistance to antimicrobial agents. *Microbiol* **2000**, *146*, 547–549. [CrossRef] [PubMed]
5. Rowe, S.; Wagner, N.J.; Li, L.; Beam, J.E.; Wilkinson, A.D.; Radlinski, L.C.; Zhang, Q.; Miao, E.A.; Conlon, B.P. Reactive oxygen species induce antibiotic tolerance during systemic *Staphylococcus aureus* infection. *Nat. Microbiol.* **2020**, *5*, 282–290. [CrossRef] [PubMed]
6. Ivanov, A.V.; Bartosch, B.; Isaguliants, M.G. Oxidative stress in infection and consequent disease. *Oxidative Med. Cell. Longev.* **2017**, *2017*, 3496043. [CrossRef]
7. Kohanski, M.A.; Dwyer, D.J.; Hayete, B.; Lawrence, C.A.; Collins, J.J. A common mechanism of cellular death induced by bactericidal antibiotics. *Cell* **2007**, *130*, 797–810. [CrossRef]
8. Keren, I.; Wu, Y.; Inocencio, J.; Mulcahy, L.R.; Lewis, K. Killing by bactericidal antibiotics does not depend on reactive oxygen species. *Science* **2013**, *339*, 1213–1216. [CrossRef]
9. Liguori, I.; Russo, G.; Curcio, F.; Bulli, G.; Aran, L.; Della-Morte, D.; Gargiulo, G.; Testa, G.; Cacciatore, F.; Bonaduce, D.; et al. Oxidative stress, aging, and diseases. *Clin. Interv. Aging* **2018**, *13*, 757–772. [CrossRef]
10. Kulkarni, R.; Antala, S.; Wang, A.; Amaral, F.E.; Rampersaud, R.; Larussa, S.J.; Planet, P.J.; Ratner, A.J. Cigarette smoke increases *Staphylococcus aureus* biofilm formation via oxidative stress. *Infect. Immun.* **2012**, *80*, 3804–3811. [CrossRef]
11. Mak, K.K.; Pichika, M.R. Artificial intelligence in drug development: Present status and future prospects. *Drug Discov. Today* **2019**, *24*, 773–780. [CrossRef]
12. Ferreira, L.L.G.; Andricopulo, A.D. ADMET modeling approaches in drug discovery. *Drug Discov. Today* **2019**, *24*, 1157–1165. [CrossRef] [PubMed]
13. Patridge, E.; Gareiss, P.; Kinch, M.S.; Hoyer, D. An analysis of FDA-approved drugs: Natural products and their derivatives. *Drug Discov. Today* **2016**, *21*, 204–207. [CrossRef] [PubMed]

14. Singh, R.; Dubey, A.K. Diversity and applications of endophytic actinobacteria of plants in special and other ecological niches. *Front. Microbiol.* **2018**, *8*, 1767. [CrossRef] [PubMed]
15. Berdy, J. Thoughts and facts about antibiotics: Where we are now and where we are heading. *J. Antibiot.* **2012**, *65*, 385–395. [CrossRef] [PubMed]
16. Singh, R.; Dubey, A.K. Endophytic actinomycetes as emerging source for therapeutic compounds. *Indo Glob. J. Pharm. Sci.* **2015**, *5*, 106–116.
17. Silver, L.L. Challenges of antibacterial discovery. *Clin. Microbiol. Rev.* **2011**, *24*, 71–109. [CrossRef]
18. Guo, X.; Liu, N.; Li, X.; Ding, Y.; Shang, F.; Gao, Y.; Ruan, J.; Huang, Y. Red soils harbor diverse culturable actinomycetes that are promising sources of novel secondary metabolites. *Appl. Environ. Microbiol.* **2015**, *81*, 3086–3103. [CrossRef]
19. Nair, D.N.; Padmavathy, S. Impact of endophytic microorganisms on plants, environment and humans. *Sci. World J.* **2014**, *2014*, 250693. [CrossRef]
20. Coombs, J.T.; Franco, C.M. Isolation and identification of actinobacteria from surface-sterilized wheat roots. *Appl. Environ. Microbiol.* **2003**, *69*, 5603–5608. [CrossRef]
21. El-Nakeeb, M.A.; Lechevalier, H.A. Selective isolation of aerobic actinomycetes. *Appl. Microbiol.* **1963**, *11*, 75–77. [CrossRef]
22. Hayakawa, M.; Nomura, S. Humic acid-vitamin agar. A new medium for the selective isolation of soil actinomycetes. *J. Ferment. Technol.* **1987**, *65*, 501–509. [CrossRef]
23. Mohseni, M.; Norouzi, H.; Hamedi, J.; Roohi, A. Screening of antibacterial producing actinomycetes from sediments of the Caspian Sea. *Int. J. Mol. Cell. Med.* **2013**, *2*, 64–71. [PubMed]
24. Du Toit, E.A.; Rautenbach, M. A sensitive standardised micro-gel well diffusion assay for the determination of antimicrobial activity. *J. Microbiol. Methods* **2000**, *42*, 159–165. [CrossRef]
25. Farris, M.H.; Olson, J.B. Detection of Actinobacteria cultivated from environmental samples reveals bias in universal primers. *Lett. Appl. Microbiol.* **2007**, *45*, 376–831. [CrossRef]
26. Dereeper, A.; Guignon, V.; Blanc, G.; Audic, S.; Buffet, S.; Chevenet, F.; Dufayard, J.F.; Guindon, S.; Lefort, V.; Lescot, M.; et al. Phylogeny.fr: Robust phylogenetic analysis for the non-specialist. *Nucleic Acids Res.* **2008**, *1*, W465–W469. [CrossRef] [PubMed]
27. Dereeper, A.; Audic, S.; Claverie, J.M.; Blanc, G. BLAST-EXPLORER helps you building datasets for phylogenetic analysis. *BMC Evol. Biol.* **2010**, *10*, 1–6. [CrossRef] [PubMed]
28. Edgar, R.C. MUSCLE: Multiple sequence alignment with high accuracy and high throughput. *Nucleic Acids Res.* **2004**, *32*, 1792–1797. [CrossRef]
29. Guindon, S.; Gascuel, O. A simple, fast, and accurate algorithm to estimate large phylogenies by maximum likelihood. *Syst. Biol.* **2003**, *52*, 696–704. [CrossRef]
30. Anisimova, M.; Gascuel, O. Approximate likelihood ratio test for branchs: A fast, accurate and powerful alternative. *Syst. Biol.* **2006**, *55*, 539–552. [CrossRef]
31. Chevenet, F.; Brun, C.; Banuls, A.L.; Jacq, B.; Chisten, R. TreeDyn: Towards dynamic graphics and annotations for analyses of trees. *BMC Bioinform.* **2006**, *7*, 439. [CrossRef]
32. Shirling, E.B.; Gottlieb, D. Methods for characterization of *Streptomyces* species. *J. Syst. Evol. Microbiol.* **1966**, *16*, 313–340. [CrossRef]
33. CLSI. Methods for dilution antimicrobial susceptibility tests for Bacteria that grow aerobically, approved standard. 9th ed. CLSI document M07-A9. *Clin. Lab. Stand. Inst.* **2012**, *32*, 1–68.
34. Sivaranjani, M.; Gowrishankar, S.; Kamaladevi, A.; Pandian, S.K.; Balamurugan, K.; Ravi, A.V. Morin inhibits biofilm production and reduces the virulence of *Listeria monocytogenes*—An in vitro and in vivo approach. *Int. J. Food Microbiol.* **2016**, *237*, 73–82. [CrossRef]
35. Bakkiyaraj, D.; Pandian, K.S.T. In vitro and in vivo antibiofilm activity of a coral associated actinomycete against drug resistant *Staphylococcus aureus* biofilms. *Biofouling* **2010**, *26*, 711–717. [CrossRef]
36. Rahman, M.M.; Islam, M.B.; Biswas, M.; Khurshid Alam, A.H.M. In vitro antioxidant and free radical scavenging activity of different parts of *Tabebuia pallida* growing in Bangladesh. *BMC Res. Notes* **2015**, *8*, 621. [CrossRef]
37. Russell, F.M.; Biribo, S.S.N.; Selvaraj, G.; Oppedisano, F.; Warren, S.; Seduadua, A.; Mulholland, E.K.; Carapetis, J.R. As a bacterial culture medium, citrated sheep blood agar is a practical alternative to citrated human blood agar in laboratories of developing countries. *J. Clin. Microbiol.* **2006**, *44*, 3346–3351. [CrossRef]

38. Ayoola, G.A.; Coker, H.A.B.; Adesegun, S.A.; Bello, A.A.A.; Obaweya, K.; Ezennia, E.C.; Atangbayila, T.O. Phytochemical screening and antioxidant activities of some selected medicinal plants used for malaria therapy in southwestern Nigeria. *Trop. J. Pharm. Res.* **2008**, *7*, 1019–1024.
39. Subramaniam, D.; Menon, T.; Elizabeth, H.L.; Swaminathan, S. Anti-HIV-1 activity of *Sargassum swartzii* a marine brown alga. *BMC Infect. Dis.* **2014**, *14*, E43. [CrossRef]
40. Savithramma, N.; Rao, M.L.; Suhrulatha, D. Screening of medicinal plants for secondary metabolites. *Middle-East J. Sci. Res.* **2011**, *8*, 579–584.
41. Suetsuna, K.; Seino, A.; Kudo, T.; Osajima, Y. Production, and biological characterization, of Dideoxygriseorhodin C by a *Streptomyces* sp. and its taxonomy. *Agric. Biol. Chem.* **1989**, *53*, 581–583. [CrossRef]
42. Panzone, G.; Trani, A.; Ferrari, P.; Gastaldo, L.; Colombo, L. Isolation and structure elucidation of 7,8-dideoxy-6-oxo-griseorhodin C produced by *Actinoplanes ianthinogenes*. *J. Antibiot.* **1997**, *50*, 665–670. [CrossRef] [PubMed]
43. Tsuge, N.; Furihata, K.; Shin-Ya, K.; Hayakawa, Y.; Seto, H. Novel antibiotics pyrisulfoxin A and B produced by *Streptomyces californicus*. *J. Antibiot.* **1999**, *52*, 505–507. [CrossRef] [PubMed]
44. Srinivasan, S.; Bhikshapathi, D.V.R.N.; Krishnaveni, J.; Veerabrahma, K. Isolation of borrelidin from *Streptomyces californicus*—An Indian soil isolate. *Indian J. Biotechnol.* **2008**, *7*, 349–355.
45. Gozari, M.; Mortazavi, M.S.; Bahador, N.; Tamadoni Jahromi, S.; Rabbaniha, M. Isolation and screening of antibacterial and enzyme producing marine actinobacteria to approach probiotics against some pathogenic vibrios in shrimp *Litopenaeus vannamei*. *Iran J. Fish Sci.* **2016**, *15*, 630–644.
46. Sharma, P.; Singh, T.A.; Bharat, B.; Bhasin, S.; Modi, H.A. Approach towards different fermentative techniques for the production of bioactive actinobacterial melanin. *Beni-Suef Univ. J. Basic Appl. Sci.* **2018**, *7*, 695–700. [CrossRef]
47. Rao, M.P.N.; Xiao, M.; Li, W.J. Fungal and bacterial pigments: Secondary metabolites with wide applications. *Front. Microbiol.* **2017**, *8*, 1113.
48. Ser, H.L.; Yin, W.F.; Chan, K.G.; Khan, T.M.; Goh, B.H.; Lee, L.H. Antioxidant and cytotoxic potentials of *Streptomyces gilvigriseus* MUSC 26T isolated from mangrove soil in Malaysia. *Prog. Microbes Mol. Biol.* **2018**, *1*, a0000002. [CrossRef]
49. Chaudhari, C.N.; Tandel, K.; Grover, N.; Bhatt, P.; Sahni, A.K.; Sen, S.; Prahraj, A.K. In vitro vancomycin susceptibility amongst methicillin resistant *Staphylococcus aureus*. *Med. J. Armed Forces India* **2014**, *70*, 215–219. [CrossRef]
50. Horiuchi, K.; Shiota, S.; Hatano, T.; Yoshida, T.; Kuroda, T.; Tsuchiya, T. Antimicrobial activity of oleanolic acid from *Salvia officinalis* and related compounds on vancomycin-resistant *Enterococci* (VRE). *Biol. Pharm. Bull.* **2007**, *30*, 1147–1149. [CrossRef]
51. Maasjost, J.; Mühldorfer, K.; Cortez de Jäckel, S.; Hafez, H.M. Antimicrobial susceptibility patterns of *Enterococcus faecalis* and *Enterococcus faecium* isolated from poultry flocks in Germany. *Avian Dis.* **2015**, *59*, 143–148. [CrossRef]
52. Gos, F.M.W.R.; Savi, D.C.; Shaaban, K.A.; Thorson, J.S.; Aluizio, R.; Possiede, Y.M.; Rohr, J.; Glienke, C. Antibacterial activity of endophytic actinomycetes isolated from the medicinal plant *Vochysia divergens* (Pantanal, Brazil). *Front. Microbiol.* **2017**, *8*, 1642. [CrossRef] [PubMed]
53. Girão, M.; Ribeiro, I.; Ribeiro, T.; Azevedo, I.C.; Pereira, F.; Urbatzka, R.; Leão, P.N.; Carvalho, M.F. Actinobacteria isolated from *Laminaria ochroleuca*: A source of new bioactive compounds. *Front. Microbiol.* **2019**, *10*, 683. [CrossRef] [PubMed]
54. Siddharth, S.; Vittal, R.R.; Wink, J.; Steinert, M. Diversity and bioactive potential of actinobacteria from unexplored regions of Western Ghats, India. *Microorganisms* **2020**, *7*, 225. [CrossRef]
55. Schillaci, D.; Petruso, S.; Raimondi, M.V.; Cusimano, M.G.; Cascioferro, S.; Scalisi, M.; La Giglia, M.A.; Vitale, M. Pyrrolomycins as potential anti-Staphylococcal biofilms agents. *Biofouling* **2010**, *26*, 433–438. [CrossRef]
56. Tote, K.; Berghe, D.V.; Deschacht, M.; de Wit, K.; Maes, L.; Cos, P. Inhibitory efficacy of various antibiotics on matrix and viable mass of *Staphylococcus aureus* and *Pseudomonas aeruginosa* biofilms. *Int. J. Antimicrob. Agents* **2009**, *33*, 525–531. [CrossRef] [PubMed]

57. Melo, T.A.; Dos Santos, T.F.; de Almeida, M.E.; Junior, L.A.; Andrade, E.F.; Rezende, R.P.; Marques, L.M.; Romano, C.C. Inhibition of *Staphylococcus aureus* biofilm by Lactobacillus isolated from fine cocoa. *BMC Microbiol.* **2016**, *16*, 250. [CrossRef]
58. Brambilla, L.Z.S.; Endo, E.H.; Cortez, D.A.G.; Filhoa, B.P.D. Anti-biofilm activity against *Staphylococcus aureus* MRSA and MSSA of neolignans and extract of *Piper regnellii*. *Rev. Bras. Farmacogn.* **2017**, *27*, 112–117. [CrossRef]
59. Yu, H.; Liu, M.; Liu, Y.; Qin, L.; Jin, M.; Wang, Z. Antimicrobial Activity and mechanism of action of *Dracocephalum moldavica* L. extracts against clinical isolates of *Staphylococcus aureus*. *Front. Microbiol.* **2019**, *10*, 1249. [CrossRef]
60. Ouyang, Y.; Li, J.; Peng, Y.; Huang, Z.; Ren, Q.; Lu, J. The role and mechanism of thiol-dependent antioxidant system in bacterial drug susceptibility and resistance. *Curr. Med. Chem.* **2020**, *27*, 1940–1954. [CrossRef]
61. Elswaifi, S.F.; Palmieri, J.R.; Hockey, K.S.; Rziqalinski, B.A. Antioxidant nanoparticles for control of infectious disease. *Infect. Disord. Drug Targets (Former. Curr. Drug Targets-Infect. Disord.)* **2009**, *9*, 445–452. [CrossRef]
62. Pellegrino, D. Antioxidants and cardiovascular risk factors. *Diseases* **2016**, *4*, 11. [CrossRef]
63. Passari, A.K.; Mishra, V.K.; Singh, G.; Singh, P.; Kumar, B.; Gupta, V.K.; Sarma, R.K.; Saikia, R.; Donovan, A.O.; Singh, B.P. Insights into the functionality of endophytic actinobacteria with a focus on their biosynthetic potential and secondary metabolites production. *Sci. Rep.* **2018**, *7*, 11809. [CrossRef] [PubMed]
64. Nimal, C.I.V.; Kumar, P.P.; Agastian, P. In vitro α-glucosidase inhibition and antioxidative potential of an endophyte species (*Streptomyces sp. loyola* UGC) isolated from *Datura stramonium* L. *Curr. Microbiol.* **2013**, *67*, 69–76. [CrossRef] [PubMed]
65. Chaudhuri, M.; Paul, A.K.; Pal, A. Isolation and assessment of metabolic potentials of bacteria endophytic to carnivorous plants *Drosera Burmannii* and *Utricularia* Spp. *Biosci. Biotechnol. Res. Asia* **2019**, *16*, 731–741. [CrossRef]
66. Garratty, G.; Arndt, P. Drugs that have been shown to cause drug-induced immune hemolytic anemia or positive direct antiglobulin tests: Some interesting findings since 2007. *Immunohematology* **2014**, *30*, 66–79.
67. Marques, M.A.; Citron, D.M.; Wang, C.C. Development of Tyrocidine A analogues with improved antibacterial activity. *Bioorganic Med. Chem.* **2007**, *15*, 6667–6677. [CrossRef]
68. Mosges, R.; Baues, C.M.; Schroder, T.; Sahin, K. Acute bacterial otitis externa: Efficacy and safety of topical treatment with an antibiotic ear drop formulation in comparison to glycerol treatment. *Curr. Med. Res. Opin.* **2011**, *27*, 871–878. [CrossRef]
69. Guleria, V.S.; Sharma, N.; Amitabh, S.; Nair, V. Ceftriaxone-induced hemolysis. *Indian J. Pharmacol.* **2013**, *45*, 530–531. [CrossRef]
70. Ghosh, T.; Biswas, M.K.; Chatterjee, S.; Roy, P. In-vitro study on the hemolytic activity of different extracts of Indian medicinal plant *Croton bonplandianum* with phytochemical estimation: A new era in drug development. *J. Drug Deliv. Ther.* **2018**, *8*, 155–160. [CrossRef]
71. Srivastava, V.; Singla, R.K.; Dubey, A.K. Inhibition of biofilm and virulence factors of *Candida albicans* by partially purified secondary metabolites of *Streptomyces chrestomyceticus* strain ADP4. *Curr. Top. Med. Chem.* **2018**, *11*, 925–945. [CrossRef]
72. Gumbi, B.P.; Moodley, B.; Birungi, G.; Ndungu, P.G. Target, suspect and non-target screening of silylated derivatives of polar compounds based on single Ion monitoring GC-MS. *Int. J. Environ. Res. Public Health* **2019**, *16*, 4022. [CrossRef] [PubMed]
73. Salem, M.Z.M.; Zayed, M.Z.; Ali, H.M.; Abd El-Kareem, M.S.M. Chemical composition, antioxidant and antibacterial activities of extracts from *Schinus molle* wood branch growing in Egypt. *J. Wood Sci.* **2016**, *62*, 548–561. [CrossRef]
74. Ghaffari, T.; Kafil, H.S.; Asnaashari, S.; Farajnia, S.; Delazar, A.; Baek, S.C.; Hamishehkar, H.; Kim, K.H. Chemical composition and antimicrobial activity of essential oils from the aerial parts of *Pinus eldarica* grown in Northwestern Iran. *Molecules* **2019**, *24*, E3203. [CrossRef]
75. Xiong, L.; Peng, C.; Zhou, Q.M.; Wan, F.; Xie, X.F.; Guo, L.; Li, X.H.; He, C.J.; Dai, O. Chemical composition and antibacterial activity of essential oils from different parts of *Leonurus japonicus* Houtt. *Molecules* **2013**, *18*, 963–973. [CrossRef]
76. Gao, Y.Q.; Guo, C.J.; Zhang, Q.; Zhou, W.M.; Wang, C.C.; Gao, J.M. Asperaculanes A and B, two sesquiterpenoids from the fungus *Aspergillus aculeatus*. *Molecules* **2014**, *20*, 325–334. [CrossRef] [PubMed]

77. Hussein, R.A.; El-Anssary, A.A. *Plants Secondary Metabolites: The Key Drivers of the Pharmacological Actions of Medicinal Plants in Herbal Medicine*; Philip, F., Ed.; IntechOpen: London, UK, 2019; pp. 11–30.
78. Karak, P. Biological activities of flavonoids: An overview. *Int. J. Pharm. Sci.* **2019**, *10*, 1567–1574.
79. Gbahou, F.; Davenas, E.; Morisset, S.; Arrang, J.M. Effects of betahistine at histamine H3 receptors: Mixed inverse agonism/agonism in vitro and partial inverse agonism in vivo. *J. Pharmacol. Exp. Ther.* **2010**, *334*, 945–954. [CrossRef]
80. Debnath, B.; Uddina, M.D.J.; Pataric, P.; Das, M.; Maitia, D.; Manna, K. Estimation of alkaloids and phenolics of five edible cucurbitaceous plants and their antibacterial activity. *Int. J. Pharm. Pharm. Sci.* **2018**, *7*, 223–227.

# 7

# Characterization of *Streptomyces sporangiiformans* sp. nov., a Novel Soil Actinomycete with Antibacterial Activity against *Ralstonia solanacearum*

**Junwei Zhao [1,†], Liyuan Han [1,†], Mingying Yu [1], Peng Cao [1], Dongmei Li [1], Xiaowei Guo [1], Yongqiang Liu [1], Xiangjing Wang [1,*] and Wensheng Xiang [1,2,*]**

1 Key Laboratory of Agricultural Microbiology of Heilongjiang Province, Northeast Agricultural University, No. 59 Mucai Street, Xiangfang District, Harbin 150030, China

2 State Key Laboratory for Biology of Plant Diseases and Insect Pests, Institute of Plant Protection, Chinese Academy of Agricultural Sciences, Beijing 100193, China

* Correspondence: wangneau2013@163.com (X.W.); xiangwensheng@neau.edu.cn (W.X.)

† These authors contributed equally to this work.

**Abstract:** *Ralstonia solanacearum* is a major phytopathogenic bacterium that attacks many crops and other plants around the world. In this study, a novel actinomycete, designated strain NEAU-SSA 1$^T$, which exhibited antibacterial activity against *Ralstonia solanacearum*, was isolated from soil collected from Mount Song and characterized using a polyphasic approach. Morphological and chemotaxonomic characteristics of the strain coincided with those of the genus *Streptomyces*. The 16S rRNA gene sequence analysis showed that the isolate was most closely related to *Streptomyces aureoverticillatus* JCM 4347$^T$ (97.9%). Phylogenetic analysis based on 16S rRNA gene sequences indicated that the strain formed a cluster with *Streptomyces vastus* JCM4524$^T$ (97.4%), *S. cinereus* DSM43033$^T$ (97.2%), *S. xiangluensis* NEAU-LA29$^T$ (97.1%) and *S. flaveus* JCM3035$^T$ (97.1%). The cell wall contained *LL*-diaminopimelic acid and the whole-cell hydrolysates were ribose, mannose and galactose. The polar lipids were diphosphatidylglycerol (DPG), phosphatidylethanolamine (PE), hydroxy-phosphatidylethanolamine (OH-PE), phosphatidylinositol (PI), two phosphatidylinositol mannosides (PIMs) and an unidentified phospholipid (PL). The menaquinones were MK-9($H_4$), MK-9($H_6$), and MK-9($H_8$). The major fatty acids were *iso*-$C_{17:0}$, $C_{16:0}$ and $C_{17:1}$ $\omega$9c. The DNA G+C content was 69.9 mol %. However, multilocus sequence analysis (MLSA) based on five other house-keeping genes (*atp*D, *gyr*B, *rec*A, *rpo*B, and *trp*B), DNA–DNA relatedness, and physiological and biochemical data showed that the strain could be distinguished from its closest relatives. Therefore, it is proposed that strain NEAU-SSA 1$^T$ should be classified as representatives of a novel species of the genus *Streptomyces*, for which the name *Streptomyces sporangiiformans* sp. nov. is proposed. The type strain is NEAU-SSA 1$^T$ (=CCTCC AA 2017028$^T$ = DSM 105692$^T$).

**Keywords:** *Streptomyces sporangiiformans* sp. nov.; antibacterial activity; multilocus sequence analysis; *Ralstonia solanacearum*

---

## 1. Introduction

*Ralstonia solanacearum* is the causal agent of bacterial wilt, one of the most devastating plant pathogenic bacteria around the world [1], which has an unusually wide host range, infecting over 200 plant species [2], including many important agricultural crops such as potato, tomato, banana and pepper. Even though different approaches have been developed to control bacterial wilt, we still lack an efficient and environmentally friendly control measure for most of the host crops [3].

Therefore, the search and discovery of novel, environmentally friendly, commercially significant, naturally bioactive compounds are in demand to control this disease at present.

The actinobacteria are known to produce biologically active secondary metabolites, including antibiotics, enzymes, enzyme inhibitors, antitumour agents and antibacterial compounds [4–6]. The genus *Streptomyces*, within the family *Streptomycetaceae*, is the largest genus of the phylum *Actinobacteria*, first proposed by Waksman and Henrici (1943) [7] and currently encompasses more than 800 species with valid published names (http://www.bacterio.net/streptomyces.html), which are widely distributed in soils throughout the world. Therefore, members of novel *Streptomyces* species are in demand as sources of novel, environmentally friendly, commercially significant, naturally bioactive compounds [8,9]. During our search for antagonistic actinobacteria from soil in Mount Song, an aerobic actinomycete, strain NEAU-SSA $1^T$ with inhibitory activity against phytopathogenic bacterium *Ralstonia solanacearum* was isolated and subjected to the polyphasic taxonomy analysis. Results demonstrated that the strain represents a novel species of the genus *Streptomyces*, for which the name *Streptomyces sporangiiformans* sp. nov. is proposed.

## 2. Materials and Methods

### *2.1. Isolation of Actinomycete Strain*

Strain NEAU-SSA $1^T$ was isolated from soil collected from Mount Song (34°29′ N, 113°2′ E), Dengfeng, Henan Province, China. The soil sample was air-dried at room temperature for 14 days before isolation for actinomycetes. After drying, the soil sample was ground into powder and then suspended in sterile distilled water, followed by a standard serial dilution technique. The diluted soil suspension was spread on humic acid-vitamin agar (HV) [10] supplemented with cycloheximide (50 mg $L^{-1}$) and nalidixic acid (20 mg $L^{-1}$). After 28 days of aerobic incubation at 28 °C, colonies were transferred and purified on the International *Streptomyces* Project (ISP) medium 3 [11], and maintained as glycerol suspensions (20%, *v*/*v*) at −80 °C for long-term preservation.

### *2.2. Morphological and Physiological and Biochemical Characteristics of NEAU-SSA $1^T$*

Gram staining was carried out by using the standard Gram stain, and morphological characteristics were observed using light microscopy (Nikon ECLIPSE E200, Nikon Corporation, Tokyo, Japan) and scanning electron microscopy (Hitachi SU8010, Hitachi Co., Tokyo, Japan) using cultures grown on ISP 3 agar at 28 °C for 6 weeks. Samples for scanning electron microscopy were prepared as described by Jin et al. [12]. Cultural characteristics were determined on the ISP 1 agar [11], ISP media 2–7 [8], Czapek's agar [13], Bennett's agar [14], and Nutrient agar [15] after 14 days at 28 °C. Color determination was done with color chips from the ISCC-NBS (Inter-Society Color Council-National Bureau of Standards) color charts [16]. Growth at different temperatures (10, 15, 20, 25, 28, 32, 35, 40, 45, and 50 °C) was determined on ISP 3 medium after incubation for 14 days. Growth tests for pH range (pH 4.0–12.0, at intervals of 1.0 pH unit) and NaCl tolerance (0, 1, 2, 3, 4, 5, 6, 7, 8, 9, 10, 15, and 20%, *w*/*v*) were tested in GY (Glucose-Yeast extract) medium [17] at 28 °C for 14 days on a rotary shaker. The buffer systems were: pH 4.0–5.0, 0.1 M citric acid/0.1 M sodium citrate; pH 6.0–8.0, 0.1 M $KH_2PO_4$/0.1 M NaOH; pH 9.0–10.0, 0.1 M $NaHCO_3$/0.1 M $Na_2CO_3$; and pH 11.0–12.0, 0.2 M $KH_2PO_4$/0.1 M NaOH. Hydrolysis of Tweens (20, 40, and 80) and production of urease were tested as described by Smibert and Krieg [18]. The utilization of sole carbon and nitrogen sources, decomposition of cellulose, hydrolysis of starch and aesculin, reduction of nitrate, coagulation and peptonization of milk, liquefaction of gelatin, and production of $H_2S$ were examined as described previously [19,20].

### *2.3. Chemotaxonomic Analysis of NEAU-SSA $1^T$*

Biomass for chemotaxonomic studies was prepared by growing the organisms in GY medium in shake flasks at 28 °C for 5 days. Cells were harvested using centrifugation, washed with distilled water, and freeze-dried. The isomer of diaminopimelic acid (DPA) in the cell wall hydrolysates was derivatized and analyzed using an HPLC (High Performance Liquid Chromatography) method [21] with an Agilent TC-$C_{18}$ Column (250 × 4.6 mm i.d. 5 µm; Agilent Technologies, Santa Clara, CA, USA) that had a mobile phase consisting of acetonitrile: 0.05 mol $L^{-1}$ phosphate buffer pH 7.2 (15:85, *v/v*) at a flow rate of 0.5 mL $min^{-1}$. The peak detection used an Agilent G1321A fluorescence detector (Agilent Technologies, Santa Clara, CA, USA) with a 365 nm excitation and 455 nm longpass emission filters. The whole-cell sugars were analyzed according to the procedures developed by Lechevalier and Lechevalier [22]. The polar lipids were examined using two-dimensional TLC (Thin-Layer Chromatography) and identified using the method of Minnikin et al. [23]. Menaquinones were extracted from the freeze-dried biomass and purified according to Collins [24]. Extracts were analyzed using a HPLC-UV method [25] with an Agilent Extend-$C_{18}$ Column (150 × 4.6 mm, i.d. 5 µm; Agilent Technologies, Santa Clara, CA, USA) at 270 nm. The mobile phase was acetonitrile-*iso*-propyl alcohol (60:40, *v/v*). To determine cellular fatty acid compositions, the strain NEAU-SSA $1^T$ was cultivated in GY medium in shake flasks at 28 °C for 4 days. Fatty acid methyl esters were extracted from the biomass as described by Gao et al. [26] and analyzed using GC-MS according to the method of Xiang et al. [27].

### 2.4. Phylogenetic Analysis of NEAU-SSA $1^T$

For DNA extraction, strain NEAU-SSA $1^T$ was cultured in GY medium for 3 days to the early stationary phase and harvested using centrifugation. The chromosomal DNA was extracted according to the method of sodium dodecyl sulfate (SDS)-based DNA extraction [28]. PCR amplification of the 16S rRNA gene sequence was carried out using the universal bacterial primers 27F (5′-AGAGTTTGATCCTGGCTCAG-3′) and 1541R (5′-AAGGAGGTGATCCAGCC-3′) under conditions described previously [29,30]. The PCR product was purified and cloned into the vector pMD19-T (Takara) and sequenced using an Applied Biosystems DNA sequencer (model 3730XL, Applied Biosystems Inc., Foster City, California, USA). The almost complete 16S rRNA gene sequence of strain NEAU-SSA $1^T$ (1412bp) was obtained and compared with type strains available at the EzBioCloud server (https://www.ezbiocloud.net/), retrieved using NCBI BLAST (National Center for Biotechnology Information, Basic Local Alignment Search Tool; https://blast.ncbi.nlm.nih.gov/Blast.cgi;) and then submitted to the GenBank database. Phylogenetic trees were constructed based on the 16S rRNA gene sequences of strain NEAU-SSA $1^T$ and related reference species. Sequences were multiply aligned in Molecular Evolutionary Genetics Analysis (MEGA) software version 7.0 using the Clustal W algorithm and trimmed manually where necessary. Phylogenetic trees were constructed with neighbor-joining [31] and maximum likelihood [32] algorithms using MEGA [33]. The stability of the topology of the phylogenetic tree was assessed using the bootstrap method with 1000 repetitions [34]. A distance matrix was generated using Kimura's two-parameter model [35]. All positions containing gaps and missing data were eliminated from the dataset (complete deletion option). 16S rRNA gene sequence similarities between strains were calculated on the basis of pairwise alignment using the EzBioCloud server [36]. To further clarify the affiliation of strain NEAU-SSA $1^T$ to its closely related strains, phylogenetic relationships of the strain NEAU-SSA $1^T$ were also confirmed using sequences of five individual housekeeping genes (*atp*D, *gyr*B, *rec*A, *rpo*B, and *trp*B) for core-genome analysis. The sequences of NEAU-SSA $1^T$ and its related strains were obtained from the genomes or GenBank/EMBL/DDBJ (European Molecular Biology Laboratory/DNA Data Bank of Japan). GenBank accession numbers of the sequences used are given in Table 1. The sequences of each locus were aligned using MEGA 7.0 software and trimmed manually at the same position before being used for further analysis. Trimmed sequences of the five housekeeping genes were concatenated head-to-tail in-frame in the order *atp*D-*gyr*B-*rec*A-*rpo*B-*trp*B. Phylogenetic analysis was performed as described above. Genome mining for bioactive secondary metabolites was performed using "antibiotics and secondary metabolite analysis shell" (antiSMASH) version 4.0 [37].

**Table 1.** GenBank Accession Numbers of the Sequences Used in MLSA.

| Strain | Type Strain | Whole Genome | *atp*D | *gyr*B | *rec*A | *rpo*B | *trp*B |
|---|---|---|---|---|---|---|---|
| *Streptomyces sporangiiformans* | NEAU-SSA 1$^{T}$ | VCHX00000000 | | – | – | – | – |
| *Streptomyces coelescens* | DSM 40421$^{T}$ | – | GU383344 | AY508508 | KT385220 | GU383768 | KT389192 |
| *Streptomyces violaceolatus* | DSM 40438$^{T}$ | – | GU383347 | AY508509 | KT385451 | GU383771 | KT389418 |
| *Streptomyces anthocyanicus* | NBRC 14892$^{T}$ | – | KT384465 | KT384814 | KT385162 | KT388784 | KT389134 |
| *Streptomyces humiferus* | DSM 43030$^{T}$ | – | KT384598 | KT384947 | KT385296 | KT388918 | KT389267 |
| *Streptomyces violaceoruber* | NBRC 12826$^{T}$ | CP020570 | KT384751 | KT385099 | KT385453 | KT389071 | KT389420 |
| *Streptomyces rubrogriseus* | LMG 20318$^{T}$ | BEWD00000000 | KT384715 | KT385065 | KT385416 | KT389036 | KT389384 |
| *Streptomyces tendae* | ATCC 19812$^{T}$ | – | KT384733 | KT385082 | KT385434 | KT389053 | KT389402 |
| *Streptomyces violaceorubidus* | LMG 20319$^{T}$ | JODM00000000 | – | – | – | – | – |
| *Streptomyces lienomycini* | LMG 20091$^{T}$ | – | KT384622 | KT384971 | KT385321 | KT388942 | KT389291 |
| *Streptomyces diastaticus* subsp. *ardesiacus* | NRRL B-1773$^{T}$ | BEWC00000000 | KT384534 | KT384883 | KT385231 | KT388853 | KT389203 |
| *Streptomyces albaduncus* | JCM 4715$^{T}$ | – | KT384449 | KT384798 | KT385146 | KJ996741 | KT389118 |
| *Streptomyces matensis* | NBRC 12889$^{T}$ | – | KT384637 | KT384986 | KT385337 | KT388957 | KT389306 |
| *Streptomyces althioticus* | NRRL B-3981$^{T}$ | – | KT384460 | KT384809 | KT385157 | KT388779 | KT389129 |
| *Streptomyces davaonensis* | JCM 4913$^{T}$ | HE971709 | – | – | – | – | – |
| *Streptomyces canus* | DSM 40017$^{T}$ | LMWO00000000 | KT384500 | KT384849 | KT385197 | KT388819 | KT389169 |
| *Streptomyces lincolnensis* | NRRL 2936$^{T}$ | CP016438 | – | – | – | – | – |
| *Streptomyces pseudovenezuelae* | DSM 40212$^{T}$ | LMWM00000000 | KT384695 | KT385045 | KT385396 | KT389016 | KT389364 |
| *Streptomyces xiangluensis* | NEAU-LA29$^{T}$ | – | MH291276 | MH345670 | MH291277 | MH291275 | MH291278 |
| *Streptomyces vastus* | NBRC 13094$^{T}$ | – | KU323834 | KT385093 | KU975607 | KT389065 | KT389414 |
| *Streptomyces cinereus* | NBRC 12247$^{T}$ | – | KT384513 | KT384862 | KT385210 | KJ996667 | KT389182 |
| *Streptomyces flaveus* | NRRL B-16074$^{T}$ | JOCU00000000 | KT384551 | KT384900 | KT385249 | KT388870 | KT389220 |
| *Streptomyces chilikensis* | RC 1830$^{T}$ | LWCC00000000 | – | – | – | – | – |
| *Streptomyces coeruleorubidus* | ISP 5145$^{T}$ | – | KT384528 | KT384877 | KT385225 | KT388847 | KT389197 |
| *Streptomyces misionensis* | DSM 40306$^{T}$ | FNTD00000000 | KT384647 | KT384996 | KT385347 | KT388967 | KT389316 |
| *Streptomyces phaeoluteichromatogenes* | NRRL 5799$^{T}$ | – | KT384680 | KT385030 | KT385381 | KT389001 | KT389350 |
| *Streptomyces tricolor* | NBRC 15461$^{T}$ | MUMF00000000 | KT384741 | KT385089 | KT385443 | KT389061 | KT389410 |
| *Streptomyces achromogenes* subsp. *achromogenes* | NBRC 12735$^{T}$ | JODT00000000 | – | – | – | – | – |
| *Streptomyces eurythermus* | ATCC 14975$^{T}$ | – | KT384544 | KT384893 | KT385242 | KT388863 | KT389213 |
| *Streptomyces nogalater* | JCM 4799$^{T}$ | – | KT384664 | KT385014 | KT385365 | KT388984 | KT389333 |
| *Streptomyces jietaisiensis* | FXJ46$^{T}$ | FNAX00000000 | KT384605 | KT384954 | KT385304 | KT388925 | KT389274 |
| *Streptomyces griseoaurantiacus* | NBRC 15440$^{T}$ | AEYX00000000 | – | – | – | – | – |
| *Streptomyces lavenduligriseus* | NRRL ISP-5487$^{T}$ | JOBD00000000 | KT384620 | AB072859 | KT385319 | KT388940 | KT389289 |
| *Streptomyces uncialis* | DCA2648$^{T}$ | LFBV00000000 | – | – | – | – | – |

**Table 1.** *Cont.*

| Strain | Type Strain | Whole Genome | *atp*D | *gyr*B | *rec*A | *rpo*B | *trp*B |
|---|---|---|---|---|---|---|---|
| *Streptomyces alboniger* | NRRL B-1832$^{T}$ | – | KT384455 | KT384804 | KT385152 | KT388774 | KT389124 |
| *Streptomyces alfalfae* | XY25$^{T}$ | CP015588 | – | – | – | – | – |
| *Streptomyces lasiicapitis* | 3H-HV17(2)$^{T}$ | – | MH651782 | KY229066 | MH651785 | MH651788 | MH651791 |
| *Streptomyces aureocirculatus* | NRRL ISP-5386$^{T}$ | JOAP00000000 | KT384476 | KT384825 | KT385173 | KT388795 | KT389145 |
| *Streptomyces aureoverticillatus* | NRRL B-3326$^{T}$ | – | KT384478 | KT384827 | KT385175 | KT388797 | KT389147 |
| *Streptomyces alboflavus* | NRRL B-2373$^{T}$ | CP021748 | – | – | – | – | – |
| *Streptomyces rutgersensis* | NBRC 12819$^{T}$ | – | KT384716 | KT385066 | KT385417 | KT389037 | KT389385 |
| *Streptomyces intermedius* | NBRC 13049$^{T}$ | – | KT384602 | KT384951 | KT385301 | KT388922 | KT389271 |
| *Streptomyces gougerotii* | NBRC 3198$^{T}$ | – | KT384572 | KT384921 | KT385270 | KT388891 | KT389241 |
| *Streptomyces diastaticus* subsp. *diastaticus* | NBRC 3714$^{T}$ | – | KT384535 | KT384884 | KT385232 | KT388854 | KT389204 |
| *Kitasatospora setae* | LM-6054$^{T}$ | AP010968 | – | – | – | – | – |

### 2.5. Draft Genome Sequencing and Assembly of NEAU-SSA 1$^T$

For draft genome sequencing and assembly, the genomic DNA of strain NEAU-SSA 1$^T$ was extracted using the method of SDS-based DNA extraction [28]. The harvested DNA was detected using agarose gel electrophoresis and quantified using Qubit® 2.0 Fluorometer (Thermo Scientific). Whole-genome sequencing was performed on the Illumina HiSeq PE150 (Illumina, San Diego, CA, USA) platform. A-tailed, ligated to paired-end adaptors, and PCR amplified samples with a 350 bp insert were used for the library construction at the Beijing Novogene Bioinformatics Technology Co., Ltd. Illumina PCR adapter reads and low-quality reads from the paired-end were filtered using a quality control step using our own compling pipeline. All good-quality paired reads were assembled using the SOAP (Short Oligonucleotide Alignment Program) denovo [38,39] (https://github.com/aquaskyline) into a number of scaffolds. Then, the filter reads were handled by the next step of the gap-closing.

### 2.6. DNA–DNA Relatedness Tests

Because of a lacking number of genome sequences of *Streptomyces aureoverticillatus* JCM4347$^T$, *Streptomyces vastus* JCM4524$^T$, *S. cinereus* DSM43033$^T$, and *S. xiangluensis* NEAU-LA29$^T$, DNA–DNA relatedness tests between strain NEAU-SSA 1$^T$ and those strains were carried out as described by De Ley et al. [40] under consideration of the modifications described by Huss et al. [41], using a model Cary 100 Bio UV/VIS-spectrophotometer (Hitachi U-3900, Hitachi Co., Tokyo, Japan) equipped with a Peltier-thermostatted 6 × 6 multicell changer and a temperature controller with in situ temperature probe (Varian). The genomic DNAs of strain NEAU-SSA 1$^T$ and its closely related species—*S. aureoverticillatus* JCM4347$^T$, *S. vastus* JCM4524$^T$, *S. cinereus* DSM43033$^T$, and *S. xiangluensis* NEAU-LA29$^T$—were extracted using the method of SDS-based DNA extraction [28]. The concentration and purity of these DNA samples were determined by measuring the optical density (OD) at 260, 280, and 230 nm. The DNA samples used for hybridization were diluted to $OD_{260}$ around 1.0 using 0.1 × SSC (saline sodium citrate buffer), then sheared using a JY92-II ultrasonic cell disruptor (ultrasonic time 3 s, interval time 4 s, 90 times; Ningbo Scientz Biotechnology Co., Ltd, Ningbo, China). The DNA renaturation rates were determined in 2 × SSC at 70 °C. The experiments were performed with three replications and the DNA–DNA relatedness value was expressed as a mean of the three values. Several genomic metrics are now available to distinguish between orthologous genes of closely related prokaryotes, including the calculation of average nucleotide identity (ANI) and digital DNA–DNA hybridization (dDDH) values [42,43]. In the present study, ANI and dDDH values were determined from the genomes of strain NEAU-SSA 1$^T$ and *S. flaveus* JCM3035$^T$ (JOCU00000000) using the ortho-ANIu algorithm from Ezbiotaxon and the genome-to-genome distance calculator (GGDC 2.0) at http://ggdc.dsmz.de.

### 2.7. In Vitro Antibacterial Activity Test

The antibacterial activity of strain NEAU-SSA 1$^T$ against two pathogenic bacteria (*Micrococcus luteus* and *Ralstonia solanacearum*) was evaluated using the agar well diffusion method [44] with the cultures growth on ISP 3 medium at 28 °C for four weeks as follows: All the spores and mycelia were collected from one ISP 3 plate (diameter, 9mm) and then extracted using 1 mL methanol with an ultrasonic step (300 W, 30–60 min). Afterwards, 200 μL methanol extract or methanol was added to the agar well, and methanol was used as the control. To further investigate the antibacterial components produced by NEAU-SSA 1$^T$, the strain was cultured in tryptone-glucose-soluble starch-yeast extract medium (tryptone 0.2%, glucose 1%, soluble starch 0.5%, yeast extract 0.2%, NaCl 0.4%, $K_2HPO_4$ 0.05%, $MgSO_4.7H_2O$ 0.05%, $CaCO_3$ 0.2%, *w*/*v*, pH 7.0–7.4), and the inhibitory activity was tested. Briefly, strain NEAU-SSA 1$^T$ was inoculated into MB medium and incubated at 28 °C for seven days in a rotary shaker. The supernatant (100 mL for this study) was obtained via centrifugation at 8000 rpm and 4 °C for 10 min and subsequently extracted by using an equal volume of ethyl acetate. Then, the extract was dried in a rotary evaporator at 40 °C and eluted with proper volume methanol (1 mL used in this study). The cell precipitate was extracted with an equal volume of methanol and also condensed as above. After that, the antibacterial activity was evaluated using the agar well diffusion method, and each well contained 200 μL of the methanol extract. To examine the effect of temperature on antibacterial activity, the ten-fold dilution methanol extract was placed in a water bath at 40, 60, 80, and 100 °C for 30 min, and then cooled to room temperature. The antibacterial activity was evaluated using the agar well diffusion method.

## 3. Result and Discussion

### 3.1. Polyphasic Taxonomic Characterization of NEAU-SSA $1^T$

The morphological characteristics of strain NEAU-SSA $1^T$ showed that the strain had the typical characteristics of the genus *Streptomyces*. Observation of 6-week cultures of strain NEAU-SSA $1^T$ grown on ISP 3 medium revealed that it formed well-developed, branched substrate hyphae and aerial mycelia. Sporangia consisted of cylindrical, and rough-surfaced spores (0.6–0.8 μm × 0.9–1.6 μm) were produced on aerial mycelia, but spore chains were not observed (Figure 1). Strain NEAU-SSA $1^T$ exhibited good growth on ISP 3, ISP 4, ISP 7, and Nutrient agar media; moderate growth on ISP 1, ISP 2, ISP 5, ISP 6, and Czapek's agar media; and poor growth on Bennett's agar medium. The cultural characteristics of strain NEAU-SSA $1^T$ is shown in Table S1. Strain NEAU-SSA $1^T$ grew well between pH 6.0 and 11.0, with an optimum pH of 7.0. The range of temperature of the strain was determined to be 15–45 °C, with the optimum growth temperature being 28 °C. The strain grew in the presence of 0–6% NaCl (*w/v*) with an optimal level of 0–1% (*w/v*). Detailed physiological characteristics are presented in the species description (Table 2 and Table S1).

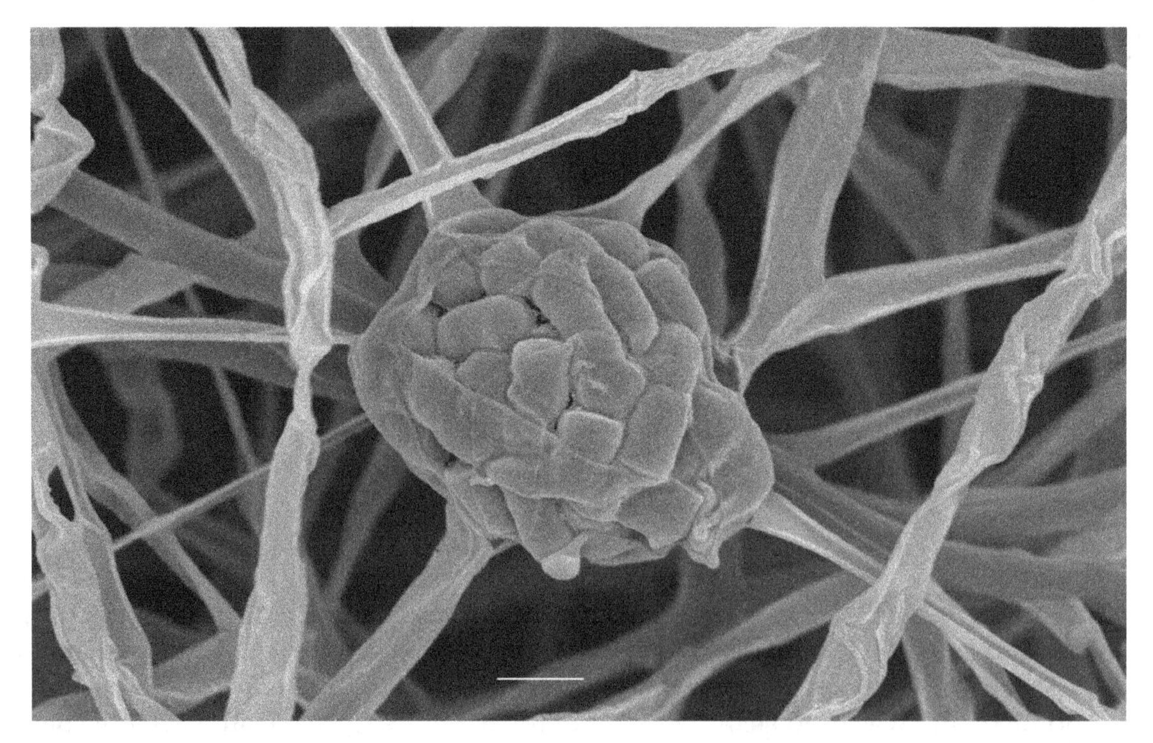

**Figure 1.** Scanning electron micrograph of strain NEAU-SSA $1^T$ grown on ISP 3 agar for 6 weeks at 28 °C; Scale bar represents 1 μm.

Chemotaxonomic analyses revealed that strain NEAU-SSA $1^T$ exhibited characteristics that are typical of representatives of the genus *Streptomyces*. The strain was found to contain *LL*-diaminopimelic acid as diamino acid. The whole-cell hydrolysates of the strain were determined to contain ribose, mannose, and galactose. The menaquinones of strain NEAU-SSA $1^T$ were MK-9($H_4$) (29.5%), MK-9($H_6$) (41.2%), and MK-9($H_8$) (29.4%). The cellular fatty acid profile of strain NEAU-SSA $1^T$ was composed of *iso*-$C_{17:0}$ (30.9%), $C_{16:0}$ (26.4%), $C_{17:1}\omega9c$ (19.9%), $C_{15:0}$ (7.8%), $C_{17:0}$ (4.4%), $C_{14:0}$ (3.3%), *iso*-$C_{16:0}$

(1.7%), *anteiso*-$C_{15:0}$ (1.7%), $C_{18:1}\omega9c$ (1.7%), $C_{16:0}$ 1-OH (1.2%), and *iso*-$C_{18:0}$ (1.1%). The polar lipids of the strain consisted of diphosphatidylglycerol (DPG), phosphatidylethanolamine (PE), hydroxy-phosphatidylethanolamine (OH-PE), phosphatidylinositol (PI), two phosphatidylinositol mannosides (PIMs), and an unidentified phospholipid (PL) (Supplementary Figure S1). All the chemotaxonomic data are consistent with the assignment of strain NEAU-SSA 1$^T$ to the genus *Streptomyces*.

**Table 2.** Differential characteristics of strain NEAU-SSA 1$^T$, *S. aureoverticillatus* JCM 4347$^T$, *S. vastus* JCM4524$^T$, *S. cinereus* DSM43033$^T$, *S. xiangluensis* NEAU-LA29$^T$, and *S. flaveus* JCM3035$^T$.

| Characteristic | 1 | 2 | 3 | 4 | 5 | 6[a] |
|---|---|---|---|---|---|---|
| Decomposition of cellulose | + | − | − | − | + | ND |
| Production of $H_2S$ | − | − | − | − | − | + |
| Tween 20 | − | + | − | − | − | ND |
| Tween 40 | + | + | − | − | + | ND |
| Tween 80 | + | + | − | − | + | ND |
| Liquefaction of gelatin | + | − | − | − | − | ND |
| Growth temperature (°C) | 15–45 | 10–45 | 20–40 | 20–40 | 20–40 | 10–37 |
| pH range for growth | 6–11 | 5–12 | 6–10 | 6–9 | 6–9 | ND |
| NaCl tolerance range (*w*/*v*, %) | 0–6 | 0–15 | 0–5 | 0–5 | 0–6 | 0–7 |
| Carbon source utilization | | | | | | |
| D-fructose | + | + | − | − | + | + |
| D-galactose | + | + | − | − | + | + |
| Lactose | + | + | − | − | + | + |
| D-maltose | + | + | − | − | + | ND |
| L-rhamnose | + | + | − | + | + | + |
| D-ribose | − | + | − | − | − | − |
| D-sorbitol | − | + | − | + | − | ND |
| D-mannose | + | + | + | − | + | + |
| Raffinose | + | + | + | − | + | + |
| L-arabinose | − | − | − | − | − | + |
| D-xylose | − | − | − | − | − | + |
| Myo-inositol | + | + | + | − | + | + |
| Nitrogen source utilization | | | | | | |
| L-glutamine | + | + | − | − | + | ND |
| Glycine | − | + | + | + | − | ND |
| L-threonine | + | + | − | − | − | + |
| L-tyrosine | − | − | − | + | + | ND |
| L-arginine | + | + | + | + | − | + |
| L-asparagine | + | + | + | + | − | ND |
| L-serine | + | + | + | + | − | + |
| L-proline | + | + | + | + | − | − |

Strains: 1—NEAU-SSA 1$^T$; 2—*S. aureoverticillatus* JCM 4347$^T$; 3—*S. vastus* JCM4524$^T$; 4—*S. cinereus* DSM43033$^T$; 5—*S. xiangluensis* NEAU-LA29$^T$; 6—*S. flaveus* JCM3035$^T$. Abbreviation: +, positive; –, negative. All data are from this study except where marked. [a] Data from Michael Goodfellow et al. [45].

Sequence analysis of the 16S rRNA gene showed that strain NEAU-SSA 1$^T$ were affiliated with the genus *Streptomyces* and most closely related to *S. aureoverticillatus* JCM 4347$^T$ (97.9%). Phylogenetic analysis based on 16S rRNA gene sequences indicated that the strain formed a cluster with *S. vastus* JCM4524$^T$ (97.4%), *S. cinereus* DSM43033$^T$ (97.2%), *S. xiangluensis* NEAU-LA29$^T$ (97.1%), and *S. flaveus* JCM3035$^T$ (97.1%) in the neighbor-joining tree (Figure 2), a relationship also recovered by the maximum-likelihood algorithm (Figure S2). Phylogenetic trees based on the neighbor-joining and maximum-likelihood algorithms were constructed from the concatenated sequence alignment of the five housekeeping genes (Figure 3 and Figure S3), and had the same topology as the 16S rRNA gene tree. Moreover, pairwise distances calculated for NEAU-SSA 1$^T$ and the related species using the concatenated sequences of *atp*D-*gyr*B-*rec*A-*rpo*B-*trp*B were well above 0.007 (Table S2), which is considered to be the threshold for species determination by Rong et al. [46]. DNA–DNA hybridization was employed to further clarify the relatedness between the strain and *S. aureoverticillatus* JCM 4347$^T$, *S. vastus* JCM4524$^T$, *S. cinereus* DSM 43033$^T$, and *S. xiangluensis* NEAU-LA29$^T$. Results showed that strain NEAU-SSA 1$^T$ shared DNA–DNA relatedness of 37.1 ± 3.4% with *S. aureoverticillatus* JCM 4347$^T$, 35.4 ± 4.3% with *S. vastus* JCM 4524$^T$, 33.1 ± 4.1% with *S. cinereus* DSM 43033$^T$, and 29.0 ± 4.9% with *S. xiangluensis* NEAU-LA29$^T$. Digital DNA–DNA hybridization was employed to clarify the relatedness between strain NEAU-SSA 1$^T$ and *S. flaveus* JCM 3035$^T$. The level of digital DNA–DNA hybridization between them was 24.9 ± 2.4%. These five values are all below the threshold value of 70% recommended by Wayne et al. [47] for assigning strains to the same genomic species. Similarly, a low ANI value of 80.99% was found between strain NEAU-SSA 1$^T$ and *S. flaveus* JCM 3035$^T$, a result well below the threshold used to delineate prokaryote species [48,49].

The assembled genome sequence of strain NEAU-SSA 1$^T$ was found to be 10,364,704 bp long and composed of 352 contigs with an N50 of 59,982 bp, a DNA G+C content of 69.9 mol % and a coverage of 200x. It was deposited into GenBank under the accession number VCHX00000000. The 16S rRNA gene sequence from the whole genome sequence shared a 100% similarity with that from PCR sequencing, suggesting that the genome sequence was not contaminated. Detailed genomic information is presented in the Table S3.

Comparison of phenotypic characteristics between strain NEAU-SSA 1$^T$ and its closely related species—*S. aureoverticillatus* JCM 4347$^T$, *S. vastus* JCM4524$^T$, *S. cinereus* DSM 43033$^T$, *S. xiangluensis* NEAU-LA29$^T$, and *S. flaveus* JCM 3035$^T$—was performed to differentiate these strains (Table 2). Differential cultural characteristics included: NaCl tolerance of the strain was up to 5.0%, which is lower than that of *S. aureoverticillatus* JCM 4347$^T$ (15%) and *S. flaveus* JCM 3035$^T$ (7%); and the strain could grow at pH 11.0, while *S. vastus* JCM4524$^T$, *S. cinereus* DSM 43033$^T$, and *S. xiangluensis* NEAU-LA29$^T$ could not. Other phenotypic differences included the production of $H_2S$; decomposition of cellulose; liquefaction of gelatin; growth temperature; hydrolysis of Tweens (20, 40, and 80); and utilization of L-arabinose, D-galactose, D-fructose, D-maltose, lactose, L-rhamnose, D-ribose, D-sorbitol, D-mannose, raffinose, D-xylose, *myo*-inositol, L-glutamine, glycine, L-threonine, L-tyrosine, L-serine, L-proline, L-asparagine, and L-arginine.

On the basis of morphological, physiological, chemotaxonomic, and phylogenetic results, strain NEAU-SSA 1$^T$ is considered to represent a novel species within the genus *Streptomyces*, for which the name *Streptomyces sporangiiformans* is proposed.

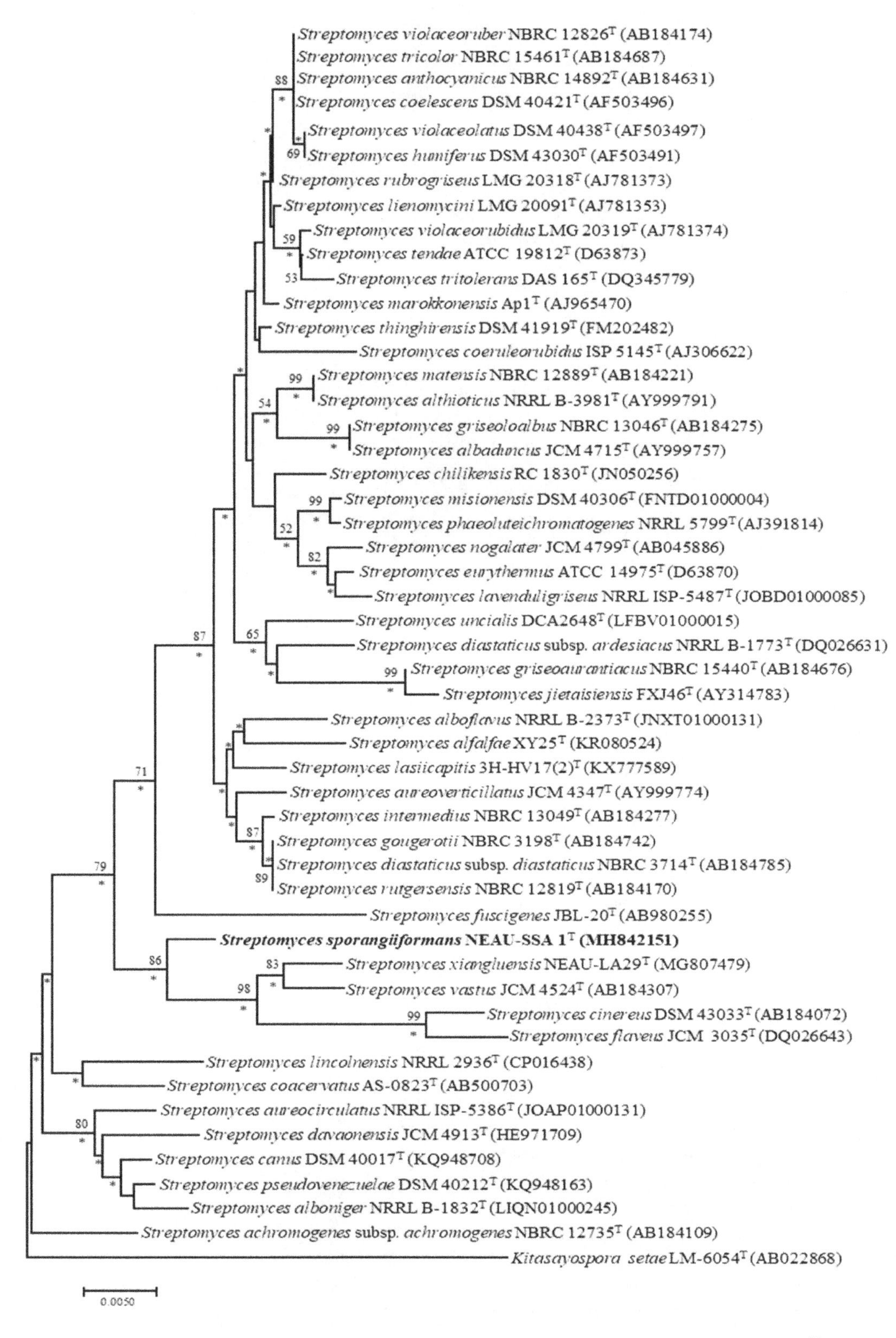


**Figure 2.** Neighbor-joining tree showing the phylogenetic position of strain NEAU-SSA $1^T$ (1412 bp) and the related species of the genus *Streptomyces* based on 16S rRNA gene sequences. The out-group used was *Kitasatospora setae* LM-6054$^T$. Only bootstrap values above 50% (percentages of 1000 replications) are indicated. Asterisks indicate branches also recovered in the maximum-likelihood tree. Scale bar represents 0.005 nucleotide substitutions per site.

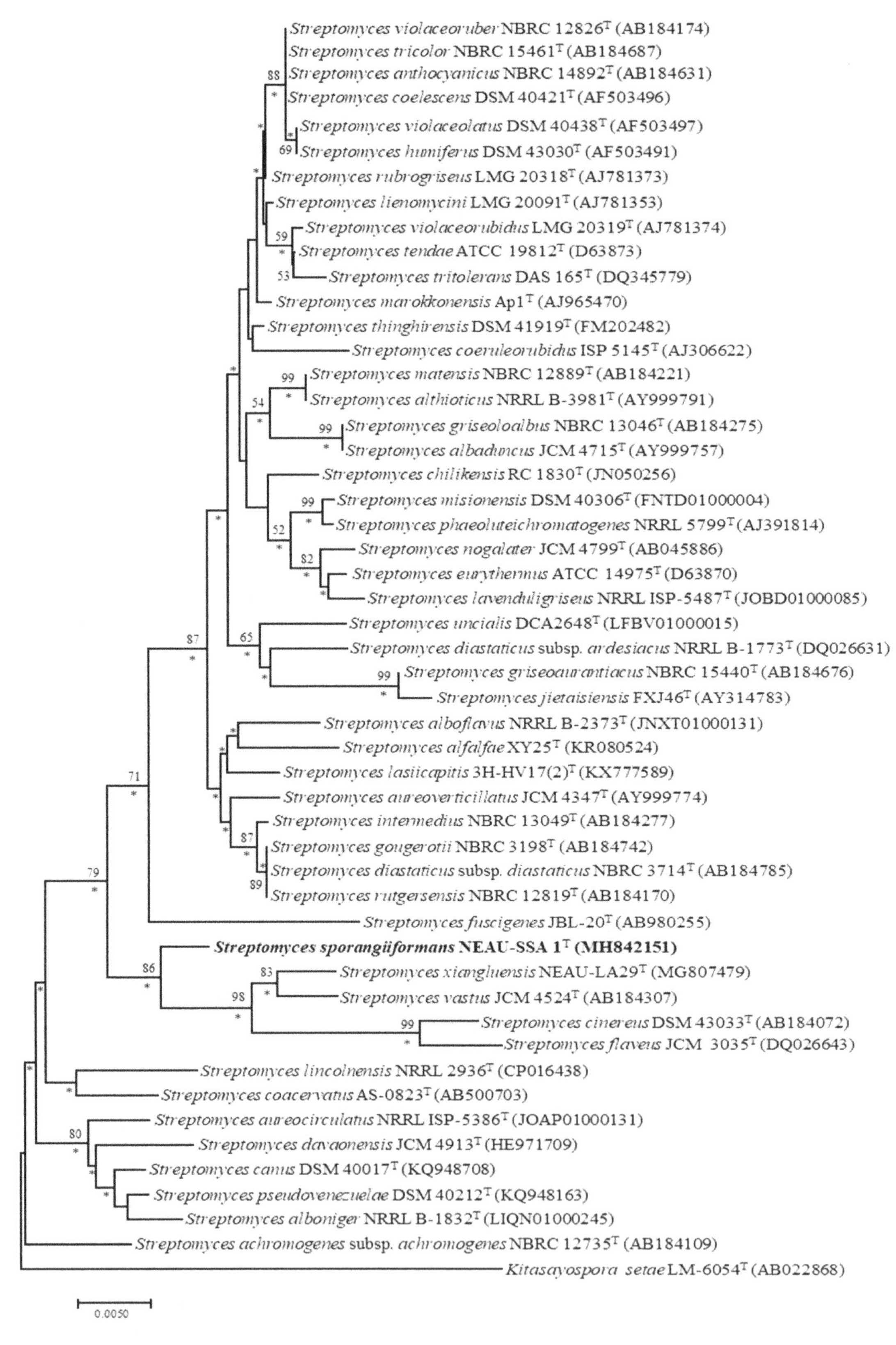


**Figure 3.** Neighbor-joining tree based on MLSA analysis of the concatenated partial sequences from five housekeeping genes (*atp*D, *gyr*B, *rec*A, *rpo*B, and *trp*B) of isolate NEAU-SSA 1$^{T}$ (in bold) and related taxa. Only bootstrap values above 50% (percentages of 1000 replications) are indicated. *Kitasatospora setae* LM-6054$^{T}$ was used as an out-group. Asterisks indicate branches also recovered in the maximum-likelihood tree. Scale bar represents 0.02 nucleotide substitutions per site.

### *3.2. Description of Streptomyces sporangiiformans sp. nov.*

*Streptomyces sporangiiformans* (spo.ran.gi.i.for'mans. N.L. neut. n. sporangium; L. pres. part. *formans* forming; N.L. part. adj. *sporangiiformans* forming sporangia).

Gram-stain-positive, aerobic actinomycete that formed well-developed, branched substrate hyphae and aerial mycelia. Sporangia consisted of cylindrical and rough surfaced spores (0.6–0.8 μm × 0.9–1.6 μm) were produced on aerial mycelia, but spore chains were not observed. Good growth on ISP 3, ISP 4, ISP 7, and Nutrient agar media; moderate growth on ISP 1, ISP 2, ISP 5, ISP 6, and Czapek's agar media; and poor growth on Bennett's agar medium. Growth occurred at pH values between 6.0 and 11.0, the optimum being pH 7.0. Tolerates up to 6.0% NaCl and grows optimally in 0–1% (*w*/*v*) NaCl. Growth was observed at temperatures between 15 and 45 °C, with an optimum temperature of 28 °C. Positive for decomposition of Tweens (40 and 80) and cellulose, hydrolysis of aesculin and starch, liquefaction of gelatin and production of urease; and negative for coagulation and peptonization of milk, hydrolysis of Tween 20, production of $H_2S$, and reduction of nitrate. D-fructose, D-galactose, D-glucose, inositol, lactose, D-maltose, D-mannose, D-raffinose, L-rhamnose, and D-sucrose were utilized as sole carbon sources, but not L-arabinose, dulcitol, D-ribose, D-sorbitol, or D-xylose. L-alanine, L-arginine, L-asparagine, L-aspartic acid, creatine, L-glutamic acid, L-glutamine, L-proline, L-serine, and L-threonine were utilized as sole nitrogen sources, but not glycine or L-tyrosine. Cell wall contained *LL*-diaminopimelic acid and the whole-cell hydrolysates were ribose, mannose, and galactose. The polar lipids contained diphosphatidylglycerol (DPG), phosphatidylethanolamine (PE), hydroxy-phosphatidylethanolamine (OH-PE), phosphatidylinositol (PI), two phosphatidylinositol mannosides (PIMs), and an unidentified phospholipid (PL). The menaquinones were MK-9($H_4$), MK-9($H_6$), and MK-9($H_8$). Major fatty acids were *iso*-$C_{17:0}$, $C_{16:0}$, and $C_{17:1}\omega9c$.

The type strain was NEAU-SSA $1^T$ (=CCTCC AA 2017028$^T$ = DSM 105692$^T$), isolated from soil collected from Mount Song, Dengfeng, Henan Province, China. The DNA G+C content of the type strain was 69.9 mol %, calculated from the assembly for the draft genome sequence. The GenBank/EMBL/DDBJ accession number for the 16S rRNA gene sequence of strain NEAU-SSA $1^T$ is MH842151. This Whole Genome Shotgun project has been deposited at DDBJ/ENA/GenBank under the accession VCHX00000000. The version described in this paper is version VCHX00000000.2.

### *3.3. Antibacterial Activity of NEAU-SSA $1^T$ against Ralstonia solanacearum*

Strain NEAU-SSA $1^T$ exhibited antibacterial activity against *Ralstonia solanacearum* with inhibitory zone diameters of 23 mm (Figure 4a). However, no inhibitory effect on the growth of *Micrococcus luteus* (Figure 4b) was observed. Comparison of the antibacterial activity of the extract of the supernatant with that of the cell pellet suggested that the antibacterial substances of strain NEAU-SSA $1^T$ were in both the supernatant and cell pellet since the extracts all showed inhibition of the growth of *Ralstonia solanacearum* with the inhibitory zone diameters of 31.5 and 26.4 mm, respectively (Figure 5a,b). The antibacterial substances in the supernatant were stable after they were placed in a water bath at 40 and 60 °C for 30 min, while they did not show antibacterial activity after 80 and 100 °C bath (Figure 6a), which indicated that they were sensitive to temperature. In contrast, the antibacterial substances in the cell pellet were insensitive to temperature (Figure 6b), which demonstrated that the antibacterial substances in the supernatant and cell pellet were different. The antiSMASH analysis led to the identification of 49 gene clusters, including 24 gene clusters that showed very low similarity to the known gene clusters of mediomycin A, cremimycin, primycin, ibomycin, naphthomycin, lasalocid, informatipeptin, polyoxypeptin, kutznerides, anisomycin, paulomycin, himastatin, desotamide, nystatin, tiacumicin B, oxazolomycin, and 4-Z-annimycin. Therefore, the relationships between the corresponding secondary metabolites produced by NEAU-SSA $1^T$ and the antibacterial activity are still ambiguous. *Streptomyces* are well known as important biological resources for their biologically active secondary metabolites, which play important roles in protecting plants against pathogens [50]. Strain NEAU-SSA $1^T$, which shows a stronger antibacterial activity against *Ralstonia solanacearum*, is a novel species of the genus *Streptomyces*, and possesses 24 lower similarity gene clusters. Therefore, it is interesting and significant to isolate and identify the secondary metabolites of the strain in further studies.

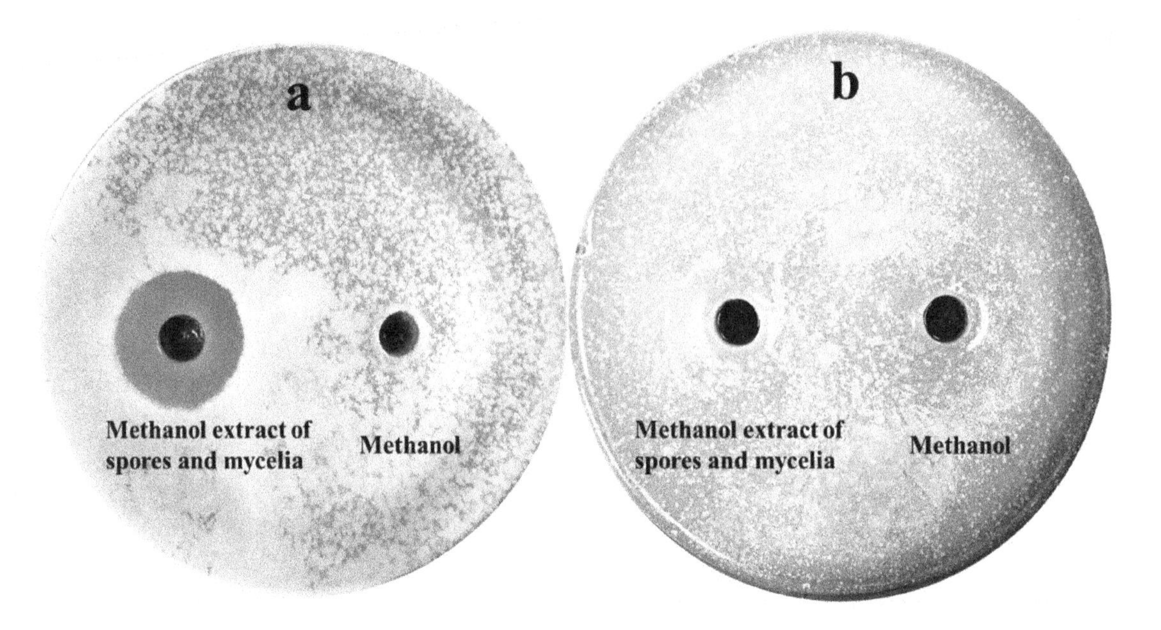


**Figure 4.** The antibacterial activity of strain NEAU-SSA $1^T$ against *Ralstonia solanacearum* (**a**) and *Micrococcus luteus* (**b**).

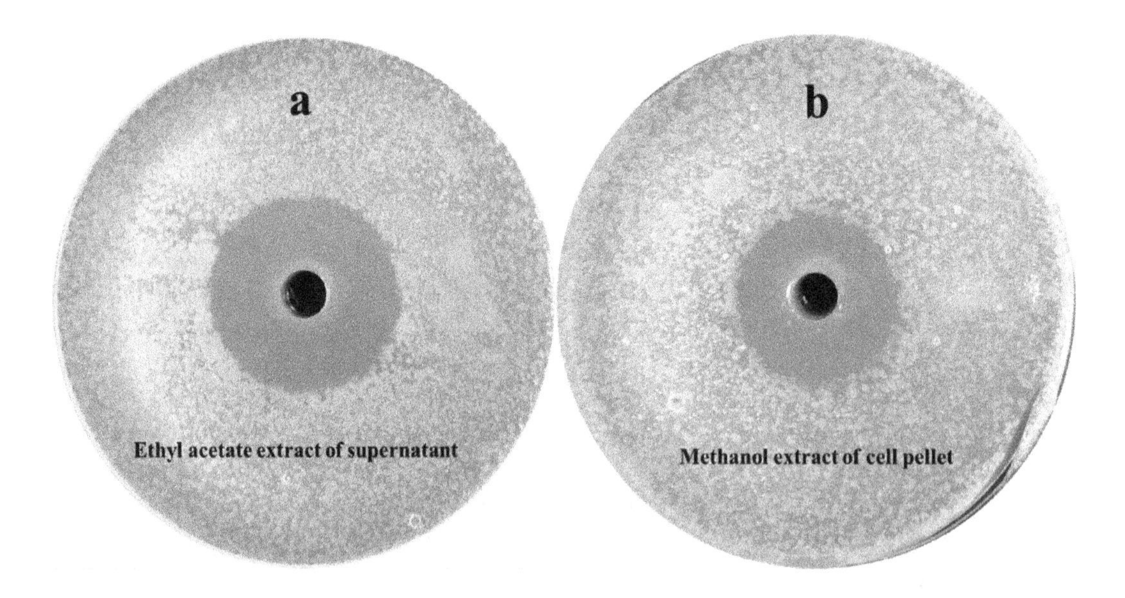


**Figure 5.** The antibacterial activity of the extract of the supernatant (**a**) and cell pellet of strain NEAU-SSA $1^T$ (**b**) against *Ralstonia solanacearum*.

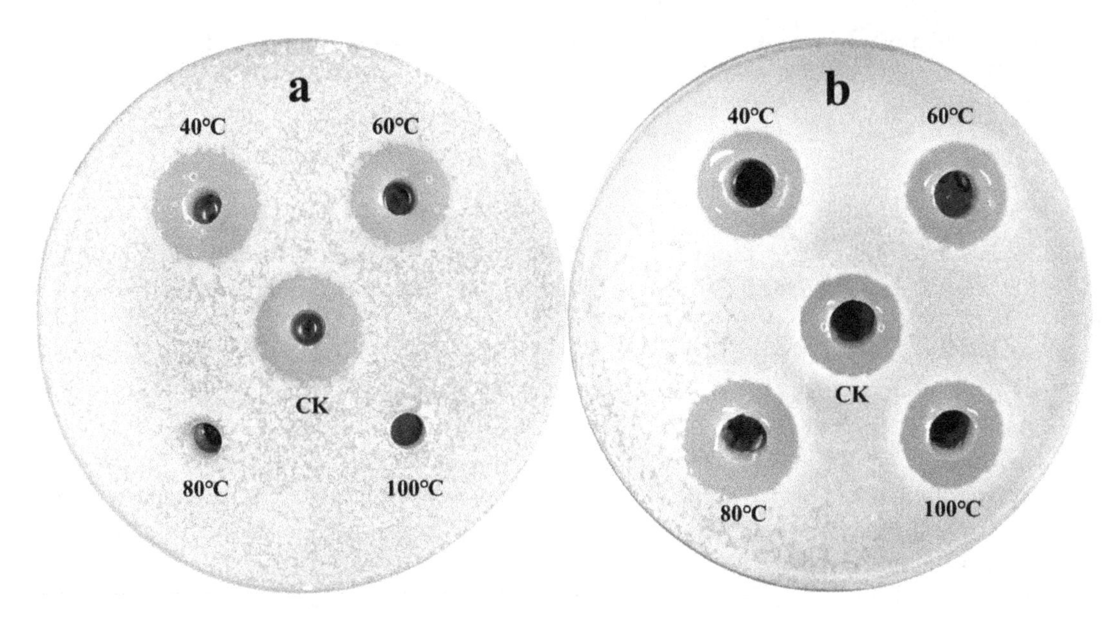


**Figure 6.** The effect of temperature on the antibacterial activity of the extract of the supernatant (**a**) and cell pellet of strain NEAU-SSA $1^T$ (**b**) against *Ralstonia solanacearum*.

## 4. Conclusions

A novel strain NEAU-SSA $1^T$ that exhibited antibacterial activity against *Ralstonia solanacearum* was isolated from a soil sample. Morphological features, phylogenetic analysis based on 16S rRNA gene sequences, and multilocus sequence analysis based on five other house-keeping genes (*atp*D, *gyr*B, *rec*A, *rpo*B, and *trp*B) suggested that strain NEAU-SSA $1^T$ belonged to the genus *Streptomyces*. Physiology and biochemical characteristics, together with DDH relatedness values and ANI values, clearly indicated that strain NEAU-SSA $1^T$ could be differentiated from the closely related strains *S. aureoverticillatus* JCM $4347^T$, *S. vastus* JCM $4524^T$, *S. cinereus* DSM $43033^T$, *S. xiangluensis* NEAU-$LA29^T$, and *S. flaveus* JCM $3035^T$. Based on the polyphasic analysis, it is proposed that strain NEAU-SSA $1^T$ should be classified as representatives of a novel species of the genus *Streptomyces*, for which the name *Streptomyces sporangiiformans* sp. nov. is proposed. The type strain is NEAU-SSA $1^T$ (=CCTCC AA $2017028^T$ = DSM $105692^T$).

**Supplementary Materials:**
Figure S1: Maximum-likelihood tree showing the phylogenetic position of strain NEAU-SSA $1^T$ (1412 bp) and the related species based on 16S rRNA gene sequences. The out-group used was *Kitasatospora setae* LM-$6054^T$. Only bootstrap values above 50% (percentages of 1000 replications) are indicated. Bar, 0.01 nucleotide substitutions per site. Figure S2: Maximum-likelihood tree based on MLSA analysis of the concatenated partial sequences from five housekeeping genes (*atp*D, *gyr*B, *rec*A, *rpo*B, and *trp*B) of isolate NEAU-SSA $1^T$ and related taxa. Only bootstrap values above 50% (percentages of 1000 replications) are indicated. *Kitasatospora setae* LM-$6054^T$ was used as an out-group. Bar, 0.05 nucleotide substitutions per site. Table S1. Growth and cultural characteristics of strain NEAU-SSA $1^T$. Table S2: MLAS distance values for selected strains in this study. Table S3: General features of the genome sequence of the type strain NEAU-SSA $1^T$.

**Author Contributions:** J.Z. and L.H. performed the isolation and morphological and biochemical characterization of strain NEAU-SSA $1^T$. M.Y. performed the antifungal test. P.C. analyzed DNA sequencing data and genomic sequencing data. D.L. performed chemotaxonomic analysis and phylogenetic analysis. X.G. prepared the figures and tables. Y.L. performed the morphological observation by transmission electron microscopy. X.W. and W.X. designed the experiments and edited the manuscript.

**Acknowledgments:** The authors would like to thank Aharon Oren (Department of Plant and Environmental Sciences, the Alexander Silberman Institute of Life Sciences, the Hebrew University of Jerusalem) for helpful advice on the specific epithet.

## References


1. Mansfield, J.; Genin, S.; Magori, S.; Citovsky, V.; Sriariyanum, M.; Ronald, P.; Dow, M.; Verdier, V.; Beer, S.V.; Machado, M.A.; et al. Top 10 plant pathogenic bacteria in molecular plant pathology. *Mol. Plant Pathol.* **2012**, *13*, 614–629. [CrossRef] [PubMed]
2. Hayward, A.C. Biology and epidemiology of bacterial wilt caused by *Pseudomonas solanacearum*. *Annu. Rev. Phytopathol.* **1991**, *29*, 65–87. [CrossRef] [PubMed]
3. Jiang, G.F.; Wei, Z.; Xu, J.; Chen, H.L.; Zhang, Y.; She, X.M.; Macho, A.P.; Ding, W.; Liao, B.S. Bacterial wilt in china: history, current status, and future perspectives. *Front. Plant Sci.* **2017**, *8*, 1549. [CrossRef] [PubMed]
4. Labeda, D.P.; Goodfellow, M.; Brown, R.; Ward, A.C.; Lanoot, B.; Vanncanneyt, M.; Swings, J.; Kim, S.B.; Liu, Z.; Chun, J.; et al. Phylogenetic study of the species within the family *Streptomycetaceae*. *Antonie Van Leeuwenhoek* **2012**, *101*, 73–104. [CrossRef]
5. Bérdy, J. Bioactive microbial metabolites, a personal view. *J. Antibiot.* **2005**, *58*, 1–26. [CrossRef] [PubMed]
6. Li, C.; He, H.R.; Wang, J.B.; Liu, H.; Wang, H.Y.; Zhu, Y.J.; Wang, X.J.; Zhang, Y.Y.; Xiang, W.S. Characterization of a LAL-type regulator NemR in nemadectin biosynthesis and its application for increasing nemadectin production in *Streptomyces cyaneogriseus*. *Sci. China Life Sci.* **2019**, *62*, 394–405. [CrossRef]
7. Waksman, S.A.; Henrici, A.T. The nomenclature and classification of the actinomycetes. *J. Bacteriol.* **1943**, *46*, 337–341. [PubMed]
8. Berdy, J. Are actinomycetes exhausted as a source of secondary metabolites? *Biotechnologia* **1995**, 13–34.
9. Fiedler, H.P.; Bruntner, C.; Bull, A.T.; Ward, A.C.; Goodfellow, M.; Potterat, O.; Puder, C.; Mihm, G. Marine actinomycetes as a source of novel secondary metabolites. *Antonie Van Leeuwenhoek* **2005**, *87*, 37–42. [CrossRef]
10. Hayakawa, M.; Nonomura, H. Humic acid-vitamin agar, a new medium for the selective isolation of soil actinomycetes. *J. Ferment Technol.* **1987**, *65*, 501–509. [CrossRef]
11. Shirling, E.B.; Gottlieb, D. Methods for characterization of *Streptomyces* species. *Int. J. Syst. Bacteriol.* **1966**, *16*, 313–340. [CrossRef]
12. Jin, L.Y.; Zhao, Y.; Song, W.; Duan, L.P.; Jiang, S.W.; Wang, X.J.; Zhao, J.W.; Xiang, W.S. *Streptomyces inhibens* sp. nov., a novel actinomycete isolated from rhizosphere soil of wheat (*Triticum aestivum* L.). *Int. J. Syst. Evol. Microbiol.* **2019**, *69*, 688–695. [CrossRef] [PubMed]
13. Waksman, S.A. *The Actinomycetes. A Summary of Current Knowledge*; The Ronald Press Co.: New York, NY, USA, 1967; p. 286.
14. Jones, K.L. Fresh isolates of actinomycetes in which the presence of sporogenous aerial mycelia is a fluctuating characteristic. *J. Bacteriol.* **1949**, *57*, 141–145. [PubMed]
15. Waksman, S.A. *The Actinomycetes, Volume 2, Classification, Identification and Descriptions of Genera and Species*; Williams and Wilkins Company: Philadelphia, PA, USA, 1961; p. 363.
16. Kelly, K.L. Color-name charts illustrated with centroid colors. In *Inter-Society Color Council-National Bureau of Standards*; U.S. National Bureau of Standards: Washington, DC, USA, 1965.
17. Jia, F.Y.; Liu, C.X.; Wang, X.J.; Zhao, J.W.; Liu, Q.F.; Zhang, J.; Gao, R.X.; Xiang, W.S. *Wangella harbinensis* gen. nov., sp. nov., a new member of the family *Micromonosporaceae*. *Antonie Van Leeuwenhoek* **2013**, *103*, 399–408. [CrossRef] [PubMed]
18. Smibert, R.M.; Krieg, N.R. Phenotypic characterization. In *Methods for General and Molecular Bacteriology*; Gerhardt, P., Murray, R.G.E., Wood, W.A., Krieg, N.R., Eds.; American Society for Microbiology: Washington, DC, USA, 1994; pp. 607–654.
19. Gordon, R.E.; Barnett, D.A.; Handerhan, J.E.; Pang, C. *Nocardia coeliaca*, *Nocardia autotrophica*, and the nocardin strain. *Int. J. Syst. Bacteriol.* **1974**, *24*, 54–63. [CrossRef]
20. Yokota, A.; Tamura, T.; Hasegawa, T.; Huang, L.H. *Catenuloplanes japonicas* gen. nov., sp. nov., nom. rev., a new genus of the order *Actinomycetales*. *Int. J. Syst. Bacteriol.* **1993**, *43*, 805–812. [CrossRef]
21. McKerrow, J.; Vagg, S.; McKinney, T.; Seviour, E.M.; Maszenan, A.M.; Brooks, P.; Seviour, R.J. A simple HPLC method for analysing diaminopimelic acid diastereomers in cell walls of Gram-positive bacteria. *Lett. Appl. Microbiol.* **2000**, *30*, 178–182. [CrossRef] [PubMed]

22. Lechevalier, M.P.; Lechevalier, H.A. The chemotaxonomy of actinomycetes. In *Actinomycete Taxonomy*; Special Publication Volume 6; Dietz, A., Thayer, D.W., Eds.; Society of Industrial Microbiology: Arlington, TX, USA, 1980; pp. 227–291.
23. Minnikin, D.E.; O'Donnell, A.G.; Goodfellow, M.; Alderson, G.; Athalye, M.; Schaal, A.; Parlett, J.H. An integrated procedure for the extraction of bacterial isoprenoid quinones and polar lipids. *J. Microbiol. Methods* **1984**, *2*, 233–241. [CrossRef]
24. Collins, M.D. Isoprenoid quinone analyses in bacterial classification and identification. In *Chemical Methods in Bacterial Systematics*; Goodfellow, M., Minnikin, D.E., Eds.; Academic Press: London, UK, 1985; pp. 267–284.
25. Qu, Z.; Ruan, J.S.; Hong, K. Application of high performance liquid chromatography and gas chromatography in the identification of actinomyces. *Biotechnol. Bull.* **2009**, *s1*, 79–82.
26. Gao, R.X.; Liu, C.X.; Zhao, J.W.; Jia, F.Y.; Yu, C.; Yang, L.Y.; Wang, X.J.; Xiang, W.S. *Micromonospora jinlongensis* sp. nov., isolated from muddy soil in China and emended description of the genus *Micromonospora*. *Antonie Van Leeuwenhoek* **2014**, *105*, 307–315. [CrossRef] [PubMed]
27. Xiang, W.S.; Liu, C.X.; Wang, X.J.; Du, J.; Xi, L.J.; Huang, Y. *Actinoalloteichus nanshanensis* sp. nov., isolated from the rhizosphere of a fig tree (*Ficus religiosa*). *Int. J. Syst. Evol. Microbiol.* **2011**, *61*, 1165–1169. [CrossRef] [PubMed]
28. Zhou, J.Z.; Bruns, M.A.; Tiedje, J.M. DNA recovery from soils of diverse composition. *Appl. Environ. Microbiol.* **1996**, *62*, 316–322. [PubMed]
29. Woese, C.R.; Gutell, R.; Gupta, R.; Noller, H.F. Detailed analysis of the higher-order structure of 16S-like ribosomal ribonucleic acids. *Microbiol. Rev.* **1983**, *47*, 621–669. [PubMed]
30. Springer, N.; Ludwig, W.; Amann, R.; Schmidt, H.J.; Görtz, H.D.; Schleifer, K.H. Occurrence of fragmented 16S rRNA in an obligate bacterial endosymbiont of *Paramecium caudatum*. *Proc. Natl. Acad. Sci. USA* **1993**, *90*, 9892–9895. [CrossRef] [PubMed]
31. Saitou, N.; Nei, M. The neighbor-joining method: a new method for reconstructing phylogenetic trees. *Mol. Biol. Evol.* **1987**, *4*, 406–425.
32. Felsenstein, J. Evolutionary trees from DNA sequences: a maximum likelihood approach. *J. Mol. Evol.* **1981**, *17*, 368–376. [CrossRef] [PubMed]
33. Kumar, S.; Stecher, G.; Tamura, K. Mega7: molecular evolutionary genetics analysis version 7.0 for bigger datasets. *Mol. Biol. Evol.* **2016**, *33*, 1870–1874. [CrossRef]
34. Felsenstein, J. Confidence limits on phylogenies: an approach using the bootstrap. *Evolution* **1985**, *39*, 83–791. [CrossRef]
35. Kimura, M. A simple method for estimating evolutionary rates of base substitutions through comparative studies of nucleotide sequences. *J. Mol. Evol.* **1980**, *16*, 111–120. [CrossRef]
36. Yoon, S.H.; Ha, S.M.; Kwon, S.; Lim, J.; Kim, Y.; Seo, H.; Chun, J. Introducing EzBioCloud: A taxonomically united database of 16S rRNA and whole genome assemblies. *Int. J. Syst. Evol. Microbiol.* **2017**, *67*, 1613–1617.
37. Blin, K.; Wolf, T.; Chevrette, M.G.; Lu, X.; Schwalen, C.J.; Kautsar, S.A.; Suarez Duran, H.G.; de Los Santos, E.L.C.; Kim, H.U.; Nave, M.; et al. antiSMASH 4.0—Improvements in chemistry prediction and gene cluster boundary identification. *Nucleic Acids Res.* **2017**, *45*, W36–W41. [CrossRef]
38. Li, R.Q.; Zhu, H.M.; Ruan, J.; Qian, W.B.; Fang, X.D.; Shi, Z.B.; Li, Y.R.; Li, S.T.; Shan, G.; Kristiansen, K.; et al. De novo assembly of human genomes with massively parallel short read sequencing. *Genome Res.* **2010**, *20*, 265–272. [CrossRef]
39. Li, R.; Li, Y.; Kristiansen, K.; Wang, J. SOAP: short oligonucleotide alignment program. *Bioinformatics* **2008**, *24*, 713–714. [CrossRef]
40. De Ley, J.; Cattoir, H.; Reynaerts, A. The quantitative measurement of DNA hybridization from renaturation rates. *Eur. J. Biochem.* **1970**, *12*, 133–142. [CrossRef]
41. Huss, V.A.R.; Festl, H.; Schleifer, K.H. Studies on the spectrometric determination of DNA hybridisation from renaturation rates. *Syst. Appl. Microbiol.* **1983**, *4*, 184–192. [CrossRef]
42. Yoon, S.H.; Ha, S.M.; Lim, J.; Kwon, S.; Chun, J. A large-scale evaluation of algorithms to calculate average nucleotide identity. *Antonie Van Leeuwenhoek* **2017**, *110*, 1281–1286. [CrossRef]
43. Meier-Kolthoff, J.P.; Auch, A.F.; Klenk, H.P.; Goker, M. Genome sequence-based species delimitation with confidence intervals and improved distance functions. *Bmc Bioinform.* **2013**, *14*, 60. [CrossRef]
44. Boyanova, L.; Gergova, G.; Nikolov, R.; Derejian, S.; Lazarova, E.; Katsarov, N.; Mitov, I.; Krastev, Z. Activity

of Bulgarian propolis against 94 *Helicobacter pylori* strains in vitro by agar-well diffusion, agar dilution and disc diffusion methods. *J. Med. Microbiol.* **2005**, *54*, 481–483. [CrossRef]
45. Goodfellow, M.; Kämpfer, P.; Busse, H.J.; Trujillo, M.E.; Suzuki, K.I.; Ludwig, W.; Whitman, W.B. *Bergey's Manual® of Systematic Bacteriology*; Springer: New York, NY, USA, 2012.
46. Rong, X.; Huang, Y. Taxonomic evaluation of the *Streptomyces hygroscopicus* clade using multilocus sequence analysis and DNA-DNA hybridization, validating the MLSA scheme for systematics of the whole genus. *Syst. Appl. Microbiol.* **2012**, *35*, 7–18. [CrossRef]
47. Wayne, L.G.; Brenner, D.J.; Colwell, R.R.; Grimont, P.A.D.; Kandler, O. International Committee on Systematic Bacteriology. Report of the ad hoc committee on reconciliation of approaches to bacterial systematics. *Int. J. Syst Bacteriol.* **1987**, *37*, 463–464. [CrossRef]
48. Richter, M.; Rossello-Mora, R. Shifting the genomic gold standard for the prokaryotic species definition. *Proc. Natl. Acad. Sci. USA* **2009**, *106*, 19126–19131. [CrossRef]
49. Chun, J.; Rainey, F.A. Integrating genomics into the taxonomy and systematics of the Bacteria and Archaea. *Int. J. Syst. Evol. Microbiol.* **2014**, *64*, 316–324. [CrossRef]
50. Ueno, M.; Quyet, N.T.; Shinzato, N.; Matsui, T. Antifungal activity of collected in subtropical region, Okinawa, against *Magnaporthe oryzae*. *Trop. Agricult. Dev.* **2016**, *60*, 48–52.

# 8

# Evaluation of Antimicrobial, Enzyme Inhibitory, Antioxidant and Cytotoxic Activities of Partially Purified Volatile Metabolites of Marine *Streptomyces* sp.S2A

**Saket Siddharth and Ravishankar Rai Vittal ***

Department of Studies in Microbiology, University of Mysore, Manasagangotri, Mysore 570006, India; saketsiddharth@gmail.com

* Correspondence: raivittal@gmail.com.

**Abstract:** In the present study, marine actinobacteria *Streptomyces* sp.S2A was isolated from the Gulf of Mannar, India. Identification was carried out by 16S rRNA analysis. Bioactive metabolites were extracted by solvent extraction method. The metabolites were assayed for antagonistic activity against bacterial and fungal pathogens, inhibition of α-glucosidase and α-amylase enzymes, antioxidant activity and cytotoxic activity against various cell lines. The actinobacterial extract showed significant antagonistic activity against four gram-positive and two gram-negative pathogens. Excellent reduction in the growth of fungal pathogens was also observed. The minimum inhibitory concentration of the partially purified extract (PPE) was determined as 31.25 μg/mL against *Klebsiella pneumoniae*, 15.62 μg/mL against *Staphylococcus epidermidis*, *Staphylococcus aureus* and *Bacillus cereus*. The lowest MIC was observed against *Micrococcus luteus* as 7.8 μg/mL. MIC against fungal pathogens was determined as 62.5 μg/mL against *Bipolaris maydis* and 15.62 μg/mL against *Fusarium moniliforme*. The α-glucosidase and α-amylase inhibitory potential of the fractions were carried out by microtiter plate method. $IC_{50}$ value of active fraction for α-glucosidase and α-amylase inhibition was found to be 21.17 μg/mL and 20.46 μg/mL respectively. The antioxidant activity of partially purified extract (PPE) (DPPH, ABTS, FRAP and Metal chelating activity) were observed and were also found to have significant cytotoxic activity against HT-29, MDA and U-87MG cell lines. The compound analysis was performed using gas chromatography-mass spectrometry (GC-MS) and resulted in three constituents; pyrrolo[1–a]pyrazine-1,4-dione,hexahydro-3-(2-methylpropyl)-, being the main component (80%). Overall, the strain possesses a wide spectrum of antimicrobial, enzyme inhibitory, antioxidant and cytotoxic activities which affords the production of significant bioactive metabolites as potential pharmacological agents.

**Keywords:** marine actinobacteria; *Streptomyces* sp.; enzyme inhibition; antimicrobial; antioxidant; cytotoxicity; GC-MS; pyrrolopyrazines

## 1. Introduction

The microbial natural products are a source of several important drugs of high therapeutic value. Going back to the history of drugs of the first choice, it suggests that novel chemical moieties forming the backbone of bioactive compounds are primarily obtained from natural sources [1]. The microbial natural products are a source of several important drugs of high therapeutic value, namely antitumor agents [2], antibiotics [3], immunosuppressive agents [4], and enzyme inhibitors [5]. The majority of commercially available pharmaceutical products are secondary metabolites or their derivatives produced by bacteria, fungi and actinobacteria [6]. Among producers of important metabolites, actinobacteria have proven to be most prolific source accounting for more than two-third of available clinical products of several medical uses [7]. Actinobacteria are filamentous gram-positive bacteria with high G + C content [8]. They are characterized by complex morphological differentiation and are considered as an intermediate group of bacteria and fungi [9]. Their presence in various ecological habitats and marine environments has enabled research communities to exploit their tremendous potential as the richest source of pharmaceutical and biologically active products [10]. Therefore, they are contemplated as the most economical and biotechnologically beneficial prokaryotes.

Secondary metabolites are organic compounds having no direct role in the vegetative growth and the development of the organism. About 40–45% of active metabolites produced by the microorganisms are contributed by various genera of actinobacteria and are currently in clinical use [11]. Over the last few decades, actinobacterial metabolites have been used as a template for the development of anticancer agents, antibiotics, enzyme inhibitors, immunomodulators and plant growth hormones [12]. Among important genera of actinobacteria, *Streptomyces* is the most dominant and prolific source of bioactive metabolites with the broad spectrum of activity. Of 10,000 known compounds, genus *Streptomyces* alone accounts for nearly 7500 compounds, while the rare actinobacterial genera including *Nocardia*, *Micromonospora*, *Streptosporangium*, *Actinomadura*, *Saccharopolyspora* and *Actinoplanes* represent 2500 compounds [13]. Although the majority of the actinobacterial bioactive metabolites come from terrestrial habitats, recent studies on actinobacteria from diverse habitats have suggested new chemical entities and bioactive compounds [14]. Moreover, the possibility of finding a novel bioactive molecule from the terrestrial habitat has diminished over the years [15]. The marine ecosystem is an untapped and underexploited source for the discovery of novel metabolites. Species isolated from marine environments have found to be different in physiological, biochemical and molecular characteristics from their terrestrial counterparts and therefore might produce novel metabolites [16]. With the increase in resistance among pathogens and unavailability of novel metabolites from terrestrial sources, marine-derived drugs could be of great importance. However, the distribution of actinobacteria in the marine ecosystem has not been explored much and the knowledge about the marine-derived metabolites remains elusive. But, recent outbreaks about the marine actinobacterial-derived bioactive metabolites with distinct lead molecules have made a significant contribution in drug discovery and may lead to the development of new drugs in future.

The present work therefore aimed to investigate the potential of secondary metabolites produced by marine actinobacteria *Streptomyces* sp.S2A and their characterization.

## 2. Materials and Methods

### 2.1. Sample Collection

Marine sediment samples were collected from Gulf of Mannar Marine National Park (Latitude 9.127823° N, Longitude 79.466155° E), Rameshwaram, India. The collected sediment samples were brought to the laboratory in sterile zip-lock plastic bags and stored at 4 °C until further use. The sediments were pre-treated with $CaCO_3$ and kept in hot air oven at 55 °C for 20 min [17].

### 2.2. Actinobacterial Isolates

Isolation of the actinobacterial strain was determined by serial dilution method on starch casein agar (Himedia, New Delhi, India) supplemented with nalidixic acid (25 μg/mL) and nystatin (50 μg/mL). The plates were incubated at 28 °C for 7 days. After incubation, individual colonies were maintained on ISP-2 slants and stored at 4 °C for further use. The ornamentation of the spore chain was analyzed by SEM.

### 2.3. Molecular Identification of Actinobacteria

Genomic DNA extraction of the strain was done using the phenol-chloroform method. The selected colony was grown in ISP-2 (International Streptomyces Project) broth on the rotary shaker

(140 rpm, Hahn-Shin, Bucheon, South Korea) at 28 °C, pH 7.2 for 14 days. The cells were harvested by centrifugation at 8000 rpm for 10 min and the pellet was washed twice with normal saline. Washed pellet was suspended in 10 mM Tris-HCl (pH 8, Merck, Burlington, VT, USA) and lysozyme (2.5 mg/Ml, Sigma, Burlington, VT, USA), incubated at 37 °C for 1 h and was re-suspended in lysis buffer (50 mM Tris, 10 mM ethylenediaminetetraacetic acid (EDTA, Qualigens Fine Chemicals Pvt. Ltd., San Diego, CA, USA), 1% sodium dodecyl sulfate (SDS, Sigma)) and proteinase K (1 mg/mL, Sigma) and incubated for 1 h at 50 °C. 400 µL of phenol (Tris-saturated, Himedia, New Delhi, India) was added and mixed vigorously for 2 min. After centrifugation, the upper aqueous layer was transferred to the fresh tube (Tarsons, Kolkata, India), followed by the addition of $CHCl_3$ (Sigma) and isoamyl alcohol (Merck) (24:1) and centrifuged at 1000 rpm for 15 min (4 °C). To the supernatant 50 µL of NaCl (5M) and twice the volume of absolute alcohol (Himedia) was added and kept for overnight incubation. Again, it was centrifuged at 14,000 rpm for 15 min and the pellet was washed with 70% alcohol. Pellet was air dried to remove traces of ethanol (EtOH, Himedia) and was suspended in 30 µL of Tris-EDTA (TE) buffer (Himedia). DNA was analyzed by 1% agarose gel electrophoresis (Bio-Rad, Hercules, CA, USA).

16S rRNA gene amplification was carried out using universal primer set, 27F (5′-AGAGTTTGA TCCTGGCTCAG-3′) and 1492R (5′-ACGGCTACCTTGTTACGACTT-3′). The PCR conditions were programmed as follows: Initial denaturation at 95 °C for 5 min; followed by 35 cycles at 95 °C for 1 min, primer annealing at 54 °C for 1 min, extension at 72 °C for 1 min. Final extension was done at 72 °C for 10 min and was kept for cooling at 10 °C. The amplified products were determined at 1.8% agarose gel electrophoresis. The sequence was compared with similar 16S rRNA sequences obtained from BLAST search in National Center for Biotechnology Information (NCBI) database and the phylogenetic tree was constructed by the neighbor joining tree algorithm using MEGA 7.0 software (Mega, Raynham, MA, USA).

### 2.4. Fermentation

Isolate *Streptomyces* sp.S2A was inoculated in ISP-2 broth (Himedia) and kept for incubation at the rotary shaker (140 rpm, 28 °C) for 14 days. The culture broth obtained was extracted thrice with ethyl acetate (EA, Fisher Scientific, Madison, WI, USA) and concentrated under the rotary evaporator at 50 °C.

### 2.5. Antimicrobial Assays

#### 2.5.1. Disc Diffusion Method

Antimicrobial activity of active fraction was assessed by disk diffusion method against *Staphylococcus epidermidis* (MTCC 435), *Staphylococcus aureus* (MTCC 740), *Bacillus cereus* (MTCC 1272), *Escherichia coli* (MTCC 40), *Klebsiella pneumoniae* (MTCC 661), *Micrococcus luteus* (MTCC 7950) *Aspergillus flavus* (MTCC 2590), *Fusarium moniliforme* (MTCC 6576), *Bipolaris maydis* and *Alternaria alternata* (MTCC 1362). The sterile discs (6 mm, Himedia) were impregnated with 30 µL of crude extract. The pathogens were inoculated in Mueller-Hinton broth (24 h for bacteria, Himedia) and Sabouraud Dextrose both (72 h for fungi, Himedia). The well-grown bacterial and fungal cultures were plated on Mueller-Hinton agar and Potato Dextrose agar respectively (Himedia). Sterile discs loaded with extract were placed on the plate. Chloramphenicol discs (Himedia) were used as positive control for antibacterial assay, while nystatin discs (Himedia) were used for the antifungal assay. Discs impregnated with dimethyl sulfoxide (DMSO, Himedia) were used as the solvent control. The plates were incubated at 37 °C and room temperature (for test bacteria and fungi respectively) and the zone of inhibition was measured.

#### 2.5.2. Determination of Minimum Inhibitory Concentration

The minimum inhibitory concentration (MIC) value of the partially purified extract (PPE) was determined by micro dilution method. Bacterial and fungal pathogens were grown in sterile broth and

10 μL of log phase culture was added into 96 well micro titre plates. Partially purified fractions were dissolved in 1% DMSO and serially diluted to give required concentrations (1 mg/mL–3.9 μg/mL). Diluted fractions and sterile broth were added into pre-coated microbial cultures, making up a total of volume of 200 μL. The plate was incubated at 37 °C and room temperature (for test bacteria and fungi respectively).

### *2.6. Antioxidant Assays*

#### 2.6.1. 2,2-diphenyl-1-picrylhydrazyl Radical Scavenging Activity (DPPH)

DPPH free radicals are highly stable and widely used to evaluate the radical scavenging activity of the antioxidants. Scavenging activity is based upon the reduction of DPPH radicals by hydrogen donating antioxidant compounds by forming DPPH-H. Radical scavenging activity of the ethyl acetate extract of the strain S2A was examined based on the previously described method by Ser et al. with minor changes [18]. Varying concentration of S2A extract was dissolved in methanol and reacted with freshly prepared DPPH solution (60 Mm, Sigma). The reaction mixture was incubated for 30 min in the dark. The absorbance was measured at 520 nm. Decreasing absorbance of DPPH solution indicates an increase in radical scavenging activity. The scavenging activity (%) was calculated using the following equation:

$$\text{DPPH scavenging activity (\%)} = [(A_o - A_1)/A_o] \times 100,$$

where $A_o$ is the absorbance of control (blank) and $A_1$ is the absorbance of the sample. Methanol was used as a blank whereas trolox (Sigma) was used as the reference compound [19].

#### 2.6.2. Metal Chelating Activity

The metal chelating activity was examined by measuring the ability of the compound to compete with ferrozine for $Fe^{2+}$, complex of which can be quantified spectrophotometrically. Metal chelating activity was measured by the method previously described by Adjimani and Asare with minor modifications [20]. Assay measures the reduction in the color intensity as a result of disruption of ferrous ion and ferrozine complexes. Briefly, varying concentration of extract was added to 0.15 mL of 2 mM $FeCl_2$. The reaction was initiated with the addition of 5 mM ferrozine (Sigma), followed by the incubation at the room temperature for 10 min. The absorbance was measured at 562 nm. The percentage of inhibition was calculated using the following equation:

$$\text{Metal chelating activity (\%)} = [(A_o - A_1)/A_o] \times 100,$$

where $A_o$ is the absorbance of control and $A_1$ is the absorbance of the sample. EDTA was used as a positive control.

#### 2.6.3. 2,2′-Azino-bis(3-ethylbenzothiazoline-6-sulfonic acid) Radical Scavenging Activity (ABTS)

ABTS (Sigma) scavenging activity is based upon the reduction of ABTS* radicals by compounds having lower redox potential than that of ABTS. The 2,2′-azino-bis(3-ethylbenzothiazoline-6-sulfonic acid) (ABTS) radical scavenging assay was carried out according to the method developed by Ser et al. [21]. Initially, ABTS stock solution (7 mM) was mixed with potassium persulfate (2.45 Mm, Himedia) to form ABTS cation complex for 12 h. The ABTS complex solution was added to varying concentrations of the extract preloaded in a 96-well microplate. The reaction was kept for incubation at room temperature for 20 min and the absorbance was measured at 734 nm. The percentage scavenging activity was calculated using the following formula:

$$\text{ABTS radical scavenging activity (\%)} = [(A_o - A_1)/A_o] \times 100,$$

where $A_o$ is the absorbance of control and $A_1$ is the absorbance of the sample. Trolox (Sigma) was used as a positive control.

2.6.4. Ferric Reducing Antioxidant Power (FRAP) Assay

This assay determined the reduction of ferric ions to ferrous ions which was monitored spectrophotometrically at 593 nm. The FRAP assay was performed according to the method previously described by Benzy and Strain with minor modification [22], based on the reduction of ferric complex to ferrous complex by the antioxidants. Initially, FRAP reagent was prepared by adding acetate buffer (pH 3.6), 10 mM TPTZ (Sigma) and 20 mM $FeCl_3$ (Himedia) at a ratio of 10:1:1. The reaction was started with the addition of varying concentration of extracts to the FRAP reagent. The mixture was then incubated at 37 °C for 10 min and absorbance was measured at 593 nm. Trolox was used as the positive control. The final FRAP values were expressed as Trolox equivalent antioxidant capacity (μM TE/g sample).

*2.7. Enzyme Inhibitory Activities*

2.7.1. Inhibition Assays for α-glucosidase Activity

The α-glucosidase inhibition was determined by the 96-well microtiter plate method based on the calorimetric assay as previously described by Vinholes et al. [23]. α-glucosidase enzyme solution (2U $mL^{-1}$, Sigma) was prepared in 100 mM phosphate buffer (pH 7.0). Ethyl acetate extracts were used in concentrations ranging from 10–100 μg $mL^{-1}$. 2 mM of *para*-nitrophenyl-α-D-glucopyranoside (Sigma) was prepared in 50 mM phosphate buffer (pH 7.0). 50 μL of the partially purified fraction was pre-incubated with an equal volume of yeast enzyme at 37 °C for 5 min, followed by the addition of 30 μL of pNPG and further incubation for 30 min. After incubation, 100 μL of stopping reagent (0.1 M $Na_2CO_3$) was added to cease the reaction. Color produced was quantified by UV spectrophotometer (Shimadzu, Kyoto, Japan) at 405 nm. Each experiment was performed in triplicate. Acarbose (Sigma) was used as a positive control, whereas purified fraction was replaced by phosphate buffer in control. Reaction mixture without enzyme was taken as blank. The percentage inhibition (%) was determined by the formula:

$$\text{percentage inhibition (\%)} = \frac{\text{absorbance of control} - \text{absorbance of sample}}{\text{absorbance of control}} \times 100$$

2.7.2. Inhibition Assays for α-amylase Activity

The α-amylase inhibition was determined by 96-well microtiter plate method based on calorimetric assay as previously described by Balasubramaniam et al. [24]. Equal volume of test samples (5 mg $mL^{-1}$) and α-amylase solution (0.5 mg $mL^{-1}$, Sigma) prepared in 30 mM phosphate buffer (pH 7.0) was pre- incubated at 37 °C for 10 min. 50 μL of 0.5% starch solution was added and incubated for 10 min at 37 °C. 120 μL of DNS reagent (Sigma) was added to stop the reaction. The reaction mixture was incubated at 95 °C for 5 min, cooled to room temperature. Absorbance was measured at 540 nm in a microplate reader. Acarbose at the concentration 2 mg $mL^{-1}$ was taken as positive control. The inhibition percentage of amylase was determined by the formula reported in the previous paragraph.

*2.8. Cytotoxicity Assay*

The human cell lines HT-29 (Colon cancer), MDA (Breast cancer) and U-87 MG (Brain cancer) were procured from National Centre for Cell Science, Pune, India. The cell lines were cultured and maintained in Dulbecco's modified Eagle's medium (DMEM, Sigma) in T-flasks in the incubator at 37 °C and internal atmosphere of 95% air and 5% $CO_2$. The cytotoxicity was determined by standard MTT dye assay, according to the method described by Carmichael et al. [25]. Briefly, varying concentration of extracts were dissolved in 1% DMSO and treated to cells seeded in 96 well tissue culture plates. The plates were kept for incubation at 37 °C for 24 h, MTT solution (Sigma) was added and incubated for 4 h at 37 °C. The amount of purple formazan crystals resulting from the reduction of MTT dye by succinic dehydrogenase in mitochondria of the viable cells was determined by measuring OD at 570 nm. The $IC_{50}$ value was calculated using graph pad prism. Each assay was performed in triplicate.

*2.9. Gas Chromatography-Mass Spectrometry (GC-MS)*

The analysis of the volatile constituents in extracts was determined by GC-MS technique (Perkin Elmer Clarus, USA). Perkin Elmer Clarus 680 employed a fused silica column, packed with Elite-5MS and the compounds were separated using helium as a carrier gas at a constant flow of 1 mL/min. The injector temperature was kept at 260 °C. Oven temperature was set as follows: 60 °C (2 min); followed by 300 °C at the rate of 10 °C $min^{-1}$. The spectrum thus obtained was compared with the database of the already known spectrum of components stored in GC-MS NIST library. The infrared spectrum of the extract was analyzed by FT-IR spectrophotometer in the range of 400–4000 $cm^{-1}$.

## 3. Results

*3.1. Isolation and Molecular Identification of the Strain*

Marine sediment from Gulf of Mannar was pre-treated with physical and chemical methods. Grown on SCA and ISP-2 medium, the cultural characteristics were identical on either of them. The aerial hyphae were white in color and substrate mycelium was colorless. SCA and ISP-2 culture plates did not show any pigment diffusion. Micromorphological studies of strain using SEM showed smooth spore ornamentation and rectiflexibilis spore morphology (Figure 1). The genomic DNA of the strain was isolated using the phenol-chloroform method and examined for 16S r-RNA sequence. The amplified sequences were subjected to BLAST analysis using the megablast tool of Genebank at NCBI under the accession number (KU921225). The BLAST search revealed that the strain belonged to *Streptomyces* sp. The highest similarity value index was found between the sequences of *Streptomyces* sp.S2A and *Streptomyces griesoruber* (100%). The neighbor-joining phylogenetic tree was drawn using MEGA 7.0 (Figure 2).

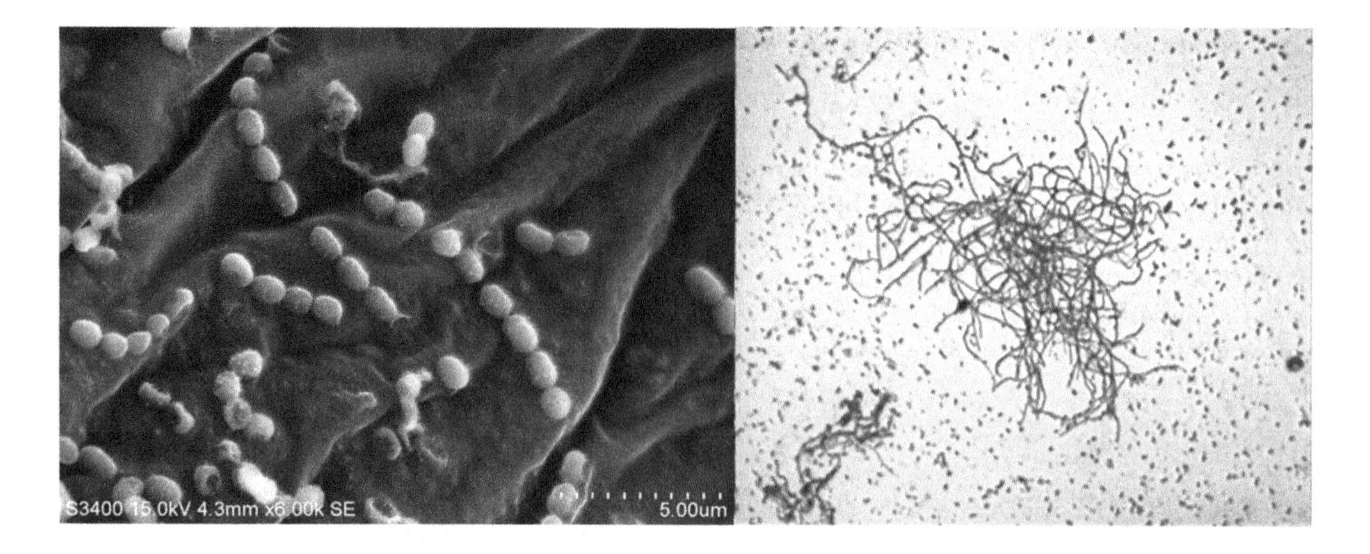


**Figure 1.** (**A**) Scanning electron micrograph showing spore ornamentation in *Streptomyces* sp.S2A; (**B**) Microscopic image of *Streptomyces* sp.S2A under 100×.

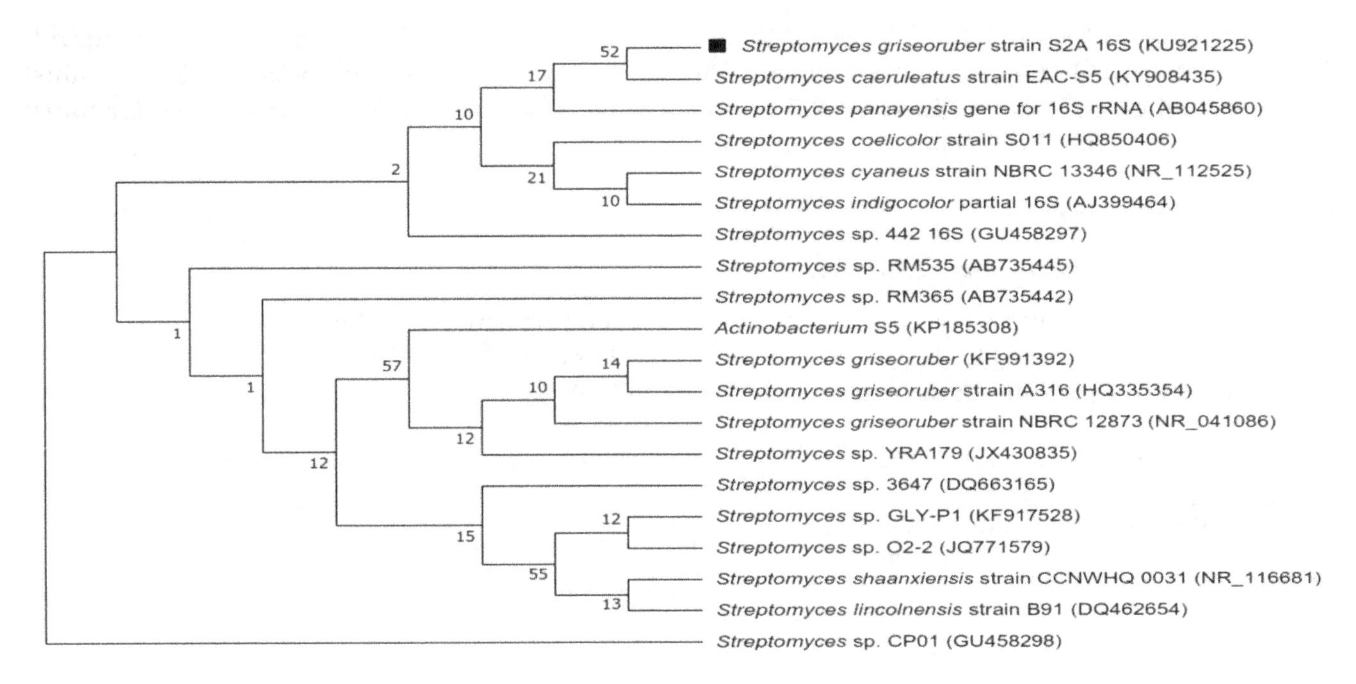


**Figure 2.** Phylogenetic tree of *Streptomyces* sp.S2A and the relationships with the closest species based on 16S rRNA gene sequencing using the neighbor-joining method.

### 3.2. Antimicrobial Assays

#### 3.2.1. Disc Diffusion Method

Antagonistic characteristics of the bioactive extract of *Streptomyces* sp.S2A showed potent antagonistic activity against bacterial and fungal pathogens (Table 1). Of six bacterial pathogens, the highest inhibition activity was manifested against *Micrococcus luteus* and *Staphylococcus epidermidis* (16 mm). Susceptibility of *Bacillus cereus, Klebsiella pneumoniae* and *Staphylococcus aureus* to bioactive compounds was highly noticeable (14 mm). *Escherichia coli* was less susceptible to the compound (10 mm). Among fungal pathogens, reduction in mycelial growth was not seen against *Aspergillus flavus* and *Alternaria alternata* whereas inhibitory activity was significantly observed against *Fusarium moniliforme* and *Bipolaris maydis* (See Supplementary Figure S1).

**Table 1.** Antimicrobial activity and MIC (μg/mL) of *Streptomyces* sp.S2A by broth dilution method.

| Test Microorganisms | Zone of Inhibition (mm) | | MIC (μg/mL) |
|---|---|---|---|
| **Bacteria** | **Extract** | **Antibiotics (Chloramphenicol)** | |
| *Klebsiella pneumoniae* MTCC 661 | 14 ± 0.4 | 30 ± 1.1 | 31.25 |
| *Micrococcus luteus* MTCC 7950 | 16 ± 0.8 | 28 ± 1.6 | 7.81 |
| *Escherichia coli MTCC* 40 | 10 ± 0.8 | 22 ± 1.9 | 15.62 |
| *Bacillus cereus* MTCC 1272 | 14 ± 1.2 | 25 ± 1.1 | 15.62 |
| *Staphylococcus epidermidis* MTCC 435 | 16 ± 0.4 | 23 ± 1.8 | 15.62 |
| *Staphylococcus aureus* MTCC 740 | 14 ± 0.8 | 24 ± 0.8 | 15.62 |
| **Fungi** | | **(Nystatin)** | |
| *Aspergillus flavus* MTCC 2590 | - | - | - |
| *Bipolaris maydis* | 14 ± 1.2 | 20±1.2 | 31.25 |
| *Alternaria alternata* MTCC 1362 | - | - | - |
| *Fusarium moniliforme* MTCC 6576 | 18 ± 1.2 | 22±1.0 | 7.81 |

#### 3.2.2. Determination of Minimum Inhibitory Concentration (MIC)

The minimum inhibitory concentration of the extract was determined as 31.25 μg/mL against *Klebsiella pneumoniae*, 15.62 μg/mL against *Staphylococcus epidermidis, Staphylococcus aureus,*

*Bacillus cereus* and *Escherichia coli*. Lowest MIC was observed against *Micrococcus luteus* as 7.8 μg/mL. The solvent DMSO (1%) had no significant inhibitory activity against pathogens. MIC against fungal pathogens was determined as 62.5 μg/mL against *Bipolaris maydis* and 15.62 μg/mL against *Fusarium moniliforme*. (Table 1).

### 3.3. *Antioxidant Assays*

#### 3.3.1. DPPH Radical Scavenging Activity

The highest inhibition concentration of radical scavenging activity of the extract was found to be 56.55 ± 3.1%, as compared to the trolox that was found to be 74.73 ± 1.13%. The $IC_{50}$ value for DPPH radical scavenging activity of the extract was 0.86 mg (Table 2).

**Table 2.** Radical scavenging activity of ethyl acetate extract of *Streptomyces* sp.S2A.

| Antioxidant Assays | Concentration of Extract (mg/mL) | % Inhibition | Absorbance | $IC_{50}$ (mg/mL) |
|---|---|---|---|---|
| DPPH | 1.0 | 56.55 ± 3.1 | - | 0.86 |
| | 0.50 | 32.33 ± 1.4 | - | |
| | 0.25 | 17.29 ± 1.6 | - | |
| Metal chelating | 2.0 | 59.98 ± 2.12 | - | 1.56 |
| | 1.0 | 37.50 ± 2.36 | - | |
| | 0.50 | 24.90 ± 2.11 | - | |
| | 0.25 | 18.40 ± 1.4 | - | |
| ABTS | 0.10 | 42.48 ± 3.1 | - | 0.011 |
| | 0.05 | 30.24 ± 3.74 | - | |
| | 0.02 | 7.29 ± 3.62 | - | |
| FRAP | 0.1 | - | 0.248 | - |
| | 0.08 | - | 0.202 | |
| | 0.06 | - | 0.145 | |
| | 0.04 | - | 0.060 | |
| | 0.02 | - | 0.028 | |

#### 3.3.2. Metal Chelating Activity

The study showed the decrease in the formation of ferrozine-$Fe^{2+}$ complex with increase in the concentration of the extract. It showed the significant chelating activity measuring from 18.40 ± 1.4% to 59.98 ± 2.12% at the concentration ranging from 0.25–2 mg/mL (Table 2).

#### 3.3.3. ABTS Radical Scavenging Activity

This assay showed the significant increase in the scavenging activity with increase in the concentration of the extract, thus decolorized the blue-green color of ABTS* back into ABTS, which is colorless. The $IC_{50}$ value for ABTS radical scavenging activity of the extract was 11.77 μg (Table 2).

#### 3.3.4. Ferric Reducing Antioxidant Power (FRAP) Assay

The increase in absorbance was observed with the increase in the concentration of the extract suggesting the significant antioxidant activity (Table 2).

### 3.4. *In Vitro Enzyme Inhibition Assay*

The EA extract exhibited α-amylase and α-glucosidase inhibitory activity in a dose-dependent manner. Acarbose was used as a standard. $IC_{50}$ value of EA extract for α-glucosidase and α-amylase inhibition was found to be 21.17 and 20.46 respectively, whereas that of acarbose was 15.47 and 18.15 μg/mL respectively (Tables 3 and 4).

**Table 3.** α-glucosidase inhibition and $IC_{50}$ values of ethyl acetate extract of *Streptomyces* sp.S2A.

| Concentration (μg/mL) | Inhibition % (EA Extract) | $IC_{50}$ (μg/mL) (EA Extract) | Inhibition % (Acarbose) | $IC_{50}$ (μg/mL) (Acarbose) |
|---|---|---|---|---|
| 6.25 | 29.12 ± 0.33 | | 36.44 ± 0.58 | |
| 12.5 | 38.54 ± 0.77 | | 45.27 ± 0.34 | |
| 25 | 55.1 ± 1.16 | 21.17 | 62.19 ± 1.10 | 15.47 |
| 50 | 68.4 ± 1.55 | | 78.52 ± 1.99 | |
| 100 | 72.31 ± 1.01 | | 86.83 ± 2.01 | |
| 200 | 81.74 ± 2.65 | | 94.22 ± 2.33 | |

**Table 4.** α-amylase inhibition and $IC_{50}$ values of ethyl acetate extract of *Streptomyces* sp.S2A.

| Concentration (μg/mL) | Inhibition % (EA Extract) | $IC_{50}$ (μg/mL) (EA Extract) | Inhibition % (Acarbose) | $IC_{50}$ (μg/mL) (Acarbose) |
|---|---|---|---|---|
| 6.25 | 16.44 ± 0.21 | | 20.19 ± 0.78 | |
| 12.5 | 34.77 ± 0.44 | | 40.05 ± 0.10 | |
| 25 | 59.29 ± 1.15 | 20.46 | 64.44 ± 1.45 | 18.15 |
| 50 | 74.32 ± 1.09 | | 87.57 ± 1.33 | |
| 100 | 81.13 ± 1.34 | | 97.03 ± 1.10 | |
| 200 | 88.67 ± 1.93 | | 97.84 ± 1.78 | |

### 3.5. Cytotoxicity Assay

The tested results of the extract against cell lines were shown in (See Supplementary Figure S2). The results revealed that the extract showed varying efficacy against cell lines. The highest activity against U-87 at 100 μg/mL was found to be 59.63 ± 1.9%. It also showed significant activity against MDA and HT-29 with cell inhibition was found to be 55.23 ± 1.09% and 52.31 ± 2.4% respectively at 100 μg/mL (Table 5). Overall, the results suggested the potential cytotoxic activity against various cell lines.

**Table 5.** Cytotoxic activity of extract of *Streptomyces* sp.S2A against HT-29, MDA and U-87 MG.

| Concentration (μg/mL) | Inhibition % | | |
|---|---|---|---|
| | U-87 MG | MDA | HT-29 |
| 5 | 13.76 ± 1.81 | 3.57 ± 1.76 | 18.51 ± 3.89 |
| 10 | 16.51 ± 2.01 | 10.71 ± 3.75 | 21.76 ± 2.32 |
| 20 | 19.26 ± 3.79 | 15.0 ± 4.10 | 23.15 ± 1.96 |
| 50 | 36.19 ± 2.11 | 30.95 ± 2.87 | 35.31 ± 2.77 |
| 100 | 59.63 ± 1.90 | 55.23 ± 1.09 | 52.31 ± 2.40 |
| $IC_{50}$ (μg/mL) | 93.32 | 80.02 | 88.68 |

### 3.6. Gas Chromatography-Mass Spectrometry (GC-MS)

Analysis of components of the active fraction with the highest activity by GC-MS analysis implied nine peaks at the retention time of (i) 17.239; (ii) 17.309; (iii) 20.811; (iv) 21.311; (v) 21.406; (vi) 21.586; (vii) 22.071; (viii) 22.126; (ix) 24.257. Further examination of MS peaks revealed $m/z$ at 168, 259, 210 and 350. According to NIST library search, peak retentions at 21.311, 21.406 and 21.586 correspond to single compound i.e., pyrrolo[1–a]pyrazine-1,4-dione,hexahydro-3-(2-methylpropyl) (Figure 3). Other tentatively identified compounds were diphenylmethane, 2-Isopropyl-1-Phenyl-3-Pyrrolidin-1-yl Propane-1,3-Dione and Benzene, 1′1-tetradecyclidenebis (The GC-MS spectrum indicated the ions at 70 and 154 corresponded to molecule $C_7H_{10}NO_2$ and $C_4H_8$ ions (See Supplementary Figures S3 and S4). The spectrum was similar to that of pyrrolo[1–a]pyrazine-1,4-dione,hexahydro-3-(2-methylpropyl) spectra in the GC-MS library and a study reported by Yang et al. [26]. The FT-IR spectrum of the partially purified metabolite showed the characteristic functional groups such as NH stretching peak of a primary amine at 3313.20 $cm^{-1}$. The functional group at 2923.91 $cm^{-1}$ corresponded to strong C-H stretching in alkanes. The peak at 1646.10 $cm^{-1}$ was assigned to C-N stretch in primary amine. The absorption peak at 1516.05 $cm^{-1}$ was assigned to C=C stretch in an alkene. The peak at 1454.69 $cm^{-1}$ was assigned to C-H bend in alkanes. The peak at 1240.27 $cm^{-1}$ was assigned to C-O stretch. The absorption peak at 1033.54 $cm^{-1}$ was assigned to the ether. The peak ranging from 605.8 $cm^{-1}$ to 701.09 $cm^{-1}$ was assigned to strong C-H bend in alkenes (Figure 4).

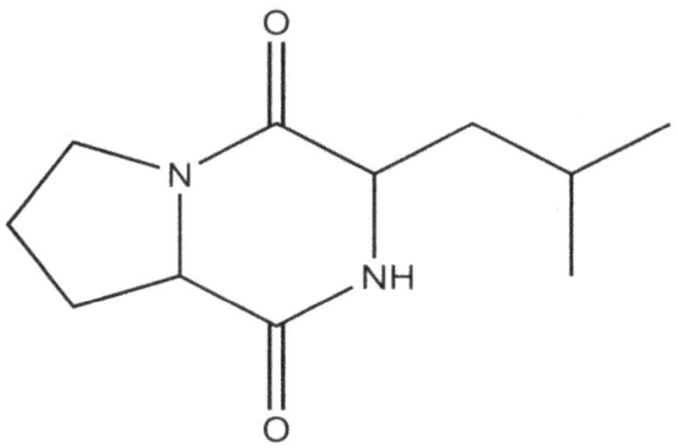

**Figure 3.** Chemical structure of the compound pyrrolo[1–a]pyrazine-1,4-dione,hexahydro-3-(2-methylpropyl).

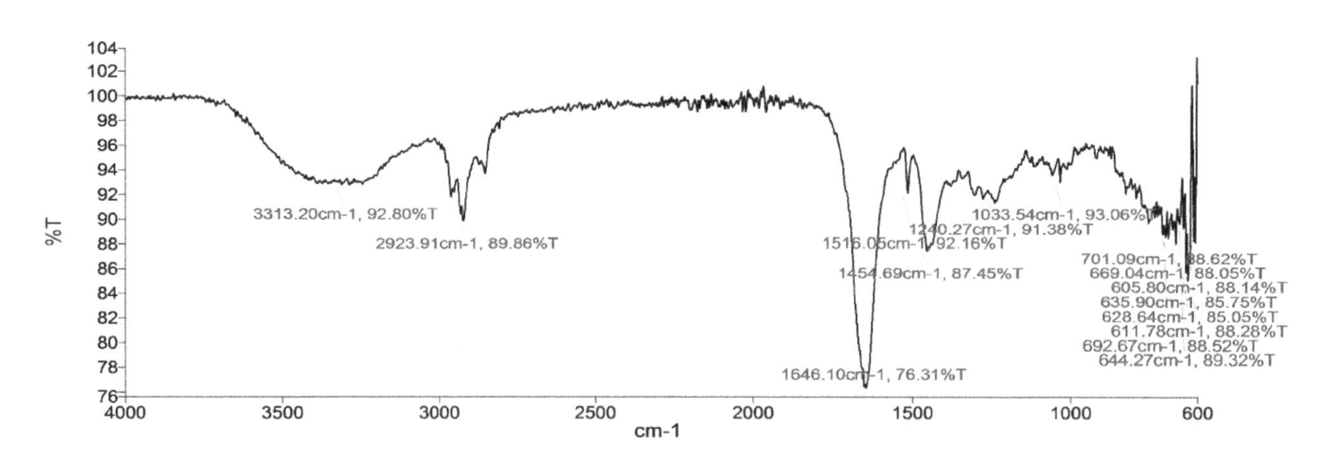


**Figure 4.** FT-IR spectrum of the active extract of *Streptomyces* sp.S2A.

## 4. Discussion

With this outlook, the present investigation was carried out to identify bioactive compound from marine actinobacteria exhibiting antagonistic activity against bacterial and fungal pathogens, enzyme inhibition activity, antioxidant and cytotoxic activity. *Streptomyces* sp.S2A isolated from marine sediment of Gulf of Mannar produced white aerial mycelium to colorless substrate mycelium on SCA medium. Shirling and Gottileb [27] reported that the pigmentation paradigm could be used for the classification and identification. However, there was no pigment pattern observed with the isolate. Extraction of metabolites with ethyl acetate yielded the dark color residue. Purification by silica gel column chromatography resulted in six fractions, of which one fraction exhibited significant activity. The antagonistic activity of the bioactive compound showed the high zone of inhibition against *Micrococcus luteus* and *Staphylococcus epidermidis* (16 mm). Moderate activity was observed against *Bacillus cereus, Klebsiella pneumoniae* and *Staphylococcus aureus* (14 mm). Susceptibility of *Escherichia coli* to the compound was found to be weaker (10 mm). Inhibitory potential of the bioactive compound against fungal pathogens showed good activity against *Fusarium moniliforme* and *Bipolaris maydis*, whereas no zone of inhibition was observed against *Aspergillus flavus* and *Alternaria aleternata*. MIC was used to determine the efficacy of the compound at different concentration. The reduction in the growth of bacterial pathogens was observed with the concentration ranging from 7.8–31.25 μg/mL. MIC of the compound showed excellent antifungal activity against *Fusarium moniliforme* and *Bipolaris maydis*. Ethyl acetate extract of *Streptomyces* sp.S2A also showed significant α-glucosidase and α-amylase inhibition activity, though $IC_{50}$ of the extract were less than that of acarbose. Antioxidant and cytotoxic activities of the extract was determined by the DPPH, ABTS, FRAP and metal chelating assays and against HT-29, MDA and U-87 MG cell lines. The results thus obtained showed the presence of potent antioxidant and anticancer agents.

The partial chemical composition relates to the metabolite was detected by GC-MS. The chromatogram of fraction A43 showed a total of nine peaks. Of all, the major constituent was pyrrolo[1–a]pyrazine-1,4-dione,hexahydro-3-(2-methylpropyl)-, constituted 80.7% and this may be the active principle compound. The other chemical compound was identified as diphenylmethane (6%), 2-Isopropyl-1-Phenyl-3-Pyrrolidin-1-yl-Propane-1,3-Dione (2%) and Benzene, 1′1-tetradecyclidenebis (2%). Pyrrolo[1–a]pyrazine-1,4-dione,hexahydro-3-(2-methylpropyl is a peptide derivative of diketopiperazine with the molecular weight as 210 and empirical formula as $C_{11}H_{18}N_2O_2$. All the bioassays mentioned in the above paragraph were confirmed with the commercially available purified compound: pyrrolo[1–a]pyrazine-1,4-dione,hexahydro-3-(2-methylpropyl). Antimicrobial and the cytotoxic activity of the purified compound were found to be significantly higher the partially purified compound, whereas the enzyme inhibition potential of the partially purified compound against α-glucosidase and α-amylase were better in comparison with commercial compound. The mass spectrum of the commercial compound corresponds to 70.0315 *m/z* and 154.0152, which is same as shown by the compound present in partially purified extract (See Supplementary Figures S5 and S6, Tables S1–S3).

Pyrrolopyrazines are known for their wide range of biological activities such as antioxidant, anti-angiogenesis, anti-tumor and antimicrobial [28]. Manimaram et al. reported the presence of antibacterial metabolite, pyrrolo[1–a]pyrazine-1,4-dione,hexahydro-3-(2-methylpropyl) in the crude extract of *Streptomyces* sp. VITMK1 isolated from mangrove soil [29]. Antifouling potential of pyrrolo[1–a]pyrazine-1,4-dione,hexahydro-3-(2-methylpropyl) against *Vibrio halioticoli* and *Loktanella honkongensis* was studied by Dash et al. [30]. Sponge-derived marine bacteria significantly inhibited the larval settlement of *Balanus amphitrite* and *Hydroides elegans*. Another marine bacteria isolated from the sponge, *Spongia officinalis* showed the potent antibacterial and antifungal activity of pyrrolo[1–a]pyrazine-1,4-dione,hexahydro-3-(2-methylpropyl) [31]. Diketopiperzines derivatives present in marine *Streptomyces* sp. had shown good anti-H1N1 activity [32]. Mithun and Rao also reported the presence of pyrrolopyrazines in *Micrococcus luteus* with anti-cancer activity against HCT-15 cell line [33]. Presence of pyrrolo[1–a]pyrazine-1,4-dione,hexahydro-3-(2-methylpropyl) was detected in *Streptomyces* sp. MUM 256 isolated from the mangrove forest in Malaysia. This compound was reported to possess antioxidant and anticancer activities [34]. Anti-cancer metabolites were also reported from *Streptomyces malaysiense* sp.MUSC 136 isolated from the mangrove ecosystem. The bioactive metabolite exhibited strong antioxidant activity and high cytotoxic activity against HCT-116 cells [35]. The first report on marine *Staphylococcus* sp. derived pyrrolo[1–a]pyrazine-1,4-dione,hexahydro-3-(2-methylpropyl) was reported by Lalitha et al. [36]. Purified metabolite was potentially active against lung (A549) and cervical (HeLa) cancer cells in a dose-dependent manner. Thus, the present study suggested that the pyrrolopyrazines derivative pyrrolo[1–a]pyrazine-1,4-dione,hexahydro-3-(2-methylpropyl) may account for the observed antagonistic, antioxidant, and cytotoxic activities in marine actinobacteria, *Streptomyces* sp.S2A. The results obtained in the current study demonstrate that bioactive metabolites produced by marine actinobacteria have tremendous potential for pharmaceutical product and are a subject of future investigation.

**Author Contributions:** Conceptualization, S.S. and R.R.V.; Methodology, S.S. and R.R.V.; Software, S.S. and R.R.V.; Validation, S.S. and R.R.V.; Formal Analysis, S.S. and R.R.V.; Investigation, S.S.; Resources, R.R.V.; Data curation, Writing—Original Draft Preparation, S.S. and R.R.V.; Writing—Review & Editing, S.S. and R.R.V.; Visualization, S.S. and R.R.V.; Supervision, R.R.V.; Project Administration, R.R.V.; Funding Acquisition, R.R.V.

**Acknowledgments:** The authors would like to thank SIF, VIT University, Vellore for carrying out GC-MS analysis. The authors would also like to acknowledge the facilities provided by DST-PURSE program, DST, New Delhi to

## References

1. Ganesan, A. The impact of natural products upon drug discovery. *Curr. Opin. Chem. Biol.* **2008**, *12*, 306–317. [CrossRef] [PubMed]
2. Chin, Y.-W.; Balunas, M.J.; Chai, H.B.; Kinghorn, A.D. Drug discovery from natural sources. *AAPS J.* **2006**, *8*, 239–253. [CrossRef]
3. Berdy, J. Bioactive microbial metabolites. *J. Antibiot.* **2005**, *58*, 1–26. [CrossRef] [PubMed]
4. Mann, J. Natural products as immunosuppressive agents. *Nat. Prod. Rep.* **2001**, *18*, 417–430. [CrossRef] [PubMed]
5. Imada, C. Enzyme inhibitors and other bioactive compounds from marine actinomycetes. *Antonie Leeuwenhoek* **2005**, *87*, 59–63. [CrossRef] [PubMed]
6. Naine, J.; Srinivasan, M.V.; Devi, S.C. Novel anticancer compounds from marine actinomycetes: A review. *J. Pharm. Res.* **2011**, *4*, 1285–1287.
7. Williams, S.T.; Goodfellow, M.; Wellington, E.M.H.; Vicker, J.C.; Alderson, G.; Sneath, P.H.A.; Sackin, M.J.; Mortimer, A.M. A probability matrix for identification of some streptomycetes. *Microbiology* **1983**, *129*, 1815–1830. [CrossRef] [PubMed]
8. Ventura, M.; Canchaya, C.; Tauch, A.; Chandra, G.; Fitzgerald, G.F.; Chater, K.F.; Van Sinderen, D. Genomics of *Actinobateria*: Tracing the evolutionary history of an ancient phylum. *Microbiol. Mol. Biol. Rev.* **2007**, *71*, 495–548. [CrossRef] [PubMed]
9. Goodfellow, M.; Williams, S.T. Ecology of actinomycetes. *Annu. Rev. Microbiol.* **1983**, *37*, 189–216. [CrossRef] [PubMed]
10. Bull, A.T.; Stach, J.E. Marine actinobacteria: New opportunities for natural product search and discovery. *Trends Microbiol.* **2007**, *15*, 491–499. [CrossRef] [PubMed]
11. Fiedler, H.P.; Bruntner, C.; Bull, A.T.; Ward, A.C.; Goodfellow, M.; Potterat, O. Marine actinomycetes as a source of novel secondary metabolites. *Antonie Leeuwenhoek* **2004**, *87*, 37–42. [CrossRef] [PubMed]
12. Shivlata, L.; Satyanarayana, T. Thermophilic and alkaliphilic *Actinobacteria*: Biology and potential applications. *Front. Microbiol.* **2015**, *6*, 1014. [CrossRef] [PubMed]
13. Miao, V.; Davies, J. Actinobacteria: The good, the bad and the ugly. *Antonie Leeuwenhoek* **2010**, *98*, 143–150. [CrossRef] [PubMed]
14. Jensen, P.R.; Mincer, T.J.; Williams, P.G.; Fenical, W. Marine actinomycete diversity and natural product discovery. *Antonie Leeuwenhoek* **2005**, *87*, 43–48. [CrossRef] [PubMed]
15. Subramani, R.; Aalbersberg, W. Marine actinomycetes: An ongoing source of novel bioactive metabolites. *Microbiol. Res.* **2012**, *167*, 571–580. [CrossRef] [PubMed]
16. Sharma, S.R.; Shah, G.S. Isolation and screening of actinomycetes for bioactive compounds from the marine coast of South-Gujarat Region. *Int. J. Res. Sci. Innov.* **2014**, *1*, 345–349.
17. Saadoun, I.; Hameed, K.M.; Moussauui, A. Characterization and analysis of antibiotic activity of some aquatic actinomycetes. *Microbios* **1999**, *99*, 173–179. [PubMed]
18. Ser, H.-L.; Palanisamy, U.D.; Yin, W.-F.; Malek, A.; Nurestri, S.; Chan, K.-G. Presence of antioxidative agent, Pyrrolo[1,2a] pyrazine-1,4-dione, hexahydro-in newly isolated *Streptomyces mangrovisoli* sp. nov. *Front. Microbiol.* **2015**, *6*, 854. [CrossRef] [PubMed]
19. Mišan, A.; Mimica-Dukić, N.; Sakač, M.; Mandić, A.; Sedej, I.; Šimurina, O.; Tumbas, V. Antioxidant activity of medicinal plant extracts in cookies. *J. Food Sci.* **2011**, *76*, 1239–1244. [CrossRef] [PubMed]
20. Adjimani, J.P.; Asare, P. Antioxidant and free radical scavenging activity of iron chelators. *Toxicol. Rep.* **2015**, *2*, 721–728. [CrossRef] [PubMed]
21. Ser, H.-L.; Tan, L.T.-H.; Palanisamy, U.D.; Abd Malek, S.N.; Yin, W.-F.; Chan, K.G. *Streptomyces antioxidans* sp. nov., a novel mangrove soil actinobacterium with antioxidative and neuroprotective potentials. *Front. Microbiol.* **2016**, *7*, 899. [CrossRef] [PubMed]

22. Benzie, I.F.F.; Strain, J.J. The ferric reducing ability of plasma (FRAP) as a measure of antioxidant power: The FRAP assay. *Anal. Biochem.* **1996**, *239*, 70–76. [CrossRef] [PubMed]
23. Vinholes, J.; Grosso, C.; Andrade, P.B.; Gil-Izquierdo, A.; Valentao, P.; de Pinho, P.G.; Ferreres, F. In vitro studies to assess the antidiabetic: Anti-cholinesterase and antioxidant potential of *Spergularia rubra*. *Food Chem.* **2011**, *129*, 454–462. [CrossRef]
24. Balasubramaniam, V.; Mustar, S.; Khalid, N.M.; Rashed, A.A.; Noh, M.F.M.; Wilcox, M.D.; Chater, P.I.; Brownlee, I.A.; Pearson, J.P. Inhibitory activities of three Malaysian edible seaweeds on lipase and alpha-amylase. *J. Appl. Phycol.* **2013**, *25*, 1405–1412. [CrossRef]
25. Carmichael, J.; DeGraff, W.G.; Gazdar, A.F.; Minna, J.D.; Mitchell, J.B. Evaluation of a tetrazolium-based semiautomated colorimetric assay, assessment of chemosensitivity testing. *Cancer Res.* **1987**, *47*, 936–942. [PubMed]
26. Guo, X.; Liu, X.; Yang, H. Synergistic algicidal effect and mechanism of two diketopiperazines produced by *Chryseobacterium* sp. strain GLY-1106 on the harmful bloom-florming *Microcystis aeruginosa*. *Sci. Rep.* **2015**, *5*, 14720. [CrossRef] [PubMed]
27. Shirling, E.B.; Gottileb, D. Methods for characterization of *Streptomyces* species. *Int. J. Syst. Bactriol.* **1966**, *16*, 312–340. [CrossRef]
28. Wang, C. Antifungal activity of volatile organic compounds from *Streptomyces alboflavus* TD-1. *FEMS Microbiol. Lett.* **2013**, *341*, 45–51. [CrossRef] [PubMed]
29. Manimaran, M.; Gopal, J.V.; Kannabiran, K. Antibacterial activity of *Streptomyces* sp. VITMK1 isolated from mangrove soil of Pichavaram, Tamil Nadu, India. *Proc. Natl. Acad. Sci. USA India Sect. B Biol. Sci.* **2015**, *87*, 499–506. [CrossRef]
30. Dash, S.; Jin, C.; Lee, O.O.; Xu, Y.; Qian, P. Antibacterial and antilarval-settlement potential and metabolite profiles of novel sponge-associated marine bacteria. *J. Ind. Microbiol. Biotechnol.* **2009**, *36*, 1047–1056. [CrossRef] [PubMed]
31. Sathiyanarayanan, G.; Gandhimathi, R.; Sabarathnam, B.; Kiran, G.S.; Selvin, J. Optimization and production of pyrrolidone antimicrobial agent from marine sponge-associated *Streptomyces* sp. MAPS15. *Bioprocess Biosyst. Eng.* **2014**, *37*, 561–573. [CrossRef] [PubMed]
32. Wang, P.; Xi, L.; Liu, P.; Wang, Y.; Wang, W.; Huang, Y.; Zhu, W. Diketopiperazine derivatives from the marine-derived actinomycete *Streptomyces* sp. FXJ7.328. *Mar. Drugs* **2013**, *11*, 1035–1049. [CrossRef] [PubMed]
33. Mithun, V.S.L.; Rao, C.S.V. Isolation and molecular characterization of anti-cancerous compound producing marine bacteria by using 16S rRNA sequencing and GC-MS techniques. *IJMER* **2012**, *2*, 4510–4515.
34. Tan, L.T.H.; Ser, H.L.; Yin, W.F.; Chan, K.G.; Lee, L.H.; Goh, B.H. Investigation of antioxidative and anticancer potentials of *Streptomyces* sp. MUM256 isolated from Malaysia mangrove soil. *Front. Microbiol.* **2015**, *6*, 1316. [CrossRef] [PubMed]
35. Ser, H.L.; Palanisamy, U.D.; Yin, W.F.; Chan, K.G.; Goh, B.H.; Lee, L.H. *Streptomyces malaysiense* sp. nov.: A novel Malaysian mangrove soil actinobacterium with antioxidative activity and cytotoxic potential against human cancer cell lines. *Sci. Rep.* **2016**, *6*, 24247. [CrossRef] [PubMed]
36. Lalitha, P.; Veena, V.; Vidhyapriya, P.; Lakshmi, P.; Krishna, R.; Sakthivel, N. Anticancer potential of pyrrole (1, 2, a) pyrazine 1, 4, dione, hexahydro 3-(2-methyl propyl) (PPDHMP) extracted from a new marine bacterium, *Staphylococcus* sp. strain MB30. *Apoptosis* **2016**, *21*, 566–577. [CrossRef] [PubMed]

# 9

# A *Streptomyces* sp. NEAU-HV9: Isolation, Identification, and Potential as a Biocontrol Agent against *Ralstonia solanacearum* of Tomato Plants

**Ling Ling [1], Xiaoyang Han [1], Xiao Li [1], Xue Zhang [1], Han Wang [1], Lida Zhang [1], Peng Cao [1], Yutong Wu [1], Xiangjing Wang [1], Junwei Zhao [1,*] and Wensheng Xiang [1,2,*]**

[1] Key Laboratory of Agricultural Microbiology of Heilongjiang Province, Northeast Agricultural University, No. 59 Mucai Street, Xiangfang District, Harbin 150030, China; LLYNL2621161093@163.com (L.L.); hanxy139251@163.com (X.H.); Lx1244070003@126.com (X.L.); zhangxue_968425@163.com (X.Z.); wanghan507555536@gmail.com (H.W.); yone910310@163.com (L.Z.); cp511@126.com (P.C.); 18103699151@163.com (Y.W.); wangneau2013@163.com (X.W.)

[2] State Key Laboratory for Biology of Plant Diseases and Insect Pests, Institute of Plant Protection, Chinese Academy of Agricultural Sciences, Beijing 100193, China

[*] Correspondence: guyan2080@126.com (J.Z.); xiangwensheng@neau.edu.cn (W.X.)

**Abstract:** *Ralstonia solanacearum* is an important soil-borne bacterial plant pathogen. In this study, an actinomycete strain named NEAU-HV9 that showed strong antibacterial activity against *Ralstonia solanacearum* was isolated from soil using an in vitro screening technique. Based on physiological and morphological characteristics and 98.90% of 16S rRNA gene sequence similarity with *Streptomyces panaciradicis* 1MR-8$^T$, the strain was identified as a member of the genus *Streptomyces*. Tomato seedling and pot culture experiments showed that after pre-inoculation with the strain NEAU-HV9, the disease occurrence of tomato seedlings was effectively prevented for *R. solanacearum*. Then, a bioactivity-guided approach was employed to isolate and determine the chemical identity of bioactive constituents with antibacterial activity from strain NEAU-HV9. The structure of the antibacterial metabolite was determined as actinomycin D on the basis of extensive spectroscopic analysis. To our knowledge, this is the first report that actinomycin D has strong antibacterial activity against *R. solanacearum* with a MIC (minimum inhibitory concentration) of 0.6 mg $L^{-1}$ (0.48 μmol $L^{-1}$). The in vivo antibacterial activity experiment showed that actinomycin D possessed significant preventive efficacy against *R. solanacearum* in tomato seedlings. Thus, strain NEAU-HV9 could be used as BCA (biological control agent) against *R. solanacearum*, and actinomycin D might be a promising candidate for a new antibacterial agent against *R. solanacearum*.

**Keywords:** antibacterial activity; *Ralstonia solanacearum*; *Streptomyces* sp. NEAU-HV9; actinomycin D

---

## 1. Introduction

Tomato is one of the world's most important vegetable crops, with a global annual yield of approximately 160 million tons [1,2]. In China, long term continuous cropping is the main planting practice for tomato, which has led to serious soilborne diseases [3]. *Ralstonia solanacearum* [4] is an important soilborne bacterial plant pathogen [5]. Bacterial wilt caused by *R. solanacearum* is a serious and common disease, which reduces the yield of tomato and many other crops in tropical, subtropical, and warm-temperature regions of the world [6]. Because of worldwide distribution and a large host range of more than 200 plant species in 50 families, including pepper, tomato, tobacco, potato, peanut, and banana, this soil bacterium has been recognized as one of the causative agents of bacterial wilt disease and is one of the leading models in pathogenicity [5]. In the absence of host plants, this bacterium can be free-living as a saprophyte in the soil or in water [7]. Plant breeding, field sanitation, crop rotation, and use of bactericides have met with only limited success for *R. solanacearum* [8]. Furthermore, pathogenic microbial multi-drug resistance is also increasing. Therefore, new natural resources and antibiotics for suppressing this soilborne disease are needed.

Various recent studies have showed that biological control of bacterial wilt disease could be achieved using antagonistic bacteria [8,9]. The suppressive effect of some antagonistic bacteria on *R. solanacearum* was reported by Toyota and Kimura [10]. Moreover, the use of antagonistic bacteria to be effective in control of *R. solanacearum* has been proved by Ciampi-Panno et al. under field conditions [8]. *Streptomycetes* are gaining interest in agriculture as plant growth promoting (PGP) bacteria and/or biological control agents (BCAs) [11,12]. The *Streptomyces* genus comprises Gram-positive bacteria which show a filamentous form; they can grow in various environments. Several *Streptomyces* species such as *S. aureofaciens*, *S. avermitilis*, *S. lividans*, *S. humidus*, *S. hygroscopicus*, *S. lydicus*, *S. plicatus*, *S. olivaceoviridis*, *S. roseoflavus*, *S. scabies* and *S. violaceusniger* have been used to control soilborne diseases due to their greatly antagonistic activities by production of various antimicrobial substances [13–15]. *Actinobacteria* are famous for producing a variety of natural bioactive metabolites. *Streptomyces* is an important source of bioactive compounds among all members of antibiotic production, accounting for two-thirds of commercially available antibiotics [16]. Actinomycins belonging to a family of chromopeptide lactones are produced by various *Streptomyces*. Among several antibiotics produced by this genus, actinomycins are prominent. More than 20 naturally-occurring actinomycins were isolated and observed to have commonality of two pentapeptidolactone moieties with an actnoyl chromophore [17]; however, they differ in functional and/or positional group. Among actinomycins, actinomycin D has been widely studied and used clinically as an anticancer drug, especially in the treatment of childhood rhabdomyosarcoma, infantile kidney tumors and several other malignant tumors [18,19]. However, no reports have been published on actinomycin D against phytopathogen *R. solanacearum*.

In the existing protocol for virulence assays, one-month old tomato plantlets are soil-inoculated with the bacterium and wilting symptoms, if any, are observed and recorded. In usual ground work, tomato seeds are sown to obtain seedlings that take 5–6 days to sprout. Seedlings are then transferred to pots containing soil and grown in a greenhouse for about one month. Following this, plants are shifted to a growth chamber where plants are inoculated with the pathogen by soil drench or the stem inoculation method [20,21]. Using this approach, it usually takes 40 days to perform a single virulence assay. The infection achieved in this way is generally not axenic as the soil conditions used are not devoid of other bacterial communities that can colonize the plant during its growth prior to the infection study. Singh et al. [22] described a simple assay to study the pathogenicity of *R. solanacearum* on freshly grown tomato seedlings instead of fully-grown tomato plants. From seed germination to completion of the infection process, the study takes around 15 to 20 days. Pathogenicity due to *R. solanacearum* was also demonstrated when there is no significant plant growth since no mineral/growth inducing factors have been added into the water [23]. Under this same condition, there are reports of the bacterium's survivability without any growth [24]. The death of tomato seedlings was actually occurring due to the presence of *R. solanacearum* in the water. On the basis of the previous study, we have discussed an approach to study biological assays in tomato seedlings.

In this study, a *Streptomyces* sp., NEAU-HV9, was isolated and showed strong antimicrobial activity against *R. solanacearum*. The taxonomic identity of NEAU-HV9 was determined by a combination of 16S rRNA gene sequence analysis with morphological and physiological characteristics. The potential control of actinomycin D produced by the strain NEAU-HV9 against *R. solanacearum* was also investigated.

## 2. Materials and Methods

### 2.1. Sample Collection

Soil samples were collected from a field situated in Bama yao Autonomous County, Hechi City, Guangxi zhuang Autonomous Region (24°15′ N, 107°26′ E). The collected soil samples were brought to the laboratory in sterile bags and kept at 4 °C until further analysis. Before isolation of actinomycetes, the soil samples were air-dried at room temperature.

### 2.2. Screening and Isolation of Actinomycetes

The soil sample (5 g) was mixed with 45 mL distilled water and followed by an ultrasonic treatment (160 W) for 3 min. The soil suspension was incubated at 28 °C and 250 rpm on a rotary shaker for 30 min. Subsequently, the supernatant was collected and subjected to serial dilutions from $10^{-2}$ to $10^{-5}$. Each dilution (200 μL) was spread on a plate of humic acid-vitamin (HV) agar [25] supplemented with cycloheximide (50 mg $L^{-1}$) and nalidixic acid (20 mg $L^{-1}$). Colonies were transferred and purified on International *Streptomyces* Project (ISP) medium 3 [26] and stored for a long time in glycerol suspensions (20%, *v*/*v*) at −80 °C after 14 days of aerobic incubation at 28 °C.

### 2.3. Screening of Antagonistic Actinobacteria Strains

The isolates were screened using the agar well diffusion method, and *R. solanacearum* was used as the indicator bacterium [27]. To further investigate the antibacterial components produced by the isolated cultures, these strains were cultured in ISP 2 medium [26] and the inhibitory activities of the supernatant and cell precipitate were tested. Initially, the isolated cultures were grown in ISP 2 medium and incubated at 28 °C on a rotary shaker. After 7 days of incubation, the supernatants were obtained by centrifugation at 8000 rpm and 4 °C for 10 min and subsequently filtrated with a 0.2 μm membrane filter. The cell precipitates were extracted with an equal volume of methanol for approximately 24 h [28]. A cell suspension (1 mL at $1 \times 10^8$ cfu $mL^{-1}$) of *R. solanacearum* was aseptically plated onto Bactoagar-glucose (BG) media supplemented with 0.5% glucose [22]. Supernatant and methanol extracts were collected from each isolate and tested initially for antimicrobial activity against *R. solanacearum*; each well contained 200 μL of supernatant or methanol extract. The plates were incubated at 37 °C for 12 h to test antibacterial activity. The diameters of inhibition zones were measured by using vernier calipers [29]. The experiments were conducted twice. The isolates that showed activities against tested organisms were collected and maintained. Among the collected isolates, the potential isolate designated as NEAU-HV9 was selected for further studies.

### 2.4. Morphological and Biochemical Characteristics of NEAU-HV9

Morphological characteristics, using cultures grown on ISP 3 medium at 28 °C for 2 weeks, were observed by light microscopy (Nikon ECLIPSE E200, Nikon Corporation, Tokyo, Japan) and scanning electron microscopy (Hitachi SU8010, Hitachi Co., Tokyo, Japan). Scanning electron microscopy samples were prepared as described by Jin et al. [30]. Cultural characteristics were determined using 2-week cultures grown at 28 °C on Czapek's agar [31], Bennett's agar [32], Nutrient agar [33], ISP 1 agar and ISP 2-7 media [26]. The color designation of substrate mycelium and aerial mycelium was done with ISCC–NBS (Inter-Society Color Council-National Bureau of Standards) Color Charts Standard Sample No. 2106 [34]. Growth at different temperatures (10 °C, 15 °C, 18 °C, 20 °C, 25 °C, 28 °C, 32 °C, 35 °C, 37 °C and 40 °C) was determined on ISP 3 medium after incubation for 14 days. Growth tests for pH range (pH 4.0–10.0, at intervals of 1.0 pH unit) using the buffer system described by Zhao et al. [35] and NaCl tolerance (0%, 1%, 2%, 3%, 4%, 5%, 6%, 7%, 8%, 9% and 10%, w/v) were tested in ISP 2 broth at 28 °C for 14 days on a rotary shaker. Biochemical testing (decomposition of adenine, casein, hypoxanthine, tyrosine, xanthine and cellulose, hydrolysis of starch, aesculin and gelatin, milk peptonization and coagulation, nitrate reduction and $H_2S$ production), the utilization of sole carbon and nitrogen sources were examined as described previously [36,37].

### 2.5. Phylogenetic Analysis of NEAU-HV9

Strain NEAU-HV9 was cultured in ISP 2 medium for 3 days at 28 °C to harvest cells. The genomic DNA was isolated using a Bacteria DNA Kit (TIANGEN Biotech, Co. Ltd., Beijing, China). The universal bacterial primers 27F and 1541R were used to carry out PCR amplification of the 16S rRNA gene sequence [38,39]. The purified PCR product cloned into the vector pMD19-T (Takara) and sequenced by using an Applied Biosystems DNA sequencer (model 3730XL, Applied Biosystems Inc., Foster City, California, USA). The almost complete 16S rRNA gene sequence (1510 bp) was uploaded to the EzBioCloud server (Available online: https://www.ezbiocloud.net/) [40] to calculate pairwise 16S rRNA gene sequence similarity between strain NEAU-HV9 and related similar species. The phylogenetic tree was reconstructed with neighbor-joining trees [41] using MEGA 7.0 software [42]. The confidence value of branches of the neighbor-joining tree was assessed using bootstrap resampling with 1000 replication [43]. A distance matrix was calculated using Kimura's two-parameter model [44]. All positions containing gaps and missing data were eliminated from the dataset (complete deletion option).

### 2.6. Fermentation

Strain NEAU-HV9 was grown and maintained for 7 days at 28 °C on ISP 3 medium agar plates. Fermentation involved the generation of a seed culture. The stock culture was transferred into two 250 mL Erlenmeyer flasks containing 50 mL of the ISP2 medium and incubated at 28 °C for 72 h on a rotary shaker at 250 rpm. All of the media were sterilized at 121 °C for 20 min. The seed culture (5%) was transferred into 75 flasks (250 mL) containing 100 mL of production medium. The production medium was composed of maltodextrin 4%, lactose 4%, yeast extract 0.5%, Mops 2% at pH 7.2–7.4. The flasks were incubated at 28 °C for 7 days, shaken at 250 rpm. The final 7.5 L fermentation broth was filtered to separate the supernatant and the mycelial cake. The supernatant was extracted with ethyl acetate three times (3 × 2 L), and the mycelial cake was extracted with MeOH (3 L). The organic phase was evaporated under reduced pressure at 55 °C to yield the red crude extract (5.2 g).

### 2.7. Isolation and Purification of Antibacterial Compounds

Crude extract from the mycelium and supernatant was combined and subjected to silica gel column chromatography (Qingdao Haiyang Chemical Group, Qingdao, China; 100–200 mesh; 100 × 3 cm column) using a gradient of ethyl acetate–MeOH (100:0–90:10) to yield three fractions (Fr.1-Fr.3) based on the TLC (thin layer chromatography) profiles. TLC was performed on silica-gel plates with solvent of ethyl acetate/MeoH (4:1). All fractions (Fr.) were screened against *R. solanacearum*. The most active, Fr.1 and Fr.2, were applied to a Sephadex LH-20 column eluted with $CH_2Cl_2$/MeOH (1:1, *v*/*v*) and then further purified by semipreparative HPLC (Agilent 1260, Zorbax SB-C18, 5 µm, 250 × 9.4 mm inner diameter; 1.5 mL/min; 220 nm; 254 nm; Agilent, Palo Alto, CA, USA) MeOH/H2O (90:10, *v*/*v*) to obtain Compound 1 (*tR* 10.928 min, 9.3 mg) and Compound 2 (*tR* 12.367 min, 60.4 mg). We chose the main product, Compound 2, for further research. NMR spectra (1H and 13C) were measured with a Bruker DRX-400 (400 MHz for $^1$H and 100 MHz for $^{13}$C) spectrometer (Bruker, Rheinstetten, Germany). The ESI-MS (electrospray ionization mass spectra) spectra were taken on a Q-TOF Micro LC-MS-MS mass spectrometer (Waters Co, Milford, MA, USA).

### 2.8. Determination of Minimum Inhibitory Concentration (MIC)

The minimum inhibitory concentration (MIC) of the antibacterial compounds was determined as described by Rathod et al. [45]. *R. solanacearum* was grown in BG medium with 0.5% glucose in shake flasks at 28 °C for 24 h. Cells were harvested by centrifugation, washed with 0.85% saline twice, then the supernatant was discarded and 0.85% saline was added to the washed cells. The suspensions were standardized to an optical density (OD) of 0.2 at 540 nm. Antibacterial compounds were two-fold serially diluted to obtain concentrations ranging from 0.2 to 12.8 mg $L^{-1}$ and one tube without drug served as a control. All of the tubes were inoculated with 1 mL of suspension of *R. solanacearum* above and incubated at 37 °C for 12 to 16 h. The turbidity of each tube with respect to the control tube was measured. The MIC value was defined as the lowest concentration of a compound that completely inhibits growth.

### 2.9. Biological Assays in Tomato Seedlings

Germination of tomato seedlings and preparation of bacterial inoculum were prepared as described by Singh et al. [22]. Freshly grown *R. solanacearum* was inoculated into 50 mL BG media broth with 0.5% glucose and incubated at 28 °C and 150 rpm for 24 h. The bacterial cultures were obtained by centrifugation at 4000 rpm and 4 °C for 15 min and were then resuspended in an equal volume of sterile distilled water to obtain a concentration of approximately $10^9$ cfu $mL^{-1}$. Strain NEAU-HV9 was cultured in ISP 2 broth on rotary shaker for 3 days at 28 °C and centrifuged at 10,000 rpm. Subsequently, cell pellets were diluted in 0.85% (*w/v*) NaCl solution and adjusted to $10^7$, $10^8$ or $10^9$ cfu $mL^{-1}$. Root inoculation of *R. solanacearum* in tomato seedlings was carried out as described by Singh et al. [22]. About 15 to 20 mL of *R. solancearum* inoculum was taken in a sterile container. Tomato seedlings (6 to 7 days old) were picked one at a time from the germinated seedling tray and then the roots of each seedling were dipped in the bacterial inoculum (up to the root-shoot junction). Four treatments were established as follows: TR 1 (tomato seedlings were pre-inoculated with suspension ($10^7$, $10^8$ or $10^9$ cfu $mL^{-1}$) of strain NEAU-HV9 and then inoculated with *R. solanacearum);* TR 2 (tomato seedlings were pre-inoculated with *R. solanacearum* and then transferred to microfuge tubes with the addition of 1 to 1.5 mL of sterile water and active fraction, where the final treatment concentrations were 1 × MIC and 2 × MIC, respectively); CK 1 (tomato seedlings were inoculated with sterile water); and CK 2 (tomato seedlings only were inoculated with *R. solanacearum*). For all of the treatments, the root-dip inoculated seedlings were transferred to an empty 1.5 mL sterile microfuge tube. After approximately 5 minutes, 1 to 1.5 mL of sterile water was added to the tube. All the inoculated seedlings, along with the controls, were transferred to a growth chamber maintained at 28 °C with 75% relative humidity (RH) and a 12-h photoperiod. Seedlings were analyzed for disease progression after 7 days. Sets of 4 seedlings were recruited in each dilution inoculation, and each assay was performed in triplicate.

### 2.10. Pot Culture Experiments

Prior to use, seed surfaces were disinfected with 2% sodium hypochlorite for 2 min [46]. Both germination and plant growth conditions followed 75–90% RH and a 12-h photoperiod at 28 °C. Four treatments were established as follows: TR 1 (one day before transplanting the test plants, strain NEAU-HV9 was added into the sterilized soil so that each gram of soil received about $1 \times 10^9$ cfu $g^{-1}$ bacterial cells. Seven day old tomato seedlings were transferred to the soil; after three weeks, plants were inoculated with a suspension ($OD_{600}$ = 0.3) of *R. solanacearum*); TR 2 (seven day old tomato seedlings were transferred to sterilized soil; after three weeks, tomato seedlings were irrigated with a solution of actinomycin D (0.6 mg $L^{-1}$). After one day, tomato plants were inoculated with a suspension ($OD_{600}$ = 0.3) of *R. solanacearum* by pouring it onto the soil of unwounded plants at a final concentration of $1 \times 10^7$ cfu $g^{-1}$ of soil [47]); CK 1 (seven day old tomato seedlings were transferred to sterilized soil; after three weeks, tomato seedlings were irrigated with sterilized water as positive control); and CK 2 (seven day old tomato seedlings were transferred to sterilized soil; after three weeks, tomato seedlings were inoculated with a suspension ($OD_{600}$ = 0.3) of *R. solanacearum* as a negative control). All plants were kept in the greenhouse at 24–28 °C and 75–90% RH with a 12-h photoperiod. Treatments were replicated three times with five plants per replication. The disease incidence was rated using the 0–4 scale [48].

## 3. Results

### 3.1. Isolation and Identification of an Antimicrobial Compound Producing Strain

More than 20 isolates from the soil samples were isolated, purified, and screened for bioactivity against *R. solanacearum*. Among them, only four isolates showed bioactivity against *R. solanacearum*. Since the methanol extract of the the cell pellet and supernatant of one isolate, designated as NEAU-HV9, revealed a higher activity (30.5 mm and 32.8 mm) against the tested bacterial strain (Table 1, Figure S2), this strain was selected for further studies.

**Table 1.** Bioactivities of the supernatant and cell pellet of NEAU-HV9 against *R. solanacearum*.

| | Methanol Extract of Cell Pellet | Supernatant |
|---|---|---|
| Inhibitory zone diameters (mm) | 30.5 | 32.8 |

Data shown are the mean of two replications.

Strain NEAU-HV9 was aerobic, Gram-stain positive and formed well-developed, branched substrate hyphae and aerial mycelium that differentiated into spiral spore chains with oval spores (Figure 1). The spore surface was wrinkled. It had good growth on ISP 1, ISP 2, ISP 3, ISP 4, ISP 5, ISP 6, ISP 7, Bennett's agar and Nutrient agar, and poor growth on Czapek's agar (Figure S1). The data on the growth characteristics of NEAU-HV9 in different media are given in Table S1.

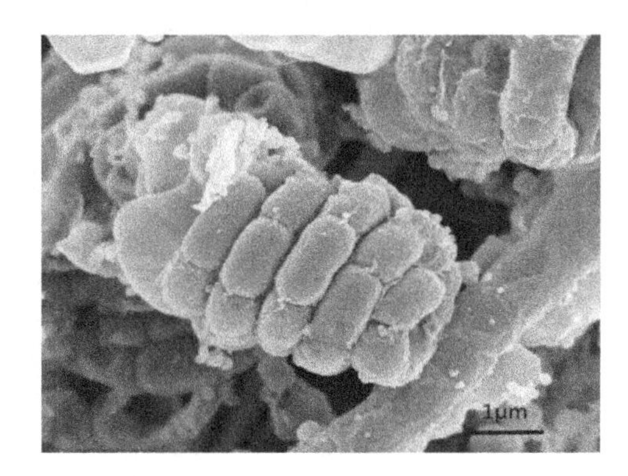


**Figure 1.** Scanning electron micrograph of strain NEAU-HV9 grown on International *Streptomyces* Project (ISP) 3 agar for 2 weeks at 28 °C.

Further characterization of NEAU-HV9 was performed by evaluating various biochemical tests (Table S2). Growth at 15 °C to 37 °C (optimum: 28 °C) and in the range of pH 5 to 9 (optimum: pH 7.0). Tolerate up to 7% (*w*/*v*) NaCl in the culture medium. Positive for hydrolysis of starch, production of $H_2S$, hydrolysis of aesculin and decomposition of adenine, hypoxanthine, tyrosine and xanthine, negative for reduction of nitrate, coagulation and peptonization of milk, liquefaction of gelatin and decomposition of casein. D-Glucose, D-maltose, D-mannitol, D-galactose, inositol, D-mannose, L-rhamnose and D-sucrose are utilized as sole carbon sources, but not L-arabinose, dulcitol, D-fructose, lactose, D-ribose, D-sorbitol or D-xylose. L-Alanine, D-arginine, L-asparagine, L-aspartic acid, L-glutamic acid, L-glutamine, glycine, L-proline, L-serine, L-threonine and L-tyrosine are utilized as sole nitrogen sources, but not creatine. The above growth data of isolate NEAU-HV9 denote that the isolate has the typical characteristics of the genus *Streptomyces*.

Recently, it has been suggested that the 16S rRNA gene can be used as a reliable molecular clock due to 16S rRNA sequences from distantly related bacterial lineages having similar functionalities [49]. Basically, the 16S rRNA gene sequence, comprising of about 1500 bp with hyper variable and conserved regions, is universal in all bacteria. According to Woese's report [50], comparing a stable part of the genetic code could determine phylogenetic relationships of bacteria. The hyper variable regions of the 16S rRNA gene sequences provide species-specific signature sequences, so it is widely used in bacterial identification all over the world. Therefore, the almost-complete 16S rRNA gene sequence (1510 bp) of strain NEAU-HV9 was obtained and has been deposited as MN578143 in the GenBank, EMBL (European Molecular Biology Laboratory) and DDBJ (DNA Data Bank of Japan) databases. BLAST sequence analysis of the 16S rRNA gene sequence indicated that strain NEAU-HV9 was related to members of the genus *Streptomyces*. The EzBioCloud analysis showed that strain NEAU-HV9 was most closely related to *Streptomyces panaciradicis* 1MR-8$^T$ and *Streptomyces sasae* JR-39$^T$ with a gene sequence similarity of 98.90% and 98.89%, respectively. In conclusion, based on the 16S rRNA gene sequence and the genetic identity of isolate NEAU-HV9, the isolated strain was further identified by neighbor-joining tree (Figure 2), and was also found to belong to the genus *Streptomyces*.

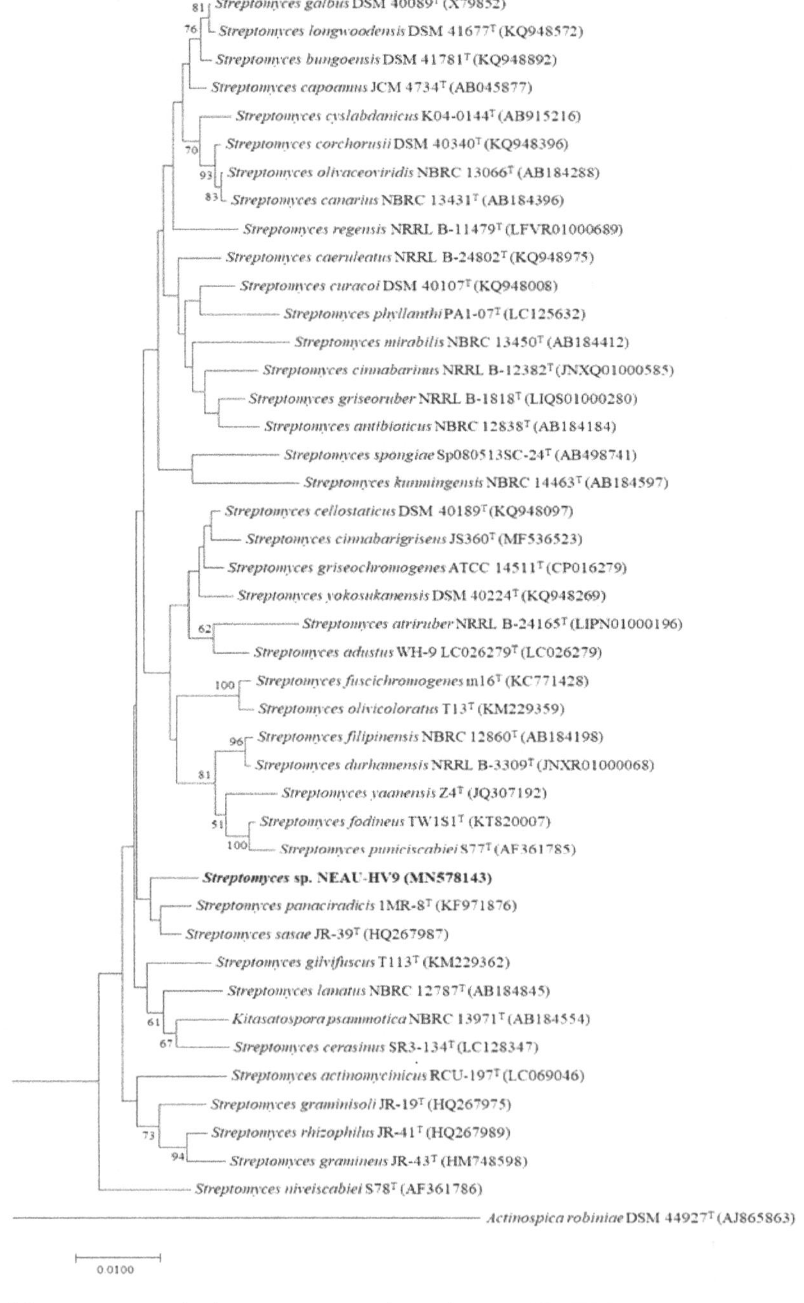


**Figure 2.** Neighbor-joining phylogenetic tree based on 16S rRNA gene sequences showing the relationships among strain NEAU-HV9 (bold) and members of the genus *Streptomyces*. Bootstrap percentages (≥50%) based on 1000 resamplings are listed at the nodes. *Actinospica robiniae* DSM 44927$^T$ was used as the out-group. Scale bar represents 0.01 nucleotide substitutions per site.

### *3.2. Structural Characterization of Compound*

The active component was isolated from fermentation medium (7.5 L) and one bioactive compound was obtained as red, amorphous powder. The compound had UV visible spectra at 215 nm, 440 nm in methanol. The compound showed absorptions at 220 nm and 254 nm with a retention time of 12.367 min (Figure S3), similar to that of actinomycin class of compounds [51,52]. The structure of the compound was further elucidated by $^{1}H$ NMR, $^{13}C$ NMR, and MS analysis as well as comparison with previously reported data. The ESI-MS of the isolated compound revealed molecular ion peaks at m/z 1277.6 $[M+Na]^{+}$ (Figure S4), which was identical to that of actinomycin D [53]; $^{1}H$ and $^{13}C$ spectra of the isolated compound in $CD_4O$ also showed great similarities to that of actinomycin D [52,53] (Figures S5 and S6). In addition, the retention time of commercial actinomycin D (Biotopped, purity: ≥98%) was 12.328 min (Figure S7), and the retention time of compound 2 was 12.367 min (Figure S3). Compound 2 and commercial actinomycin D have similar activity against *R. solanacearum* (Figure S8). The above results showed that the structure of the main active compound was confirmed to be actinomycin D (Figure 3).

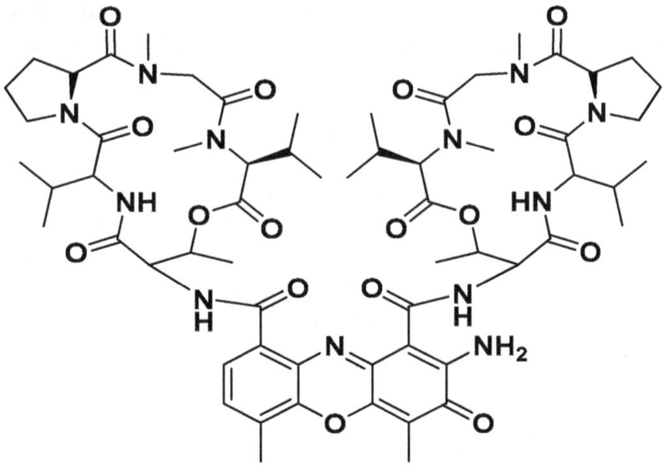

**Figure 3.** Chemical structure of actinomycin D.

### *3.3. Bioactivity of Isolated Compound*

#### 3.3.1. Minimum Inhibitory Concentration (MIC)

The minimum inhibitory concentration (MIC) of the antibacterial compound was determined as described by Rathod et al. [45]. The minimum inhibitory concentration of actinomycin D was determined as 0.6 mg $L^{-1}$ (0.48 µmol $L^{-1}$) against *R. solanacearum* (Table 2).

**Table 2.** Minimum inhibitory concentration (MIC) values of actinomycin D against *R. solanacearum*.

| Pathogen | MIC (mg/L) |
|---|---|
| *R. solanacearum* | 0.6 ± 0.2 |

Data shown are the mean of three replications.

#### 3.3.2. Biological Assays in Tomato Seedlings

The efficacy of the selected antagonist for the control of *R. solanacearum* was evaluated on tomato seedlings (Figure 4 and Figure S9, Table 3). The disease assessment was carried out using the method described in Kumar [23]. For the plants in the TR 1 group, the $10^{9}$ cfu $mL^{-1}$ suspension of NEAU-HV9 was effective against *R. solanacearum* when compared with the control (CK 2); all seedlings were as healthy as the CK 1 group. The $10^{8}$ cfu $mL^{-1}$ suspension of NEAU-HV9 showed very weak bioactivity against *R. solanacearum* compared with the control (CK 2); only one seedling was healthy and others were wilted. The $10^7$ cfu $mL^{-1}$ suspension of NEAU-HV9 exhibited no bioactivity against *R. solanacearum*; all seedlings were wilted, the same as the seedlings that were dried.

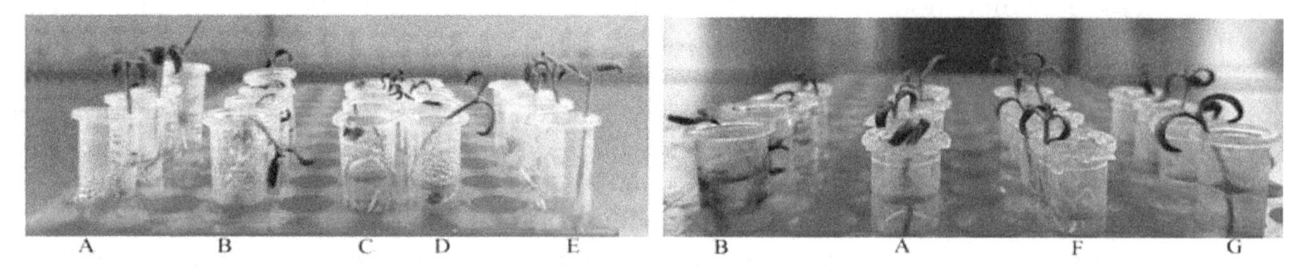


**Figure 4.** Control efficiency of strain NEAU-HV9 against *R. solanacearum*. A, tomato seedlings were inoculated with sterile water (CK 1); B, tomato seedlings only were inoculated with *R. solanacearum* (CK 2); C, tomato seedlings were pre-inoculated with suspension ($10^7$ cfu $mL^{-1}$) of NEAU-HV9 and then inoculated with *R. solanacearum* (TR 1); D, tomato seedlings were pre-inoculated with suspension ($10^8$ cfu $mL^{-1}$) of NEAU-HV9 and then inoculated with *R. solanacearum* (TR 1); E, tomato seedlings were pre-inoculated with suspension ($10^9$ cfu $mL^{-1}$) of NEAU-HV9 and then inoculated with *R. solanacearum* (TR 1); F, actinomycin D at the concentration 1 × MIC (TR 2); G, actinomycin D at the concentration 2 × MIC (TR 2).

**Table 3.** Effect of the strain NEAU-HV9 and actinomycin D on the incidence and control of tomato bacterial wilt in tomato seedlings.

| Treatment | Wilt Incidence (%) | Control Efficacy (%) |
|---|---|---|
| NEAU-HV9 ($10^7$ cfu $mL^{-1}$) | $100 \pm 0$ [a] | $0 \pm 0$ [a] |
| NEAU-HV9 ($10^8$ cfu $mL^{-1}$) | $93 \pm 11.6$ [b] | $6.6 \pm 11.5$ [b] |
| NEAU-HV9 ($10^9$ cfu $mL^{-1}$) | $0 \pm 0$ [c] | $100 \pm 0$ [c] |
| Actinomycin D (1 × MIC) | $0 \pm 0$ [c] | $100 \pm 0$ [c] |
| Actinomycin D (2 × MIC) | $0 \pm 0$ [c] | $100 \pm 0$ [c] |
| Control | $100 \pm 0$ [a] | ... |

Data shown are the mean of three replications. Means within the same column followed by the same letter are not significantly different ($p = 0.05$) according to Fisher's least significant difference test.

Actinomycin D was highly effective against *R. solanacearum* in tomato seedlings (Figure 4 and Figure S9, Table 3). It was notable that all seedlings of the control (CK 2) were wilted, however after treatment with actinomycin D at the concentration 1 × MIC and 2 × MIC, none of the seedlings exhibited disease symptoms; the control efficacy of the formulation was 100%.

### 3.4. Pot Culture Experiments

In the pot culture experiments, NEAU-HV9 and actinomycin D effectively suppressed the development of bacterial wilt caused by *R. solanacearum* (Table 4, Figure S10). The negative control treatment had 73.9% relative disease incidence. For strain NEAU-HV9 and actinomycin D, the control efficacies of the formulations were 82% and 100%, respectively.

**Table 4.** Effect of the strain NEAU-HV9 and actinomycin D on the incidence and control of tomato bacterial wilt in pot culture experiments.

| Treatment | Wilt Incidence (%) | Control Efficacy (%) |
|---|---|---|
| NEAU-HV9 | $13.3 \pm 5.8$ [b] | $82 \pm 6$ [a] |
| Actinomycin D | $0 \pm 0$ [b] | $100 \pm 0$ [a] |
| Control | $73.9 \pm 6.6$ [a] | |

Wilt incidence (WI) was calculated as the percentage of leaves that were completely wilted. Control efficacy was calculated using the following formula: control efficacy (%) = 100 × (WI of control − WI of treatment)/WI of control. Data shown are the mean of three replications. Means within the same column followed by the same letter are not significantly different ($p = 0.05$) according to Fisher's least significant difference test.

## 4. Discussion

Soil-borne diseases have caused a significant decline in yield in the monoculture tomato field [3]. *Ralstonia solanacearum* is an important soil-borne bacterial plant pathogen which is distributed all over the world [5]. Recently, the biological control of soil-borne diseases has attracted more attention due to its environmental friendliness and high efficiency [54]. Therefore, isolation, screening and application of highly efficient antagonistic microorganisms is a key factor in biological control. With this outlook, a *Streptomyces* sp. strain NEAU-HV9 was isolated and found to exhibit antibacterial activities against *R. solanacearum* in the present study. By using 16S rRNA gene sequence analysis, combined with morphological, cultural and physiological characteristics, the results showed that strain NEAU-HV9 belongs to members of the genus *Streptomyces* and was most closely related to *Streptomyces panaciradicis* 1MR-8$^T$ and *Streptomyces sasae* JR-39$^T$ with gene sequence similarities of 98.90% and 98.89%, respectively.

Actinobacteria, particularly *Streptomyces*, are ubiquitous in the rhizosphere soil and can protect plant from pathogenic fungi/bacteria [55], so they have always been used in agriculture [56]. For instance, several *Streptomyces* species such as strains CAI-24, CAI-121, CAI-127, KAI-32 and KAI-90 have been used as BCAs against *Fusarium* wilt in chickpea plants [57]. The *Streptomyces* sp. CB-75, selected from banana rhizosphere soil, showed antifungal activity against 11 plant pathogenic fungi [54]. In this study, the *Streptomyces* sp. NEAU-HV9 exhibited strong antagonistic activity against *R. solanacearum*. According to the study of Singh et al., susceptibility of early stages of tomato seedlings toward the pathogen was confirmed by root-inoculation of *R. solanacearum* in early stages of tomato seedlings [22]. The antagonistic strains should reach a certain amount to demonstrate a significant biocontrol effect [58,59]. In this study, we inoculated very high numbers of *R. solanacearum* and very high levels of *Streptomyces* sp. NEAU-HV9 ($10^9$ cfu mL$^{-1}$) in small tubes in the TR 1 group. After culturing for seven days, all tomato seedlings were as healthy as the CK 1 control group (Table 3). There are only *R. solanacearum* and *Streptomyces* sp. NEAU-HV9 in this artificial system, which can better prove that a single NEAU-HV9 was able to be effective against *R. solanacearum*. On the seventh day, more than 90% of seedlings inoculated with *R. solanacearum* were found to be killed, but water-inoculated control seedlings were not wilted/dried [23]. Freshly grown tomato seedlings are too small to carry out detailed disease assessment, and can only be described as healthy, healthy wilted or dried. In the tests of this study, tomato seedlings inoculated with suspension ($10^7$ or $10^8$ cfu mL$^{-1}$) of NEAU-HV9 and *R. solanacearum* showed healthy wilted and dried disease phenotypes at different levels, while all tomato seedlings inoculated with suspension ($10^9$ cfu mL$^{-1}$) of NEAU-HV9 and *R. solanacearum* were healthy (Figure 4). The results indicated that there are only *R. solanacearum* and *Streptomyces* sp. NEAU-HV9 in this artificial system, which can better prove that a single NEAU-HV9 was able to be effective against *R. solanacearum*. In addition, strain NEAU-HV9 effectively controlled *R. solanacearum* on larger plants in pot culture experiments (Table 4). Thus, the test presented in this study is viable for a preliminary screening of antagonistic actinobacterial strains against *R. solanacearum* and has important aspects with respect to reduced time, space consumption and economics. Meanwhile, the results showed the possibility of using *Streptomyces* sp. NEAU-HV9 as bioinoculant for *R. solanacearum*.

A wide range of bioactive secondary metabolites with anti-inflammatory, antibacterial, antifungal, antialgal, antimalarial and anticancer activities were produced by actinomycetes. Actinomycetes have produced about two-thirds of available antibiotics that have great practical value [60,61]. For example, *Streptomyces* TP-A0595 produced an antagonist that was determined as 6-prenylindole and effective against *Alternaria brassicicola* by inhibiting the formation of infection hyphae [62]. *Streptomyces griseus* H7602 produced a monomer compound that has suppressive effect on infection by *Phytophthora capsici* [63]. Some well-known antibiotics have been isolated from *Streptomyces* and used as fungicides. Many types of antibiotics with high antibacterial activity were produced from *Streptomyces spectabilis*, including streptovaricin [64], desertomycin [65] and spectinomycin [66], and they have high application value in the pharmaceutical industry [67]. Actinomycin D (or Dactinomycin) is a proverbial antitumor-antibiotic drug, which belongs to the actinomycin family and was isolated from *Streptomyces*. Actinomycin D has been demonstrated to have various biological activities.

Gram-negative bacteria were largely inhibited by using 10–100 mg per liter concentrations [68]. Actinomycin D produced by the bacterium *Streptomyces hydrogenans* IB310 was effective against both bacterial and fungal phytopathogens [51], and the authors proposed that actinomycin D might be developed as an antibacterial agent used in agriculture. However, there are no reports on antibacterial activities against *R. solanacearum* and it is not currently used in agriculture. In this study, *Streptomyces* sp. NEAU-HV9, which showed strong antibacterial activity against *R. solanacearum*, was isolated and identified. To learn more about the chemical nature of the antibacterial activity of the culture filtrate, the active compound actinomycin D was finally obtained. In this paper, we tested the in vitro antibacterial activity of actinomycin D against *R. solanacearum* and obtained a MIC value of 0.6 mg $L^{-1}$ (0.48 μmol $L^{-1}$), which was many fold lower than other reported new natural antibacterial agents [69], synthesized antibacterial agents and those of commercial fungicides including gentamicin and streptomycin [70]. The antibacterial activity of actinomycin D against *R. solanacearum* tomato seedlings treated with 1 × MIC and 2 × MIC were determined. None of the seedlings inoculated with actinomycin D exhibited disease symptoms and the phytotoxic rating of actinomycin D was similar to that of a water control. Thus, actinomycin D was not phytotoxic at a concentration of 0.6 mg $L^{-1}$ (0.48 μmol $L^{-1}$). The results suggest that actinomycin D might be useful as a candidate pesticide for the treatment of *Ralstonia solanacearum* in tomato.

## 5. Conclusions

In summary, this study found that *Streptomyces* sp. NEAU-HV9 exerted significant antibacterial activity against *R. solanacearum*, and actinomycin D, which was produced by *Streptomyces* sp. NEAU-HV9, exhibited a minimum inhibitory concentration (MIC) against *R. solanacearum* of 0.6 mg $L^{-1}$ (0.48 μmol $L^{-1}$). In addition, *Streptomyces* sp. NEAU-HV9 and actinomycin D can effectively inhibit the occurrence of *R. solanacearum*. From the results, it is obvious that *Streptomyces* sp. NEAU-HV9 is an important microbial resource as a biological control against *R. solanacearum* and actinomycin D is a promising candidate for the development of potential antibacterial biocontrol agents.

**Supplementary Materials:**
Table S1: Growth and cultural characteristics of strain NEAU-HV9 after 2 weeks at 28 °C; Table S2: Physiological and biochemical characteristics of strain NEAU-HV9; Figure S1: Cultural characteristics of strain NEAU-HV9 observed on ISP 1, ISP 2, ISP 3, ISP 4, ISP 5, ISP6, ISP7, Nutrient agar, Bennett's agar and Czapek's agar after being incubated at 28 °C for 2 weeks; Figure S2: Bioactivities of the supernatant and cell pellet of NEAU-HV9 against *R. solanacearum*; Figure S3: The HPLC profiles of crude extract produced by *Streptomyces* NEAU-HV9; Figure S4: Mass Spectrometry of actinomycin D ($C_{62}H_{86}N_{12}O_{16}Na$: 1277.6); Figure S5: $^{1}H$ NMR of actinomycin D; Figure S6: $^{13}C$ NMR of actinomycin D; Figure S7: The HPLC profiles of commercial actinomycin D; Figure S8: Bioactivities of commercial actinomycin D and the main product of NEAU-HV9 against *R. solanacearum*; Figure S9: Control efficiency of strain NEAU-HV9 against *R. solanacearum*; Figure S10: Control efficiency of the actinomycin D and strain NEAU-HV9 against *R. solanacearum*.

**Author Contributions:** L.L., X.H., X.L., H.W. and Y.W. performed the experiments. X.Z., L.Z. and P.C. analyzed the data. L.L. wrote the paper. X.W. and X.H. prepared the figures and tables. J.Z. and W.X. designed the experiments and reviewed the manuscript. All authors have read and agreed to the published version of the manuscript.

## References

1. Lee, C.G.; Lida, T.; Uwagaki, Y.; Otani, Y.; Nakaho, K.; Ohkuma, M. Comparison of prokaryotic and eukaryotic communities in soil samples with and without tomato bacterial wilt collected from different fields. *Microbes Eniviron.* **2017**, *32*, 376–385. [CrossRef] [PubMed]
2. Singh, V.K.; Singh, A.K.; Kumar, A. Disease management of tomato through PGPB: Current trends and future perspective. *3Biotech* **2017**, *7*, 255. [CrossRef] [PubMed]
3. Zhou, X.G.; Wu, F.Z. Dynamics of the diversity of fungal and Fusarium communities during continuous cropping of cucumber in the green house. *FEMS Microbiol. Ecol.* **2012**, *802*, 469–478. [CrossRef] [PubMed]
4. Yabuuchi, E.;-Kosako, Y.; Yano, I.; Hota, H.; Nishiuchi, Y. Transfer of two *Burkholderia* and an *Alcaligenes species* to *Ralstonia* gen. nov., *Ralstonia solanacearum* (Smith, 1986) comb. nov. *Microbiol. Immun.* **1995**, *39*, 897–904. [CrossRef]

5. Hayward, A.C. Biology and epidemiology of bacterial wilt caused by *Pseudomonas solanacearum*. *Ann. Rev. Phytopathol.* **1991**, *29*, 65–87. [CrossRef]
6. Vu, T.T.; Kim, J.C.; Choi, Y.H.; Choi, G.J.; Jang, K.S.; Choi, T.H.; Yoon, T.M.; Lee, S.W. Effect of Gallotannins Derived from *Sedum takesimense* on Tomato Bacterial Wilt. *Plant Dis.* **2013**, *97*, 1593–1598. [CrossRef]
7. Genin, S.; Boucher, C. Lessons learned from the genome analysis of *Ralstonia solanacearum*. *Ann. Rev. Phytopathol.* **2004**, *42*, 107–134. [CrossRef]
8. Ciampi-Panno, L.; Fernandez, C.; Bustamante, P.; Andrade, N.; Ojeda, S.; Conteras, A. Biological control of bacterial wilt of potatoes caused by *Pseudomonas solanacearum*. *Am. Potato J.* **1989**, *66*, 315–332. [CrossRef]
9. McLaughlin, R.J.; Sequeira, L.; Weingartner, D.P. Biocontrol of bacterial wilt of potato with an avirulent strain of *Pseudomonas solanacearum*: Interactions with root-knot nematodes. *Am. Potato J.* **1990**, *67*, 93–107. [CrossRef]
10. Toyota, K.; Kimura, M. Suppresision of *Ralstonia solanacearum* in soil following colonization by other strains of *R. solanacearum*. *Soil Sci. Plant Nutr.* **2000**, *46*, 449–459.
11. Viaene, T.; Langendries, S.; Beirinckx, S.; Maes, M.; Goormachtig, S. *Streptomyces* as a plant's best friend? *FEMS Microbiol. Ecol.* **2016**, *92*. [CrossRef] [PubMed]
12. Dias, M.P.; Bastos, M.S.; Xavier, V.B.; Cassel, E.; Astarita, L.V.; Santarémm, E.R. Plant growth and resistance promoted by *Streptomyces* spp. in tomato. *Plant Physiol. Biochem.* **2017**, *118*, 479–493. [CrossRef] [PubMed]
13. Xio, K.; Kinkel, L.L.; Samac, D.A. Biological control of *Phytophthora* root rots on alfalfa and soybean with *Streptomyces*. *Biol. Control* **2002**, *23*, 285–295. [CrossRef]
14. El-Tarabily, K.A.; Sivasithamparam, K. Non-streptomycete Actinobacteria as biocontrol agents of soil-borne fungal plant pathogens and as plant growth promoters. *Soil Biol. Biochem.* **2006**, *38*, 1505–1520. [CrossRef]
15. Taechowisan, T.; Chuaychot, N.; Chanaphat, S.; Wanbanjob, A.; Tantiwachwutikul, P. Antagonistic effects of Streptomyces sp. SRM1 on colletotrichum musae. *Biotechnology* **2009**, *8*, 86–92. [CrossRef]
16. Bentley, S.D.; Chater, K.F.; Cerden, A.M.; Challis, G.L.; Thomson, N.R.; James, K.D.; Harris, D.E.; Quail, M.A.; Kieser, H.; Harper, D.; et al. Complete genome sequence of the model actinomycete *Streptomyces coelicolor* A3. *Nature* **2002**, *417*, 141–147. [CrossRef]
17. Brockmann, H. Die actinomycine. *Angew. Chem.* **1960**, *87*, 1767–1947.
18. Farber, S.; D'Angio, G.; Evans, A.; Mitus, A. Clinical studies of actinomycin D with special reference to Wilms' tumor in children. *J. Urol.* **2002**, *168*, 2560–2562. [CrossRef]
19. Womer, R.B. Soft tissue sarcomas. *Eur. J. Cancer* **1997**, *33*, 2230–2234. [CrossRef]
20. Monteiro, F.; Genin, S.; van, D.I.; Valls, M. A luminescent reporter evidences active expression of *Ralstonia solanacearum* type III secretion system genes throughout plant infection. *Microbiology* **2012**, *158*, 2107–2116. [CrossRef]
21. Guidot, A.; Jiang, W.; Ferdy, J.B.; Thébaud, C.; Barberis, P.; Gouzy, J.; Genin, S. Multihost experimental evolution of the pathogen *Ralstonia solanacearum* unveils genes involved in adaptation to plants. *Mol. Biol. Evol.* **2014**, *11*, 2913–2928. [CrossRef] [PubMed]
22. Singh, N.; Phukan, T.; Sharma, P.L.; Kabyashree, K.; Barman, A.; Kumar, R.; Sonti, R.V.; Genin, S.; Ray, S.K. An Innovative Root Inoculation Method to Study *Ralstonia solanacearum* Pathogenicity in Tomato Seedlings. *Phytopathology* **2018**, *108*, 436–442. [CrossRef]
23. Kumar, R. Studying Virulence Functions of *Ralstonia solanacearum*, the Causal Agent of Bacterial Wilt in Plants. Ph.D. Thesis, Tezpur University, Tezpur, India, 2014.
24. van Elsas, J.D.; Kastelein, P.; de Vries, P.M.; van Overbeek, L.S. Effects of ecological factors on the survival and physiology of *Ralstonia solanacearum* bv 2 in irrigation water. *Can. J. Microbiol.* **2001**, *47*, 842–854. [CrossRef] [PubMed]
25. Hayakawa, M.; Nonomura, H. Humic acid-vitamin agar, a new medium for selective isolation of soil actinomycetes. *J. Ferment. Technol.* **1987**, *65*, 501–509. [CrossRef]
26. Shirling, E.B.; Gottlieb, D. Methods for characterization of *Streptomyces* species. *Int. J. Syst. Bacteriol.* **1966**, *16*, 313–340. [CrossRef]
27. Zhang, D.C.; Brouchkov, A.; Griva, G.; Schinner, F.; Margesin, R. Isolation and characterization of bacteria from ancient siberian permafrost sediment. *Biology* **2013**, *2*, 85–106. [CrossRef]
28. Fu, Y.S.; Yan, R.; Liu, D.; Zhao, J.W.; Song, J.; Wang, X.J.; Cui, L.; Zhang, J.; Xiang, W.S. Characterization of *Sinomonas gamaensis* sp. nov., a Novel Soil Bacterium with Antifungal Activity against *Exserohilum turcicum*. *Microorganisms* **2019**, *7*, 170. [CrossRef] [PubMed]

29. Gao, F.; Wu, Y.; Wang, M. Identification and antifungal activity of an actinomycete strain against *Alternaria* spp. *Span. J. Agric. Res.* **2014**, *12*, 1158–1165. [CrossRef]
30. Jin, L.Y.; Zhao, Y.; Song, W.; Duan, L.P.; Jiang, S.W.; Wang, X.J.; Zhao, J.W.; Xiang, W.S. *Streptomyces inhibens* sp. nov., a novel actinomycete isolated from rhizosphere soil of wheat (*Triticum aestivum* L.). *Int. J. Syst. Evol. Microbiol.* **2019**, *69*, 688–695. [CrossRef]
31. Waksman, S.A. *The Actinomycetes. A Summary of Current Knowledge*; Ronald: New York, NY, USA, 1967.
32. Jones, K.L. Fresh isolates of actinomycetes in which the presence of sporogenous aerial mycelia is a fluctuating characteristic. *J. Bacteriol.* **1949**, *57*, 141–145. [CrossRef]
33. Waksman, S.A. Classification, identification and descriptions of genera and species. In *The Actinomycetes*; Williams and Wilkins: Baltimore, MD, USA, 1961; Volume 2.
34. Kelly, K.L. *Inter-Society Color Council-National Bureau of Standards Color-Name Charts Illustrated with Centroid Colors*; Government Printing Office: Washington, DC, USA, 1964.
35. Zhao, J.W.; Han, L.Y.; Yu, M.Y.; Cao, P.; Li, D.M.; Guo, X.W.; Liu, Y.Q.; Wang, X.J.; Xiang, W.S. Characterization of *Streptomyces sporangiiformans* sp. nov., a Novel Soil Actinomycete with Antibacterial Activity against *Ralstonia solanacearum*. *Microorganisms.* **2019**, *7*, 360. [CrossRef] [PubMed]
36. Gordon, R.E.; Barnett, D.A.; Handerhan, J.E.; Pang, C. *Nocardia coeliaca, Nocardia autotrophica,* and the nocardin strain. *Int. J. Syst. Bacteriol.* **1974**, *24*, 54–63. [CrossRef]
37. Yokota, A.; Tamura, T.; Hasegawa, T.; Huang, L.H. *Catenuloplanes japonicas* gen. nov., sp. nov., nom. rev., a new genus of the order *Actinomycetales*. *Int. J. Syst. Bacteriol.* **1993**, *43*, 805–812. [CrossRef]
38. Woese, C.R.; Gutell, R.; Gupta, R.; Noller, H.F. Detailed analysis of the higher-order structure of 16S-like ribosomal ribonucleic acids. *Microbiol. Rev.* **1983**, *47*, 621–669. [CrossRef] [PubMed]
39. Springer, N.; Ludwig, W.; Amann, R.; Schmidt, H.J.; Görtz, H.D.; Schleifer, K.H. Occurrence of fragmented 16S rRNA in an obligate bacterial endosymbiont of *Paramecium caudatum*. *Proc. Natl. Acad. Sci. USA* **1993**, *90*, 9892–9895. [CrossRef]
40. Yoon, S.H.; Ha, S.M.; Kwon, S.; Lim, J.; Kim, Y.; Seo, H.; Chun, J. Introducing EzBioCloud: A taxonomically united database of 16S rRNA and whole genome assemblies. *Int. J. Syst. Evol. Microbiol.* **2017**, *67*, 1613–1617. [CrossRef]
41. Saitou, N.; Nei, M. The neighbor-joining method: A new method for reconstructing phylogenetic trees. *Mol. Biol. Evol.* **1987**, *4*, 406–425.
42. Kumar, S.; Stecher, G.; Tamura, K. Mega7: Molecular evolutionary genetics analysis version 7.0 for bigger datasets. *Mol. Biol. Evol.* **2016**, *33*, 1870–1874. [CrossRef]
43. Felsenstein, J. Confidence limits on phylogenies: An approach using the bootstrap. *Evolution* **1985**, *39*, 783–791. [CrossRef]
44. Kimura, M. A simple method for estimating evolutionary rates of base substitutions through comparative studies of nucleotide sequences. *J. Mol. Evol.* **1980**, *16*, 111–120. [CrossRef] [PubMed]
45. Rathod, B.B.; Korasapati, R.; Sripadi, P.; Reddy, S.P. Novel actinomycin group compound from newly isolated *Streptomyces* sp. RAB12: Isolation, characterization, and evaluation of antimicrobial potential. *Appl. Microbiol. Biotechnol.* **2018**, *102*, 1241–1250. [CrossRef]
46. Guo, J.H.; Qi, H.Y.; Guo, Y.H.; Ge, H.; Gong, L.Y.; Zhang, L.X.; Sun, P.H. Biocontrol of tomato wilt by growth-promoting rhizobacteria. *Biol. Control* **2004**, *29*, 66–72. [CrossRef]
47. Roy, N.; Choi, K.; Khan, R.; Lee, S.W. Culturing Simpler and Bacterial Wilt Suppressive Microbial Communities from Tomato Rhizosphere. *Plant Pathol. J.* **2019**, *35*, 362–371. [PubMed]
48. Roberts, D.P.; Denny, T.P.; Schell, M.A. Cloning of the egl gene of *Pseudomonas solanacearum* and analysis of its role in phytopathogenicity. *J. Bacteriol.* **1988**, *170*, 1445–1451. [CrossRef] [PubMed]
49. Tsukuda, M.; Kitahara, K.; Miyazaki, K. Comparative RNA function analysis reveals high functional similarity between distantly related bacterial 16 S rRNAs. *Sci. Rep.* **2017**, *7*, 9993. [CrossRef]
50. Woese, C.R. Bacterial evolution. *Microbiol. Rev.* **1987**, *51*, 221–2711. [CrossRef]
51. Kulkarni, M.; Gorthi, S.; Banerjee, G.; Chattopadhyay, P. Production, characterization and optimization of actinomycin D from Streptomyces hydrogenans IB310, a(n antagonistic bacterium against phytopathogens. *Biocatal. Agric. Biotechnol.* **2017**, *10*, 69–74. [CrossRef]
52. Zhang, L.L.; Wan, C.X.; Luo, X.X.; Lv, L.L.; Wang, X.P.; Xia, Z.F. Manufacture of Actinomycin D with Streptomyces mutabilis for Controlling Plant Pathogenic Bacteria, Cow Mastitis, and Female Colpitis. China Patent CN104450580A, 25 March 2015.

53. Chen, C.; Song, F.; Wang, Q.; Abdel-Mageed, W.M.; Guo, H.; Fu, C.; Hou, W.; Dai, H.; Liu, X.; Yang, N.; et al. A marine-derived Streptomyces sp. MS449 produces high yield of actinomycin X2 and actinomycin D with potent anti-tuberculosis activity. *Appl. Microbiol. Biotechnol.* **2012**, *95*, 919–927. [CrossRef]
54. Chen, Y.F.; Zhou, D.B.; Qi, D.F.; Gao, Z.F.; Xie, J.H.; Luo, Y.P. Growth promotion and disease suppression ability of a *streptomyces* sp. CB-75 from banana rhizosphere soil. *Front. Microbiol.* **2018**, *8*, 2704. [CrossRef]
55. Crawford, D.L.; Lynch, J.M.; Whipps, J.M.; Ousley, M.A. Isolation and characterization of actionomycete antagonists of a fungal root pathogen. *Appl. Environ. Microb.* **1993**, *59*, 3889–3905. [CrossRef]
56. Gao, L.; Qiu, Z.; You, J.; Tan, H.; Zhou, S. Isolation and characterization of endophytic *Streptomyces* strains from surface-sterilized tomato (*Lycopersicon esculentum*) roots. *Lett. Appl. Microbiol.* **2004**, *39*, 425–430.
57. Gopalakrishnan, S.; Pande, S.; Sharma, M.; Humayun, P.; Kiran, B.K.; Sandeep, D.; Vidya, M.S.; Deepthi, K.; Rupela, O. Evaluation of actinomycete isolates obtained from herbal vermicompost for biological control of *Fusarium* wilt of chickpea. *Crop Protect.* **2011**, *30*, 1070–1078. [CrossRef]
58. Bull, C.T. Relationship between root colonization and suppression of *Gaeumannomyces graminis* var tritici by *Pseudomonas fluorescens* strain 2–79. *Phytopathology* **1991**, *81*, 954–959. [CrossRef]
59. Raaijmakers, J.M.; Leeman, M.; van Oorschot, M.M.P.; van der Sluis, I.; Schippers, B.; Bakker, P.A.H.M. Dose-response relationships in biological-control of Fusarium-wilt of radish by *Pseudomonas* spp. *Phytopathology* **1995**, *85*, 1075–1081. [CrossRef]
60. Wang, J.J.; Zhao, Y.; Ruan, Y.Z. Effects of bio-organic fertilizers produced by four *Bacillus amyloliquefaciens* strains on banana *Fusarium* wilt Disease. *Compost Sci. Util.* **2015**, *23*, 185–198. [CrossRef]
61. Hong, K.; Gao, A.H.; Xie, Q.Y.; Gao, H.; Zhuang, L.; Lin, H.P.; Yu, H.P.; Li, J.; Yao, X.S.; Goodfellow, M.; et al. Actino-mycetes for marine drug discovery isolated from man-grove soils and plants in China. *Mar. Drugs* **2009**, *7*, 24–44. [CrossRef]
62. Sasaki, T.; Igarashi, Y.; Ogawa, M.; Furumai, T. Identification of 6-prenylindole as an antifungal metabolite of *Streptomyces* sp. TP-A0595 and synthesis and bioactivity of 6-substituted indoles. *J. Antibiot.* **2002**, *55*, 1009–1012. [CrossRef] [PubMed]
63. Nguyen, X.H.; Naing, K.W.; Lee, Y.S.; Kim, Y.H.; Moon, J.H.; Kim, K.Y. Antagonism of antifungal metabolites from *Streptomyces* griseus H7602 against *Phytophthora capsici*. *J. Basic Microbiol.* **2015**, *55*, 45–53. [CrossRef] [PubMed]
64. Kakinuma, K.; Hanson, C.A.; Rinehart, K.L., Jr. Spectinabilin, a new nitro-containing metabolite isolated from *Streptomyces spectabilis*. *Tetrahedron* **1976**, *32*, 217–222. [CrossRef]
65. Ivanova, V. New macrolactone of the desertomycin family from *Streptomyces spectabilis*. *Prep. Biochem. Biotechnol.* **1997**, *27*, 19–38. [CrossRef]
66. Kim, K.R.; Kim, T.J.; Suh, J.W. The gene cluster for spectinomycin biosynthesis and the aminoglycoside-resistance function of spcm in *Streptomyces spectabilis*. *Curr. Microbiol.* **2008**, *57*, 371–374. [CrossRef] [PubMed]
67. Selvakumar, J.; Chandrasekaran, S.; Vaithilingam, M. Bio prospecting of marine-derived *Streptomyces spectabilis* VITJS10 and exploring its cytotoxicity against human liver cancer celllines. *Pharmacogn. Mag.* **2015**, *11*, 469–473.
68. Waksman, S.A.; Woodruff, H.B. Bacteriostatic and bactericidal substances produced by a soil actinomyces. *Proc. Soc. Exp. Biol. Med.* **1940**, *45*, 609–614. [CrossRef]
69. Vu, T.T.; Kim, H.; Tran, V.K.; Vu, H.D.; Hoang, T.X.; Han, J.W.; Choi, Y.H.; Jang, K.S.; Choi, G.J.; Kim, J.J. Antibacterial activity of tannins isolated from *Sapium baccatum* extract and use for control of tomato bacterial wilt. *PLoS ONE* **2017**, *12*, e0181499. [CrossRef]
70. Zheng, S.J.; Zhu, R.; Tang, B.; Chen, L.Z.; Bai, H.J.; Zhang, J.W. Synthesis and biological evaluations of a series of calycanthaceous analogues as antifungal agents. *Nat. Prod. Res.* **2019**, 1–9. [CrossRef] [PubMed]

# 10

# Genome Mining Coupled with OSMAC-Based Cultivation Reveal Differential Production of Surugamide A by the Marine Sponge Isolate *Streptomyces* sp. SM17 When Compared to Its Terrestrial Relative *S. albidoflavus* J1074

**Eduardo L. Almeida [1], Navdeep Kaur [2], Laurence K. Jennings [2], Andrés Felipe Carrillo Rincón [1], Stephen A. Jackson [1,3], Olivier P. Thomas [2] and Alan D.W. Dobson [1,3,*]**

[1] School of Microbiology, University College Cork, T12 YN60 Cork, Ireland; e.leaodealmeida@umail.ucc.ie (E.L.A.); andres-felipe.carrillo@alumnos.unican.es (A.F.C.R.); stevejackson71@hotmail.com (S.A.J.)

[2] Marine Biodiscovery, School of Chemistry and Ryan Institute, National University of Ireland Galway (NUI Galway), University Road, H91 TK33 Galway, Ireland; navdeep.kaur@nuigalway.ie (N.K.); laurence.jennings@nuigalway.ie (L.K.J.); olivier.thomas@nuigalway.ie (O.P.T.)

[3] Environmental Research Institute, University College Cork, T23 XE10 Cork, Ireland

* Correspondence: a.dobson@ucc.ie

**Abstract:** Much recent interest has arisen in investigating *Streptomyces* isolates derived from the marine environment in the search for new bioactive compounds, particularly those found in association with marine invertebrates, such as sponges. Among these new compounds recently identified from marine *Streptomyces* isolates are the octapeptidic surugamides, which have been shown to possess anticancer and antifungal activities. By employing genome mining followed by an one strain many compounds (OSMAC)-based approach, we have identified the previously unreported capability of a marine sponge-derived isolate, namely *Streptomyces* sp. SM17, to produce surugamide A. Phylogenomics analyses provided novel insights on the distribution and conservation of the surugamides biosynthetic gene cluster (*sur* BGC) and suggested a closer relatedness between marine-derived *sur* BGCs than their terrestrially derived counterparts. Subsequent analysis showed differential production of surugamide A when comparing the closely related marine and terrestrial isolates, namely *Streptomyces* sp. SM17 and *Streptomyces albidoflavus* J1074. SM17 produced higher levels of surugamide A than *S. albidoflavus* J1074 under all conditions tested, and in particular producing >13-fold higher levels when grown in YD and 3-fold higher levels in SYP-NaCl medium. In addition, surugamide A production was repressed in TSB and YD medium, suggesting that carbon catabolite repression (CCR) may influence the production of surugamides in these strains.

**Keywords:** genome mining; OSMAC; phylogenomics; secondary metabolites; surugamides; surugamide A; marine sponge-associated bacteria; *Streptomyces*; *albidoflavus* phylogroup

---

## 1. Introduction

Members of the *Streptomyces* genus are widely known to be prolific producers of natural products. Many of these compounds have found widespread use in the pharmaceutical industry as antibiotics, immunosuppressant, antifungal, anticancer, and anti-parasitic drugs [1]. However, there continues to be an urgent need to discover new bioactive compounds, and especially antibiotics, primarily due to the emergence of antibiotic resistance in clinically important bacterial pathogens [2,3]. In particular, the increase in multi-resistant ESKAPE pathogens (*Enterococcus faecium*, *Staphylococcus aureus*, *Klebsiella pneumoniae*, *Acinetobacter baumannii*, *Pseudomonas aeruginosa*, and *Enterobacter* species) has focused research efforts to develop new antibiotics to treat these priority antibiotic-resistant bacteria [4].

Up until relatively recently, marine ecosystems had largely been neglected as a potential source for the discovery of novel bioactive compounds, in comparison to terrestrial environments, primarily due to issues of accessibility [5]. Marine sponges are known to host a variety of different bacteria and fungi, which produce a diverse range of natural products, including compounds with antiviral, antifungal, antiprotozoal, antibacterial, and anticancer activities [5,6]. Marine sponge-associated *Streptomyces* spp. are a particularly important source of bioactive compounds, with examples including *Streptomyces* sp. HB202, isolated from the sponge *Halichondria panicea,* which produces mayamycin, a compound with activity against *Staphylococcus aureus* [7]; and streptophenazines G and K, with activity against *Bacillus subtilis* [8]; together with *Streptomyces* sp. MAPS15, which was isolated from *Spongia officinalis,* which produces 2-pyrrolidine, with activity against *Klebsiella pneumoniae* [9]. Additionally, our group has reported the production of antimycins from *Streptomyces* sp. SM8 isolated from the sponge *Haliclona simulans*, with antifungal and antibacterial activities [10,11]. In further work, we genetically characterised 13 *Streptomyces* spp. that were isolated from both shallow and deep-sea sponges, which displayed antimicrobial activities against a number of clinically relevant bacterial and yeast species [12,13]. Amongst these strains, the *Streptomyces* sp. SM17 demonstrated an ability to inhibit the growth of *E. coli* NCIMB 12210, methicillin-resistant *S. aureus* (MRSA), and *Candida* spp., when employing deferred antagonism assays [12,13].

Among other clinically relevant natural products derived from marine *Streptomyces* isolates are the recently identified surugamides family of molecules. The cyclic octapeptide surugamide A and its derivatives were originally identified in the marine-derived *Streptomyces* sp. JAMM992 [14], and have been shown to belong to a particularly interesting family of compounds due not only to their relevant bioactivity, but also due to their unusual metabolic pathway involving D-amino acids [14–16]. Since their discovery, concerted efforts have been employed in order to chemically characterise these compounds and determine the genetic mechanisms involved in their production [14,15,17–20]. The surugamides and their derivatives have been shown to possess a number of bioactivities, with the surugamides A–E and the surugamides G–J being shown to possess anticancer activity by inhibiting bovine cathepsin B, a cysteine protease reported to be involved in the invasion of metastatic tumour cells [14,16]; while another derivative, namely acyl-surugamide A, has been shown to possess anti-fungal activity [16]. It has been determined that the non-ribosomal peptide synthase-encoding *surABCD* genes are the main biosynthetic genes involved in the biosynthesis of surugamides and their derivatives [19], with these genes being involved in the production of at least 20 different compounds [16]. Surugamides A–E have been reported to be produced by the *surA* and *surD* genes, while the linear decapeptide surugamide F has been shown to be produced by the *surB* and *surC* genes, involving a unique pattern of intercalation of the biosynthetic genes [19]. Further metabolic pathways studies have reported that the expression of the *surABCD* gene cluster is strongly regulated by the *surR* transcriptional repressor [16], while the cyclisation of the cyclic surugamides has been shown to involve a penicillin binding protein (PBP)-like thioesterase encoded by the *surE* gene [17,18,21].

Although apparently widespread in marine-derived *Streptomyces* isolates [18,19], the production of surugamides has also been reported in the *S. albidoflavus* strain J1074 [16,22], a derivative of the soil isolate *S. albus* G [23,24]. The *S. albidoflavus* strain J1074 is a well-characterised *Streptomyces* isolate, which is frequently used as a model for the genus and has commonly been successfully employed in the heterologous expression of biosynthetic gene clusters (BGCs) [25–29]. This strain was originally classified as an *S. albus* isolate, however, due to more recent taxonomy studies, it has been reclassified as a *S. albidoflavus* species isolate [30,31]. Interestingly, surugamides and their derivatives have been shown to only be produced by *S. albidoflavus* J1074 under specific conditions, such as when employing chemical stress elicitors [16], and more recently when cultivating the strain in a soytone-based liquid-based medium SG2 [22].

In a previous study [32], we reported that the *S. albidoflavus* J1074 and *Streptomyces* sp. SM17 possessed morphological and genetic similarities. Differences were observed, however, when both strains were exposed to high salt concentrations using culture media, such as TSB or ISP2, in which the marine sponge-derived strain SM17 grew and differentiated more rapidly in comparison with the soil strain *S. albidoflavus* J1074, which appeared to have trouble growing and differentiating when salts were present in the growth medium [32]. Genome mining based on the prediction of secondary metabolites BGCs also showed many similarities between the two strains [32]. Among these predicted BGCs, both the *S. albidoflavus* J1074 and *Streptomyces* sp. SM17 isolates appeared to possess the *sur* BGC, encoding for the production of surugamides A/D. Due to the fact that marine-derived *Streptomyces* isolates have been shown to produce good levels of surugamides when grown under standard conditions [18,19], and that production of surugamides and derivatives can be induced in the presence of chemical stress elicitors in *S. albidoflavus* J1074 [16], it appears likely that marine-derived *Streptomyces* isolates and their *sur* BGCs could share genetic similarities that might help to optimise production of the compound. To investigate this possibility we 1) employed genome mining approaches together with phylogenomics in order to better characterise the SM17 strain, and to investigate the distribution and differences/similarities between marine- (or aquatic saline-) and terrestrial-derived *sur* BGCs and *sur* BGC-harbouring microorganisms; and 2) experimentally compared the metabolic profiles of surugamide A production between a marine (SM17) and a terrestrial (J1074) *Streptomyces* isolate. With respect to the latter, we employed an "one strain many compounds" (OSMAC)-based approach, which has been shown to be a useful strategy in eliciting production of natural products from silent gene clusters by employing different culture conditions [33,34]; together with analytical chemistry methods such as liquid chromatography–mass spectrometry to monitor production of surugamide A in both *S. albidoflavus* J1074 and *Streptomyces* sp. SM17.

## 2. Materials and Methods

### 2.1. Bacterial Strains and Nucleotide Sequences

The *Streptomyces* sp. SM17 strain was isolated from the marine sponge *Haliclona simulans*, from the Kilkieran Bay, Galway, Ireland, as previously described [13]. The *Streptomyces albidoflavus* J1074 strain was provided by Dr Andriy Luzhetskyy (Helmholtz Institute for Pharmaceutical Research Saarland, Saarbrücken, Germany). Their complete genome sequences are available from the GenBank database [35] under the accession numbers NZ_CP029338 and NC_020990, for *Streptomyces* sp. SM17 and *S. albidoflavus* J1074, respectively. The surugamides biosynthetic gene cluster (*sur* BGC) sequence used as a reference for this study was the one previously described in *Streptomyces albidoflavus* LHW3101 (GenBank accession number: MH070261) [18]. Other genomes used in this study's analyses were obtained from the GenBank RefSeq database [35].

### 2.2. Phylogenetic Analyses

The NCBI BLASTN tool [36,37] was used to determine the closest 30 *Streptomyces* strains with complete genome available in the GenBank RefSeq database [35] to the *Streptomyces* sp. SM17. Then, phylogeny analysis was performed with the concatenated sequences of the 16S rRNA, and the housekeeping genes *atpD*, *gyrB*, *recA*, *rpoB*, and *trpB*. The sequences were aligned using the MAFFT program [38], and the phylogeny analysis was performed using the MrBayes program [39]. In MrBayes, the general time reversible (GTR) model of nucleotide substitution was used [40], with gamma-distributed rates across sites with a proportion of invariable sites, with 1 million generations sampled every 100 generations. Final consensus phylogenetic tree generated by MrBayes was processed using MEGA X [41], with a posterior probability cut-off of 95%.

Phylogeny analysis of the surugamides biosynthetic gene cluster (*sur* BGC) was performed by using the *S. albidoflavus* LHW3101 *sur* BGC nucleotide sequence as reference [18] and searching for similar sequences on the GenBank RefSeq database using the NCBI BLASTN tool [35–37], only taking into account complete genomes. The genome regions with similarity to the *S. albidoflavus* LHW3101 *sur* BGC undergone phylogeny analysis using the same aforementioned tools and parameters.

*2.3. Prediction of Secondary Metabolites Biosynthetic Gene Clusters*

In order to assess the similarities and differences between the *Streptomyces* isolates belonging to the *albidoflavus* phylogroup, in regard to their potential to produce secondary metabolites, BGCs were predicted in their genomes, using the antiSMASH (version 5 available at https://docs.antismash.secondarymetabolites.org/) program [42]. The predicted BGCs were then processed using the BiG-SCAPE program (version 20190604, available at https://git.wageningenur.nl/medema-group/BiG-SCAPE) [43], with the MiBIG database (version 1.4 available at https://mibig.secondarymetabolites.org/) as reference [44], and similarity clustering of gene cluster families (GCFs) was performed. The similarity network was processed using Cytoscape (version 3.7.1, available at https://cytoscape.org/) [45].

*2.4. Gene Synteny Analysis*

The genome regions previously determined to share similarities with the *S. albidoflavus* LHW3101 *sur* BGC were manually annotated, for the known main biosynthetic genes (*surABCD*), the penicillin binding protein (PBP)-like peptide cyclase and hydrolase *surE* gene, and the gene with regulatory function *surR* [15–19,21]. This was performed using the UniPro UGENE toolkit (version 1.32.0, available at http://ugene.net/) [46], the GenBank database, and the NCBI BLASTN tool (available at https://blast.ncbi.nlm.nih.gov/Blast.cgi, accessed on June 2019) [35–37]. The gene synteny and reading frame analysis was performed using the UniPro UGENE toolkit [46] and the Artemis genome browser (version 18.0.0, available at https://www.sanger.ac.uk/science/tools/artemis) [47].

*2.5. Diagrams and Figures*

All the Venn diagrams presented in this study were generated using the Venn package in R [48,49], and RStudio [50]. All the images presented in this study were edited using the Inkscape program (available from https://inkscape.org/).

*2.6. Strains Culture, Maintenance, and Secondary Metabolites Production*

The same culture media and protocols were employed for both isolates *Streptomyces* sp. SM17 and *Streptomyces albidoflavus* J1074. Glycerol stocks were prepared from spores collected from soya-mannitol (SM) medium after 8 days of cultivation at 28 °C and preserved at −20 °C. To verify the secondary metabolites production profile, spores were cultivated for 7 days on SM agar medium at 28 °C, then pre-inoculated in 5 mL TSB medium, and cultivated at 28 °C and 220 rpm for 2 days. Then, 10% (*v*/*v*) of the pre-inoculum was transferred to 30 mL of the following media: TSB; SYP-NaCl (1% starch, 0.4% yeast extract, 0.2% peptone, and 0.1% NaCl); YD (0.4% yeast extract, 1% malt extract, and 4% dextrin pH 7.0); P1 (2% glucose, 1% soluble starch, 0.1% meat extract, 0.4% yeast extract, 2.5% soy flour, 0.2% NaCl, and 0.005% $K_2HPO_4$ pH 7.3); P2 (1% glucose, 0.6% glycerol, 0.1% yeast extract, 0.2% malt extract, 0.6% $MgCl_2.6H_2O$, 0.03% $CaCO_3$, and 10% sea water); P3 (2.5% soy flour, 0.75% starch, 2.25% glucose, 0.35% yeast extract, 0.05% $ZnSO_4 \times 7H_2O$, and 0.6% $CaCO_3$ pH 6.0); CH-F2 (2% soy flour, 0.5% yeast extract, 0.2% $CaCO_3$, 0.05% citric acid, 5% glucose, and pH 7.0); SY (2.5% soluble starch, 1.5% soy flour, 0.2% yeast extract, and 0.4% $CaCO_3$ pH 7.0); Sporulation medium (2% soluble starch and 0.4 yeast extract); and Oatmeal medium (2% oatmeal). These were cultivated at 28 °C and 220 rpm for 4 days in TSB; and for 8 days in SYP-NaCl, YD, SY, P1, P2, P3, CH-F2, Sporulation, and Oatmeal media. Once the bioprocess was completed, the broth was frozen at −20 °C for further chemical analysis.

*2.7. Metabolic Profiling, Compound Isolation, and Chemical Structure Analysis*

The *Streptomyces* broth of TSB, SYP-NaCl, and YD medium cultures (180 mL) was exhaustively extracted using a solvent mixture of 1:1 MeOH:DCM yielding a crude extract (3.89 g). This crude extract was first separated using SPE on $C_{18}$ bonded silica gel (Polygoprep C18 (Fisher Scientific, Dublin, Ireland) 12%C, 60 Å, 40–63 μm), eluting with varying solvent mixtures to produce five fractions: $H_2O$ (743.62 mg), 1:1 $H_2O$:MeOH (368.6 mg), MeOH (15.4 mg), 1:1 MeOH:DCM (10.9 mg), DCM (8.2 mg). The final three fractions (MeOH, 1:1 MeOH:DCM, DCM, 34.5 mg) were then combined and subject to analytical reverse phase HPLC on a Waters Symmetry (VWR, Dublin, Ireland) C18 5 μm, 4.6 × 250 mm column. The column was eluted with 10% MeCN (0.1% TFA)/90% H2O (0.1% TFA) for 5 min, then a linear gradient to 100% MeCN (0.1% TFA) over 21 min was performed. The column was further eluted with 100% MeCN (0.1% TFA) for 6 min. After the HPLC was complete, a linear gradient back to 10% MeCN (0.1% TFA)/90% $H_2O$ (0.1% TFA) over 1 min and then further elution of 10% MeCN (0.1% FA)/90% $H_2O$ (0.1% FA) for 4 min was performed. This yielded pure surugamide A (0.8 mg). Surugamide A was characterised using MS and NMR data to confirm the structure for use as an analytical standard.

Surugamide A was quantified in the broth using LC-MS analysis on an Agilent UHR-qTOF 6540 (Agilent Technologies, Cork, Ireland) mass spectrometer. The column used for separation was Waters equity UPLC BEH (Apex Scientific, Kildare, Ireland) C18 1.7 μm 2.1 × 75 mm. The column was eluted with 10% MeCN (0.1% FA)/90% $H_2O$ (0.1% FA) for 2 min, then a linear gradient to 100% MeCN (0.1% FA) over 6 min was performed. The column was further eluted with 100% MeCN (0.1% FA) for 4 min. After the UPLC was complete, a linear gradient back to 10% MeCN (0.1% FA)/90% $H_2O$ (0.1% FA) over 1 min and then further elution of 10% MeCN (0.1% FA)/90% $H_2O$ (0.1% FA) for 3 min was performed before the next run. The MS detection method was positive ion. A calibration curve was produce using the LC-MS method above and injecting the pure surugamide A at seven concentrations (100, 25, 10, 2, 1, 0.2, 0.1 mg/L). Thirty millilitres of each *Streptomyces* strain in broth were extracted using a solvent mixture of 1:1 MeOH:DCM three times to yield a crude extract. These extracts were resuspended in MeOH and filtered through PTFE 0.2 μm filters (Sigma Aldrich, Arklow, Ireland) before being subject to the above LC-MS method.

The surugamide A calibration standards 1–7 and the six extracts were analysed using the Agilent MassHunter Quantification software package. This allowed the quantification of surugamide A in the extracts based on the intensity of peaks in the chromatogram with matching retention time and exact mass.

## 3. Results and Discussion

### 3.1. Multi-locus Sequence Analysis and Taxonomy Assignment of the Streptomyces sp. SM17 Isolate

In order to taxonomically characterise the *Streptomyces* sp. SM17 isolate based on genetic evidence, multi-locus sequence analysis (MLSA) [51] employing the 16S rRNA sequence, in addition to five housekeeping genes, namely *atpD* (ATP synthase subunit beta), *gyrB* (DNA gyrase subunit B), *recA* (recombinase RecA), *rpoB* (DNA-directed RNA polymerase subunit beta), and *trpB* (tryptophan synthase beta chain) was performed, in a similar manner to a previous report [32]. A similarity search was performed in the GenBank database [35], using the NCBI BLASTN tool [36,37], based on the 16S rRNA nucleotide sequence of the SM17 isolate. The top 30 most similar *Streptomyces* species for which complete genome sequences were available in GenBank were selected for further phylogenetic analysis.

The concatenated nucleotide sequences [51,52] of the 16S rRNA and the aforementioned five housekeeping genes, were first aligned using the MAFFT program [38], and the phylogeny analysis was performed using the MrBayes program [39]. The general time reversible (GTR) model of nucleotide substitution with gamma-distributed rates across sites with a proportion of invariable sites was applied [40], with 1 million generations sampled every 100 generations. The final phylogenetic tree was then processed using MEGA X [41], with a posterior probability cut-off of 95% (Figure 1).

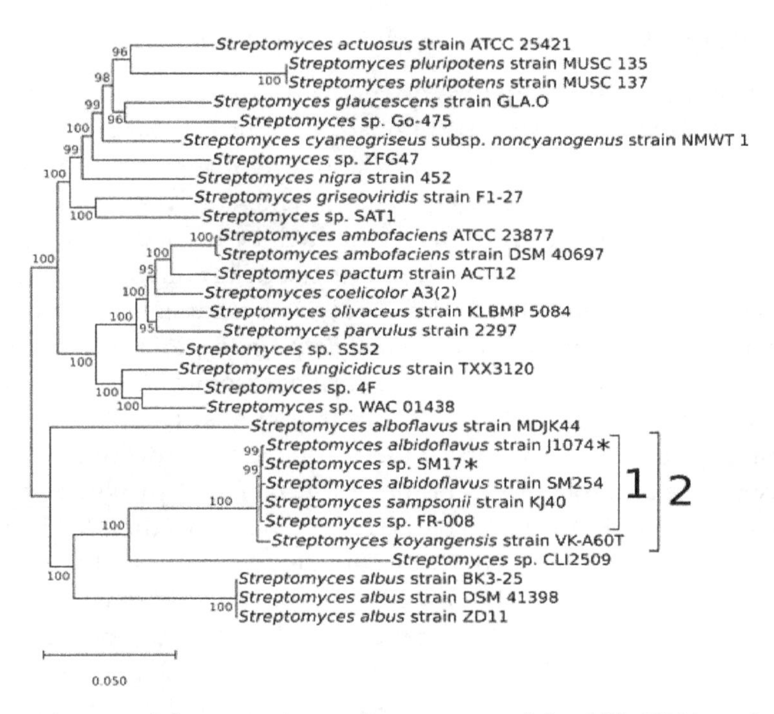


**Figure 1.** Phylogenetic tree of the concatenated sequences of the 16S rRNA and the housekeeping genes *atpD*, *gyrB*, *recA*, *rpoB*, and *trpB*, from the *Streptomyces* sp. SM17 together with 30 *Streptomyces* isolates for which complete genome sequences were available in the GenBank database. Analysis was performed using MrBayes, with a posterior probability cut-off of 95%. 1) *albidoflavus* phylogroup. 2) Clade including the neighbour isolate *Streptomyces koyangensis* strain VK-A60T. The strains SM17 and J1074 are indicated with asterisks.

The resulting phylogenetic tree clearly indicates the presence of a clade that includes the isolates *Streptomyces albidoflavus* strain J1074; *Streptomyces* sp. SM17; *Streptomyces albidoflavus* strain SM254; *Streptomyces sampsonii* strain KJ40; *Streptomyces* sp. FR-008; and *Streptomyces koyangensis* strain VK-A60T (clade 2 in Figure 1). In addition, this larger clade contains a sub-clade (clade 1 in Figure 1) that includes *Streptomyces* isolates similar to the strain *Streptomyces albidoflavus* J1074. The J1074 strain is a well-studied *Streptomyces* isolate widely used as a model for the genus and for various biotechnological applications, including the heterologous expression of secondary metabolites biosynthetic gene clusters (BGCs) [25–29]. This isolate was originally classified as "*Streptomyces albus* J1074", but due to recent taxonomy data, it has been reclassified as *Streptomyces albidoflavus* J1074 [30,31]. Hence, in this study, this strain will be referred to as *Streptomyces albidoflavus* J1074, and this clade will from now on be referred to as the *albidoflavus* phylogroup (Figure 1).

Interestingly, members of the *albidoflavus* phylogroup were all isolated from quite different environments. The *Streptomyces albidoflavus* strain J1074 stems from the soil isolate *Streptomyces albus* G [23,24]. The *Streptomyces sampsonii* strain KJ40 was isolated from rhizosphere soil in a poplar plantation [53]. The *Streptomyces* sp. strain FR-008 is a random protoplast fusion derivative of two *Streptomyces hygroscopicus* isolates [54]. On the other hand, two of these strains were isolated from aquatic saline environments, with *Streptomyces* sp. SM17 being isolated from the marine sponge *Haliclona simulans* [13]; while the *Streptomyces albidoflavus* strain SM254 strain was isolated from copper-rich subsurface fluids within an iron mine, following growth on artificial sea water (ASW) [55]. The fact that these isolates, although derived from quite distinct environmental niches, simultaneously share significant genetic similarities is interesting, and raises questions about their potential evolutionary relatedness.

### 3.2. *Analysis of Groups of Orthologous Genes in the Albidoflavus Phylogroup*

In an attempt to provide further genetic evidence with respect to the similarities shared among the members of the *albidoflavus* phylogroup (Figure 1), a pan-genome analysis was performed to determine the number of core genes, accessory genes, and unique genes present in this group of isolates. The Roary program was employed for this objective [56], which allowed the identification of groups of orthologous and paralogous genes (which from now on will be referred to simply as "genes") present in the set of *albidoflavus* genomes, with a protein identity cut-off of 95%, which is the identity value recommended by the Roary program manual when analysing organisms belonging to the same species.

A total of 7565 genes were identified in the *albidoflavus* pan-genome, and among these a total of 5177 were determined to be shared among all the *albidoflavus* isolates (i.e., the core genome) (Figure 2). This represents a remarkably high proportion of genes that appear to be highly conserved between all the isolates, representing approximately 68.4% of the pan-genome. Additionally, when considering the genomes individually (Table S1), the core genome accounts for approximately 84.5% of the FR-008 genome; 88.5% of J1074; 85.5% of KJ40; 86.7% of SM17; and 83.7% of the SM254 genome. On the other hand, the accessory genome (i.e., genes present in at least two isolates) was determined to consist of 1055 genes (or ~13.9% of the pan-genome); while the unique genome (i.e., genes present in only one isolate) was determined to consist of 1333 genes (or ~17.6% of the pan-genome). This strikingly high conservation of genes present in their genomes together with the previous multi-locus phylogeny analysis are very strong indicators that these microorganisms may belong to the same species.

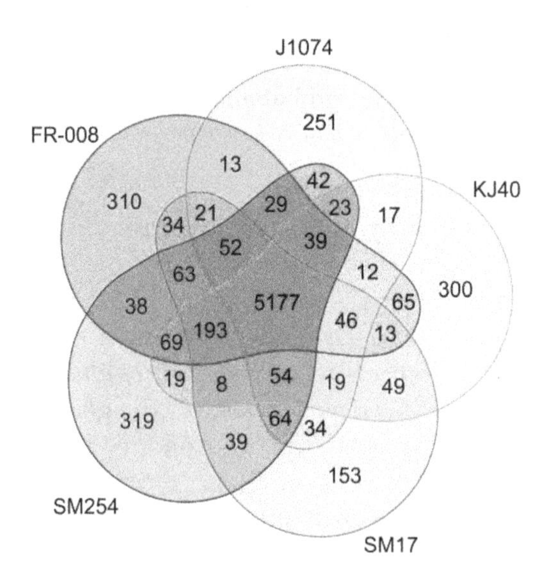


**Figure 2.** Venn diagram representing the presence/absence of groups of orthologous genes in the organisms belonging to the *albidoflavus* phylogroup.

An additional pan-genome analysis similar to the aforementioned analysis was also performed including the *Streptomyces koyangensis* strain VK-A60T in the dataset (Figure S1), which was an isolate shown to be a closely related neighbour to the *albidoflavus* phylogroup (Figure 1, clade 2). When compared to the previous analysis, the pan-genome analysis including the VK-A60T isolate showed significant changes in the values representing the core genome, which changed from 5177 genes (Figure 2) to 3912 genes (Figure S1), with an additional 1273 genes also shared among all of the *albidoflavus* isolates (Figure S1). The results also showed a much larger number of genes uniquely present in the VK-A60T genome than in the other genomes, with 2059 unique genes identified from a total of 6245 CDSs present in the VK-A60T genome in total, or approximately a third of its total number of genes (Figure S1). This proportion of unique genes present in the VK-A60T genome is considerably higher than the proportions of unique genes observed in the other *albidoflavus* phylotype genomes (Figure 2), which accounted for approximately only 2.5% of the total number of genes in

SM17; 4.2% in J1074; 4.9% in KJ40; 5% in FR-008; and 5.1% in SM254. Taken together, these results further demonstrate the similarities between the isolates belonging to the *albidoflavus* phylogroup, while the VK-A60T isolate is clearly more distantly related.

Thus, from previous studies [30,31] and in light of the phylogeny analysis and further genomic evidence presented in this study, it is likely that all the isolates belonging to the *albidoflavus* phylogroup are in fact members of the same species. It is reasonable to infer that, for example, the isolates in the *albidoflavus* phylogroup that possess no species assignment thus far (i.e., strains SM17 and FR-008) are indeed members of the *albidoflavus* species. Also, it is possible that the *Streptomyces sampsonii* KJ40 has been misassigned, and possibly requires reclassification as an *albidoflavus* isolate.

Misassignment and reclassification of *Streptomyces* species is a common issue, and an increase in the quantity and the quality of available data from these organisms (e.g., better-quality genomes available in the databases) will provide better support for taxonomy claims, or correction of these when new information becomes available [31,57–59].

### *3.3. Prediction of Secondary Metabolites Biosynthetic Gene Clusters in the Albidoflavus Phylogroup*

Isolates belonging to the *albidoflavus* phylogroup have been reported to produce bioactive compounds of pharmacological relevance, such as antibiotics. As mentioned previously, the *Streptomyces albidoflavus* strain J1074 is the best described member of the *albidoflavus* phylogroup to date. As such, several of secondary metabolites produced by this isolate have been identified, including acyl-surugamides and surugamides with antifungal and anticancer activities, respectively [16]; together with paulomycin derivatives with antibacterial activity [60]. The *Streptomyces* sp. FR-008 isolate has been shown to produce the antimicrobial compound FR-008/candicidin [61,62]; while the *Streptomyces sampsonii* KJ40 isolate has been shown to produce a chitinase that possesses anti-fungal activity against plant pathogens [53]. On the other hand, although no bioactive compounds have been characterised from *Streptomyces albidoflavus* SM254, this isolate has been shown to possess anti-fungal activity, specifically against the fungal bat pathogen *Pseudogymnoascus destructans*, which is responsible for the White-nose Syndrome [55,63]. The *Streptomyces* sp. SM17 isolate has also previously been shown to possess antibacterial and antifungal activities against clinically relevant pathogens, including methicillin-resistant *Staphylococcus aureus* (MRSA) [13]. However, no natural products derived from this strain have been identified and isolated until now.

In order to further *in silico* assess the potential of these *albidoflavus* phylogroup isolates to produce secondary metabolites, and also to determine how potentially similar or diverse they are within this phylogroup, prediction of secondary metabolites biosynthetic gene clusters (BGCs) was performed using the antiSMASH (version 5) program [42]. The antiSMASH prediction was processed using the BiG-SCAPE program [43], in order to cluster the BGCs into gene cluster families (GCFs), based on sequence and Pfam [64] protein families similarity, and also by comparing them to the BGCs available from the minimum information about a biosynthetic gene cluster (MiBIG) repository [44] (Figure 3). When compared to known BGCs from the MiBIG database, a significant number of BGCs predicted to be present in the *albidoflavus* phylogroup genomes could potentially encode for the production of novel compounds, including those belonging to the non-ribosomal peptide synthetase (NRPS) and bacteriocin families of compounds (Figure 3). The presence/absence of homologous BGCs in the *albidoflavus* isolates' genomes was determined using BiG-SCAPE and is represented in Figure 4. Interestingly, the vast majority of the BGCs predicted in the *albidoflavus* phylogroup are shared among all of its members (15 BGCs); while another large portion (8 BGCs) are present in at least two isolates (Figure 4). Among the five members of the *albidoflavus* phylogroup, only the J1074 strain and the SM17 strain appeared to possess unique BGCs when compared to the other strains. Three unique BGCs were predicted to be present in the J1074 genome: a predicted type I polyketide synthase (T1PKS)/NRPS without significant similarity to the BGCs from the MiBIG database; a predicted bacteriocin, which also did not show any significant similarity to the BGCs from the MiBIG database; and a BGC predicted to encode for the production of the antibiotic paulomycin, with similarity to the paulomycin-encoding

BGCs from *Streptomyces paulus* and *Streptomyces* sp. YN86 [65], which has also been experimentally shown to be produced by the J1074 strain [60]. One BGC predicted to encode a type III polyketide synthase (T3PKS)—with no significant similarity to the BGCs from the MiBIG database—was also identified as being unique to the SM17 genome.

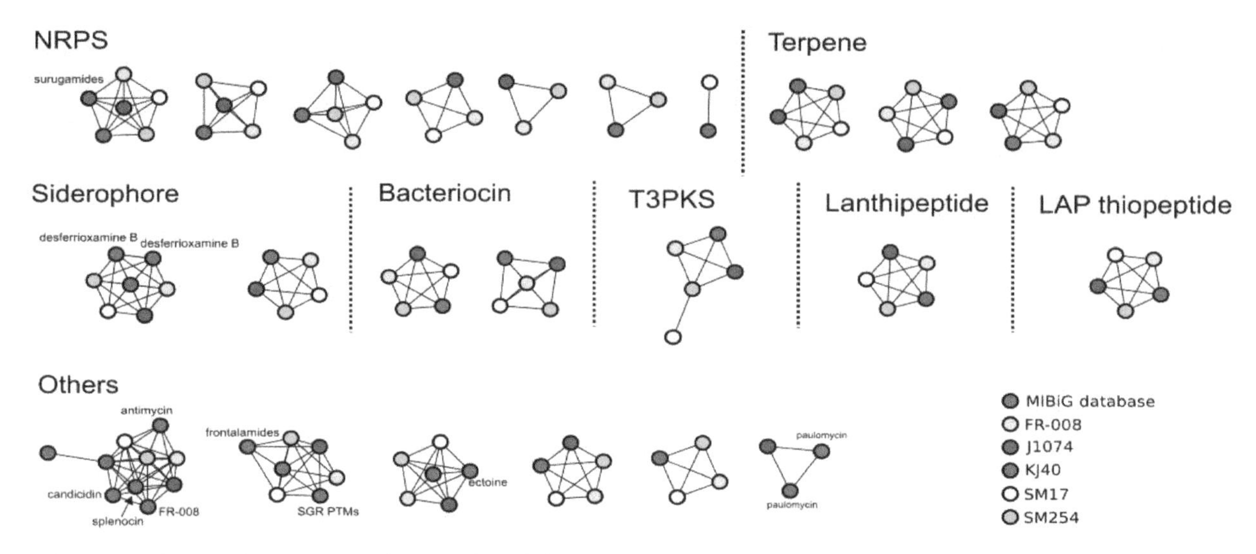


**Figure 3.** Biosynthetic gene clusters (BGCs) similarity clustering using BiG-SCAPE. Singletons, i.e., BGCs without significant similarity with the BGCs from the minimum information about a biosynthetic gene cluster (MiBIG) database or with the BGCs predicted in other genomes, are not represented.

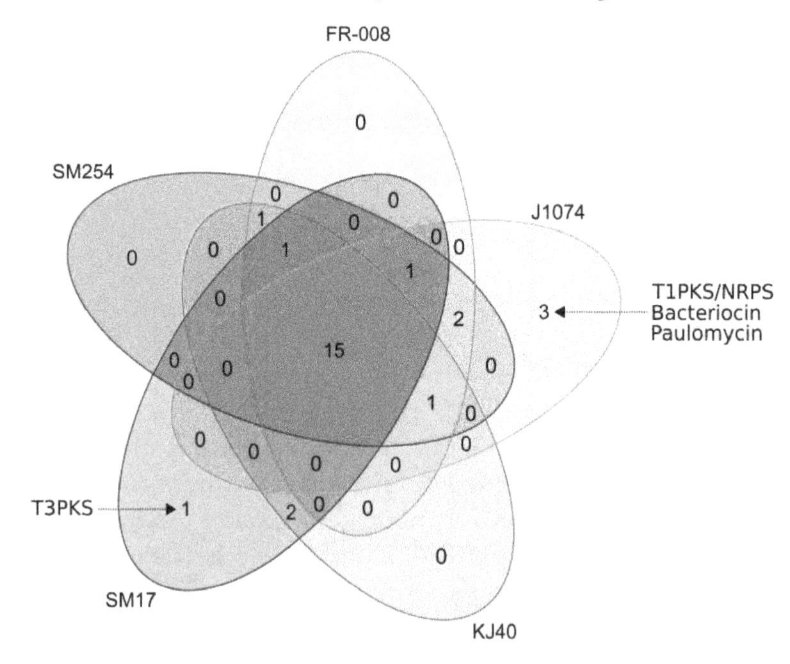


**Figure 4.** Venn diagram representing BGCs presence/absence in the genomes of the members of the *albidoflavus* phylogroup, determined using antiSMASH and BiG-SCAPE.

Importantly, BGCs with similarity to the surugamide A/D BGC from "*Streptomyces albus* J1074" (now classified as *S. albidoflavus*) from the MiBIG database [16] were identified in all the other genomes of the members of the *albidoflavus* phylogroup. This raises the possibility that this BGC may be commonly present in *albidoflavus* species isolates. However, as only a few complete genomes of isolates belonging to this phylogroup are currently available, further data will be required to support this hypothesis. Nevertheless, these results further highlight the genetic similarities of the isolates belonging to the *albidoflavus* phylogroup, even with respect to their potential to produce secondary metabolites.

### 3.4. Phylogeny and Gene Synteny Analysis of Sur BGC Homologs

In parallel to the previous phylogenomics analysis performed with the *albidoflavus* phylogroup isolates, sequence similarity and phylogenetic analyses were performed, using the previously described and experimentally characterised *Streptomyces albidoflavus* LHW3101 surugamides biosynthetic gene cluster (*sur* BGC, GenBank accession number: MH070261) as a reference [18]. The aim was to assess how widespread in nature the *sur* BGC might be, and the degree of genetic variation, if any; that might be present in *sur* BGCs belonging to different microorganisms.

Nucleotides sequence similarity to the *sur* BGC was performed in the GenBank database [35], using the NCBI BLASTN tool [36,37]. It is important to note that, since the quality of the data is crucial for sequence similarity, homology, and phylogeny inquiries, only complete genome sequences were employed in this analysis. For this reason, for example, the marine *Streptomyces* isolate in which surugamides and derivatives were originally identified, namely *Streptomyces* sp. JAMM992 [14], was not included, since its complete genome is not available in the GenBank database.

The sequence similarity analysis identified five microorganisms that possessed homologs to the *sur* BGC and had their complete genome sequences available in the GenBank database: *Streptomyces* sp. SM17; *Streptomyces albidoflavus* SM254; *Streptomyces* sp. FR-008; *Streptomyces albidoflavus* J1074; and *Streptomyces sampsonii* KJ40. Notably, these results overlapped with the isolates belonging to the previously discussed *albidoflavus* phylogroup (Figure 1), further highlighting the possibility that the *sur* BGC may be commonly present in and potentially exclusive to the *albidoflavus* species.

Phylogenetic analysis was performed in the genomic regions determined to be homologs to the *Streptomyces albidoflavus* LHW3101 *sur* BGC, using the MrBayes program [39] (Figure 5). Although a larger number of sequences should ideally be employed in this type of analysis, these results suggest the possibility of a clade with aquatic saline environment-derived *sur* BGCs (Figure 5). Thus, these aquatic saline environment-derived *sur* BGCs are likely to share more genetic similarities amongst each other, rather than with those derived from terrestrial environments. Since this analysis took into consideration the whole genome regions that contained the *sur* BGCs of each isolate, it is likely that the similarities and differences present in these regions involve not only coding sequences (CDSs) for biosynthetic genes and/or transcriptional regulators, but also could include promoter regions and other intergenic sequences.

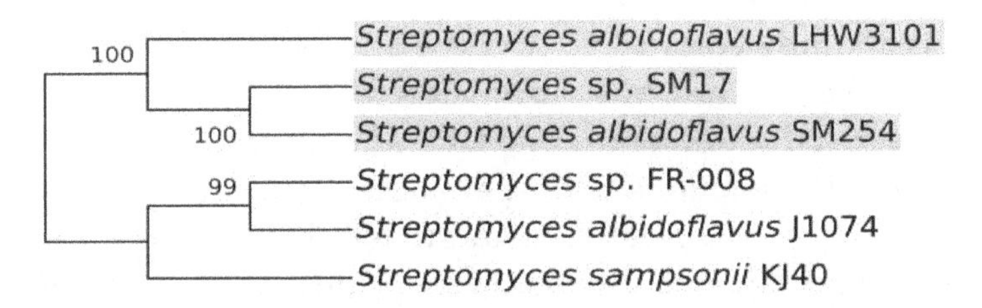


**Figure 5.** Consensus phylogenetic tree of the *sur* BGC region of the *S. albidoflavus* LHW3101 reference *sur* BGC sequence, plus five *Streptomyces* isolates determined to have *sur* BGC homologs, generated using MrBayes and Mega X, with a 95% posterior probability cut-off. Aquatic saline environment-derived isolates are highlighted in cyan.

With this in mind, the genomic regions previously determined to share homology with the *sur* BGC from *S. albidoflavus* LHW3101 were further analysed, with respect to the genes present in the surrounding region, the organisation of the BGCs, together with the overall gene synteny (Figure 6). Translated CDSs predicted in the region were manually annotated using the NCBI BLASTP tool [36,37], together with GenBank [35] and the CDD [66] databases. These included the main biosynthetic genes, namely *surABCD*, the transcriptional regulator *surR*, and the thioesterase *surE*—all of which had previously been reported to have roles in the biosynthesis of surugamides and their derivatives [15–19] (Figure 6).

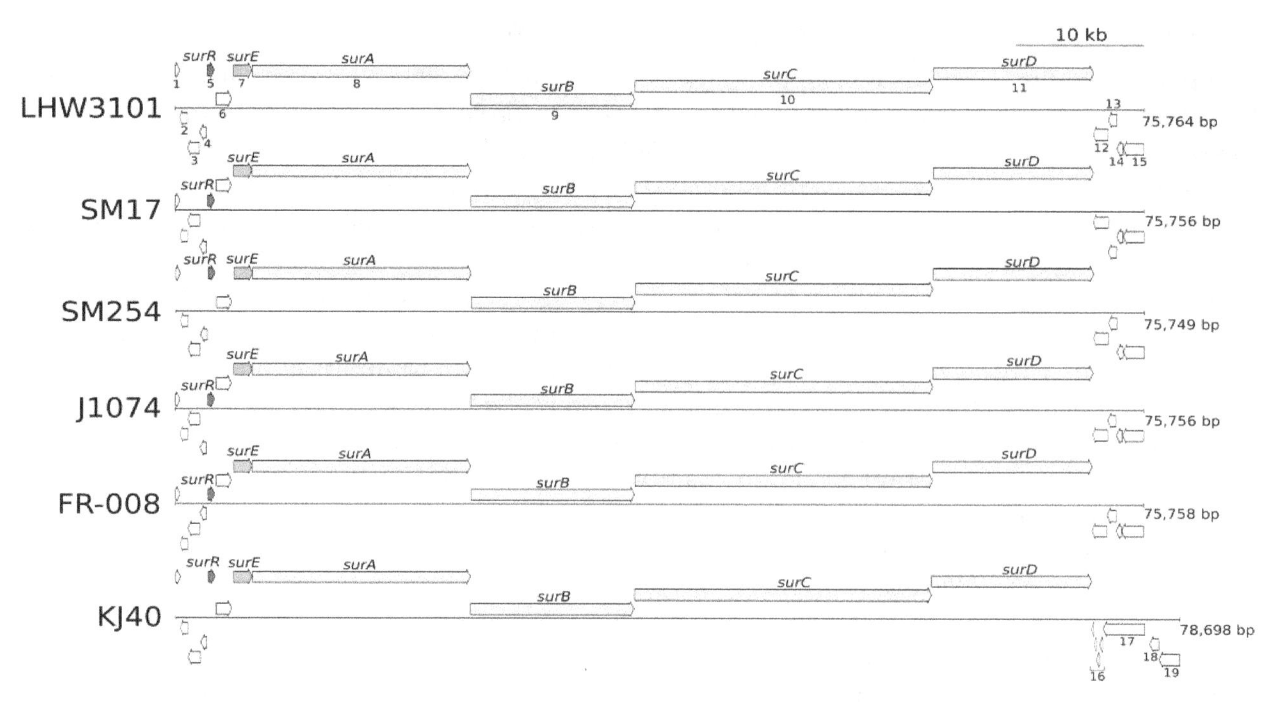


**Figure 6.** Gene synteny of the *sur* BGC region, including the reference *sur* BGC nucleotide sequence (LHW3101) and each of the *albidoflavus* phylogroup genomes. Arrows at different positions represent genes transcribed in different reading frames.

Interestingly, this result indicated that the gene synteny of the biosynthetic genes as well as the flanking genes is highly conserved, with the exception to the 3′ flanking region of the BGC from *S. sampsonii* KJ40. Notably, even the reading frames of the *surE* gene and the *surABCD* genes are conserved amongst all the genomes. As indicated by the numbers in Figure 6, the 5′ region in all the genomic regions consisted of: 1) A MbtH-like protein, which have been reported to be involved in the synthesis of non-ribosomal peptides, antibiotics, and siderophores, in *Streptomyces* species [67,68]; 2) a putative ABC transporter, which is a family of proteins with varied biological functions, including conferring resistance to drugs and other toxic compounds [69,70]; 3) a BcrA family ABC transporter, which is a family commonly involved in peptide antibiotics resistance [71,72]; 4) a hypothetical protein; followed by 5) the transcriptional repressor SurR, which has been experimentally demonstrated to repress the production of surugamides [16]; 6) a hypothetical membrane protein; 7) the thioesterase SurE, which is homologous to the penicillin binding protein, reported to be responsible for the cyclisation of surugamides molecules [21]; and finally 8–11) the main surugamides biosynthetic genes *surABCD*, all of which encode non-ribosomal peptide synthetase (NRPS) proteins [19]. The 3′ flanking region consisted of: 12) A predicted multi-drug resistance (MDR) transporter belonging to the major facilitator superfamily (MFS) of membrane transport proteins [73,74]; 13) a predicted TetR/AcrR transcriptional regulator, which is a family of regulators reported to be involved in antibiotic resistance [75]; 14) a hypothetical protein; and 15) another predicted MDR transporter belonging to the MFS superfamily. In contrast, the 3′ flanking region of the KJ40 strain *sur* BGC, consisted of: 16) A group of four hypothetical proteins, which may represent pseudogene versions of the first MDR transporter identified in the other isolates (gene number 12 in Figure 6); 17) a predicted rearrangement hotspot (RHS) repeat protein, which is a family of proteins reported to be involved in mediating intercellular competition in bacteria [76]; 18) a hypothetical protein; and 19) a MDR transporter belonging to the MFS superfamily, which, interestingly, is a homolog of protein number 15, which is present in all the other isolates.

The conserved gene synteny observed in the *sur* BGC genomic region, particularly those positioned upstream of the main biosynthetic *surABCD* genes, together with the observation that even the reading frames of the *surE* and the *surABCD* genes are conserved among all the genomes analysed, coupled with the previous phylogenetic and pan-genome analyses, suggest the following. Firstly, it is very likely that these strains share a common ancestry and that the *sur* BGC genes had a common origin. Secondly, there is a strong evolutionary pressure ensuring the maintenance of not only gene synteny, but also of the reading frames of the main biosynthetic genes involved in the production of surugamides. The latter raises the question of which other genes in this region may be involved in the production of these compounds, or potentially conferring mechanisms of self-resistance to surugamides in the isolates, particularly since many of the genes have predicted functions that are compatible with the transport of small molecules and with multi-drug resistance. These observations are particularly interesting considering that these strains are derived from quite varied environments and geographic locations.

### *3.5. Growth, Morphology, Phenotype, and Metabolism Assessment of Streptomyces sp. SM17 in Complex Media*

In order to assess the metabolic potential of the SM17 strain [77], particularly with respect to the production of surugamide A, the isolate was cultivated in a number of different growth media, within an OSMAC-based approach [33,34]. While the SM17 strain was able to grow in SYP-NaCl, YD, SY, P1, P2, P3, and CH-F2 liquid media, the strain was unable to grow in Oatmeal and Sporulation media. The latter indicated an inability to metabolise oat and starch when nutrients other than yeast extract are not present. Morphologically, the SM17 strain formed cell aggregates or pellets in TSB, YD, and SYP-NaCl, while this differentiation was not observed in the other media. Preliminary chemical analyses of these samples, employing liquid chromatography–mass spectrometry (UPLC-DAD-HRMS), indicated that secondary metabolism in SM17 was not very active when the strain was cultivated in SY, P1, P2, P3, and CH-F2 media. In contrast, significant production of surugamides was evidenced in the extracts from TSB, SYP-NaCl, and YD media, with characteristic ions at *m/z* 934.6106 (surugamide A) and 920.5949 (surugamide B) $[M + Na]^+$, which correlated with the formation of cell pellets and the production of natural products, as previously described in other *Streptomyces* strains [77,78].

### *3.6. Differential Production of Surugamide A by Streptomyces sp. SM17 and S. albidoflavus J1074*

To confirm the production of surugamide A by the SM17 isolate, extracts from the TSB, SYP-NaCl, and YD media were combined and purified using high-performance liquid chromatography (HPLC). The structures of the major compounds of the extract were subsequently analysed using nuclear magnetic resonance (NMR) spectroscopy, which allowed for the identification of the chemical structure of the surugamide A molecule as major metabolite by comparison with reference NMR data (Figure 7) [14].

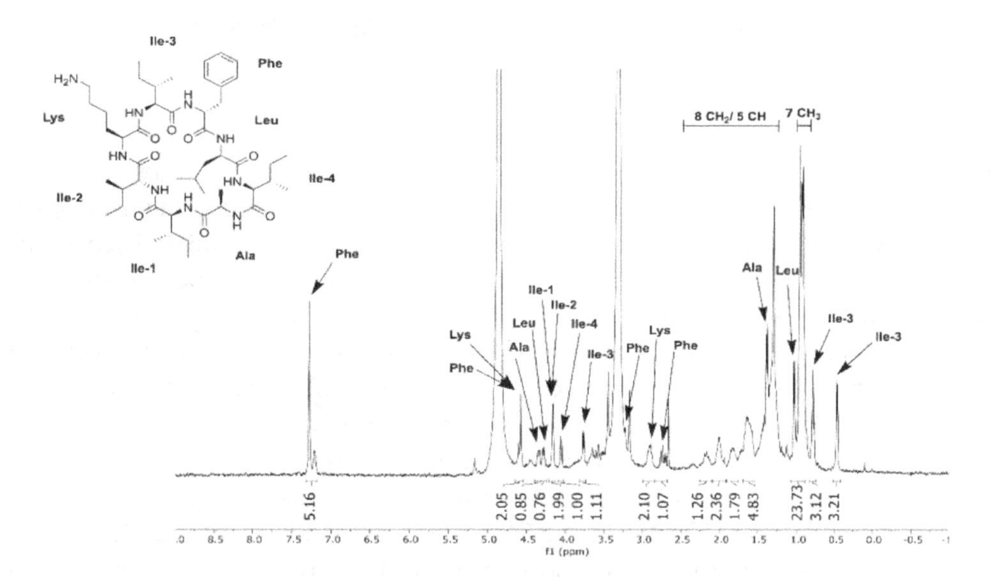


**Figure 7.** Structure of surugamide A isolated from SM17 grown in TSB, SYP-NaCl, and YD medium with annotated $^1$H NMR spectrum obtained in $CD_3OD$ at 500 MHz.

The isolates *Streptomyces* sp. SM17 and *S. albidoflavus* J1074 were subsequently cultivated in the aforementioned media in which the SM17 strain had been shown to be metabolically active, namely the TSB, SYP-NaCl, and YD media. This was performed in order to assess whether there were any significant differences in the production of surugamide A when different growth media are employed for the production of this compound, and to compare the levels of surugamide A produced by the SM17 and the J1074 isolates. The MeOH/DCM (1:1) extracts from the aforementioned cultures of SM17 and J1074 were subjected to liquid chromatography–mass spectrometry (UPLC-HRMS) to quantify the levels of surugamide A being produced under each condition (Table 1), using a surugamide A standard calibration curve (Figure S2).

**Table 1.** Surugamide A production by SM17 and J1074 measured using different media.

| Strain | Media | Percent (*w*/*w*) of Extract | Concentration of Surugamide A (mg/L) Corrected in 5 mg/mL of Extract |
|---|---|---|---|
| SM17 | TSB | 2.44% | 122.01 |
| SM17 | SYP-NaCl | 10.60% | 530.15 |
| SM17 | YD | 1.13% | 56.27 |
| J1074 | TSB | 0.27% | 13.32 |
| J1074 | SYP-NaCl | 3.55% | 176.82 |
| J1074 | YD | 0.09% | 4.26 |

The LC-MS quantification analysis (Table 1) indicated that both strains were capable of producing surugamide A in all the conditions tested. However, the SM17 strain appeared to produce considerably higher yields of the compound when compared to J1074, in all the conditions analysed. In addition, the *S. albidoflavus* J1074 isolate appeared to produce quite low levels of surugamide A when grown in TSB and YD media, accounting for less than 1% (*w*/*w*) of the extracts from these media. Interestingly, higher yields of surugamide A were produced in the SYP-NaCl medium in both strains, when compared with the levels of surugamide A produced by these strains when grown in TSB and the YD media (Table 1). In the SM17 culture in SYP-NaCl, surugamide A accounted for 10.60% (*w*/*w*) of the extract, compared to 2.44% and 1.13% from TSB and YD, respectively; while in J1074 it accounted for 3.55% (*w*/*w*) of the extract from the SYP-NaCl culture, compared to 0.27% and 0.09% from TSB and YD, respectively (Table 1). These results provide further insights into factors that are potentially involved in regulation the biosynthesis of surugamide A, in the *albidoflavus* phylogroup and in *Streptomyces* sp. SM17 in particular.

Firstly, it appears likely that surugamide A biosynthesis may be regulated, at least in part, by carbon catabolite repression (CCR). Carbon catabolite repression is a well-described regulatory mechanism in bacteria that controls carbon metabolism [79–82], and which has also been reported to regulate the biosynthesis of secondary metabolites in a number of different bacterial species, including in *Streptomyces* isolates [83–86]. While the TSB and the YD media contain glucose and dextrins as carbon sources, respectively; the complex polysaccharide starch is the carbon source in the SYP-NaCl medium. Therefore, it is reasonable to infer that glucose and dextrin may repress the production of surugamide A in *Streptomyces* sp. SM17 and in *Streptomyces albidoflavus* J1074, while starch does not. Further evidence for this can be found when considering the different production media previously employed in the production of surugamides by different *Streptomyces* isolates. For example, in the original research that led to the discovery of surugamides in *Streptomyces* sp. JAMM992 [14], the PC-1 medium (1% starch, 1% polypeptone, 1% meat extract, 1% molasses, pH 7.2) was employed for production of these compounds. Similar to the SYP-NaCl medium employed in our study, the PC-1 medium also contains starch as the carbon source, together with another complex carbon source, namely molasses. Likewise, for the production of surugamides in *S. albidoflavus* strain LHW3101 [18], the TSBY medium (3% tryptone soy broth, 10.3% sucrose, 0.5% yeast extract) was employed, which utilises sucrose as its main carbon source. In contrast, when elicitors were employed to induce the production of surugamides and their derivatives in the J1074 strain [16], by activating the *sur* BGC, which appeared to be silent in this isolate, the R4 medium (0.5% glucose, 0.1% yeast extract, among other non-carbon related components) was employed, which utilises glucose as its main carbon supply, and, as shown in this study, it potentially represses the production of surugamide A. Thus, from these previous reports and from our observations, it appears likely that CCR plays an important role in regulating the biosynthesis of surugamides.

Secondly, it is important to note the presence of salts in the form of NaCl in the SYP-NaCl medium. As previously mentioned, genetic and phylogenetic analyses of the *sur* BGC indicated similarities between those BGCs belonging to aquatic saline-derived *Streptomyces* isolates (Figure 5), together with the likelihood that these *sur* BGCs might have had a common origin. Thus, it is plausible that this origin may have been marine, and hence the presence of salts in the growth medium may also have an influence on the biosynthesis of surugamide A. Different concentrations of salts in the form of NaCl in the culture medium have also previously been shown to impact on the chemical profile of metabolites produced in the marine-obligate bacteria *Salinispora arenicola* [87].

Nevertheless, it is interesting to observe that, despite the repression/induction of the biosynthesis of surugamide A observed when different media were employed, the SM17 isolate clearly produces considerably higher amounts of surugamide A when compared to *S. albidoflavus* J1074—reaching yields up to >13-fold higher in the YD medium, and around 3-fold higher when grown in the SYP-NaCl medium (Table 1).

## 4. Conclusions

Marine-derived bacteria, particularly those isolated in association with marine invertebrates, such as sponges, have been shown to be reservoirs of bioactive molecules, including those with antibacterial, antifungal, and anticancer activities. Among these newly identified bioactive compounds, the surugamides and their derivatives are of particular interest due to their clinically relevant bioactivities, i.e., anticancer and antifungal, and their original metabolic pathway.

Based on genome mining, this study identified the previously unreported capability of the marine sponge-derived isolate *Streptomyces* sp. SM17 to produce surugamide A and also sheds new light on factors such as the carbon catabolite repression (CCR) that may be involved in regulating production of this molecule. Phylogenomics analysis indicated that the *sur* BGC is commonly present in members of the proposed *albidoflavus* phylogroup, and that the *sur* BGCs present in different isolates derived from varied environmental niches may possess a common ancestry. Although high quality genomic data from this proposed *albidoflavus* phylogroup are still lacking, results presented here suggest that the *sur* BGCs derived from *Streptomyces* isolated from aquatic saline environment are more similar to each other, when compared to those isolated from terrestrial environments.

Chemical analysis was performed in order to assess differential production of surugamide A when comparing a marine *Streptomyces* isolate with a terrestrial *Streptomyces* isolate, namely SM17 and J1074 strains, respectively, following an OSMAC-based approach employing different culture media. This analysis showed that not only the marine-derived isolate SM17 was capable of producing more surugamide A when compared to J1074 under all the conditions tested, but also that the biosynthesis of surugamide A is likely to be influenced by the CCR, and potentially by the presence of salts in the growth medium. These results also highlight the importance of employing an OSMAC-based approach even when analysing the production of known compounds, since there is a clear difference in the yields of surugamide A obtained when employing different culture media. Thus, it is possible to gain further insights into the production of bacterial types of compounds by 1) discovering strains that possess a higher capability to produce these compounds; 2) establishing optimal conditions for the biosynthesis of their production; and 3) providing a better understanding of the genetic and regulatory mechanisms potentially underpinning the production of these compounds.

**Author Contributions:** Conceived and designed the experiments: E.L.A., A.F.C.R, O.P.T, A.D.W.D. Performed the experiments: E.L.A., A.F.C.R., N.K., L.K.J. Analysed the data: E.L.A., A.F.C.R., N.K., L.K.J., O.P.T., S.A.J., A.D.W.D. Wrote the paper: E.L.A., O.P.T., A.D.W.D.

## References

1. Hwang, K.-S.; Kim, H.U.; Charusanti, P.; Palsson, B.Ø.; Lee, S.Y. Systems biology and biotechnology of *Streptomyces* species for the production of secondary metabolites. *Biotechnol. Adv.* **2014**, *32*, 255–268. [CrossRef] [PubMed]
2. Thabit, A.K.; Crandon, J.L.; Nicolau, D.P. Antimicrobial resistance: Impact on clinical and economic outcomes and the need for new antimicrobials. *Expert Opin. Pharmacother.* **2015**, *16*, 159–177. [CrossRef] [PubMed]
3. Tommasi, R.; Brown, D.G.; Walkup, G.K.; Manchester, J.I.; Miller, A.A. ESKAPEing the labyrinth of antibacterial discovery. *Nat. Rev. Drug Discov.* **2015**, *14*, 529–542. [CrossRef] [PubMed]
4. Demers, D.; Knestrick, M.; Fleeman, R.; Tawfik, R.; Azhari, A.; Souza, A.; Vesely, B.; Netherton, M.; Gupta, R.; Colon, B.; et al. Exploitation of Mangrove Endophytic Fungi for Infectious Disease Drug Discovery. *Mar. Drugs* **2018**, *16*, 376. [CrossRef] [PubMed]
5. Indraningrat, A.; Smidt, H.; Sipkema, D. Bioprospecting Sponge-Associated Microbes for Antimicrobial Compounds. *Mar. Drugs* **2016**, *14*, 87. [CrossRef] [PubMed]
6. Calcabrini, C.; Catanzaro, E.; Bishayee, A.; Turrini, E.; Fimognari, C. Marine Sponge Natural Products with Anticancer Potential: An Updated Review. *Mar. Drugs* **2017**, *15*, 310. [CrossRef]
7. Schneemann, I.; Kajahn, I.; Ohlendorf, B.; Zinecker, H.; Erhard, A.; Nagel, K.; Wiese, J.; Imhoff, J.F. Mayamycin, a Cytotoxic Polyketide from a *Streptomyces* Strain Isolated from the Marine Sponge *Halichondria panicea*. *J. Nat. Prod.* **2010**, *73*, 1309–1312. [CrossRef] [PubMed]
8. Kunz, A.; Labes, A.; Wiese, J.; Bruhn, T.; Bringmann, G.; Imhoff, J. Nature's Lab for Derivatization: New and Revised Structures of a Variety of Streptophenazines Produced by a Sponge-Derived *Streptomyces* Strain. *Mar. Drugs* **2014**, *12*, 1699–1714. [CrossRef]
9. Sathiyanarayanan, G.; Gandhimathi, R.; Sabarathnam, B.; Seghal Kiran, G.; Selvin, J. Optimization and production of pyrrolidone antimicrobial agent from marine sponge-associated *Streptomyces* sp. MAPS15. *Bioprocess Biosyst. Eng.* **2014**, *37*, 561–573. [CrossRef]
10. Almeida, E.L.; Margassery, L.M.; Kennedy, J.; Dobson, A.D.W. Draft Genome Sequence of the Antimycin-Producing Bacterium *Streptomyces* sp. Strain SM8, Isolated from the Marine Sponge *Haliclona simulans*. *Genome Announc.* **2018**, *6*, e01535-17. [CrossRef]
11. Viegelmann, C.; Margassery, L.; Kennedy, J.; Zhang, T.; O'Brien, C.; O'Gara, F.; Morrissey, J.; Dobson, A.; Edrada-Ebel, R. Metabolomic Profiling and Genomic Study of a Marine Sponge-Associated *Streptomyces* sp. *Mar. Drugs* **2014**, *12*, 3323–3351. [CrossRef] [PubMed]
12. Jackson, S.; Crossman, L.; Almeida, E.; Margassery, L.; Kennedy, J.; Dobson, A. Diverse and Abundant Secondary Metabolism Biosynthetic Gene Clusters in the Genomes of Marine Sponge Derived *Streptomyces* spp. Isolates. *Mar. Drugs* **2018**, *16*, 67. [CrossRef] [PubMed]
13. Kennedy, J.; Baker, P.; Piper, C.; Cotter, P.D.; Walsh, M.; Mooij, M.J.; Bourke, M.B.; Rea, M.C.; O'Connor, P.M.; Ross, R.P.; et al. Isolation and Analysis of Bacteria with Antimicrobial Activities from the Marine Sponge *Haliclona simulans* Collected from Irish Waters. *Mar. Biotechnol.* **2009**, *11*, 384–396. [CrossRef] [PubMed]
14. Takada, K.; Ninomiya, A.; Naruse, M.; Sun, Y.; Miyazaki, M.; Nogi, Y.; Okada, S.; Matsunaga, S. Surugamides A–E, Cyclic Octapeptides with Four d-Amino Acid Residues, from a Marine *Streptomyces* sp.: LC–MS-Aided Inspection of Partial Hydrolysates for the Distinction of d- and l-Amino Acid Residues in the Sequence. *J. Org. Chem.* **2013**, *78*, 6746–6750. [CrossRef] [PubMed]
15. Matsuda, K.; Kuranaga, T.; Sano, A.; Ninomiya, A.; Takada, K.; Wakimoto, T. The Revised Structure of the Cyclic Octapeptide Surugamide A. *Chem. Pharm. Bull.* **2019**, *67*, 476–480. [CrossRef] [PubMed]
16. Xu, F.; Nazari, B.; Moon, K.; Bushin, L.B.; Seyedsayamdost, M.R. Discovery of a Cryptic Antifungal Compound from *Streptomyces albus* J1074 Using High-Throughput Elicitor Screens. *J. Am. Chem. Soc.* **2017**, *139*, 9203–9212. [CrossRef] [PubMed]
17. Thankachan, D.; Fazal, A.; Francis, D.; Song, L.; Webb, M.E.; Seipke, R.F. A trans-Acting Cyclase Offloading Strategy for Nonribosomal Peptide Synthetases. *ACS Chem. Biol.* **2019**, *14*, 845–849. [CrossRef] [PubMed]

18. Zhou, Y.; Lin, X.; Xu, C.; Shen, Y.; Wang, S.-P.; Liao, H.; Li, L.; Deng, H.; Lin, H.-W. Investigation of Penicillin Binding Protein (PBP)-like Peptide Cyclase and Hydrolase in Surugamide Non-ribosomal Peptide Biosynthesis. *Cell Chem. Biol.* **2019**, *26*, 737–744.e4. [CrossRef] [PubMed]
19. Ninomiya, A.; Katsuyama, Y.; Kuranaga, T.; Miyazaki, M.; Nogi, Y.; Okada, S.; Wakimoto, T.; Ohnishi, Y.; Matsunaga, S.; Takada, K. Biosynthetic Gene Cluster for Surugamide A Encompasses an Unrelated Decapeptide, Surugamide F. *ChemBioChem* **2016**, *17*, 1709–1712. [CrossRef]
20. Kuranaga, T.; Matsuda, K.; Sano, A.; Kobayashi, M.; Ninomiya, A.; Takada, K.; Matsunaga, S.; Wakimoto, T. Total Synthesis of the Nonribosomal Peptide Surugamide B and Identification of a New Offloading Cyclase Family. *Angew. Chemie Int. Ed.* **2018**, *57*, 9447–9451. [CrossRef] [PubMed]
21. Matsuda, K.; Kobayashi, M.; Kuranaga, T.; Takada, K.; Ikeda, H.; Matsunaga, S.; Wakimoto, T. SurE is a trans-acting thioesterase cyclizing two distinct non-ribosomal peptides. *Org. Biomol. Chem.* **2019**, *17*, 1058–1061. [CrossRef] [PubMed]
22. Koshla, O.T.; Rokytskyy, I.V.; Ostash, I.S.; Busche, T.; Kalinowski, J.; Mösker, E.; Süssmuth, R.D.; Fedorenko, V.O.; Ostash, B.O. Secondary Metabolome and Transcriptome of *Streptomyces albus* J1074 in Liquid Medium SG2. *Cytol. Genet.* **2019**, *53*, 1–7. [CrossRef]
23. Chater, K.F.; Wilde, L.C. Restriction of a bacteriophage of *Streptomyces albus* G involving endonuclease *Sal*I. *J. Bacteriol.* **1976**, *128*, 644–650. [PubMed]
24. Chater, K.F.; Wilde, L.C. *Streptomyces albus* G Mutants Defective in the *Sal*GI Restriction-Modification System. *Microbiology* **1980**, *116*, 323–334. [CrossRef] [PubMed]
25. Huang, C.; Yang, C.; Zhang, W.; Zhu, Y.; Ma, L.; Fang, Z.; Zhang, C. Albumycin, a new isoindolequinone from *Streptomyces albus* J1074 harboring the fluostatin biosynthetic gene cluster. *J. Antibiot. (Tokyo)* **2019**, *72*, 311–315. [CrossRef] [PubMed]
26. Zaburannyi, N.; Rabyk, M.; Ostash, B.; Fedorenko, V.; Luzhetskyy, A. Insights into naturally minimised *Streptomyces albus* J1074 genome. *BMC Genomics* **2014**, *15*, 97. [CrossRef] [PubMed]
27. Myronovskyi, M.; Rosenkränzer, B.; Nadmid, S.; Pujic, P.; Normand, P.; Luzhetskyy, A. Generation of a cluster-free *Streptomyces albus* chassis strains for improved heterologous expression of secondary metabolite clusters. *Metab. Eng.* **2018**, *49*, 316–324. [CrossRef]
28. Bilyk, O.; Sekurova, O.N.; Zotchev, S.B.; Luzhetskyy, A. Cloning and Heterologous Expression of the Grecocycline Biosynthetic Gene Cluster. *PLoS ONE* **2016**, *11*, e0158682. [CrossRef]
29. Jiang, G.; Zhang, Y.; Powell, M.M.; Zhang, P.; Zuo, R.; Zhang, Y.; Kallifidas, D.; Tieu, A.M.; Luesch, H.; Loria, R.; et al. High-Yield Production of Herbicidal Thaxtomins and Thaxtomin Analogs in a Nonpathogenic *Streptomyces* Strain. *Appl. Environ. Microbiol.* **2018**, *84*, e00164-18. [CrossRef]
30. Labeda, D.P.; Doroghazi, J.R.; Ju, K.-S.; Metcalf, W.W. Taxonomic evaluation of *Streptomyces albus* and related species using multilocus sequence analysis and proposals to emend the description of *Streptomyces albus* and describe *Streptomyces pathocidini* sp. nov. *Int. J. Syst. Evol. Microbiol.* **2014**, *64*, 894–900. [CrossRef]
31. Labeda, D.P.; Dunlap, C.A.; Rong, X.; Huang, Y.; Doroghazi, J.R.; Ju, K.-S.; Metcalf, W.W. Phylogenetic relationships in the family Streptomycetaceae using multi-locus sequence analysis. *Antonie Van Leeuwenhoek* **2017**, *110*, 563–583. [CrossRef] [PubMed]
32. Almeida, E.L.; Carillo Rincón, A.F.; Jackson, S.A.; Dobson, A.D. Comparative genomics of marine sponge-derived *Streptomyces* spp. isolates SM17 and SM18 with their closest terrestrial relatives provides novel insights into environmental niche adaptations and secondary metabolite biosynthesis potential. *Front. Microbiol.* **2019**, *10*, 1713. [CrossRef] [PubMed]
33. Romano, S.; Jackson, S.; Patry, S.; Dobson, A. Extending the "One Strain Many Compounds" (OSMAC) Principle to Marine Microorganisms. *Mar. Drugs* **2018**, *16*, 244. [CrossRef] [PubMed]
34. Pan, R.; Bai, X.; Chen, J.; Zhang, H.; Wang, H. Exploring Structural Diversity of Microbe Secondary Metabolites Using OSMAC Strategy: A Literature Review. *Front. Microbiol.* **2019**, *10*, 294. [CrossRef] [PubMed]
35. Benson, D.A.; Cavanaugh, M.; Clark, K.; Karsch-Mizrachi, I.; Ostell, J.; Pruitt, K.D.; Sayers, E.W. GenBank. *Nucleic Acids Res.* **2018**, *46*, D41–D47. [CrossRef] [PubMed]
36. Johnson, M.; Zaretskaya, I.; Raytselis, Y.; Merezhuk, Y.; McGinnis, S.; Madden, T.L. NCBI BLAST: A better web interface. *Nucleic Acids Res.* **2008**, *36*, W5–W9. [CrossRef] [PubMed]
37. Camacho, C.; Coulouris, G.; Avagyan, V.; Ma, N.; Papadopoulos, J.; Bealer, K.; Madden, T.L. BLAST+: Architecture and applications. *BMC Bioinform.* **2009**, *10*, 421. [CrossRef] [PubMed]

38. Katoh, K.; Standley, D.M. MAFFT Multiple Sequence Alignment Software Version 7: Improvements in Performance and Usability. *Mol. Biol. Evol.* **2013**, *30*, 772–780. [CrossRef] [PubMed]
39. Ronquist, F.; Teslenko, M.; van der Mark, P.; Ayres, D.L.; Darling, A.; Höhna, S.; Larget, B.; Liu, L.; Suchard, M.A.; Huelsenbeck, J.P. MrBayes 3.2: Efficient Bayesian Phylogenetic Inference and Model Choice Across a Large Model Space. *Syst. Biol.* **2012**, *61*, 539–542. [CrossRef] [PubMed]
40. Waddell, P.J.; Steel, M. General Time-Reversible Distances with Unequal Rates across Sites: Mixing Γ and Inverse Gaussian Distributions with Invariant Sites. *Mol. Phylogenet. Evol.* **1997**, *8*, 398–414. [CrossRef] [PubMed]
41. Kumar, S.; Stecher, G.; Li, M.; Knyaz, C.; Tamura, K. MEGA X: Molecular Evolutionary Genetics Analysis across Computing Platforms. *Mol. Biol. Evol.* **2018**, *35*, 1547–1549. [CrossRef] [PubMed]
42. Blin, K.; Shaw, S.; Steinke, K.; Villebro, R.; Ziemert, N.; Lee, S.Y.; Medema, M.H.; Weber, T. antiSMASH 5.0: Updates to the secondary metabolite genome mining pipeline. *Nucleic Acids Res.* **2019**, *47*, 81–87. [CrossRef] [PubMed]
43. Navarro-Muñoz, J.C.; Selem-Mojica, N.; Mullowney, M.W.; Kautsar, S.; Tryon, J.H.; Parkinson, E.I.; Santos, E.L.C.D.L.; Yeong, M.; Cruz-Morales, P.; Abubucker, S.; et al. A computational framework for systematic exploration of biosynthetic diversity from large-scale genomic data. *bioRxiv* **2018**, *10*, 445270.
44. Medema, M.H.; Kottmann, R.; Yilmaz, P.; Cummings, M.; Biggins, J.B.; Blin, K.; de Bruijn, I.; Chooi, Y.H.; Claesen, J.; Coates, R.C.; et al. Minimum Information about a Biosynthetic Gene cluster. *Nat. Chem. Biol.* **2015**, *11*, 625–631. [CrossRef]
45. Shannon, P.; Markiel, A.; Ozier, O.; Baliga, N.S.; Wang, J.T.; Ramage, D.; Amin, N.; Schwikowski, B.; Ideker, T. Cytoscape: A software environment for integrated models of biomolecular interaction networks. *Genome Res.* **2003**, *13*, 2498–2504. [CrossRef] [PubMed]
46. Okonechnikov, K.; Golosova, O.; Fursov, M. Unipro UGENE: A unified bioinformatics toolkit. *Bioinformatics* **2012**, *28*, 1166–1167. [CrossRef] [PubMed]
47. Rutherford, K.; Parkhill, J.; Crook, J.; Horsnell, T.; Rice, P.; Rajandream, M.-A.; Barrell, B. Artemis: Sequence visualization and annotation. *Bioinformatics* **2000**, *16*, 944–945. [CrossRef] [PubMed]
48. Dusa, A. Venn: Draw Venn Diagrams 2018. Available online: https://cran.r-project.org/web/packages/venn/index.html (accessed on 25 September 2019).
49. *R Core Team R: A Language and Environment for Statistical Computing*, version 3.5.3; Vienna, Austria, 2018. Available online: https://www.R-project.org/ (accessed on 25 September 2019).
50. RStudio Team. *RStudio Team RStudio: Integrated Development Environment for RStudio Inc.*; RStudio Inc.: Boston, MA, USA, 2015; Volume 14.
51. Glaeser, S.P.; Kämpfer, P. Multilocus sequence analysis (MLSA) in prokaryotic taxonomy. *Syst. Appl. Microbiol.* **2015**, *38*, 237–245. [CrossRef]
52. Gadagkar, S.R.; Rosenberg, M.S.; Kumar, S. Inferring species phylogenies from multiple genes: Concatenated sequence tree versus consensus gene tree. *J. Exp. Zool. Part B Mol. Dev. Evol.* **2005**, *304B*, 64–74. [CrossRef]
53. Li, S.; Zhang, B.; Zhu, H.; Zhu, T. Cloning and Expression of the Chitinase Encoded by *ChiKJ406136* from *Streptomyces sampsonii* (Millard & Burr) Waksman KJ40 and Its Antifungal Effect. *Forests* **2018**, *9*, 699.
54. Liu, Q.; Xiao, L.; Zhou, Y.; Deng, K.; Tan, G.; Han, Y.; Liu, X.; Deng, Z.; Liu, T. Development of *Streptomyces* sp. FR-008 as an emerging chassis. *Synth. Syst. Biotechnol.* **2016**, *1*, 207–214. [CrossRef] [PubMed]
55. Badalamenti, J.P.; Erickson, J.D.; Salomon, C.E. Complete Genome Sequence of *Streptomyces albus* SM254, a Potent Antagonist of Bat White-Nose Syndrome Pathogen *Pseudogymnoascus destructans*. *Genome Announc.* **2016**, *4*, e00290-16. [CrossRef] [PubMed]
56. Page, A.J.; Cummins, C.A.; Hunt, M.; Wong, V.K.; Reuter, S.; Holden, M.T.G.; Fookes, M.; Falush, D.; Keane, J.A.; Parkhill, J. Roary: Rapid large-scale prokaryote pan genome analysis. *Bioinformatics* **2015**, *31*, 3691–3693. [CrossRef] [PubMed]
57. Rong, X.; Doroghazi, J.R.; Cheng, K.; Zhang, L.; Buckley, D.H.; Huang, Y. Classification of *Streptomyces* phylogroup *pratensis* (Doroghazi and Buckley, 2010) based on genetic and phenotypic evidence, and proposal of *Streptomyces pratensis* sp. nov. *Syst. Appl. Microbiol.* **2013**, *36*, 401–407. [CrossRef] [PubMed]
58. Ward, A.; Allenby, N. Genome mining for the search and discovery of bioactive compounds: The *Streptomyces* paradigm. *FEMS Microbiol. Lett.* **2018**, *365*, fny240. [CrossRef]
59. Li, Y.; Pinto-Tomás, A.A.; Rong, X.; Cheng, K.; Liu, M.; Huang, Y. Population Genomics Insights into Adaptive Evolution and Ecological Differentiation in Streptomycetes. *Appl. Environ. Microbiol.* **2019**, *85*, e02555-18. [CrossRef]

60. Hoz, J.F.-D.L.; Méndez, C.; Salas, J.A.; Olano, C. Novel Bioactive Paulomycin Derivatives Produced by *Streptomyces albus* J1074. *Molecules* **2017**, *22*, 1758. [CrossRef]
61. Chen, S.; Huang, X.; Zhou, X.; Bai, L.; He, J.; Jeong, K.J.; Lee, S.Y.; Deng, Z. Organizational and Mutational Analysis of a Complete FR-008/Candicidin Gene Cluster Encoding a Structurally Related Polyene Complex. *Chem. Biol.* **2003**, *10*, 1065–1076. [CrossRef]
62. Zhao, Z.; Deng, Z.; Pang, X.; Chen, X.-L.; Zhang, P.; Bai, L.; Li, H. Production of the antibiotic FR-008/candicidin in *Streptomyces* sp. FR-008 is co-regulated by two regulators, FscRI and FscRIV, from different transcription factor families. *Microbiology* **2015**, *161*, 539–552.
63. Hamm, P.S.; Caimi, N.A.; Northup, D.E.; Valdez, E.W.; Buecher, D.C.; Dunlap, C.A.; Labeda, D.P.; Lueschow, S.; Porras-Alfaro, A. Western Bats as a Reservoir of Novel *Streptomyces* Species with Antifungal Activity. *Appl. Environ. Microbiol.* **2017**, *83*, e03057-16. [CrossRef]
64. El-Gebali, S.; Mistry, J.; Bateman, A.; Eddy, S.R.; Luciani, A.; Potter, S.C.; Qureshi, M.; Richardson, L.J.; Salazar, G.A.; Smart, A.; et al. The Pfam protein families database in 2019. *Nucleic Acids Res.* **2019**, *47*, D427–D432. [CrossRef] [PubMed]
65. Li, J.; Xie, Z.; Wang, M.; Ai, G.; Chen, Y. Identification and Analysis of the Paulomycin Biosynthetic Gene Cluster and Titer Improvement of the Paulomycins in *Streptomyces paulus* NRRL 8115. *PLoS ONE* **2015**, *10*, e0120542. [CrossRef] [PubMed]
66. Marchler-Bauer, A.; Derbyshire, M.K.; Gonzales, N.R.; Lu, S.; Chitsaz, F.; Geer, L.Y.; Geer, R.C.; He, J.; Gwadz, M.; Hurwitz, D.I.; et al. CDD: NCBI's conserved domain database. *Nucleic Acids Res.* **2015**, *43*, D222–D226. [CrossRef] [PubMed]
67. Quadri, L.E.; Sello, J.; Keating, T.A.; Weinreb, P.H.; Walsh, C.T. Identification of a *Mycobacterium tuberculosis* gene cluster encoding the biosynthetic enzymes for assembly of the virulence-conferring siderophore mycobactin. *Chem. Biol.* **1998**, *5*, 631–645. [CrossRef]
68. Lautru, S.; Oves-Costales, D.; Pernodet, J.-L.; Challis, G.L. MbtH-like protein-mediated cross-talk between non-ribosomal peptide antibiotic and siderophore biosynthetic pathways in *Streptomyces coelicolor* M145. *Microbiology* **2007**, *153*, 1405–1412. [CrossRef]
69. Glavinas, H.; Krajcsi, P.; Cserepes, J.; Sarkadi, B. The role of ABC transporters in drug resistance, metabolism and toxicity. *Curr. Drug Deliv.* **2004**, *1*, 27–42. [CrossRef]
70. Polgar, O.; Bates, S.E. ABC transporters in the balance: Is there a role in multidrug resistance? *Biochem. Soc. Trans.* **2005**, *33*, 241–245. [CrossRef]
71. Podlesek, Z.; Comino, A.; Herzog-Velikonja, B.; Zgur-Bertok, D.; Komel, R.; Grabnar, M. *Bacillus licheniformis* bacitracin-resistance ABC transporter: Relationship to mammalian multidrug resistance. *Mol. Microbiol.* **1995**, *16*, 969–976. [CrossRef]
72. Ohki, R.; Tateno, K.; Okada, Y.; Okajima, H.; Asai, K.; Sadaie, Y.; Murata, M.; Aiso, T. A Bacitracin-Resistant *Bacillus subtilis* Gene Encodes a Homologue of the Membrane-Spanning Subunit of the *Bacillus licheniformis* ABC Transporter. *J. Bacteriol.* **2003**, *185*, 51–59. [CrossRef]
73. Kumar, S.; He, G.; Kakarla, P.; Shrestha, U.; Ranjana, K.C.; Ranaweera, I.; Willmon, T.M.; Barr, S.R.; Hernandez, A.J.; Varela, M.F. Bacterial Multidrug Efflux Pumps of the Major Facilitator Superfamily as Targets for Modulation. *Infect. Disord. Drug Targets* **2016**, *16*, 28–43. [CrossRef]
74. Yan, N. Structural Biology of the Major Facilitator Superfamily Transporters. *Annu. Rev. Biophys.* **2015**, *44*, 257–283. [CrossRef] [PubMed]
75. Cuthbertson, L.; Nodwell, J.R. The TetR Family of Regulators. *Microbiol. Mol. Biol. Rev.* **2013**, *77*, 440–475. [CrossRef] [PubMed]
76. Koskiniemi, S.; Lamoureux, J.G.; Nikolakakis, K.C.; de Roodenbeke, C.T.K.; Kaplan, M.D.; Low, D.A.; Hayes, C.S. Rhs proteins from diverse bacteria mediate intercellular competition. *Proc. Natl. Acad. Sci. USA* **2013**, *110*, 7032–7037. [CrossRef] [PubMed]
77. Manteca, Á.; Yagüe, P. *Streptomyces* as a Source of Antimicrobials: Novel Approaches to Activate Cryptic Secondary Metabolite Pathways. In *Antimicrobials, Antibiotic Resistance, Antibiofilm Strategies and Activity Methods*; IntechOpen: London, UK, 2019. [CrossRef]
78. Manteca, A.; Alvarez, R.; Salazar, N.; Yague, P.; Sanchez, J. Mycelium Differentiation and Antibiotic Production in Submerged Cultures of *Streptomyces coelicolor*. *Appl. Environ. Microbiol.* **2008**, *74*, 3877–3886. [CrossRef] [PubMed]

79. Stülke, J.; Hillen, W. Carbon catabolite repression in bacteria. *Curr. Opin. Microbiol.* **1999**, *2*, 195–201. [CrossRef]
80. Kremling, A.; Geiselmann, J.; Ropers, D.; de Jong, H. Understanding carbon catabolite repression in *Escherichia coli* using quantitative models. *Trends Microbiol.* **2015**, *23*, 99–109. [CrossRef]
81. Deutscher, J. The mechanisms of carbon catabolite repression in bacteria. *Curr. Opin. Microbiol.* **2008**, *11*, 87–93. [CrossRef] [PubMed]
82. Brückner, R.; Titgemeyer, F. Carbon catabolite repression in bacteria: Choice of the carbon source and autoregulatory limitation of sugar utilization. *FEMS Microbiol. Lett.* **2002**, *209*, 141–148. [CrossRef]
83. Romero-Rodríguez, A.; Ruiz-Villafán, B.; Tierrafría, V.H.; Rodríguez-Sanoja, R.; Sánchez, S. Carbon Catabolite Regulation of Secondary Metabolite Formation and Morphological Differentiation in *Streptomyces coelicolor*. *Appl. Biochem. Biotechnol.* **2016**, *180*, 1152–1166. [CrossRef]
84. Inoue, O.O.; Schmidell Netto, W.; Padilla, G.; Facciotti, M.C.R. Carbon catabolite repression of retamycin production by *Streptomyces olindensis* ICB20. *Braz. J. Microbiol.* **2007**, *38*, 58–61. [CrossRef]
85. Gallo, M.; Katz, E. Regulation of secondary metabolite biosynthesis: Catabolite repression of phenoxazinone synthase and actinomycin formation by glucose. *J. Bacteriol.* **1972**, *109*, 659–667. [PubMed]
86. Magnus, N.; Weise, T.; Piechulla, B. Carbon Catabolite Repression Regulates the Production of the Unique Volatile Sodorifen of *Serratia plymuthica* 4Rx13. *Front. Microbiol.* **2017**, *8*, 2522. [CrossRef] [PubMed]
87. Bose, U.; Hewavitharana, A.; Ng, Y.; Shaw, P.; Fuerst, J.; Hodson, M. LC-MS-Based Metabolomics Study of Marine Bacterial Secondary Metabolite and Antibiotic Production in *Salinispora arenicola*. *Mar. Drugs* **2015**, *13*, 249–266. [CrossRef] [PubMed]

# 11

# Survey of Biosynthetic Gene Clusters from Sequenced Myxobacteria Reveals Unexplored Biosynthetic Potential

**Katherine Gregory, Laura A. Salvador, Shukria Akbar, Barbara I. Adaikpoh and D. Cole Stevens ***

Department of BioMolecular Sciences, School of Pharmacy, University of Mississippi, University, MS 38677, USA; kcgregor@go.olemiss.edu (K.G.); lasalvad@go.olemiss.edu (L.A.S.); sakbar@go.olemiss.edu (S.A.); biadaikp@go.olemiss.edu (B.I.A.)

* Correspondence: stevens@olemiss.edu.

**Abstract:** Coinciding with the increase in sequenced bacteria, mining of bacterial genomes for biosynthetic gene clusters (BGCs) has become a critical component of natural product discovery. The order Myxococcales, a reputable source of biologically active secondary metabolites, spans three suborders which all include natural product producing representatives. Utilizing the BiG-SCAPE-CORASON platform to generate a sequence similarity network that contains 994 BGCs from 36 sequenced myxobacteria deposited in the antiSMASH database, a total of 843 BGCs with lower than 75% similarity scores to characterized clusters within the MIBiG database are presented. This survey provides the biosynthetic diversity of these BGCs and an assessment of the predicted chemical space yet to be discovered. Considering the mere snapshot of myxobacteria included in this analysis, these untapped BGCs exemplify the potential for natural product discovery from myxobacteria.

**Keywords:** myxobacteria; biosynthetic gene clusters; natural product discovery

---

## 1. Introduction

Ubiquitous to soils and marine sediments, bacteriovorous myxobacteria display organized social behaviors and predation strategies [1–4]. Perhaps intrinsic to their role as predators, myxobacteria are a critical source of diverse secondary metabolites that exhibit unique modes-of-action across a broad range of biological activities [5]. Distinct from other bacterial sources, the vast majority of the 60 species within the order Myxococcales produce natural products [5,6]. This gifted diversity of secondary metabolite producing representatives has established myxobacteria as a prolific resource for drug discovery efforts perhaps only second to Actinomycetales [7,8]. Bolstered by the observed lack of overlap between actinomycetal and myxobacterial drug-like metabolites, the potential to discover novel specialized metabolites from myxobacteria remains considerably high [7,8]. Herein, we report a survey of all myxobacterial natural product biosynthetic gene clusters (BGCs) deposited in the antiSMASH database and provide an account of all BGCs with and without characterization and assigned metabolites in an effort to observe the capacity for discovery from readily cultivable, sequenced myxobacteria [9,10]. Such analysis provides an assessment of the potential associated with the continued discovery efforts as well as development and application of methodologies to activate situational or cryptic secondary metabolism not functional during axenic cultivation [11,12]. A homology network of 994 BGCs from 36 sequenced myxobacterial genomes was constructed using the combined BiG-SCAPE-CORe Analysis of Syntenic Orthologues to prioritize Natural products biosynthetic gene clusters (CORASON) platform [13]. BiG-SCAPE facilitates the exploration of calculated BGC sequence similarity networks and provides the opportunity to visualize biosynthetic diversity across datasets [13]. Gene cluster families (GCFs) rendered by BiG-SCAPE are connected by edges that indicate shared domain types, sequence similarity, and similarity of domain pair-types amongst input BGCs [13]. Comparative analysis against the Minimum Information about a Biosynthetic Gene Cluster (MIBiG) repository (v1.4) indicates an untapped reservoir of BGCs that encompasses a broad range of biosynthetic diversity [14]. The 36 Myxococcales within the antiSMASH database currently span all 3 suborders with 26 Cystobacterineae, 7 Sorangineae, and 3 Nannocystineae included. Considering that the myxobacteria within the antiSMASH database minimally represent the breadth of the order Myxococcales, these observations not only support thorough investigation of identified myxobacteria and the presented biosynthetic space but also continued efforts for the identification and subsequent exploration of new myxobacteria [1,3].

## 2. Materials and Methods

Dataset. All BGCs associated with the order Myxococcales, a total of 994 BGCs from 36 myxobacteria, were downloaded as .gbk files from the publicly available antiSMASH database (https://antismash-db.secondarymetabolites.org) [9]. The original genome sequence data for all included myxobacteria are also publicly available and can be accessed at the National Center for Biotechnology Information, U.S. National Library of Medicine (https://www.ncbi.nlm.nih.gov/genome/browse#!/prokaryotes/myxobacteria).

BIG-SCAPE-CORASON analysis. BiG-SCAPE version 20181005 (available at: https://git.wageningenur.nl/medema-group/BiG-SCAPE) was utilized locally to analyse the 994 BGCs as individual .gbk files downloaded from the antiSMASH database (1/30/2019) [9,13]. BiG-SCAPE analysis was supplemented with Pfam database version 31 [15]. The singleton parameter in BiG-SCAPE was selected to ensure that BGCs with distances lower than the default cutoff distance of 0.3 were included in the corresponding output data. The MIBiG parameter in BiG-SCAPE was set to include the MIBiG repository version 1.4 of annotated BGCs [14]. The hybrids-off parameter was selected to prevent hybrid BGC redundancy. Generated network files separated by BiG-SCAPE class were combined for visualization using Cytoscape version 3.7.1; annotations associated with each BGC were included into Cytoscape networks by importing curated tables generated by BiG-SCAPE [16]. Phylogenetic trees provided by CORASON were generated during BiG-SCAPE analysis. Annotated network and table files including GCF associations are provided as Supplementary files. All BGCs with sequence similarities to deposited MIBiG clusters ≥75% were indicated and annotated using Cytoscape. An annotated .cys Cytoscape file is included as Supplementary Material. All associated .network and .tsv files are provided as Supplementary Materials. All histograms were generated GraphPad Prism version 7.0d for Mac OS X, GraphPad Software, San Diego, California, USA, www.graphpad.com.

## 3. Results

### 3.1. *BiG-SCAPE Analysis of BGCs from Sequenced Myxobacteria*

A sequence similarity network calculated using BiG-SCAPE consisted of 994 total BGCs as unique nodes from 36 myxobacteria and included 1035 edges (included self-looped nodes) representing homology across 753 GCFs (Figure 1). Of these 994 BGCs from the antiSMASH database, a total of 124 were determined to be located on contig edges by antiSMASH. Clusters determined to be on contig edges could contribute to redundancy within our analysis. While no 2 BGCs from an individual myxobacterium were found within a GCF, this does not preclude a single BGC split across multiple contigs from being included multiple times. A total of 613 singletons without homology using a similarity cutoff of 0.30 were also included in the network to appropriately depict all myxobacterial BGCs within the antiSMASH database [9,13]. Predicted BGC classes included 64 type I or modular polyketide synthases (t1PKS), 57 PKS categorized by antiSMASH as "PKSother" that includes all non-modular categories of PKSs, 125 nonribosomal peptide synthetases (NRPS), 166 hybrid PKS-NRPS, 245 ribosomally synthesized and post-translationally modified peptides (RiPPs), 149 terpene clusters, 3 saccharide clusters, and 185 clusters not belonging to any of the aforementioned classes that antiSMASH categorizes as "Others" clusters [9,10].

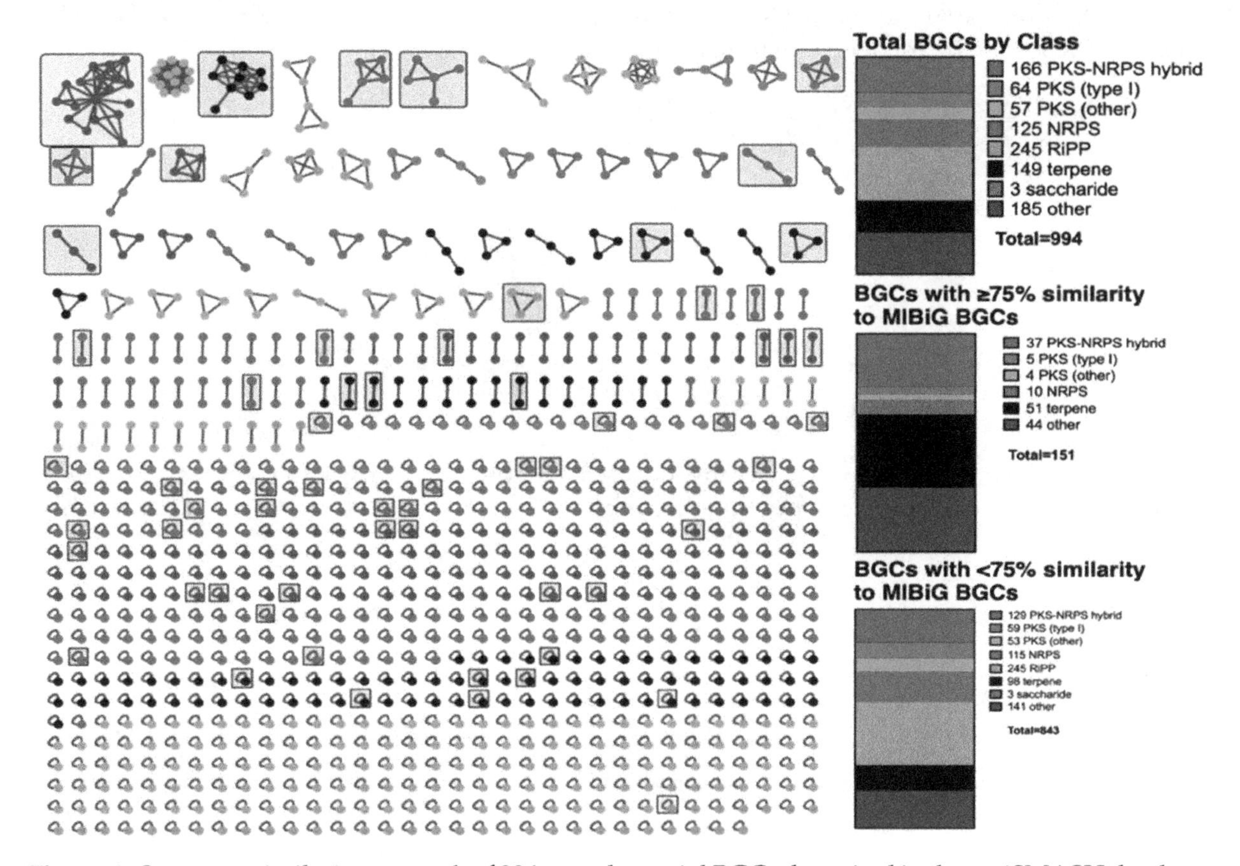


**Figure 1.** Sequence similarity network of 994 myxobacterial BGCs deposited in the antiSMASH database generated by BiG-SCAPE and rendered with Cytoscape [9,10,13,14,16]. All GCFs that include at least 1 BGC with sequence similarity greater than ≥75% to a characterized cluster deposited in the MIBiG repository are boxed in grey (excluding 25 geosmin BGCs) [9,14]. Totals for BGC class diversity and BGCs (including 25 geosmin BGCs identified as 22 Terpene and 3 Other clusters) with and without homology to MIBiG clusters as well as color reference provided (right).

While hybrid PKS-NRPS pathways that include both PKS and NRPS domains are organized into a specific separate grouping, all other hybrid pathways that include more than one BGC are categorized in the Others class [9,13]. The Others-associated BGCs included clusters with 133 predicted products as well as 52 hybrid BGCs (Figure 2). This breadth of biosynthetic diversity from just 36 myxobacteria includes 23 out of 52 BGC-types currently designated by antiSMASH [9,10].

### 3.2. *Discovered Metabolites from Myxobacteria and Associated BGCs*

Of the 994 BGCs analysed, 151 possess sequence similarities ≥75% with annotated BGCs in the MIBiG repository (v 1.4) [14]. Sequence similarities from the antiSMASH database are provided by KnownClusterBlast analysis of BGCs within the database against characterized pathways within the MIBiG repository [9,14,17,18]. As these BGCs produce characterized metabolites or potentially analogues thereof (Figure 3), a total of 85% of the BGCs within the network might produce yet to be discovered metabolites [19–43]. Considering the range in quality across the 36 total genomes and draft genomes incorporated in the antiSMASH database, we also considered additional BGCs with similarity scores lower than 75% that had similarities with MIBiG clusters reported from myxobacteria identified by antiSMASH. This analysis provided an additional 23 BGCs that might produce metabolites with overlapping chemical diversities to the products delineated within the MIBiG repository (Figure 4) [44–58]. Of these 23 BGCs omitted from our original analysis, only 10 would have been included if our sequence similarity cutoff had been lowered to 67% sequence similarity. Including this inference, 82% of the BGCs within the network lack any association with a reported myxobacterial metabolite. The biosynthetic diversity of these mapped BGCs includes 5 t1PKS, 10 NRPS, 37 hybrid PKS-NRPS, 4 PKSother, 51 terpene clusters, and 44 Others (Figure 1).

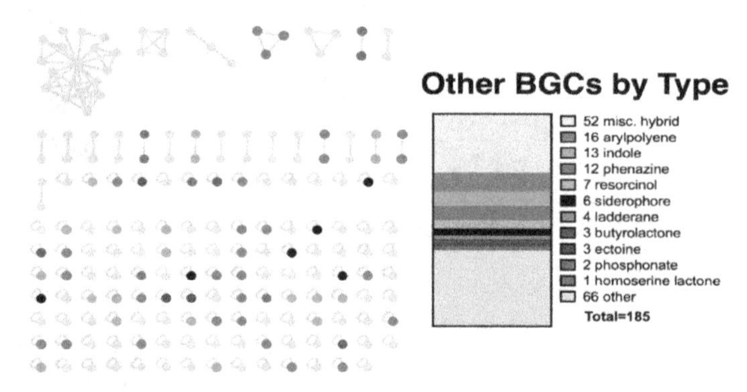


**Figure 2.** Sequence similarity network of myxobacterial BGCs classified as Others in the antiSMASH database with predicted product type and totals (right) [9].

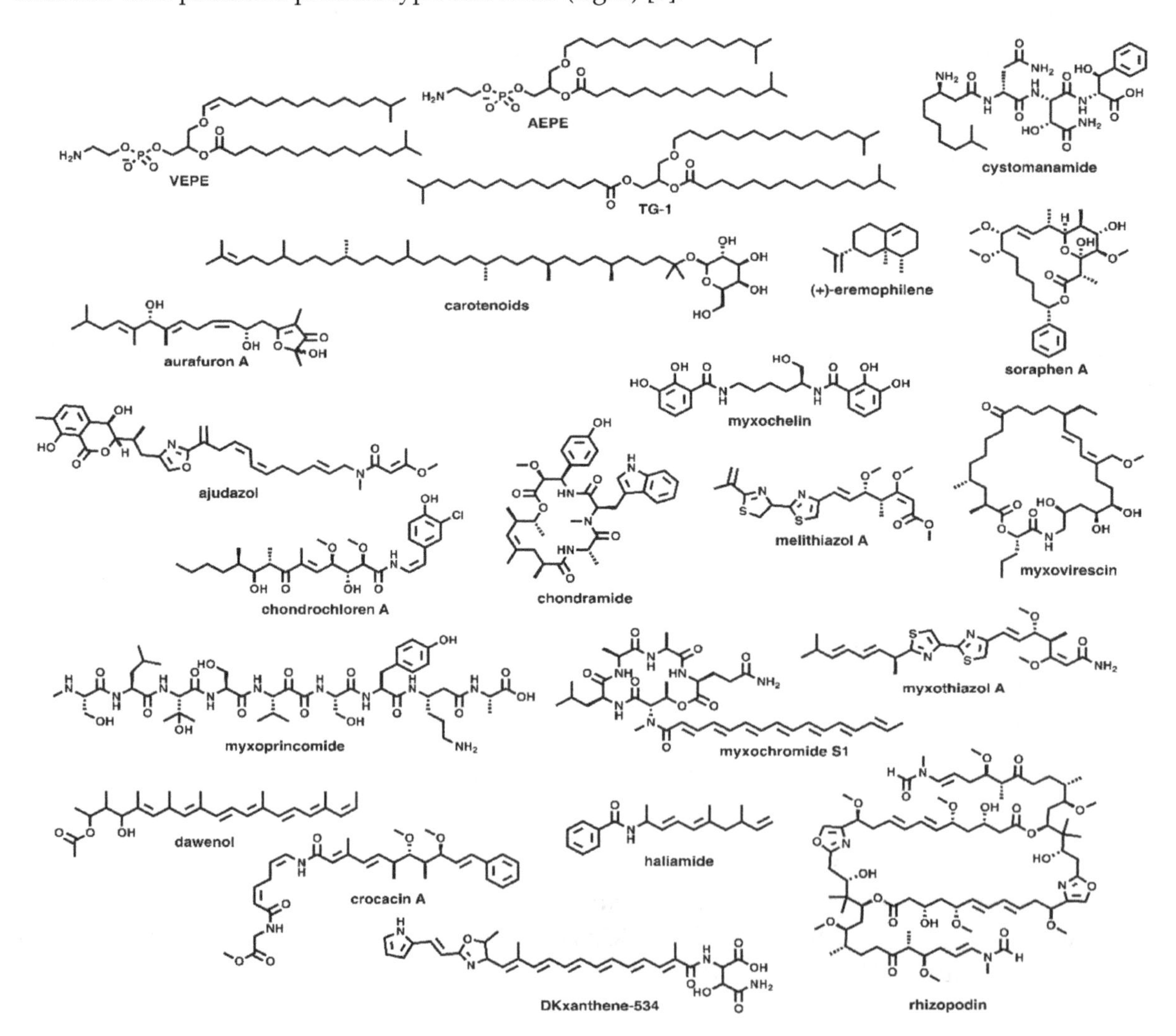


**Figure 3.** Secondary metabolites associated with BGCs determined to possess ≥75% sequence similarity to characterized clusters in MIBiG [17–41].

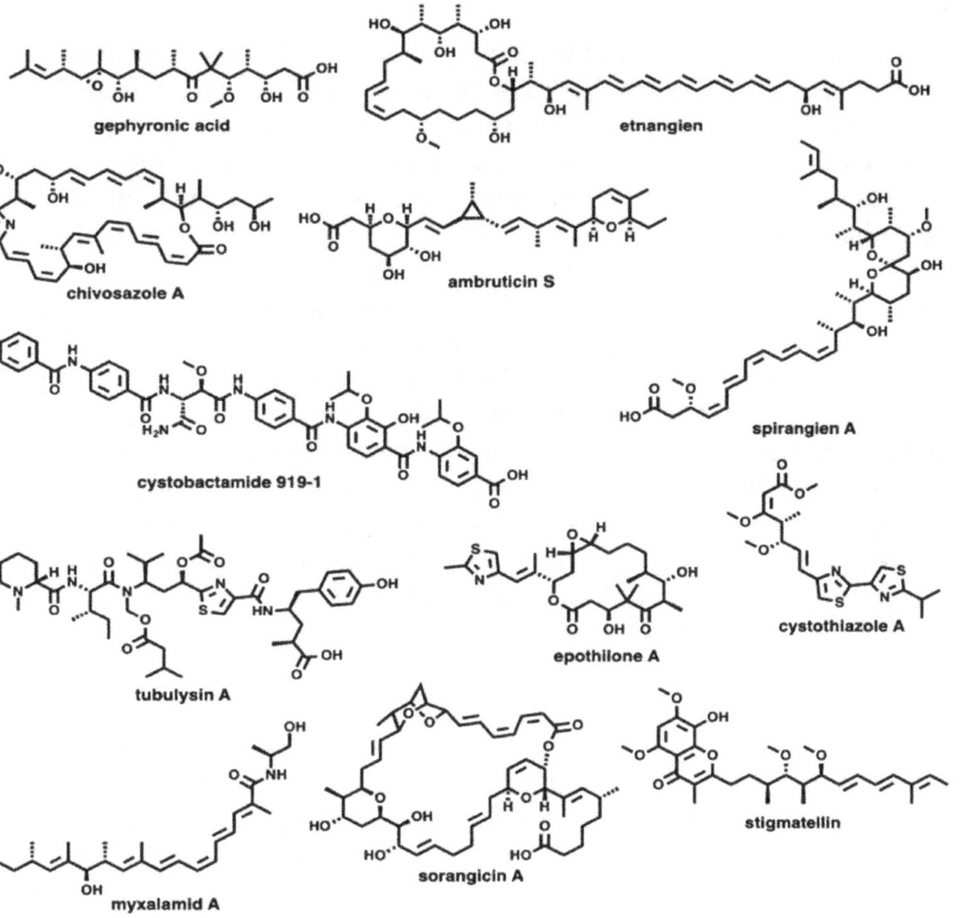

**Figure 4.** Secondary metabolites associated with known BGCs from myxobacteria with sequence similarity to BGCs included in the MIBiG dataset below the 75% similarity cutoff [42–56].

While the vast majority of BGCs were considered singletons or unclustered individual nodes without sequence similarity to other analysed BGCs, GCFs with more than 1 member BGC often shared sequence similarities with characterized MIBiG clusters. Interestingly, BGCs with high sequence similarity to specific MIBiG clusters were not always assigned the same cluster class nor were they included within an individual GCF. For example, 9 GCFs that include a single BGC with high homology to the myxochelin BGC were assigned as NRPS, hybrid PKS-NRPS, and Others type clusters [33,39]. Trees generated by CORASON provide the phylogenetic diversity associated with these myxochelin BGCs (Figure S1) [13]. Analysis of these trees indicated that such wholesale affiliation with each of these GCFs led to inclusion of BGCs that were in fact not related to the myxochelin BGC but instead shared proximal similarity to a BGC within the family that also included a neighbouring myxochelin-like BGC (Figure S1) [33,39]. While this omits unexplored BGCs and demonstrates the limitations of our totals, this only supports our conclusion that a vast wealth of biosynthetic space from myxobacteria remains unexplored. Other BGCs observed across multi-member GCFs included: 26 BGCs within 11 GCFs homologous to a carotenoid cluster from *Myxococcus xanthus*, 24 BGCs and 4 GCFs associated with the characterized VEPE/AEPE/TG-1 biosynthetic pathway from *M. xanthus* DK1622, and 11 BGCs across 5 GCFs with similarity to the hybrid PKS-NRPS DKxanthene cluster [19,20,24,30,31]. While all of the BGCs included in this charted biosynthetic space might not correlate to the corresponding metabolites associated with each MIBiG cluster, we consider this a rigorous assessment that provides a conservative estimate of uncharacterized BGCs and remaining opportunity for natural product discovery.

## 4. Discussion

This survey assesses the potential to discover novel metabolites from these myxobacteria and depicts unexplored biosynthetic space. Perhaps the most obvious absence in the 151 BGCs associated with characterized BGCs was that no RiPP clusters with sequence similarity to MIBiG clusters were observed [59–61]. However, there are no myxobacterial RiPP BGCs currently deposited in the MIBiG database, and crocagin A produced by *Chondromyces crocatus* is the only myxobacterial RiPP discovered to date [62]. Considering the 245 uncharacterized BGCs predicted to produce RiPPs within our network, myxobacteria are an excellent resource for the discovery of RiPPs. Also, with respect to notable outliers, no sequence similarities were observed for the 3 saccharide BGCs that include the aminoglycoside and aminocyclitol subtypes [63–65]. All other BGCs considered unexplored accounted for the vast majority of BGCs within each cluster class, including the following: 92% of t1PKS, 98% of PKSother, 92% of NRPS, 81% of terpene clusters, 78% of hybrid PKS-NRPS, and 76% of Others. Interestingly, within the BGCs assigned to the Others class, 3 butyrolactone and 1 homoserine lactone clusters were identified. Specialized metabolites belonging to these types of clusters are typically quorum-signaling molecules produced by Streptomyces and numerous non-myxobacterial Proteobacteria respectively [66–69]. Although putative quorum signal receptors are present within myxobacterial genomes and exogenous homoserine lactones increase the predatory behavior of *M. xanthus*, no metabolite associated with these quorum signaling systems has been reported from a myxobacteria [70,71].

## 5. Conclusions

The continued discovery of novel, biologically active bacterial metabolites is required to address the need for antimicrobials and anticancer therapeutics. Assessment of biosynthetic space within the growing amount of genome data from myxobacteria can provide insight to direct responsible discovery efforts [72–75]. This survey likely underestimates the unexplored biosynthetic space from myxobacteria. However, the vast discrepancies between BGCs with and without sequence similarity to characterized pathways suggests continued discovery of novel metabolites from this subset of 36 myxobacteria and exemplifies the outstanding potential associated with the Myxococcales at large.

**Author Contributions:** Conceptualization, supervision, and administration, D.C.S.; methodology, formal analysis, data curation, and validation, K.G., L.A.S., and D.C.S.; writing, K.G., L.A.S., S.A., B.I.A., and D.C.S.

**Acknowledgments:** The authors would like to acknowledge the University of Mississippi School of Pharmacy for startup support and the Sally Barksdale Honors College for encouraging undergraduate research.

## References

1. Brinkhoff, T.; Fischer, D.; Vollmers, J.; Voget, S.; Beardsley, C.; Thole, S.; Mussmann, M.; Kunze, B.; Wagner-Dobler, I.; Daniel, R.; et al. Biogeography and phylogenetic diversity of a cluster of exclusively marine myxobacteria. *ISME J.* **2012**, *6*, 1260–1272. [CrossRef] [PubMed]
2. Cao, P.; Dey, A.; Vassallo, C.N.; Wall, D. How Myxobacteria Cooperate. *J. Mol. Biol.* **2015**, *427*, 3709–3721. [CrossRef] [PubMed]
3. Mohr, K.I. Diversity of Myxobacteria-We Only See the Tip of the Iceberg. *Microorganisms* **2018**, *6*, 84. [CrossRef] [PubMed]
4. Munoz-Dorado, J.; Marcos-Torres, F.J.; Garcia-Bravo, E.; Moraleda-Munoz, A.; Perez, J. Myxobacteria: Moving, Killing, Feeding, and Surviving Together. *Front. Microbiol.* **2016**, *7*, 781. [CrossRef] [PubMed]
5. Herrmann, J.; Fayad, A.A.; Müller, R. Natural products from myxobacteria: Novel metabolites and bioactivities. *Nat. Prod. Rep.* **2017**, *34*, 135–160. [CrossRef] [PubMed]
6. Landwehr, W.; Wolf, C.; Wink, J. Actinobacteria and Myxobacteria-Two of the Most Important Bacterial Resources for Novel Antibiotics. *Curr. Top. Microbiol. Immunol.* **2016**, *398*, 273–302. [CrossRef] [PubMed]
7. Baltz, R.H. Natural product drug discovery in the genomic era: Realities, conjectures, misconceptions, and opportunities. *J. Ind. Microbiol. Biotechnol.* **2019**, *46*, 281–299. [CrossRef]

8. Liu, R.; Deng, Z.; Liu, T. Streptomyces species: Ideal chassis for natural product discovery and overproduction. *Metab. Eng.* **2018**, *50*, 74–84. [CrossRef] [PubMed]
9. Blin, K.; Pascal Andreu, V.; de Los Santos, E.L.C.; Del Carratore, F.; Lee, S.Y.; Medema, M.H.; Weber, T. The antiSMASH database version 2: A comprehensive resource on secondary metabolite biosynthetic gene clusters. *Nucleic Acids Res.* **2019**, *47*, D625–D630. [CrossRef]
10. Blin, K.; Shaw, S.; Steinke, K.; Villebro, R.; Ziemert, N.; Lee, S.Y.; Medema, M.H.; Weber, T. antiSMASH 5.0: Updates to the secondary metabolite genome mining pipeline. *Nucleic Acids Res.* **2019**. [CrossRef]
11. Baral, B.; Akhgari, A.; Metsa-Ketela, M. Activation of microbial secondary metabolic pathways: Avenues and challenges. *Synth. Syst. Biotechnol.* **2018**, *3*, 163–178. [CrossRef] [PubMed]
12. Mao, D.; Okada, B.K.; Wu, Y.; Xu, F.; Seyedsayamdost, M.R. Recent advances in activating silent biosynthetic gene clusters in bacteria. *Curr. Opin. Microbiol.* **2018**, *45*, 156–163. [CrossRef] [PubMed]
13. Navarro-Muñoz, J.C.; Selem-Mojica, N.; Mullowney, M.W.; Kautsar, S.; Tryon, J.H.; Parkinson, E.I.; de los Santos, E.L.C.; Yeong, M.; Cruz-Morales, P.; Abubucker, S.; et al. A computational framework for systematic exploration of biosynthetic diversity from large-scale genomic data. *Biorxiv* **2018**. [CrossRef]
14. Medema, M.H.; Kottmann, R.; Yilmaz, P.; Cummings, M.; Biggins, J.B.; Blin, K.; de Bruijn, I.; Chooi, Y.H.; Claesen, J.; Coates, R.C.; et al. Minimum Information about a Biosynthetic Gene cluster. *Nat. Chem. Biol.* **2015**, *11*, 625–631. [CrossRef] [PubMed]
15. Richardson, L.J.; Rawlings, N.D.; Salazar, G.A.; Almeida, A.; Haft, D.R.; Ducq, G.; Sutton, G.G.; Finn, R.D. Genome properties in 2019: A new companion database to InterPro for the inference of complete functional attributes. *Nucleic Acids Res.* **2019**, *47*, D564–D572. [CrossRef] [PubMed]
16. Shannon, P.; Markiel, A.; Ozier, O.; Baliga, N.S.; Wang, J.T.; Ramage, D.; Amin, N.; Schwikowski, B.; Ideker, T. Cytoscape: A software environment for integrated models of biomolecular interaction networks. *Genome Res.* **2003**, *13*, 2498–2504. [CrossRef] [PubMed]
17. Blin, K.; Wolf, T.; Chevrette, M.G.; Lu, X.; Schwalen, C.J.; Kautsar, S.A.; Suarez Duran, H.G.; de Los Santos, E.L.C.; Kim, H.U.; Nave, M.; et al. antiSMASH 4.0-improvements in chemistry prediction and gene cluster boundary identification. *Nucleic Acids Res.* **2017**, *45*, W36–W41. [CrossRef] [PubMed]
18. Weber, T.; Blin, K.; Duddela, S.; Krug, D.; Kim, H.U.; Bruccoleri, R.; Lee, S.Y.; Fischbach, M.A.; Müller, R.; Wohlleben, W.; et al. antiSMASH 3.0-a comprehensive resource for the genome mining of biosynthetic gene clusters. *Nucleic Acids Res.* **2015**, *43*, W237–W243. [CrossRef]
19. Bhat, S.; Ahrendt, T.; Dauth, C.; Bode, H.B.; Shimkets, L.J. Two lipid signals guide fruiting body development of Myxococcus xanthus. *MBio* **2014**, *5*, e00939-13. [CrossRef]
20. Botella, J.A.; Murillo, F.J.; Ruiz-Vazquez, R. A cluster of structural and regulatory genes for light-induced carotenogenesis in Myxococcus xanthus. *Eur. J. Biochem* **1995**, *233*, 238–248. [CrossRef]
21. Buntin, K.; Weissman, K.J.; Müller, R. An unusual thioesterase promotes isochromanone ring formation in ajudazol biosynthesis. *Chembiochem* **2010**, *11*, 1137–1146. [CrossRef] [PubMed]
22. Cervantes, M.; Murillo, F.J. Role for vitamin B(12) in light induction of gene expression in the bacterium Myxococcus xanthus. *J. Bacteriol.* **2002**, *184*, 2215–2224. [CrossRef] [PubMed]
23. Cortina, N.S.; Krug, D.; Plaza, A.; Revermann, O.; Müller, R. Myxoprincomide: A natural product from Myxococcus xanthus discovered by comprehensive analysis of the secondary metabolome. *Angew. Chem. Int. Ed. Engl.* **2012**, *51*, 811–816. [CrossRef] [PubMed]
24. Etzbach, L.; Plaza, A.; Garcia, R.; Baumann, S.; Müller, R. Cystomanamides: Structure and biosynthetic pathway of a family of glycosylated lipopeptides from myxobacteria. *Org. Lett.* **2014**, *16*, 2414–2417. [CrossRef] [PubMed]
25. Frank, B.; Wenzel, S.C.; Bode, H.B.; Scharfe, M.; Blocker, H.; Müller, R. From genetic diversity to metabolic unity: Studies on the biosynthesis of aurafurones and aurafuron-like structures in myxobacteria and streptomycetes. *J. Mol. Biol.* **2007**, *374*, 24–38. [CrossRef] [PubMed]
26. Gaitatzis, N.; Kunze, B.; Müller, R. In vitro reconstitution of the myxochelin biosynthetic machinery of Stigmatella aurantiaca Sg a15: Biochemical characterization of a reductive release mechanism from nonribosomal peptide synthetases. *Proc. Natl. Acad. Sci. USA* **2001**, *98*, 11136–11141. [CrossRef] [PubMed]
27. Li, Y.; Weissman, K.J.; Müller, R. Myxochelin biosynthesis: Direct evidence for two- and four-electron reduction of a carrier protein-bound thioester. *J. Am. Chem. Soc.* **2008**, *130*, 7554–7555. [CrossRef] [PubMed]

28. Ligon, J.; Hill, S.; Beck, J.; Zirkle, R.; Molnar, I.; Zawodny, J.; Money, S.; Schupp, T. Characterization of the biosynthetic gene cluster for the antifungal polyketide soraphen A from Sorangium cellulosum So ce26. *Gene* **2002**, *285*, 257–267. [CrossRef]
29. Lopez-Rubio, J.J.; Elias-Arnanz, M.; Padmanabhan, S.; Murillo, F.J. A repressor-antirepressor pair links two loci controlling light-induced carotenogenesis in Myxococcus xanthus. *J. Biol. Chem.* **2002**, *277*, 7262–7270. [CrossRef]
30. Lorenzen, W.; Ahrendt, T.; Bozhuyuk, K.A.; Bode, H.B. A multifunctional enzyme is involved in bacterial ether lipid biosynthesis. *Nat. Chem. Biol.* **2014**, *10*, 425–427. [CrossRef]
31. Meiser, P.; Weissman, K.J.; Bode, H.B.; Krug, D.; Dickschat, J.S.; Sandmann, A.; Müller, R. DKxanthene biosynthesis–understanding the basis for diversity-oriented synthesis in myxobacterial secondary metabolism. *Chem. Biol.* **2008**, *15*, 771–781. [CrossRef] [PubMed]
32. Muller, S.; Rachid, S.; Hoffmann, T.; Surup, F.; Volz, C.; Zaburannyi, N.; Müller, R. Biosynthesis of crocacin involves an unusual hydrolytic release domain showing similarity to condensation domains. *Chem. Biol.* **2014**, *21*, 855–865. [CrossRef] [PubMed]
33. Osswald, C.; Zaburannyi, N.; Burgard, C.; Hoffmann, T.; Wenzel, S.C.; Müller, R. A highly unusual polyketide synthase directs dawenol polyene biosynthesis in Stigmatella aurantiaca. *J. Biotechnol.* **2014**, *191*, 54–63. [CrossRef] [PubMed]
34. Perez-Marin, M.C.; Padmanabhan, S.; Polanco, M.C.; Murillo, F.J.; Elias-Arnanz, M. Vitamin B12 partners the CarH repressor to downregulate a photoinducible promoter in Myxococcus xanthus. *Mol. Microbiol.* **2008**, *67*, 804–819. [CrossRef] [PubMed]
35. Pistorius, D.; Müller, R. Discovery of the rhizopodin biosynthetic gene cluster in Stigmatella aurantiaca Sg a15 by genome mining. *Chembiochem* **2012**, *13*, 416–426. [CrossRef] [PubMed]
36. Rachid, S.; Scharfe, M.; Blocker, H.; Weissman, K.J.; Müller, R. Unusual chemistry in the biosynthesis of the antibiotic chondrochlorens. *Chem. Biol.* **2009**, *16*, 70–81. [CrossRef] [PubMed]
37. Schifrin, A.; Ly, T.T.; Gunnewich, N.; Zapp, J.; Thiel, V.; Schulz, S.; Hannemann, F.; Khatri, Y.; Bernhardt, R. Characterization of the gene cluster CYP264B1-geoA from Sorangium cellulosum So ce56: Biosynthesis of (+)-eremophilene and its hydroxylation. *Chembiochem* **2015**, *16*, 337–344. [CrossRef] [PubMed]
38. Silakowski, B.; Schairer, H.U.; Ehret, H.; Kunze, B.; Weinig, S.; Nordsiek, G.; Brandt, P.; Blocker, H.; Hofle, G.; Beyer, S.; et al. New lessons for combinatorial biosynthesis from myxobacteria. The myxothiazol biosynthetic gene cluster of Stigmatella aurantiaca DW4/3-1. *J. Biol. Chem.* **1999**, *274*, 37391–37399. [CrossRef]
39. Simunovic, V.; Zapp, J.; Rachid, S.; Krug, D.; Meiser, P.; Müller, R. Myxovirescin A biosynthesis is directed by hybrid polyketide synthases/nonribosomal peptide synthetase, 3-hydroxy-3-methylglutaryl-CoA synthases, and trans-acting acyltransferases. *Chembiochem* **2006**, *7*, 1206–1220. [CrossRef]
40. Sun, Y.; Tomura, T.; Sato, J.; Iizuka, T.; Fudou, R.; Ojika, M. Isolation and Biosynthetic Analysis of Haliamide, a New PKS-NRPS Hybrid Metabolite from the Marine Myxobacterium Haliangium ochraceum. *Molecules* **2016**, *21*, 59. [CrossRef]
41. Weinig, S.; Hecht, H.J.; Mahmud, T.; Müller, R. Melithiazol biosynthesis: Further insights into myxobacterial PKS/NRPS systems and evidence for a new subclass of methyl transferases. *Chem. Biol.* **2003**, *10*, 939–952. [CrossRef] [PubMed]
42. Wenzel, S.C.; Kunze, B.; Hofle, G.; Silakowski, B.; Scharfe, M.; Blocker, H.; Müller, R. Structure and biosynthesis of myxochromides S1-3 in Stigmatella aurantiaca: Evidence for an iterative bacterial type I polyketide synthase and for module skipping in nonribosomal peptide biosynthesis. *Chembiochem* **2005**, *6*, 375–385. [CrossRef] [PubMed]
43. Rachid, S.; Krug, D.; Kunze, B.; Kochems, I.; Scharfe, M.; Zabriskie, T.M.; Blocker, H.; Müller, R. Molecular and biochemical studies of chondramide formation-highly cytotoxic natural products from Chondromyces crocatus Cm c5. *Chem. Biol.* **2006**, *13*, 667–681. [CrossRef] [PubMed]
44. Baumann, S.; Herrmann, J.; Raju, R.; Steinmetz, H.; Mohr, K.I.; Huttel, S.; Harmrolfs, K.; Stadler, M.; Müller, R. Cystobactamids: Myxobacterial topoisomerase inhibitors exhibiting potent antibacterial activity. *Angew. Chem. Int. Ed. Engl.* **2014**, *53*, 14605–14609. [CrossRef] [PubMed]
45. Beyer, S.; Kunze, B.; Silakowski, B.; Müller, R. Metabolic diversity in myxobacteria: Identification of the myxalamid and the stigmatellin biosynthetic gene cluster of Stigmatella aurantiaca Sg a15 and a combined polyketide-(poly)peptide gene cluster from the epothilone producing strain Sorangium cellulosum So ce90. *Biochim. Biophys. Acta* **1999**, *1445*, 185–195. [PubMed]

46. Feng, Z.; Qi, J.; Tsuge, T.; Oba, Y.; Kobayashi, T.; Suzuki, Y.; Sakagami, Y.; Ojika, M. Construction of a bacterial artificial chromosome library for a myxobacterium of the genus Cystobacter and characterization of an antibiotic biosynthetic gene cluster. *Biosci. Biotechnol. Biochem.* **2005**, *69*, 1372–1380. [CrossRef] [PubMed]
47. Frank, B.; Knauber, J.; Steinmetz, H.; Scharfe, M.; Blocker, H.; Beyer, S.; Müller, R. Spiroketal polyketide formation in Sorangium: Identification and analysis of the biosynthetic gene cluster for the highly cytotoxic spirangienes. *Chem. Biol.* **2007**, *14*, 221–233. [CrossRef]
48. Irschik, H.; Kopp, M.; Weissman, K.J.; Buntin, K.; Piel, J.; Müller, R. Analysis of the sorangicin gene cluster reinforces the utility of a combined phylogenetic/retrobiosynthetic analysis for deciphering natural product assembly by trans-AT PKS. *Chembiochem* **2010**, *11*, 1840–1849. [CrossRef]
49. Julien, B.; Shah, S.; Ziermann, R.; Goldman, R.; Katz, L.; Khosla, C. Isolation and characterization of the epothilone biosynthetic gene cluster from Sorangium cellulosum. *Gene* **2000**, *249*, 153–160. [CrossRef]
50. Julien, B.; Tian, Z.Q.; Reid, R.; Reeves, C.D. Analysis of the ambruticin and jerangolid gene clusters of Sorangium cellulosum reveals unusual mechanisms of polyketide biosynthesis. *Chem. Biol.* **2006**, *13*, 1277–1286. [CrossRef]
51. Menche, D.; Arikan, F.; Perlova, O.; Horstmann, N.; Ahlbrecht, W.; Wenzel, S.C.; Jansen, R.; Irschik, H.; Müller, R. Stereochemical determination and complex biosynthetic assembly of etnangien, a highly potent RNA polymerase inhibitor from the myxobacterium Sorangium cellulosum. *J. Am. Chem. Soc.* **2008**, *130*, 14234–14243. [CrossRef] [PubMed]
52. Molnar, I.; Schupp, T.; Ono, M.; Zirkle, R.; Milnamow, M.; Nowak-Thompson, B.; Engel, N.; Toupet, C.; Stratmann, A.; Cyr, D.D.; et al. The biosynthetic gene cluster for the microtubule-stabilizing agents epothilones A and B from Sorangium cellulosum So ce90. *Chem. Biol.* **2000**, *7*, 97–109. [CrossRef]
53. Perlova, O.; Gerth, K.; Kaiser, O.; Hans, A.; Müller, R. Identification and analysis of the chivosazol biosynthetic gene cluster from the myxobacterial model strain Sorangium cellulosum So ce56. *J. Biotechnol.* **2006**, *121*, 174–191. [CrossRef] [PubMed]
54. Sandmann, A.; Sasse, F.; Müller, R. Identification and analysis of the core biosynthetic machinery of tubulysin, a potent cytotoxin with potential anticancer activity. *Chem. Biol.* **2004**, *11*, 1071–1079. [CrossRef] [PubMed]
55. Silakowski, B.; Nordsiek, G.; Kunze, B.; Blocker, H.; Müller, R. Novel features in a combined polyketide synthase/non-ribosomal peptide synthetase: The myxalamid biosynthetic gene cluster of the myxobacterium Stigmatella aurantiaca Sga15. *Chem. Biol.* **2001**, *8*, 59–69. [CrossRef]
56. Tang, L.; Shah, S.; Chung, L.; Carney, J.; Katz, L.; Khosla, C.; Julien, B. Cloning and heterologous expression of the epothilone gene cluster. *Science* **2000**, *287*, 640–642. [CrossRef] [PubMed]
57. Young, J.; Stevens, D.C.; Carmichael, R.; Tan, J.; Rachid, S.; Boddy, C.N.; Müller, R.; Taylor, R.E. Elucidation of gephyronic acid biosynthetic pathway revealed unexpected SAM-dependent methylations. *J. Nat. Prod.* **2013**, *76*, 2269–2276. [CrossRef] [PubMed]
58. Zhu, L.P.; Li, Z.F.; Sun, X.; Li, S.G.; Li, Y.Z. Characteristics and activity analysis of epothilone operon promoters from Sorangium cellulosum strains in Escherichia coli. *Appl. Microbiol. Biotechnol.* **2013**, *97*, 6857–6866. [CrossRef]
59. Hetrick, K.J.; van der Donk, W.A. Ribosomally synthesized and post-translationally modified peptide natural product discovery in the genomic era. *Curr. Opin. Chem. Biol.* **2017**, *38*, 36–44. [CrossRef]
60. Hudson, G.A.; Mitchell, D.A. RiPP antibiotics: Biosynthesis and engineering potential. *Curr. Opin. Microbiol.* **2018**, *45*, 61–69. [CrossRef]
61. Ortega, M.A.; van der Donk, W.A. New Insights into the Biosynthetic Logic of Ribosomally Synthesized and Post-translationally Modified Peptide Natural Products. *Cell Chem. Biol.* **2016**, *23*, 31–44. [CrossRef] [PubMed]
62. Viehrig, K.; Surup, F.; Volz, C.; Herrmann, J.; Abou Fayad, A.; Adam, S.; Kohnke, J.; Trauner, D.; Müller, R. Structure and Biosynthesis of Crocagins: Polycyclic Posttranslationally Modified Ribosomal Peptides from Chondromyces crocatus. *Angew. Chem. Int. Ed. Engl.* **2017**, *56*, 7407–7410. [CrossRef] [PubMed]
63. Flatt, P.M.; Mahmud, T. Biosynthesis of aminocyclitol-aminoglycoside antibiotics and related compounds. *Nat. Prod. Rep.* **2007**, *24*, 358–392. [CrossRef] [PubMed]
64. Kudo, F.; Eguchi, T. Aminoglycoside Antibiotics: New Insights into the Biosynthetic Machinery of Old Drugs. *Chem. Rec.* **2016**, *16*, 4–18. [CrossRef] [PubMed]
65. Yu, Y.; Zhang, Q.; Deng, Z. Parallel pathways in the biosynthesis of aminoglycoside antibiotics. *F1000Res* **2017**, *6*. [CrossRef]

66. Biarnes-Carrera, M.; Breitling, R.; Takano, E. Butyrolactone signalling circuits for synthetic biology. *Curr. Opin. Chem. Biol.* **2015**, *28*, 91–98. [CrossRef] [PubMed]
67. Camilli, A.; Bassler, B.L. Bacterial small-molecule signaling pathways. *Science* **2006**, *311*, 1113–1116. [CrossRef]
68. Papenfort, K.; Bassler, B.L. Quorum sensing signal-response systems in Gram-negative bacteria. *Nat. Rev. Microbiol.* **2016**, *14*, 576–588. [CrossRef]
69. Polkade, A.V.; Mantri, S.S.; Patwekar, U.J.; Jangid, K. Quorum Sensing: An Under-Explored Phenomenon in the Phylum Actinobacteria. *Front. Microbiol.* **2016**, *7*, 131. [CrossRef]
70. Brotherton, C.A.; Medema, M.H.; Greenberg, E.P. luxR Homolog-Linked Biosynthetic Gene Clusters in Proteobacteria. *mSystems* **2018**, *3*. [CrossRef]
71. Lloyd, D.G.; Whitworth, D.E. The Myxobacterium Myxococcus xanthus Can Sense and Respond to the Quorum Signals Secreted by Potential Prey Organisms. *Front. Microbiol.* **2017**, *8*, 439. [CrossRef] [PubMed]
72. Amiri Moghaddam, J.; Crusemann, M.; Alanjary, M.; Harms, H.; Davila-Cespedes, A.; Blom, J.; Poehlein, A.; Ziemert, N.; Konig, G.M.; Schaberle, T.F. Analysis of the Genome and Metabolome of Marine Myxobacteria Reveals High Potential for Biosynthesis of Novel Specialized Metabolites. *Sci. Rep.* **2018**, *8*, 16600. [CrossRef] [PubMed]
73. Bouhired, S.; Rupp, O.; Blom, J.; Schaberle, T.F.; Schiefer, A.; Kehraus, S.; Pfarr, K.; Goesmann, A.; Hoerauf, A.; Konig, G. Complete Genome Sequence of the Corallopyronin A-Producing Myxobacterium Corallococcus coralloides B035. *Microbiol. Resour. Announc.* **2019**, *8*. [CrossRef] [PubMed]
74. Garcia, R.; Müller, R. Simulacricoccus ruber gen. nov., sp. nov., a microaerotolerant, non-fruiting, myxospore-forming soil myxobacterium and emended description of the family Myxococcaceae. *Int. J. Syst. Evol. Microbiol.* **2018**, *68*, 3101–3110. [CrossRef] [PubMed]
75. Livingstone, P.G.; Morphew, R.M.; Whitworth, D.E. Genome Sequencing and Pan-Genome Analysis of 23 Corallococcus spp. Strains Reveal Unexpected Diversity, With Particular Plasticity of Predatory Gene Sets. *Front. Microbiol.* **2018**, *9*, 3187. [CrossRef] [PubMed]

# 12

# Identification and Heterologous Expression of the Albucidin Gene Cluster from the Marine Strain *Streptomyces Albus* Subsp. *Chlorinus* NRRL B-24108

**Maksym Myronovskyi [1], Birgit Rosenkränzer [1], Marc Stierhof [1], Lutz Petzke [2], Tobias Seiser [2] and Andriy Luzhetskyy [1,3,*]**

[1] Pharmazeutische Biotechnologie, Universität des Saarlandes, 66123 Saarbrücken, Germany; maksym.myronovskyi@uni-saarland.de (M.M.); b.rosenkraenzer@mx.uni-saarland.de (B.R.); s8mcstie@stud.uni-saarland.de (M.S.)

[2] BASF SE, 67056 Ludwigshafen, Germany; lutz.petzke@basf.com (L.P.); tobias.seiser@basf.com (T.S.)

[3] Helmholtz-Institut für Pharmazeutische Forschung Saarland, 66123 Saarbrücken, Germany

[*] Correspondence: a.luzhetskyy@mx.uni-saarland.de.

**Abstract:** Herbicides with new modes of action and safer toxicological and environmental profiles are needed to manage the evolution of weeds that are resistant to commercial herbicides. The unparalleled structural diversity of natural products makes these compounds a promising source for new herbicides. In 2009, a novel nucleoside phytotoxin, albucidin, with broad activity against grass and broadleaf weeds was isolated from a strain of *Streptomyces albus* subsp. *chlorinus* NRRL B-24108. Here, we report the identification and heterologous expression of the previously uncharacterized albucidin gene cluster. Through a series of gene inactivation experiments, a minimal set of albucidin biosynthetic genes was determined. Based on gene annotation and sequence homology, a model for albucidin biosynthesis was suggested. The presented results enable the construction of producer strains for a sustainable supply of albucidin for biological activity studies.

**Keywords:** albucidin; herbicide; nucleoside; biosynthetic gene cluster; heterologous expression; *Streptomyces albus* Del14

---

## 1. Introduction

Pesticides play an important role in modern agriculture. Among all chemicals being produced, pesticides are in second place after fertilizers in their extent of use. A total of 2.4 billion kilograms of pesticides were applied worldwide in 2007 [1]. Nevertheless, lack of weed control is still the most topical issue. Among all pests, weeds have the largest negative effect on crop productivity [2,3]. In light of the rapidly increasing evolution of herbicide resistance, the need for new herbicides with new modes of action (MOAs) and safer ecological profiles is growing [4,5].

From all new pesticide active ingredients registered by the Environmental Protection Agency from 1997 to 2010, almost 70% have origins in natural products. Interestingly, only 8% of conventional herbicides are natural product-derived [3,6]. The wide structural diversity of natural products and their small amount of overlap with synthetic compounds imply their potential as lead structures for the development of new pesticides [7–9]. This is further confirmed by the phytotoxin literature, which suggests that natural products have many more MOAs than the commercial herbicides currently possess [3].

A novel bleaching herbicide, albucidin, from the strain *Streptomyces albus* subsp. *chlorinus* NRRL B-24108 was discovered in 2009 [10]. In this paper, we present the identification, heterologous expression, and engineering of the albucidin gene cluster. We also propose the biosynthetic route that leads to the production of albucidin. The identified minimal set of biosynthetic genes allows for the straightforward construction of overproducing strains for a high yield albucidin supply for biological activity studies.

## 2. Materials and Methods

### 2.1. General Experimental Procedures

All strains, plasmids and BACs used in this work are listed in Tables S1 and S2. *Escherichia coli* strains were cultured in LB medium [11]. *Streptomyces* strains were grown on soya flour mannitol agar (MS agar) [12] and in liquid tryptic soy broth (TSB; Sigma-Aldrich, St. Louis, MO, USA). For albucidin production, liquid SG medium [13] was used. The antibiotics kanamycin, apramycin, hygromycin, ampicillin and nalidixic acid were supplemented when required.

### 2.2. Isolation and Manipulation of DNA

The previously constructed BAC library of *Streptomyces albus* subsp. *chlorinus* NRRL B-24108 was used [14]. DNA manipulation, *E. coli* transformation and *E. coli/Streptomyces* intergeneric conjugation were performed according to standard protocols [11,12,15]. BAC DNA was purified with the BACMAX™ DNA purification kit (Lucigen, Middleton, WI, USA). Restriction endonucleases were used according to the manufacturer's recommendations (New England Biolabs, Ipswich, MA, USA). All the strains and plasmids are listed in the Tables S1 and S2, respectively.

### 2.3. Metabolite Extraction and Analysis

For metabolite extraction, *Streptomyces* strains were grown in 15 mL of TSB in a 100 mL baffled flask for 1 day, and 1 mL of seed culture was used to inoculate 100 mL of SG production medium in a 500 mL baffled flask. Cultures were grown for 7 days at 28 °C and 180 rpm in an Infors multitron shaker. Albucidin was extracted from the culture supernatant with an equal amount of butanol, evaporated, and dissolved in methanol. Albucidin production was analysed on a Bruker Amazon Speed mass spectrometer coupled to UPLC Thermo Dionex Ultimate 3000 RS. Analytes were separated either on a Waters ACQUITY BEH C18 column (1.7 μm, 2.1 mm × 30 mm) or on a Waters ACQUITY BEH C18 column (1.7 μm, 2.1 mm × 100 mm). Water + 0.1% formic acid and methanol + 0.1% formic acid were used as the mobile phases. For the determination of high-resolution mass, analytes were analysed with a Thermo LTQ Orbitrap XL coupled to UPLC Thermo Dionex Ultimate 3000 RS. Analytes were separated on a Waters ACQUITY BEH C18 column (1.7 μm, 2.1 mm × 100 mm) with water + 0.1% formic acid and methanol + 0.1% formic acid as the mobile phase.

### 2.4. Chemical Mutagenesis

One millilitre of spore suspension of *Streptomyces albus* subsp. *chlorinus* NRRL B-24108 was inoculated into 100 mL of SG medium in a 500 mL baffled flask and cultivated overnight at 28 °C and 180 rpm. The pH of the culture was adjusted to 8.5 with 1 M NaOH. Ten millilitres of culture was transferred into three 50 mL falcon tubes. Then, 64 mg of wet NTG was dissolved in 16 mL of water. Next, 6.666 mL, 4.285 mL and 1.765 mL of NTG stock solution were added to the tubes containing culture to reach final NTG concentrations of 800 μg/mL, 600 μg/mL and 300 μg/mL, respectively. The samples were incubated at 28 °C for 30 min in the overhead shaker. The mycelium was precipitated by centrifugation, and the supernatant was discarded. The mycelium was washed twice with 5% sodium thiosulfate solution. The treated samples were plated on MS agar plates and cultivated for 14 days at 28 °C. The spores were washed with water and plated in dilutions on MS agar. The plates with spore dilutions were incubated for 10 days at 28 °C. Single colonies were picked on 30 mm plates with SG agar. The plates were incubated for 14 days at 28 °C. Agar blocks were cut out from the plates and transferred into 2 mL tubes. Albucidin was extracted from the agar blocks with 500 μL of butanol for 48 h. The extracts were analysed using HPLC-MS.

### 2.5. Albucidin Isolation and $^1$H-NMR Spectroscopy

*Streptomyces albus* 1K1 was grown in 10 L of SG medium, and albucidin was extracted with butanol. The dry extract was dissolved in 50 mL of water containing 5% acetonitrile and 0.1% formic acid. The extract was loaded onto 3 C18 SPE columns (Discovery DSC-18 SPE 52607-U) equilibrated with 5% acetonitrile in water with 0.1% formic acid. The flowthrough was collected. The columns were washed twice with 12 mL of 5% acetonitrile containing 0.1% formic acid. The flowthrough and the wash fractions were combined and evaporated in a rotary evaporator. The presence of albucidin was detected by HPLC-MS.

The dry material after the SPE purification step was dissolved in methanol and used for size-exclusion chromatography. Separation was performed on a glass column (30 mm × 1000 mm) packed with Sephadex LH-20 and methanol as the mobile phase. Fractions containing albucidin were identified by HPLC-MS. Albucidin-containing fractions were combined and evaporated. The dry extract was dissolved in 5 mL of 5% methanol in water containing 10 mM potassium phosphate buffer pH 6.4.

HPLC separation was performed on a preparative HPLC Thermo Dionex Ultimate 3000 equipped with a Macherey Nagel Nucleodur HTec C18 column (5 μm, 21 mm × 150 mm). A 10 mM potassium phosphate buffer (pH 6.4) was used as solvent A, and 50% methanol in 10 mM phosphate buffer (pH 6,4) was used as solvent B. The following gradient at a flowrate of 15 mL/min was used for separation: 0 min–13% B, 20 min—25% B, 21 min–25% B, 24 min–100% B, 25 min–100% B, 28 min–13% B, 29 min–13% B. Albucidin eluted at 18 min. The albucidin-containing fractions were pooled and evaporated.

For the final purification step, the dry material after HPLC purification was dissolved in 5 mL of water and loaded onto a Sephadex LH-20 column (30 mm × 450 mm) previously equilibrated with water. Water was used as the mobile phase. The fractions containing albucidin were identified by HPLC-MS, pooled and evaporated.

The $^1$H-NMR spectra were recorded on a Bruker Avance 500 spectrometer (Bruker, BioSpin GmbH, Rheinstetten, Germany) at 300 K equipped with a 5 mm BBO probe using deuterated trifluoroacetic acid (Deutero, Kastellaun, Germany) as the solvent containing tetramethylsilane (TMS) as a reference. Albucidin was measured in deuterated water (Deutero, Kastellaun, Germany). The chemical shifts are reported in parts per million (ppm) relative to TMS. All spectra were recorded with the standard $^1$H pulse program using 128 scans. The structure of albucidin was confirmed by comparison of the recorded $^1$H NMR data (Figure S1) with published data [10].

### 2.6. Construction of the 1K1 BAC Derivatives

The derivatives of 1K1 BAC with gene deletions were constructed using the RedET approach. For this, the antibiotic resistance marker was amplified by PCR with primers harbouring overhang regions complementary to the boundaries of the DNA to be deleted. The amplified fragment was used for recombineering of the BAC. The recombinant BACs were analysed by restriction mapping and sequencing. The primers used for recombineering purposes are listed in Table S3.

For the construction of BAC 1K1_LS, the ampicillin marker from pUC19 was amplified with the primers LS-F/LS-R. For the construction of the BAC 1K1_RS, the hygromycin marker from pACS-hyg [16] was amplified with the primers RS-F/RS-R. For the construction of the BACs 1K1_KO14, 1K1_KO15 and 1K1_KO16, the ampicillin cassette was amplified with the pairs of primers KO14-F/KO14-R, KO15-F/KO15-R and KO16-F/KO16-R, respectively. For the construction of the BACs 1K1_KO7, 1K1_KO8, 1K1_KO9, 1K1_KO10, 1K1_KO11, 1K1_KO12 and 1K1_KO13, the ampicillin cassette was amplified with the pairs of primers KO7-F/KO7-R, KO8-F/KO8-R, KO9-F/KO9-R, KO10-F/KO10-R, KO11-F/KO11-R, KO12-F/KO12-R and KO13-F/KO13-R, respectively.

BAC 1K1_alb_act was constructed in two steps. First, 1K1_RS2 BAC was constructed from 1K1 using an ampicillin marker amplified with primers RS2-F/RS2-R. Then BAC 1K1_alb_act was constructed by recombineering the BAC 1K1_RS2 using a hygromycin marker from pACS-hyg amplified with the primers ACT-F/ACT-R.

### *2.7. Genome Mining and Bioinformatics Analysis*

The *S. albus* subsp. *chlorinus* genome was screened for secondary metabolite biosynthetic gene clusters using the antiSMASH online tool [17] and the software Geneious [18]. The genomic sequence of the albucidin producer *S. albus* subsp. *chlorinus* NRRL B-24108 was deposited in GenBank under accession number VJOK00000000 [14].

## 3. Results and Discussion

### *3.1. Identification of the Albucidin Biosynthetic Gene Cluster*

The aim of this study was to identify the biosynthetic genes leading to the production of the nucleoside phytotoxin albucidin. For this purpose, the genome sequence of the producer strain of *Streptomyces albus* subsp. *chlorinus* NRRL B-24108 was analysed by genome-mining software [17]. This analysis led to the identification of several putative nucleoside clusters. To prove the involvement of these candidate clusters in albucidin production, they were heterologously expressed in a genetically engineered cluster-free strain *Streptomyces albus* Del14 [19] and in *Streptomyces lividans* TK24 [20]. No albucidin production was detected in the extracts of the obtained strains, indicating that either the expressed clusters were not involved in the biosynthesis of albucidin or they were not expressed in the heterologous host environment. The inactivation of the candidate clusters in the natural albucidin producer was not feasible because the strain is refractory to genetic manipulation. Considering the difficulties in identifying the albucidin gene cluster using conventional methods, an alternative approach using chemical mutagenesis was chosen.

For chemical mutagenesis of the albucidin-producing strain *S. albus* subsp. *chlorinus* NRRL B-24108, 1-methyl-3-nitro-1-nitrosoguanidine (NTG) was used. The strain in the exponential growth stage was treated with various NTG concentrations (800 μg/mL, 600 μg/mL and 300 μg/mL) for 30 min. After mutagenesis, the cells were washed with 5% thiosulfate solution and plated in dilutions on MS-agar medium for segregation of mutations. The spores of the obtained mutant populations were washed and plated on MS-agar plates in dilutions to obtain single colonies. Altogether, 4000 individual mutants were analysed for albucidin production. The mutants were cultivated on individual plates with the production medium SG agar. The metabolites were extracted with butanol, and albucidin production was assayed by HPLC-MS. Eight mutants that lost the ability to produce albucidin were identified in the course of this screening: 6-238, 6-260, 6-389, 6-444, 6-612, 6-892, 8-610 and 8-639. The genomic DNA of the obtained zero mutants was sequenced using Illumina technology. The point mutations in the genomes of the mutants were detected by mapping the sequencing reads to the reference genome of the wild type albucidin producer. Up to 100 transition mutations were identified in the genomes of the mutant strains. By comparing the mutation patterns of the separate mutants, a short genomic region was identified that was affected by point mutations in all analysed zero mutants, implying its potential involvement in albucidin production (Figure S2). The identified region contains two genes, *SACHL2_05525* and *SACHL2_05524*, which encode putative radical SAM proteins and were named *albA* and *albB* (Table 1, Figure 1b). The genes constitute a putative operon with the third gene *SACHL2_05523*, which was named *albC*. The *albC* gene encodes a putative ribonucleoside-triphosphate reductase and was not affected by point mutations in the analysed zero mutants of *S. albus* subsp. *chlorinus* NRRL B-24108. The identified genes *albA* and *albB* were not a part of the nucleoside gene clusters previously identified by genome mining and analysed in this study. Interestingly, these genes were located within the DNA fragment annotated by genome mining software as a putative NRPS gene cluster.

**Table 1.** Genes encoded within the chromosomal fragment cloned in BAC 1K1.

| Gene | Locus Tag [1] | Putative Function |
|---|---|---|
| 1 | *SACHL2_05539* | Hypothetical protein |
| 2 | *SACHL2_05538* | ABC transporter |
| 3 | *SACHL2_05537* | Transcriptional regulatory protein LiaR |
| 4 | *SACHL2_05536* | Hypothetical protein |
| 5 | *SACHL2_05535* | Hypothetical protein |
| 6 | *SACHL2_05534* | beta-lactamase/D-alanine carboxypeptidase |
| 7 | *SACHL2_05533* | Chondramide synthase |
| 8 | *SACHL2_05532* | Hypothetical protein |
| 9 | *SACHL2_05531* | Hypothetical protein |
| 10 | *SACHL2_05530* | Thymidylate kinase |
| 11 | *SACHL2_05529* | Pyruvate, phosphate dikinase |
| 12 | *SACHL2_05528* | Hypothetical protein |
| 13 | *SACHL2_05527* | Hypothetical protein |
| 14 | *SACHL2_05526* | Hypothetical protein |
| 15; *albA* | *SACHL2_05525* | Biotin synthase, radical SAM protein |
| 16; *albB* | *SACHL2_05524* | Radical SAM protein |
| 17; *albC* | *SACHL2_05523* | Ribonucleoside-triphosphate reductase |
| 18 | *SACHL2_05522* | Tyrocidine synthase 3 |
| 19 | *SACHL2_05521* | Plipastatin synthase, subunit A |
| 20 | *SACHL2_05520* | Acyl carrier protein |
| 21 | *SACHL2_05519* | Demethylmenaquinone methyltransferase |
| 22 | *SACHL2_05518* | Linear gramicidin synthase, subunit D |
| 23 | *SACHL2_05517* | Hypothetical protein |
| 24 | *SACHL2_05516* | Acyl carrier protein |
| 25 | *SACHL2_05515* | Fatty-acid–CoA ligase |
| 26 | *SACHL2_05514* | Ribonucleotide-diphosphate reductase |
| 27 | *SACHL2_05513* | Hypothetical protein |
| 28 | *SACHL2_05512* | Tyrocidine synthase 3 |
| 29 | *SACHL2_05511* | Hypothetical protein |

[1] The locus tags refer to the genome sequence of *S. albus* subsp. *chlorinus* NRRL B-24108 available under GenBank accession number VJOK00000000.

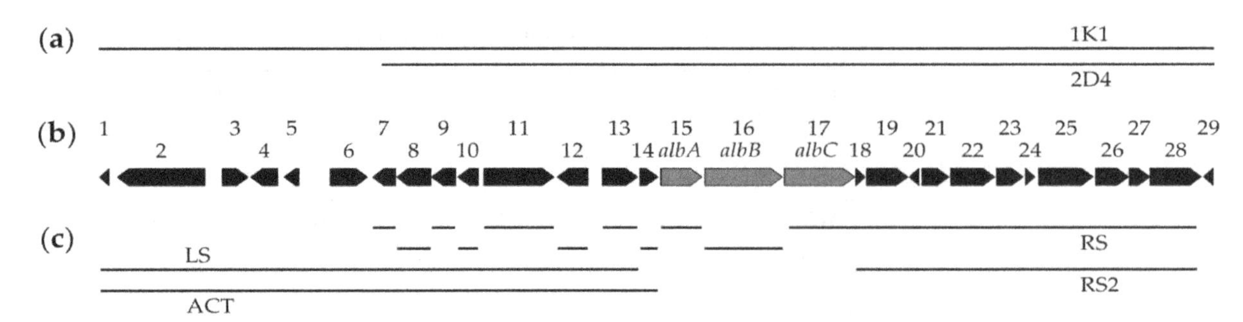


**Figure 1.** Chromosomal fragment of *S. albus* subsp. *chlorinus* NRRL B-24108 with the albucidin biosynthetic genes. (**a**) Schematic representations of DNA fragments cloned in BACs 1K1 and 2D4; (**b**) The genes encoded within the fragment cloned in BAC 1K1. The *albA–C* operon is marked in red; and (**c**) Overview of the performed deletions within BAC 1K1.

To determine whether the identified genes *albA* and *albB* encode albucidin biosynthetic enzymes, a BAC 1K1 containing the abovementioned genes (Figure 1a) was selected from the genomic library of *S. albus* subsp. *chlorinus* NRRL B-24108. BAC 1K1 was transferred into the heterologous host strains *S. albus* Del14 and *S. lividans* TK24 by conjugation, and the production profile of the obtained strains *S. albus* 1K1 and *S. lividans* 1K1 was analysed by HPLC-MS. The production of the compound with a high-resolution mass corresponding to albucidin could be detected in the extracts of *S. albus* 1K1 (Figure S3). No production could be detected in *S. lividans* 1K1.

Due to the lack of an albucidin standard, we set out to purify the compound identified in the extracts of *S. albus* 1K1 for structure elucidation studies by NMR spectroscopy. The *S. albus* 1K1 strain was cultivated in 10 L of SG medium for 7 days. The culture supernatant was extracted with equal amount of butanol, and the obtained extract was concentrated under vacuum. Four milligrams of the compound was purified using size exclusion and reverse phase chromatography and used for subsequent NMR studies. Analysis of the recorded NMR spectra of the purified compound unequivocally demonstrated its identity as albucidin (Figure S1 and Figure 2a).

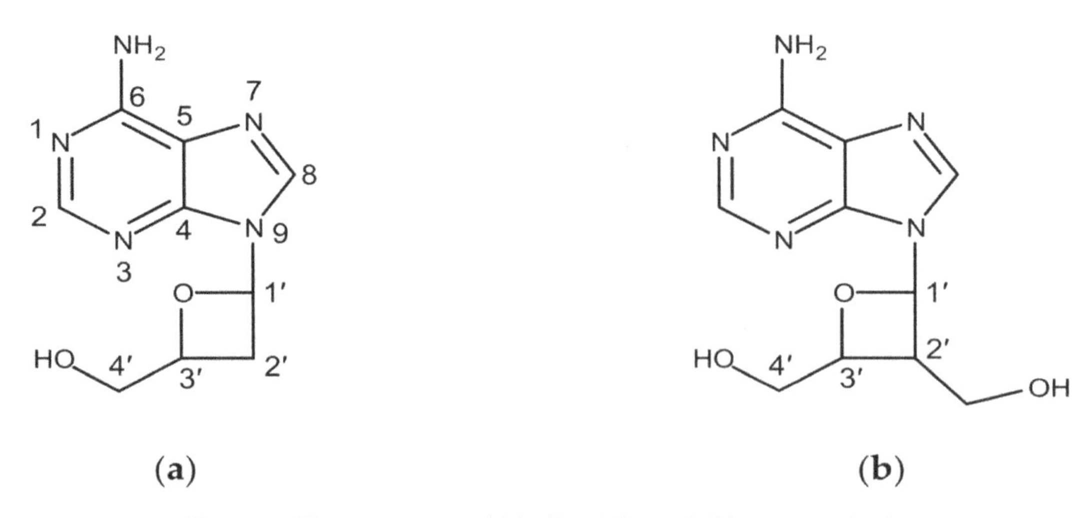


**Figure 2.** The structures of (**a**) albucidin and (**b**) oxetanocin A.

The production of albucidin by *S. albus* 1K1 gives evidence that the genes *albA* and *albB* identified by chemical mutagenesis encode albucidin biosynthetic genes. The lack of albucidin production by *S. lividans* 1K1 can be explained by differences in regulatory networks of the *S. albus* Del14 and *S. lividans* TK24 strains.

### 3.2. Identification of the Minimal Set of Albucidin Biosynthetic Genes

BAC 1K1, which leads to the production of albucidin under expression in the heterologous host *S. albus* Del14, contains a 32 kb chromosomal fragment from the natural albucidin producer *S. albus* subsp. *chlorinus* NRRL B-24108. Twenty-nine open reading frames were annotated in this 32 kb region (Table 1, Figure 1b). Of these genes, only two, *albA* and *albB*, were affected by point mutations in albucidin zero mutants identified in the course of the chemical mutagenesis studies. These two genes constitute a putative operon with the gene *albC*, implying that either only the genes *albA* and *albB* are necessary for albucidin production or that all three gene within the operon are required. To experimentally determine the minimal set of albucidin biosynthetic genes, a series of gene deletions was performed within the cloned region of 1K1 BAC.

The genes *albA–C* are located in the middle part of the chromosomal fragment cloned in 1K1 BAC. The *alb* operon is preceded by the genes *SACHL2_05539–SACHL2_05526*, followed by the genes *SACHL2_05522–SACHL2_05511* (Table 1). For the sake of simplicity, the 29 genes *SACHL2_05539–SACHL2_05511* cloned in the BAC 1K1 will be designated in the text according to their sequence number (1 to 29). (Figure 1b) To determine which genes within the 1K1 cloned fragment are essential for albucidin production, five deletions (LS, KO14, KO15, KO16 and RS) were performed in the 1K1 BAC yielding the BACs 1K1_LS, 1K1_KO14, 1K1_KO15, 1K1_KO16 and 1K1_RS (Figure 1c). In BAC 1K1_LS, the left shoulder encompassing genes 1–13 was substituted by an ampicillin resistance marker (Figure 1c). The genes 14, *albA* (gene 15) and *albB* (gene 16) were substituted with the ampicillin resistance gene in BACs 1K1_KO14, 1K1_KO15 and 1K1_KO16, respectively (Figure 1c). In the BAC 1K1_RS, the right shoulder encompassing genes *albC* (gene 17)-28 was substituted by the hygromycin resistance marker (Figure 1c). The constructed BACs were transferred separately into the *S. albus* Del14 strain by conjugation, and the albucidin production of the resulting strains was assayed by HPLC-MS.

The deletion of gene 14 did not affect albucidin production in *S. albus* 1K1_KO14 (Figure S4B). As expected from the results of chemical mutagenesis, inactivation of the genes *albA* (gene 15) and *albB* (gene 16) completely abolished albucidin production in the strains *S. albus* 1K1_KO15 and *S. albus* 1K1_KO16 (Figure S4C,D). This unambiguously demonstrates the essential role of the genes *albA* and *albB* in albucidin biosynthesis.

Deletion of the genes *albC* (gene 17)-28 did not affect albucidin production by the *S. albus* RS strain (Figure S4F). It was expected from the gene annotation and results of chemical mutagenesis that the genes 18–28 do not participate in albucidin biosynthesis. However, the dispensability of the gene *albC* (gene 17) is surprising since it belongs to the same operon as the essential genes *albA* (gene 15) and *albB* (gene 16). The deletion of the *albC* gene (gene 17) might be cross-complemented by an unidentified gene in the genome of the host strain *S. albus* Del14.

Albucidin production was heavily abolished in the strain *S. albus* 1K1_LS (Figure S4E), implying that at least one of the genes 1–13 that were deleted in the BAC 1K1_LS might be essential for albucidin biosynthesis. No genes encoding regulatory proteins or structural enzymes that might participate in nucleoside biosynthesis were identified in close proximity to the *albA–C* operon. To identify the genes within the deleted LS region that influence albucidin production, a BAC 2D4 was isolated from the genomic library of *S. albus* subsp. *chlorinus* NRRL B-24108. The chromosomal fragment cloned in BAC 2D4 overlaps with the fragment cloned in BAC 1K1 and covers the *albA–C* operon (Figure 1a). In contrast to 1K1, 2D4 BAC lacks genes 1–6, which are present in the deleted LS region. BAC 2D4 was transferred into *S. albus* Del14. Albucidin production could be detected in the extracts of the obtained strain *S. albus* 2D4 by HPLC-MS (Figure S5). This indicates that the genes 1–6 within the LS region are not involved in albucidin production and that one of the genes among 7–13 is responsible for the abolishment of albucidin production in *S. albus* 1K1_LS.

To identify which of the genes 7–13 is involved in albucidin biosynthesis, each of them was individually substituted by an ampicillin resistance marker in 1K1 BAC yielding 1K1_KO7, 1K1_KO8, 1K1_KO9, 1K1_KO10, 1K1_KO11, 1K1_KO12 and 1K1_KO13 (Figure 1b,c). The constructed BACs were transferred into the *S. albus* Del14 strain, and the albucidin production was analysed. The albucidin production levels of all obtained strains, except *S. albus* 1K1_KO12, were in the range of *S. albus* 1K1 harbouring the unmodified BAC (Figure S6). Albucidin production was abolished in *S. albus* 1K1_KO12 (Figure S6G), indicating that gene 12 is responsible for the detrimental effect of the LS deletion on albucidin biosynthesis. No enzymatic activity could be assigned to the peptide product of gene 12 using blast analysis. The product also did not show homology to any known regulatory protein. Considering this, it was proposed that only the genes *albA* and *albB* encode structural enzymes essential for albucidin production in the heterologous host *S. albus* Del14 and that the product of the gene 12 elicits a regulatory effect on transcription of the *albA–C* operon through a mechanism that is not understood. To prove this, a BAC 1K1_alb_act was constructed containing only *albA–C* genes under the control of a strong promoter. The genes downstream of the *albA–C* operon (genes 18–28) were substituted with the ampicillin resistance gene and the genes upstream of the operon (genes 1–14) were substituted with the hygromycin resistance gene (Figure 1c). The hygromycin resistance gene used was under the control of the strong synthetic promoter TS81 and did not contain a terminator at its 3′-end [21]. The insertion of the hygromycin marker in front of the *albA–C* genes was performed in the orientation, which enabled their read-through from the TS81 promoter and their transcriptional activation. The constructed BAC 1K1_alb_act was transferred into the heterologous host strain *S. albus* Del14. The production of albucidin was detected in the extracts of the obtained strain *S. albus* 1K1_alb_act by HPLC-MS (Figure S7). Three times increase of albucidin production was observed in the strain *S. albus* 1K1_alb_act compared to *S. albus* 1K1 containing non-modified albucidin cluster (Figure S7). Taking into account that the total recovered albucidin yield from the *S. albus* 1K1 strain was approximately 0.4 mg/L, the calculated albucidin production by *S. albus* 1K1_alb_act corresponded to 1.2 mg/L. The albucidin production rate of 2 mg/L was reported for the original producer *Streptomyces albus* subsp. *chlorinus* NRRL B-24108 [10].

Albucidin production by the strain *S. albus* 1K1_alb_act clearly demonstrates that the genes *albA* and *albB* constitute the minimal set of the genes required for albucidin biosynthesis in heterologous host *S. albus* Del14. The role of the gene *albC* in albucidin biosynthesis is not completely understood. Because *albC* constitutes a single operon with *albA* and *albB* and its product shows homology to nucleotide biosynthetic enzymes, it cannot be completely excluded that the *albC* is involved in albucidin production in the natural producer. However, the deletion of *albC* has no effect on albucidin production in heterologous host.

The identification of the minimal set of albucidin biosynthetic genes allows its expression in various heterologous chassis strains as well as rational construction of albucidin overproducers. The engineering of the albucidin biosynthetic genes can be performed in *E. coli* and the obtained constructs can be heterologously expressed in *Streptomyces* hosts. In contrast to the genetically intractable original albucidin producer *Streptomyces albus* subsp. *chlorinus* NRRL B-24108, commonly used heterologous strains possess a well-established toolkit for their genetic manipulation. This opens the possibility to engineer their metabolic network to increase the intracellular levels of biosynthetic precursors and therefore to increase the production yields. The heterologous strains are often characterized by the simplified metabolic background which provides better detection limits for heterologously expressed compounds than the original producers, higher product yields and simplified downstream processing. Construction of the albucidin overproducers based on heterologous expression hosts is not necessarily limited to a rational approach. The chassis strains expressing heterologous cluster may be also subjected to classical mutagenesis and screening for overproducing clones.

### *3.3. Proposed Biosynthetic Pathway of Albucidin*

Structurally, albucidin is closely related to oxetanocin A (Figure 2b), which has been isolated from the culture of *Bacillus megaterium* NK84-0218 [22]. Both compounds are the only known naturally occurring nucleosides featuring four membered oxetane rings in their structure. From a structural view, albucidin is 2′-dehydroxymethyl oxetanocin A. Two genes, *oxsA* and *oxsB*, encoding a putative HD domain phosphohydrolase and a cobalamin-dependent S-adenosylmethionine radical enzyme have been reported to be responsible for oxetanocin biosynthesis [23]. dAMP, dADP and dATP were identified as direct oxetanocin precursors [24]. The product of *oxsB* catalyses the contraction of the deoxyribose ring, while the product of *oxsA* is responsible for the removal of one or multiple phosphates from a phosphorylated 2′-deoxyadenosine derivative [24,25]. Through the simultaneous actions of OxsA and OxsB, the phosphorylated 2′-deoxyadenosine is converted to the oxetanocin A precursor, its aldehyde form, which must be reduced to complete biosynthesis [24]. This reaction is not encoded by the genes within the oxetanocin A cluster and is likely to be carried by an unidentified enzyme of *B. megaterium* NK84-0218.

Gene inactivation studies have given evidence that two genes, *albA* and *albB*, are required for the production of albucidin. Both genes encode putative SAM radical proteins. At the protein level, the *albA* gene shows homology to biotin synthases and the *albB* gene shows homology to the product of the oxetanocin biosynthetic gene *oxsB*. Despite the high structural similarity of albucidin and oxetanocin A, the homologue of the second oxetanocin biosynthetic gene *oxsA* cannot be found within the albucidin cluster or in the genome of albucidin producer *S. albus* subsp. *chlorinus* NRRL B-24108. The homology of the *albB* gene to *oxsB* implies that the product of *albB* might also be responsible for the ring contraction reaction in albucidin biosynthesis. However, the structural differences between albucidin and oxetanocin and the absence of an *oxsA* homologue imply substantial differences in biosynthetic routes leading to the biosynthesis of the nucleosides. Due to the lack of an *oxsA* homologue that is responsible for the dephosphorylation of adenine deoxyribonucleotides during oxetanocin biosynthesis, we propose that deoxyadenosine is used instead of dAMP, dADP or dATP as a precursor for albucidin production. The product of *albB* is likely responsible for the contraction of the deoxyribose ring of deoxyadenosine (Figure 3) in a similar manner as its homologue OxsB catalyses oxetane ring formation in oxetanocin A biosynthesis [24]. As a result of this reaction, the aldehyde form of oxetanocine A is formed. The conversion of the latter into albucidin is likely to be catalysed by the product of *albA*, which removes the aldehyde group from the 2′-position (Figure 3).

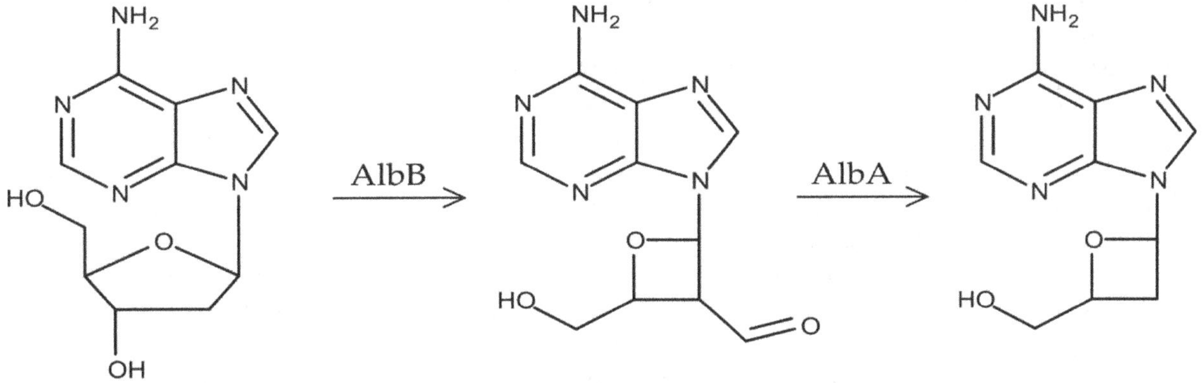

**Figure 3.** The proposed scheme of albucidin biosynthesis. 2′-Deoxyadenosine (**1**) is converted into the aldehyde form of oxetanocin A (**2**) by the product of *albB*. The latter is then converted into albucidin (**3**) by the product of *albA*.

In this paper, we report the identification, cloning, and heterologous expression of the albucidin biosynthetic gene cluster from *Streptomyces albus* subsp. *chlorinus* NRRL B-24108. Albucidin is a nucleoside phytotoxin featuring a rare oxetane ring in its structure. This metabolite shows herbicidal activity against a broad spectrum of grass and broadleaf weeds. In treated plants, albucidin induces metabolic perturbation, chlorosis, and bleaching [10]. The exact MOA of the compound remains unknown. The identification of the albucidin cluster presented in this paper enables biosynthetic studies of albucidin, optimization of its production as well as albucidin supply for the determination of its MOA.

## 4. Patents

(WO2018224939) Gene cluster for the biosynthesis of albucidin.

**Supplementary Materials:**
Figure S1: $^1$H NMR spectrum of albucidin (500 MHz), Figure S2: The genes *albA* and *albB* with the mapped point mutations identified in the course of NTG-mutagenesis, Figure S3: High resolution HPLC-MS analysis of albucidin production by *Streptomyces albus* 1K1, Figure S4: HPLC-MS analysis of albucidin production by *Streptomyces albus* strain harboring 1K1 BAC and its derivatives with gene deletions, Figure S5: HPLC-MS analysis of albucidin production by *Streptomyces albus* 1K1 BAC (**a**) and *Streptomyces albus* 2D4 BAC (**b**), Figure S6: HPLC-MS analysis of albucidin production by *Streptomyces albus* strain harboring 1K1 BAC and its derivatives with gene deletions, Figure S7: HPLC-MS analysis of albucidin production by *Streptomyces albus* strain harboring full length 1K1 BAC (**a**) and minimized 1K1_alb_act BAC, containing only transcriptionally activated *albA–C* operon (**b**), Table S1: Bacterial strains used in the study, Table S2: Plasmids and BACs used in the study, Table S3: Primers used in this study. References [26,27] are cite d in the supplementary materials.

**Author Contributions:** L.P., T.S., M.M., and A.L. designed the experiments; M.M. and B.R. performed the experiments; M.S. performed structure elucidation studies; M.M., B.R., and A.L. analysed the data and wrote the manuscript, and all authors reviewed the manuscript. All authors have read and agreed to the published version of the manuscript.

## References


1. Malakof, D.; Stokstad, E. Infographic: Pesticide Planet. *Science* **2013**, *341*, 730–731.
2. Pimentel, D.; Zuniga, R.; Morrison, D. Update on the environmental and economic costs associated with alien-invasive species in the United States. *Ecol. Econ.* **2005**, *52*, 273–288. [CrossRef]
3. Dayan, F.E.; Duke, S.O. Natural compounds as next-generation herbicides. *Plant Physiol.* **2014**, *166*, 1090–1105. [CrossRef]
4. Duke, S.O. Why have no new herbicide modes of action appeared in recent years? *Pest Manag. Sci.* **2012**, *68*, 505–512. [CrossRef]
5. Heap, I. Global perspective of herbicide-resistant weeds. *Pest Manag. Sci.* **2014**, *70*, 1306–1315. [CrossRef]
6. Cantrell, C.L.; Dayan, F.E.; Duke, S.O. Natural products as sources for new pesticides. *J. Nat. Prod.* **2012**, *75*, 1231–1242. [CrossRef]
7. Duke, S.O.; Stidham, M.A.; Dayan, F.E. A novel genomic approach to herbicide and herbicide mode of action discovery. *Pest Manag. Sci.* **2019**, *75*, 314–317. [CrossRef]
8. Harvey, A.L. Natural products as a screening resource. *Curr. Opin. Chem. Biol.* **2007**, *11*, 480–484. [CrossRef]
9. Koch, M.A.; Schuffenhauer, A.; Scheck, M.; Wetzel, S.; Casaulta, M.; Odermatt, A.; Ertl, P.; Waldmann, H. Charting biologically relevant chemical space: A structural classification of natural products (SCONP). *Proc. Natl. Acad. Sci. USA* **2005**, *102*, 17272–17277. [CrossRef]
10. Hahn, D.R.; Graupner, P.R.; Chapin, E.; Gray, J.; Heim, D.; Gilbert, J.R.; Gerwick, B.C. Albucidin: A novel bleaching herbicide from *Streptomyces albus* subsp. *chlorinus* NRRL B-24108. *J. Antibiot.* **2009**, *62*, 191–194. [CrossRef]
11. Green, M.R.; Sambrook, J. *Molecular Cloning: A Laboratory Manual*, 4th ed.; Cold Spring Harbor Laboratory Press: New York, NY, USA, 2012.
12. Kieser, T.; Bibb, M.J.; Buttner, M.J.; Chater, K.F.; Hopwood, D.A. *Practical Streptomyces Genetics*; John Innes Foundation: Norwich, UK, 2000.
13. Rebets, Y.; Ostash, B.; Luzhetskyy, A.; Hoffmeister, D.; Brana, A.; Mendez, C.; Salas, J.A.; Bechthold, A.; Fedorenko, V. Production of landomycins in *Streptomyces globisporus* 1912 and *S. cyanogenus* S136 is regulated by genes encoding putative transcriptional activators. *FEMS Microbiol. Lett.* **2003**, *222*, 149–153. [CrossRef]
14. Rodríguez Estévez, M.; Myronovskyi, M.; Gummerlich, N.; Nadmid, S.; Luzhetskyy, A. Heterologous Expression of the Nybomycin Gene Cluster from the Marine Strain *Streptomyces albus* subsp. *chlorinus* NRRL B-24108. *Mar. Drugs* **2018**, *16*, 435.
15. Mazodier, P.; Petter, R.; Thompson, C. Intergeneric conjugation between Escherichia coli and *Streptomyces* species. *J. Bacteriol.* **1989**, *171*, 3583–3585. [CrossRef] [PubMed]
16. Myronovskyi, M.; Brötz, E.; Rosenkränzer, B.; Manderscheid, N.; Tokovenko, B.; Rebets, Y.; Luzhetskyy, A. Generation of new compounds through unbalanced transcription of landomycin A cluster. *Appl. Microbiol. Biotechnol.* **2016**, *100*, 9175–9186. [CrossRef] [PubMed]

17. Blin, K.; Medema, M.H.; Kottmann, R.; Lee, S.Y.; Weber, T. The antiSMASH database, a comprehensive database of microbial secondary metabolite biosynthetic gene clusters. *Nucleic Acids Res.* **2017**, *45*, D555–D559. [CrossRef]
18. Kearse, M.; Moir, R.; Wilson, A.; Stones-Havas, S.; Cheung, M.; Sturrock, S.; Buxton, S.; Cooper, A.; Markowitz, S.; Duran, C.; et al. Geneious Basic: An integrated and extendable desktop software platform for the organization and analysis of sequence data. *Bioinformatics* **2012**, *28*, 1647–1649. [CrossRef] [PubMed]
19. Myronovskyi, M.; Rosenkränzer, B.; Nadmid, S.; Pujic, P.; Normand, P.; Luzhetskyy, A. Generation of a cluster-free *Streptomyces albus* chassis strains for improved heterologous expression of secondary metabolite clusters. *Metab. Eng.* **2018**, *49*, 316–324. [CrossRef]
20. Rückert, C.; Albersmeier, A.; Busche, T.; Jaenicke, S.; Winkler, A.; Friðjónsson, Ó.H.; Hreggviðsson, G.Ó.; Lambert, C.; Badcock, D.; Bernaerts, K.; et al. Complete genome sequence of *Streptomyces lividans* TK24. *J. Biotechnol.* **2015**, *199*, 21–22. [CrossRef]
21. Siegl, T.; Tokovenko, B.; Myronovskyi, M.; Luzhetskyy, A. Design, construction and characterisation of a synthetic promoter library for fine-tuned gene expression in actinomycetes. *Metab. Eng.* **2013**, *19*, 98–106. [CrossRef]
22. Shimada, N.; Hasegawa, S.; Harada, T.; Tomisawa, T.; Fujii, A.; Takita, T. Oxetanocin, a novel nucleoside from bacteria. *J. Antibiot.* **1986**, *39*, 1623–1625. [CrossRef]
23. Morita, M.; Tomita, K.; Ishizawa, M.; Takagi, K.; Kawamura, F.; Takahashi, H.; Morino, T. Cloning of oxetanocin A biosynthetic and resistance genes that reside on a plasmid of *Bacillus megaterium* strain NK84-0128. *Biosci. Biotechnol. Biochem.* **1999**, *63*, 563–566. [CrossRef] [PubMed]
24. Bridwell-Rabb, J.; Zhong, A.; Sun, H.G.; Drennan, C.L.; Liu, H.-W. A B12-dependent radical SAM enzyme involved in oxetanocin A biosynthesis. *Nature* **2017**, *544*, 322–326. [CrossRef] [PubMed]
25. Bridwell-Rabb, J.; Kang, G.; Zhong, A.; Liu, H.-W.; Drennan, C.L. An HD domain phosphohydrolase active site tailored for oxetanocin-A biosynthesis. *Proc. Natl. Acad. Sci. USA* **2016**, *113*, 13750–13755. [CrossRef] [PubMed]
26. Grant, S.G.; Jessee, J.; Bloom, F.R.; Hanahan, D. Differential plasmid rescue from transgenic mouse DNAs into Escherichia coli methylation-restriction mutants. *Proc. Natl. Acad. Sci. USA* **1990**, *87*, 4645–4649. [CrossRef]
27. Flett, F.; Mersinias, V.; Smith, C.P. High efficiency intergeneric conjugal transfer of plasmid DNA from Escherichia coli to methyl DNA-restricting streptomycetes. *FEMS Microbiol. Lett.* **1997**, *155*, 223–229. [CrossRef]

# 13

# Taxonomic Characterization and Secondary Metabolite Analysis of NEAU-wh3-1: An *Embleya* Strain with Antitumor and Antibacterial Activity

**Han Wang [1,†], Tianyu Sun [1,†], Wenshuai Song [1], Xiaowei Guo [1], Peng Cao [1], Xi Xu [1], Yue Shen [1,2,*] and Junwei Zhao [1,*]**

[1] Key Laboratory of Agricultural Microbiology of Heilongjiang Province, Northeast Agricultural University, No. 600 Changjiang Road, Xiangfang District, Harbin 150030, China; wanghan507555536@gmail.com (H.W.); sty1561214024@163.com (T.S.); wenshuaisong@163.com (W.S.); guoweizi@hotmail.com (X.G.); cp511@126.com (P.C.); xuxi1758899581@126.com (X.X.)

[2] College of Science, Northeast Agricultural University, No. 600 Changjiang Road, Xiangfang District, Harbin 150030, China

* Correspondence: shenyuelele@163.com (Y.S.); guyan2080@126.com (J.Z.)

† These authors contributed equally to this work.

**Abstract:** Cancer is a serious threat to human health. With the increasing resistance to known drugs, it is still urgent to find new drugs or pro-drugs with anti-tumor effects. Natural products produced by microorganisms have played an important role in the history of drug discovery, particularly in the anticancer and anti-infective areas. The plant rhizosphere ecosystem is a rich resource for the discovery of actinomycetes with potential applications in pharmaceutical science, especially *Streptomyces*. We screened *Streptomyces*-like strains from the rhizosphere soil of wheat (*Triticum aestivum* L.) in Hebei province, China, and thirty-nine strains were obtained. Among them, the extracts of 14 isolates inhibited the growth of colon tumor cell line HCT-116. Strain NEAU-wh-3-1 exhibited better inhibitory activity, and its active ingredients were further studied. Then, 16S rRNA gene sequence similarity studies showed that strain NEAU-wh3-1 with high sequence similarities to *Embleya scabrispora* DSM 41855$^T$ (99.65%), *Embleya hyalina* MB891-A1$^T$ (99.45%), and *Streptomyces lasii* 5H-CA11$^T$ (98.62%). Moreover, multilocus sequence analysis based on the five other house-keeping genes (*atp*D, *gyr*B, *rpo*B, *rec*A, and *trp*B) and polyphasic taxonomic approach comprising chemotaxonomic, phylogenetic, morphological, and physiological characterization indicated that the isolate should be assigned to the genus *Embleya* and was different from its closely related strains, therefore, it is proposed that strain NEAU-wh3-1 may be classified as representatives of a novel species of the genus *Embleya*. Furthermore, active substances in the fermentation broth of strain NEAU-wh-3-1 were isolated by bioassay-guided analysis and identified by nuclear magnetic resonance (NMR) and mass spectrometry (MS) analyses. Consequently, one new Zincophorin analogue together with seven known compounds was detected. The new compound showed highest antitumor activity against three human cell lines with the 50% inhibition ($IC_{50}$) values of 8.8–11.6 μg/mL and good antibacterial activity against four pathogenic bacteria, the other known compounds also exhibit certain biological activity.

**Keywords:** Rhizosphere soil; *Embleya*; NEAU-wh-3-1; compound; antitumor activity; antibacterial activity

---

## 1. Introduction

Tumor, especially malignant tumor, has become one of the major diseases, which is a serious threat to the health of people around the world [1,2]. According to records of the World Health Organization

(WHO) in 2018, more than 9 million people died of cancer, which was the second leading cause of death worldwide [3]. This figure will further rise because of aging, intensification of industrialization and urbanization, lifestyle modifications, etc. [2]. Thus, the burden of cancer cannot be ignored and the search for effective anticancer drugs is urgent [4]. On the other hand, the severe cancer incidence is also an invisible spur to the development of anti-tumor drugs throughout the world. So far, chemotherapy is still an important method for cancer treatment. Among the chemotherapeutics used, antitumor antibiotics derived from natural products account for a large proportion [5–7]. Natural product antibiotics were derived from various source materials including terrestrial plants, terrestrial microorganisms, marine organisms, and some invertebrates [8].

Microbial natural products have, in fact, been an excellent resource for drug discovery, particularly in the anticancer and anti-infective areas [9–11]. The phylum Actinobacteria accounts for a high proportion of soil microbial biomass and contains the most economically significant prokaryotes, producing more than half of the bioactive compounds in a literature survey, including antibiotics, antitumor agents, and enzymes [8,12–14]. Many famous antibiotics, such as bleomycin (BLM), mitomycin, anthracyclines [15], actinomycin D (ActD), polyether ionophore antibiotics, tetracyclines, quinolones, and so on, are derived from actinomycetes, which played an important role in the drug market [16]. Zincophorin, also referred to as M144255 or griseochellin, is a polyoxygenated ionophoric antibiotic [17], and has been reported to possess strong activity against Gram-positive bacteria and have strong cytotoxicity against human lung carcinoma cells A549 and Madin-Darby canine kidney cells MDCK [17,18], which was also isolated from actinomycetes. As the main genus of Actinobacteria, *Streptomyces* is the largest antibiotic producer. More than 70% of nearly 10,000 microbial origin compounds are produced by *Streptomyces* while some rare actinobacterial genera only accounted for less than 30% [19–22]. As abundant resources of larger number and wider variety of new antibiotics, *Streptomyces* strains have been continuously noted rather than any other actinomycete genera [19,23]. *Streptomyces* are widely distributed in terrestrial ecosystems, especially in the soil [24,25]. However, as time goes on, the possibility of finding novel compounds from *Streptomyces* in conventional soil has decreased and the rediscovery rate is high [22,26]. In recent years, studies on actinomycetes from diverse habitats have suggested new chemical structures and bioactive compounds [27,28]. Rhizosphere soil, the thin layer of soil around the roots of plants, has been a potential region for the discovery of functional microbes due to its special ecological environment. As early as the beginning of the last century, Hiltner proposed that there are more microorganisms in rhizosphere soil than surrounding soil [29–31]. There is a close relationship between rhizosphere microorganisms and plants. Plants can release organic compounds and signal molecules through root secretions to recruit microbial flora that are beneficial to their own growth. Microbes can control plant pathogens and pests by synthesizing multiple antibiotics, thereby indirectly promoting plant growth [32–34]. In recent years, many biologically active microorganisms and active substances produced by their secondary metabolism have been isolated from plant rhizosphere soil [35–38].

The genus *Embleya*, was very recently transferred from genus *Streptomyces* and established by Nouioui et al [39] and is a new member of the family *Streptomycetaceae* in the order *Streptomycetales* [39,40]. *Embleya* forms well-branched substrate mycelia with long aerial hyphae in open spirals and contains LL-diaminopimelic acid in the cell wall peptidoglycan, MK-9($H_4$) or MK-9($H_6$) as the major isoprenoid quinone and phosphatidylethanolamine (PE) as the predominant phospholipid [41], which is very similar to that of *Streptomyces* [42]. At present, the genus comprises only two species: *Embleya scabrispora* and *Embleya hyaline*. *Embleya scabrispora* was originally proposed as *Streptomyces scabrisporus* sp. nov. [43], and it has been reclassified to the genus *Embleya* as the type species [39,40], it could produce hitachimycin with antitumor, antibacterial, and antiprotozoal activities [44–46]; and *Embleya hyaline* was first described as *Streptomyces hyalinum* [41,47], and it has been reported to produce nybomycin which is an effective agent against antibiotic-resistant *Staphylococcus aureus* and it was called a reverse antibiotic [48].

In this study, an *Embleya* strain, NEAU-wh-3-1, with better antitumor activity was isolated from the wheat rhizosphere soil. The taxonomic identity of strain NEAU-wh3-1 was determined by a combination of 16S rRNA gene sequence and five other house-keeping genes (*atp*D, *gyr*B, *rpo*B, *rec*A, and *trp*B) analysis with morphological and physiological characteristics. The active substances of strain NEAU-wh-3-1 were also isolated, identified, and determined. Furthermore, the cytotoxicity and antimicrobial activity of the isolated compounds were tested.

## 2. Materials and Methods

### 2.1. Isolation of Streptomyces-Like Strains

Rhizosphere soil of wheat (*Triticum aestivum* L.) was collected from Langfang, Hebei Province, Central China (39°32′ N, 116°40′ E). The soil sample should be protected from light and air-dried at room temperature for 14 days before isolation for *Streptomyces*-like strains. After drying, the soil sample was ground into powder and then suspended in sterile distilled water followed by a standard serial dilution technique. The diluted soil suspension was spread on humic acid-vitamin agar (HV) [49] supplemented with cycloheximide (50 mg $L^{-1}$) and nalidixic acid (20 mg $L^{-1}$). After 28 days of aerobic incubation at 28 °C, colonies were transferred and purified on the International *Streptomyces* Project (ISP) medium 3 [50], and maintained as glycerol suspensions (20%, *v/v*) at −80 °C for long-period preservation.

### 2.2. Screening of Strains with Antitumor Activity

All the isolated were cultured on ISP medium 2 (yeast extract-malt extract agar) and incubated at 28 °C for 7 days. The spores of the strains were transferred into 250 mL Erlenmeyer flasks containing 30 mL of the production broth containing maltodextrin 4%, lactose 4%, yeast extract 0.5%, and MOPS 2%, at pH 7.2–7.4. on a rotary shaker at 250 r.p.m at 28 °C. After seven days, the production broth was extracted with an equal volume of methanol for approximately 24 h. After filtration, the filtrate substances were evaporated under reduced pressure at 50 °C to yield the crude extract and then dissolved in DMSO (dimethyl sulfoxide) at concentrations of 20 μg/mL and 100 μg/mL. The HCT-116 (human colorectal carcinoma) cell lines were maintained in Dulbecco's modified Eagle's medium supplemented with 10% (*w/v*) fetal bovine serum in a humidified incubator at 37 °C of 5% $CO_2$ incubator. The antitumor activities of extracts with two concentrations were investigated by the SRB (Sulforhodamine B) colorimetric method. Briefly, treated cells were harvested and seeded at a density of $5 \times 10^4$ cells/well into a sterile flat bottom 96-well plate for 24 h, the cells were treated with different concentrations of the extracts for 48 h and growth inhibition was measured by determining the optical density at 510 nm, and the assay was performed basing on an established method [51].

### 2.3. Morphological and Physiological and Biochemical Characteristics of NEAU-wh3-1

Gram staining was carried out by using the standard method and morphological characteristics were observed by light microscopy (Nikon ECLIPSE E200, Nikon Corporation, Tokyo, Japan) and scanning electron microscopy (Hitachi SU8010, Hitachi Co., Tokyo, Japan) using cultures grown on ISP 3 agar at 28 °C for 2 weeks. Samples for scanning electron microscopy were prepared as described by Jin et al. [52]. Growth at different temperatures (4, 10, 15, 20, 25, 28, 32, 37, 40, and 45 °C) was determined on ISP 3 medium after incubation for 14 days. Growth tests for pH range (pH 4.0–12.0, at intervals of 1.0 pH unit) using the buffer system described by Zhao et al. [53] and tolerance of various NaCl concentrations (0–10%, with an interval of 1%, *w/v*) were tested in GY (Glucose-yeast extract powder) medium (glucose 1%, yeast extract 1%, $K_2HPO_4$ $3H_2O$ 0.05%, $MgSO_4$ $7H_2O$ 0.05%, *w/v*, pH 7.2) at 28 °C for 14 days on a rotary shaker. Hydrolysis of Tweens (20, 40, and 80) and production of urease were tested as described by Smibert and Krieg [54]. The utilization of sole carbon and nitrogen sources, decomposition of cellulose, hydrolysis of starch and aesculin, reduction of nitrate, coagulation and peptonization of milk, liquefaction of gelatin, and production of $H_2S$ were examined as described previously [55,56].

### 2.4. Chemotaxonomic Analysis of NEAU-wh3-1

Biomass for chemotaxonomic characterization was prepared by growing strain NEAU-wh3-1 in ISP 2 broth in shake flasks at 28 °C for 7 days. Cells were harvested by centrifugation, washed twice with distilled water, and freeze-dried. The whole-cell sugars were analyzed according to the procedures developed by Lechevalier and Lechevalier [57]. The polar lipids were examined by two-dimensional TLC (thin layer chromatography) and identified using the method of Minnikin et al. [58]. Menaquinones were extracted from freeze-dried biomass and purified according to Collins [59]. *Streptomyces lutosisoli* DSM 42165$^T$ [60] was used as the reference strain for identification of menaquinones. Extracts were analyzed by a HPLC-UV method [61] using an Agilent Extend-C18 Column (150 × 4.6 mm, i.d. 5 μm) (Agilent Corp., Santa Clara, CA, USA) at 270 nm.

### 2.5. Phylogenetic Analysis of NEAU-wh3-1

Extraction of genomic DNA, PCR amplification of the 16S rRNA gene sequence and sequencing of PCR products were carried out using a standard procedure [62]. The PCR product was purified and cloned into the vector pMD19-T (Takara Bio Inc., Dalian, China) and sequenced using an Applied Biosystems DNA sequencer (model 3730XL, Applied Biosystems Inc., Foster City, California, USA). The almost complete 16S rRNA gene sequence of strain NEAU-wh3-1, comprising 1487 bp, was obtained and compared with type strains available in the EzBioCloud server [63] and retrieved using NCBI BLAST (https://blast.ncbi.nlm.nih.gov/Blast.cgi;), and then submitted to the GenBank database. The phylogenetic tree was constructed based on the 16S rRNA gene sequences of strain NEAU-wh3-1 and related reference species. Sequences were multiply aligned in Molecular Evolutionary Genetics Analysis (MEGA) using the Clustal W algorithm and trimmed manually where necessary. Phylogenetic trees were generated with the neighbor-joining [64] and maximum-likelihood [65] algorithms using MEGA software version MEGA 7.0 [66]. The stability of the topology of the phylogenetic tree was assessed using the bootstrap method with 1000 replicates [67]. A distance matrix was generated using Kimura's two-parameter model [68]. All positions containing gaps and missing data were eliminated from the dataset (complete deletion option). The *gyr*B gene was amplified with primers PF-1 and PR-2 [69] under the PCR program for 16S rRNA gene. PCR of the *atp*D, *rec*A, *rpo*B, and *trp*B genes were performed using primers and amplification conditions described by Guo et al. [70]. The sequence data were exported as single gene alignments or a concatenated five-gene alignment for subsequent analysis as described above. Trimmed sequences of the five housekeeping genes were concatenated head-to-tail in-frame in the order *atp*D (430 bp)-*gyr*B (354 bp)-*rec*A (431 bp)-*rpo*B (208 bp)-*trp*B (556 bp). Phylogenetic analysis was performed as described above.

### 2.6. Production

The strain *Embleya* sp. NEAU-wh3-1 was grown on the ISP medium 2 (yeast extract-malt extract agar) and incubated for 6–7 days at 28 °C. The spores of the strain were transferred into two 1.0 L Erlenmeyer flasks containing 250 mL of the seed medium and incubated at 28 °C for 48 h on a rotary shaker at 250 r.p.m. All of the media were sterilized at 121 °C for 30 min. The seed culture (8%) was transferred into 60 flasks (1.0 L) containing 250 mL of production broth. The production broth was composed of maltodextrin 4%, lactose 4%, yeast extract 0.5%, MOPS 2%, at pH 7.2–7.4. The flasks were incubated at 28 °C for 7 days, shaken at 250 r.p.m.

### 2.7. Extraction and Isolation

The final 15.0 L production broth was filtered to separate supernatant and mycelial cake. The supernatant was subjected to a Diaion HP-20 resin column and eluted with 95% EtOH. The mycelial cake was washed with water (3 L) and subsequently extracted with MeOH (3 L) to obtain soluble material. The MeOH extract and the EtOH eluents were evaporated under reduced pressure at 50 °C to yield the crude extract (24 g). The crude extract was chromatographed on a silica gel column and eluted with a stepwise gradient of $CH_2Cl_2$/MeOH (95:5/90:10/85:15/80:20/70:30/65:35, *v/v*) and giving three fractions (Fr.1–Fr.3) based on the TLC profiles, which was performed on silica-gel plates with solvent system of $CHCl_3$/MeOH (9:1, *v/v*). The Fr.1 was subjected to a Sephadex LH-20 column eluted with $CH_2Cl_2$/MeOH (1:1, *v/v*) and detected by TLC to give two subfractions (Fr.1-1-Fr.1-2). The Fr.1-1 was further isolated by semi-preparative HPLC (Agilent 1100, Zorbax SB-C18, 5 μm, 250 × 9.4 mm inner diameter; 1.5 mL $min^{-1}$; 254 nm; Agilent, PaloAlto, CA, USA) eluting with $CH_3CN/H_2O$ (90:10, *v/v*) to give compound **1** ($t_R$ 25.06 min, 10.5 mg), the Fr.1-2 was further isolated by preparative HPLC (Shimadzu LC-8 A, Shimadzu-C18, 5 μm, 250 × 20 mm inner diameter; 20 mL $min^{-1}$; 220 /254 nm; Shimadzu, Kyoto, Japan) eluting with a stepwise gradient MeOH/$H_2O$ (30–80%, *v/v* 30 min), and giving compound **2** ($t_R$ 12.7 min, 7.5 mg), compound **3** ($t_R$ 17.5 min, 12.7 mg) and compound **4** ($t_R$ 22.6 min, 16.3 mg). The Fr.2 was subjected to another silica gel column eluted with n-hexane/acetone (95:5-60:40, *v/v*) and further purified by semi-preparative HPLC (Agilent 1100, Zorbax SB-C18, 5 μm, 250 × 9.4 mm inner diameter; 1.5 mL$min^{-1}$; 254 nm; Agilent, PaloAlto, CA, USA) eluting with $CH_3CN/H_2O$ (75:25, *v/v*) to give compound **5** ($t_R$ 15.1 min, 13.5 mg) and compound **6** ($t_R$ 24.3 min, 18.5 mg). Fr.3 was treated by an another silica gel column and eluted with a stepwise gradient of n-hexane/acetone (100:0-40:60, *v/v*) to give three fractions Fr.3-1–Fr.3-3 according to their TLC profiles, which was observed on silica-gel plates with solvent system of n-hexane/acetone (1:3, *v/v*). The Fr.3-3 was further purified by semi-preparative HPLC (Agilent 1260, Zorbax SB-C18, 5 μm, 250 × 9.4 mm inner diameter; 1.5 mL $min^{-1}$; 220nm; 254 nm; Agilent, PaloAlto, CA, USA) eluting with $CH_3CN/H_2O$ (45:55, *v/v*) to obtain compounds **7** ($t_R$ 25.1 min, 13.0 mg) and **8** ($t_R$ 30.1 min, 7.5 mg).

### *2.8. General Experimental Procedures*

IR spectra were recorded on a Thermo Nicolet Avatar FT-IR-750 spectrophotometer (Thermo, Tokyo, Japan) using KBr disks. Optical rotations were measured on a Perkin-Elmer 341 polarimeter (PerkinElmer, Inc. Suzhou, China). UV spectra were recorded on a Varian CARY 300 BIO spectrophotometer (Varian, Cary, NC, USA). The HR-ESI-MS and ESI-MS were taken on a Q-TOF Micro LC-MS-MS mass spectrometer (Waters Co, Milford, MA, U.S.A.). Nuclear magnetic resonance (NMR) spectra (400 MHz for $^1H$ and 100 MHz for $^{13}C$) were measured with a Bruker DRX-400 spectrometer (Bruker, Rheinstetten, Germany). HPLC analysis was performed on a preparative HPLC (Shimadzu LC-8 A, Shimadzu-C18, 5 μm, 250 × 20 mm inner diameter; 20 mL $min^{-1}$; 220/254 nm; Shimadzu, Kyoto, Japan) as well as a semipreparative HPLC (Agilent 1100, Zorbax SB-C18, 5 μm, 250 × 9.4 mm inner diameter; 1.5 mL/min; 220/254 nm; Agilent, Palo Alto, CA, USA). Column chromatography were consisted of silica gel (100–200 mesh, Qingdao Haiyang Chemical Group Co., Qingdao, China) as well as Sephadex LH-20 gel (GE Healthcare, Glies, UK), which were analyzed by thin-layer chromatography (TLC). TLC was performed on silica-gel plates (HSGF254, Yantai Chemical Industry Research Institute, Yantai, China) and the developed plates were observed under a UV lamp at 254 nm or by heating after spraying with sulfuric acid-ethanol, 5:95 (*v/v*).

### *2.9. Biological Assays*

The cytotoxicity of the eight compounds was assayed by cell counting kit-8 (CCK-8) colorimetric method [71] in vitro against the human leukemia cells K562, hepatocellular liver carcinoma cell line HepG2, and the human colon tumor cell line HCT-116. The cell lines were routinely in Dulbecco's Modified Eagle's Medium (DMEM) containing 10% calf serum at 37 °C for 4 h in a humidified atmosphere of 5% $CO_2$ incubator. The adherent cells at logarithmic phase were digested by pancreatic enzymes and inoculated onto 96-well culture plate at a density of $1.0 \times 10^4$ cells per/well. Test samples and control were dissolved in DMSO (dimethyl sulfoxide) and then added to the medium, incubated for 72 h. Then, the cell counting kit-8 (CCK-8, Dojindo, Kumamoto, Japan) reagent was added to the medium followed by further incubation for 3 h. Absorbance at 450 nm with a 600 nm reference was measured thereafter using a SpectraMax M5 microplate reader (Molecular Devices Inc., Sunnyvale, CA, USA). The inhibitory rate of cell proliferation was expressed as $IC_{50}$ values and calculated by the following formula:

$$\text{Growth inhibition (\%)} = [\text{ODcontrol-ODtreated}]/\text{ODcontrol}\times 100$$

Doxorubicin was tested as a positive control, and cell solutions containing 0.5% DMSO were tested as a negative control.

The antibacterial activities of the isolated compounds were tested against Gram-positive bacteria *Staphylococcus aureus*, *Bacillus subtilis*, and *Sarcina lutea* and Gram-negative bacteria *Klebsiella pneumoniae* and *Escherichia coli* with the minimum inhibitory concentration (MIC) method recommended by the Clinical and Laboratory Standards Institute [72].

## 3. Results

### *3.1. Isolation and Screening of an Antitumor Compound Producing Strains*

Thirty-nine strains belonging to actinomycetes were isolated from the soil samples. The crude extracts of these isolates were examined for their cytotoxic activity at dilution concentrations of 100 μg/mL and 20 μg/mL. As a result of primary screening, fourteen strains showed cytotoxic activity to human colon tumor cell line HCT-116 (Figure 1). Due to the superior cytotoxic activity of strain NEAU-wh3-1, which inhibition rate was greater than 80% at both concentrations, further chemical investigations were performed on this strain.

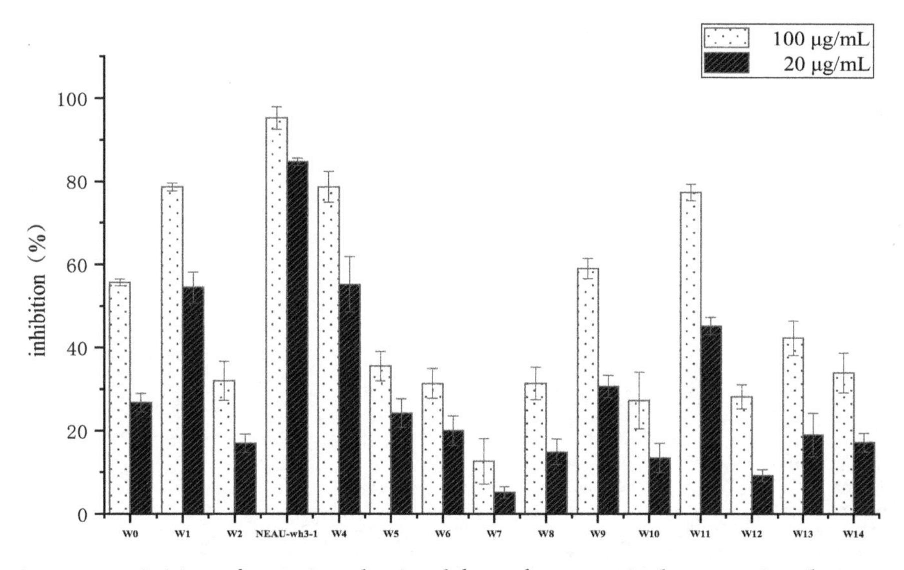


**Figure 1.** Antitumor activities of extracts obtained from fourteen isolates against human colon tumor cell line HCT-116.

### *3.2. Polyphasic Taxonomic Characterization of NEAU-wh3-1*

Morphological observation of 2-week-old cultures of strain NEAU-wh3-1 grown on ISP 3 medium revealed that the strain has the typical characteristics of genus *Embleya* and formed well-developed, branched substrate hyphae and aerial mycelium that differentiated into spiral spore chains consisted of cylindrical spores (0.6–0.8μm × 0.9–1.3μm), the spores were rough-surfaced and non-motile (Figure 2). Strain NEAU-wh3-1 was found to grow at a temperature range of 4 to 37 °C (optimum temperature 28 °C), pH 5 to 12 (optimum pH 7), and NaCl tolerance of 0% to 3% (optimum NaCl of 1%). The physiological and biochemical properties of strain NEAU-wh3-1, *Embleya scabrispora* DSM 41855$^{T}$, *Embleya hyalina* MB891-A1$^{T}$, and *Streptomyces lasii* 5H-CA11$^{T}$ are given in Table 1.

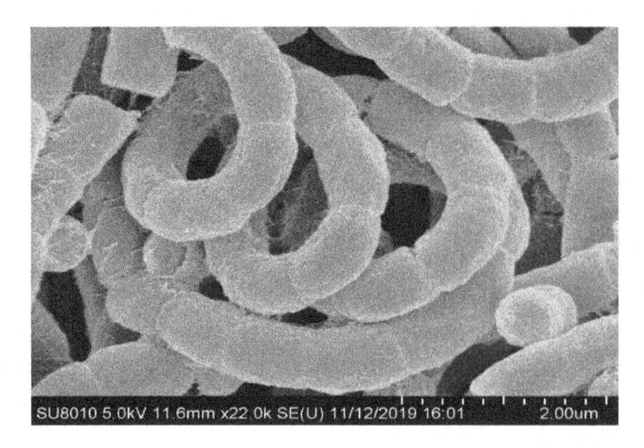


**Figure 2.** Scanning electron micrograph of spore chains of strain NEAU-wh3-1 grown on ISP 3 agar for 2 weeks at 28 °C.

**Table 1.** The physiological and biochemical properties of strain NEAU-wh3-1, *Embleya scabrispora* DSM 41855$^T$, *Embleya hyalina* MB891-A1$^T$, *Streptomyces lasii* 5H-CA11$^T$.

| Characteristic | 1 | 2[a,c] | 3[b] | 4[c] |
|---|---|---|---|---|
| Decomposition of | | | | |
| Cellulose | − | − | ND | + |
| Tween 20 | − | + | ND | + |
| Tween 40 | + | + | ND | + |
| Tween 80 | − | + | ND | + |
| Liquefaction of gelatin | − | − | + | + |
| Growth temperature (°C) | 4–37 | 18–36 | 10–28 | 15–38 |
| pH range for growth | 5–12 | 4–10 | 6–11 | 5–11 |
| NaCl tolerance range (*w/v*, %) | 0–3 | 0–3 | 0–1 | 0–2.5 |
| Milk coagulation | − | w | − | + |
| Nitrate reduction | − | + | − | − |
| Starch hydrolysis | − | w | − | − |
| Carbon source utilization | | | | |
| L-arabinose | + | ± | − | − |
| Dulcitol | + | w | − | W |
| D-Fructose | + | − | + | − |
| D-Galactose | + | − | + | − |
| D-Glucose | + | + | + | + |
| Inositol | − | + | + | − |
| Lactose | + | + | − | + |
| D-Maltose | + | − | ± | − |
| D-Mannitol | + | − | − | − |
| D-Mannose | + | − | + | − |
| D-Raffinose | + | − | − | + |
| D-Ribose | − | + | ND | − |
| D-Sorbitol | + | − | − | − |
| D-Sucrose | + | ± | W | + |
| D-Xylose | + | + | − | − |
| L-Rhamnose | − | + | + | − |
| Nitrogen source utilization | | | | |
| L-Alanine | + | W | ND | + |
| L-Arginine | + | W | ND | + |
| L-Asparagine | + | W | ND | − |
| L-Aspartic acid | + | + | ND | + |
| Creatine | + | w | ND | − |
| L-Glutamic acid | + | w | ND | + |
| L-Glutamine | + | w | ND | + |
| Glycine | + | w | ND | + |
| L-Proline | − | + | ND | + |
| L-Serine | + | − | ND | − |
| L-Threonine | + | w | ND | + |
| L-Tyrosine | + | + | ND | + |
| Phospholipids | DPG, PE, PI, UL | PE, PGL | DPG, PE, PI | DPG, PME, PI, PIM, GL |
| Menaquinones | MK-9($H_4$), MK-9($H_6$), MK-9($H_8$) | MK-9($H_2$), MK-9($H_4$), MK-9($H_6$) | MK-9($H_4$), MK-9($H_6$), MK-9($H_8$) | MK-9($H_4$), MK-9($H_6$), MK-9($H_8$) |
| Whole cell-wall sugars | Arabinose, glucose, ribose | Arabinose | Arabinose, glucose | Glucose, ribose |

Strains: 1, NEAU-wh3-1; 2, *Embleya scabrispora* DSM 41855$^T$; 3, *Embleya hyalina* MB891-A1$^T$; 4, *Streptomyces lasii* 5H-CA11$^T$. Abbreviation: +, positive; −, negative; ±, ambiguous; ND, not determined; w, weak; DPG, diphosphatidylglycerol; PME, phosphatidylmonomethylethanolamine; PE, phosphatidylethanolamine; PI, phosphatidylinositol; PIM, phosphatidylinositolmannoside; UL, unidentified lipid; GL, glucosamine-containing lipid; PGL, phospholipid containing glucosamine. All data are from this study except where marked. [a] Data from Ping et al. [43]; [b] Data from Komaki et al. [41]; [c] Data from Liu et al. [73].

Chemotaxonomic analyses revealed that strain NEAU-wh3-1 contained LL-diaminopimelic acid as cell wall diamino acid. The whole-cell sugar was found to contain arabinose, glucose, and ribose. The phospholipid profile consisted of diphosphatidylglycerol (DPG), phosphatidylethanolamine (PE), phosphatidylinositol (PI), and two unidentified lipids (ULs) (Supplementary Figure S1). The menaquinones detected were MK-9($H_4$) (46.5%), MK-9($H_6$) (45.8%), and MK-9($H_8$) (7.7%).

The almost complete 16S rRNA gene sequence of strain NEAU-wh3-1 (1487 bp) was determined and deposited with the accession number MN928616 in the GenBank/EMBL (European Molecular Biology Laboratory)/DDBJ (DNA Data Bank of Japan) databases. EzBioCloud analysis suggests that strain NEAU-wh3-1 shared the highest 16S rRNA gene sequence similarities with *Embleya scabrispora* DSM 41855$^T$ (99.65%), *Embleya hyalina* MB891-A1$^T$ (99.45%), and *Streptomyces lasii* 5H-CA11$^T$ (98.62%). Phylogenetic analysis based on the 16S rRNA gene sequences indicated that the strain formed a stable cluster with *E. scabrispora* DSM 41855$^T$, *E. hyalina* MB891-A1$^T$, and *S. lasii* 5H-CA11$^T$ based on neighbor-joining algorithm (Figure 3) and also supported by the maximum-likelihood algorithm (Supplementary Figure S2). To further clarify the affiliation of strain NEAU-wh3-1 to its closely related strains, partial sequences of housekeeping genes including *atp*D, *gyr*B, *rec*A, *rpo*B, and *trp*B were obtained. GenBank accession numbers of the sequences are displayed in Table S1. The phylogenetic tree based on the neighbor-joining tree constructed from the concatenated sequence alignment (1979 bp) of five housekeeping genes (Figure 4) suggested that the isolate clustered with *E. scabrispora* DSM 41855$^T$ and *E. hyalina* MB891-A1$^T$, and also supported by the maximum-likelihood algorithm (*Streptomyces lasii* 5H-CA11$^T$ lacks housekeeping genes; Supplementary Figure S3). Moreover, pairwise distances calculated for NEAU-wh3-1 and the related species using concatenated sequences of *atp*D-*gyr*B-*rec*A-*rpo*B-*trp*B were well above 0.007 (Table S2) for the related species, which was considered to be the threshold for species determination [74].

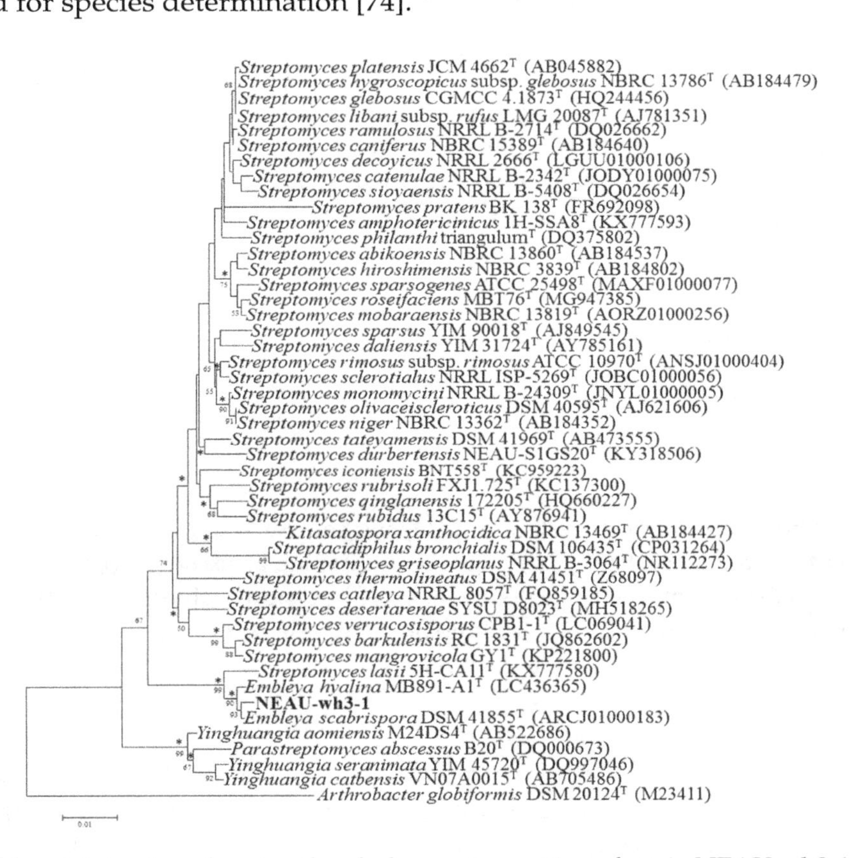


**Figure 3.** Neighbor-joining tree showing the phylogenetic position of strain NEAU-wh3-1 (1487 bp) and related taxa based on 16S rRNA gene sequences. Bootstrap values > 50% (based on 1000 replications) are shown at branch points. *Arthrobacter globiformis* DSM 20124$^T$ (M23411) was used as an outgroup. *Asterisks* indicate branches also recovered in the maximum-likelihood tree; Bar, 0.01 substitutions per nucleotide position.

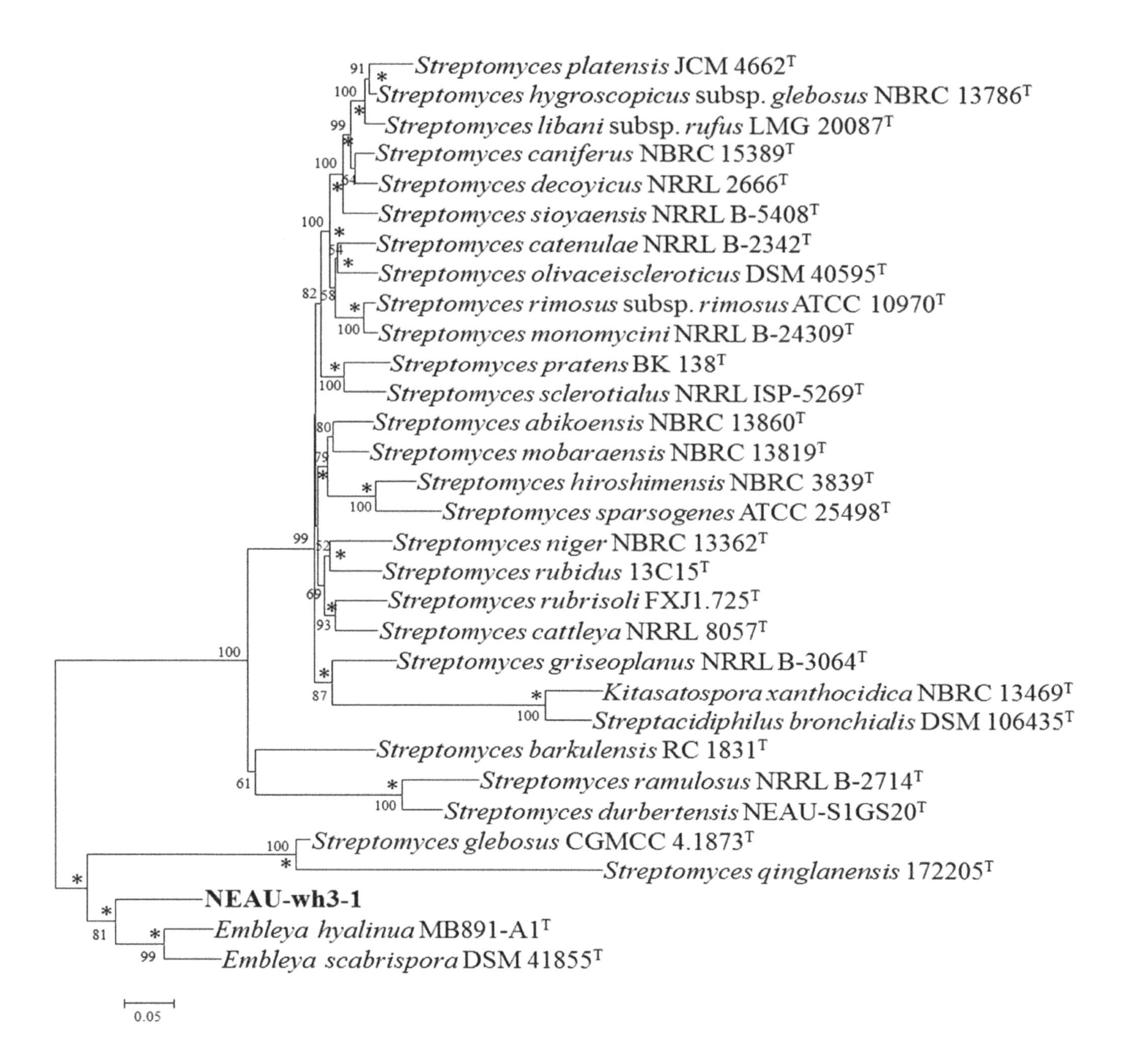


**Figure 4.** Neighbor-joining tree based on multilocus sequence analysis (MLSA) analysis of the concatenated partial sequences (1979 bp) from five housekeeping genes (*atp*D, *gyr*B, *rec*A, *rpo*B, and *trp*B) of strain NEAU-wh3-1 (in bold) with related taxa. Only bootstrap values above 50% (percentages of 1000 replications) are indicated. Asterisks indicate branches also recovered in the maximum-likelihood tree; Bar, 0.05 substitutions per nucleotide position.

### 3.3. Structural Elucidation

The strain NEAU-wh3-1 was grown preparative scale in 15.0 L of production broth for 7 days. Bioassay-guided isolation of the active components of the strain yielded eight main bioactive compounds. Compounds **2–8** are known compounds, which structures were elucidated as conglobatin (**2**) [75], piericidin C1 (**3**) [76], piericidin C5 (**4**) [77], piericidin A1 (**5**) [78], piericidin A3 (**6**) [76], Mer-A 2026 A (**7**) [79], and BE-52211 D (**8**) [80] by analysis of their spectroscopic data and comparison with literature values (Figure 5, Figures S12–S27). Compound **1** is a new zincophorin analogue (Figure 6, Figures S4–S11) [17].

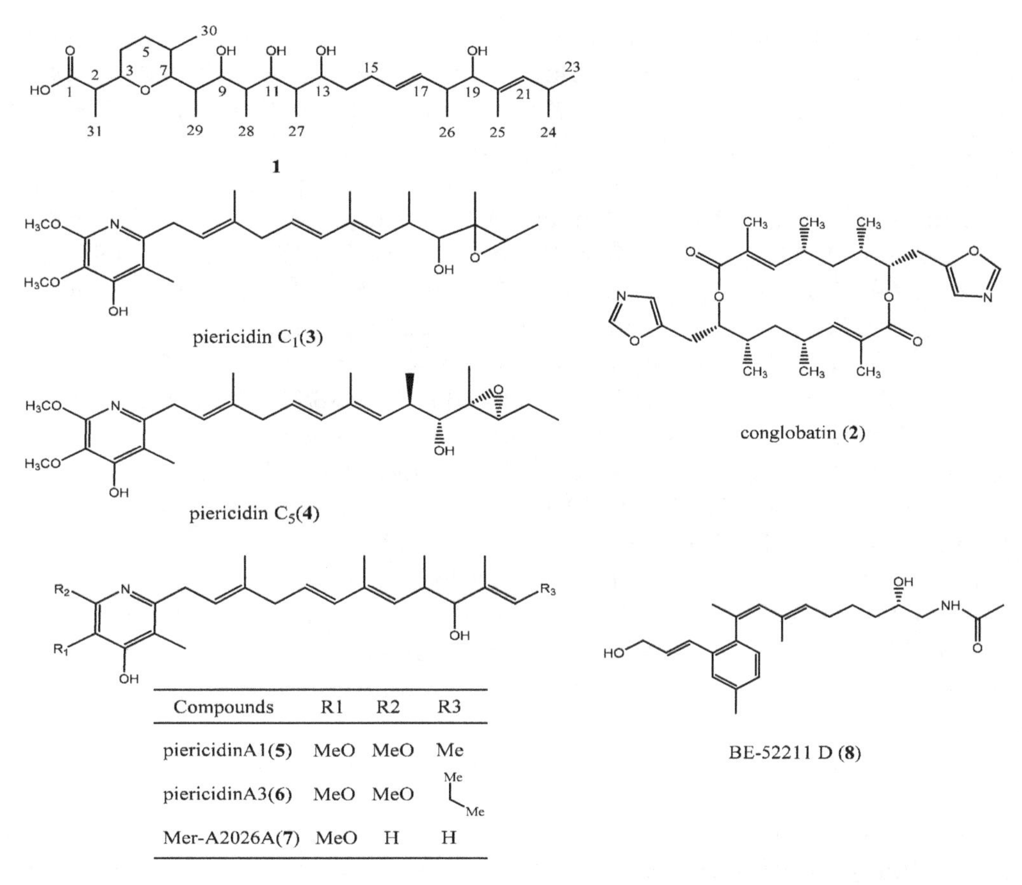


| Compounds | R1 | R2 | R3 |
|---|---|---|---|
| piericidinA1(5) | MeO | MeO | Me |
| piericidinA3(6) | MeO | MeO | |
| Mer-A2026A(7) | MeO | H | H |

**Figure 5.** The structures of compounds **1–8**. Compound **1** was isolated as white solid with $[\alpha]_D^{25}$ + 15 (c 0.043, EtOH) and UV (EtOH) λmax nm (log ε): 202 (4.53). Its molecular formula was established as $C_{31}H_{56}O_7$ by HR-ESI-MS at *m/z* 539.3942 $[M-H]^-$ (calcd 539.3953 as $C_{31}H_{55}O_7$). The IR spectrum revealed hydroxyl absorption at 3320 $cm^{-1}$ and carbonyl absorption at 1735 $cm^{-1}$, as well as methyl and methylene absorptions at 2953 $cm^{-1}$ and 2924 $cm^{-1}$.

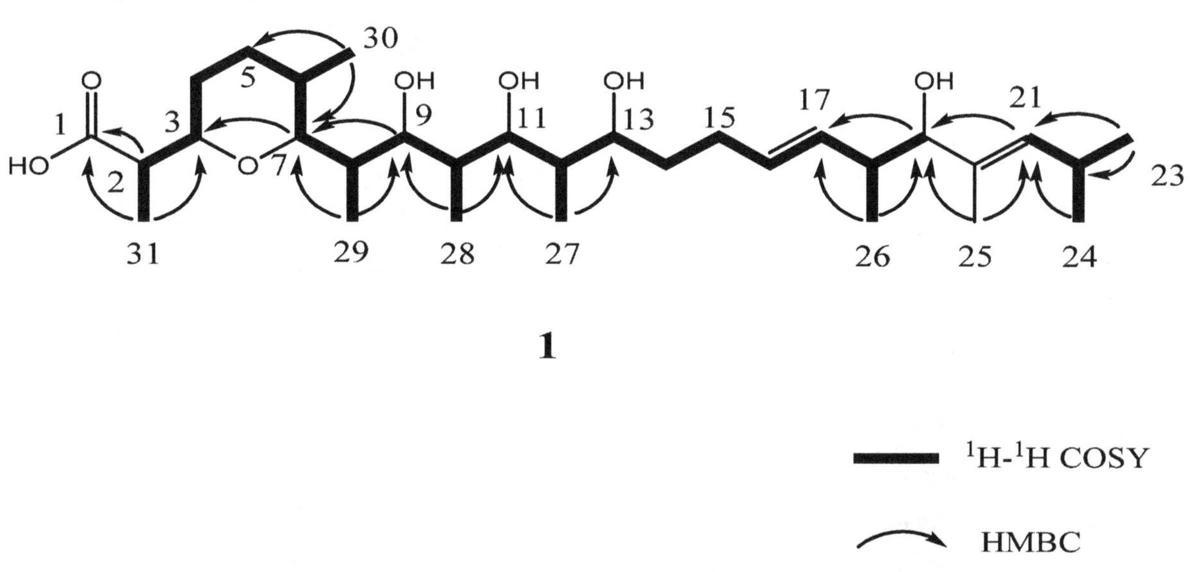


**Figure 6.** 2D nuclear magnetic resonance (NMR) correlations of compound **1**.

Analysis of $^1H$ NMR spectrum of **1** revealed the presence of three olefinic protons at $\delta_H$ 5.52 (1H, m), 5.36 (1H, m), 5.20 (1H, d, $J$ = 8.9 Hz), seven aliphatic methine protons at $\delta_H$ 4.07 (1H, m), 4.06 (1H, m), 3.77 (1H, d, $J$ = 8.9 Hz), 3.72 (1H, dd, $J$ = 9.6, 2.1 Hz), 3.58 (1H, d, $J$ = 9.2 Hz), 3.49 (1H, m), 3.28 (1H, m), seven methylene protons at $\delta_H$ 2.18 (1H, m), 2.13 (1H, m), 1.77 (1H, m), 1.67 (2H, m), 1.40 (1H, m),

1.35 (1H, m), 1.28 (1H, m), one singlet methyl at $\delta_H$ 1.63 s, in addition to eight doublet methyl protons at $\delta_H$ 1.18 (3H, d, $J$ = 7.1 Hz), 1.15 (3H, d, $J$ = 7.2 Hz), 1.11 (3H, d, $J$ = 7.0 Hz), 0.98 (3H, d, $J$ = 6.5 Hz), 0.97 (3H, d, $J$ = 6.6 Hz), 0.86 (3H, d, $J$ = 6.7 Hz), 0.81 (3H, d, $J$ = 6.5 Hz), 0.70 (3H, d, $J$ = 6.7 Hz). The $^{13}$C NMR and DEPT135 spectra (Table 2) of 1 showed 31 resonances attributable to a carbonyl carbon at $\delta_C$ 175.6, one $sp^2$ quaternary carbon at $\delta_C$ 132.3, three $sp^2$ methines at $\delta_C$ 136.6, 134.5, 132.4. In the sp$^3$-carbon region, the spectrum showed six oxygenated methines at $\delta_C$ 84.3, 83.3, 81.7, 76.1, 74.0, 69.4, six methines at $\delta_C$ 42.2, 37.1, 36.5, 36.0, 33.0, 26.1, four methylenes at $\delta_C$ 34.2, 28.6, 27.1, 24.9 and nine methyl carbons at $\delta_C$ 22.9, 22.9, 17.1, 16.7, 15.4, 12.3, 11.3, 10.8, 10.1. The $^1$H–$^1$H COSY correlations (Figure 6) of H-2/H-3/H$_2$-4/H$_2$-5/H-6/H-7/H-8/H-9/H-10/H-11/H-12/H-13/H$_2$-14/H$_2$-15/H-16/H-17/H-18/H-19 established connectivity from H-2 atom along the chain through to C-19 atom. The correlations between H-21/H-22/H$_3$-23/H$_3$-24, H-12/H$_3$-27, H-10/H$_3$-28, H-18/H$_3$-26, H-8/H$_3$-29 protons in the $^1$H–$^1$H COSY spectrum (Figure 6) indicated the five structural units of C-21–C-24, C-18–C-26, C-12–C-27, C-10-C-28, C-8-C-29. The observed HMBC (heteronuclear multiple bond correlation) correlations (Figure 6) from H$_3$-23, H$_3$-24 to C-21, C-22, from H$_3$-25 to C-19, C-21, from H$_3$-26 to C-17, C-18, and C-19, from H$_3$-27 to C-11, C-12, and C-13, from H$_3$-28 to C-9, C-10, C-11, from H$_3$-29 to C-7, C-8, C-9, from H$_3$-30 to C-5, C-6, C-7, from H$_3$-31 to C-2 and C-3, from H-21 to C-19, H-20 to C-18, from H-19 to C-17, from H$_3$-23 to C-21 established the linkage of C-2–C-22. The carbonyl group was connected with C-2 by the HMBC corrections from H-2 and H$_3$-31 to C-1 ($\delta_C$ 175.6). The correlations from H-7 ($\delta_H$ 3.77 d, $J$ = 8.9 Hz) to C-3 ($\delta_C$ 74.0) indicated the linkage of C-3 and C-7 through an oxygen atom to form a tetrahydropyran ring. Taking the molecular formula of $C_{31}H_{56}O_7$ into account, four hydroxyl groups were situated at C-9, C-11, C-13, C-19, respectively, and a carboxyl group was situated at C-1. Comparison the NMR data of **1** with Zincophorin [17], a mocarboxylic acid ionophore contains one single tetrahydropyran ring, which was isolated from a strain of *Streptomyces griseus*, implied that **1** was identified to be an analogue of Zincophorin, the difference between two compounds was that the terminal ethyl group in Zincophorin was replaced by a H proton in compound **1**. On the basis of the above spectroscopic data, a gross structure of **1** was established and named Zincophorin B, and the $^1$H and $^{13}$C resonances in **1** were assigned (Table 2).

**Table 2.** $^1$H and $^{13}$C NMR data of compound 1 in $CDCl_3$.

| No. | $\delta_H$ ($J$ in Hz) | $\delta_C$ (p.p.m) | No. | $\delta_H$ ($J$ in Hz) | $\delta_C$ (p.p.m) |
|---|---|---|---|---|---|
| 1 | | 175.6 | 16 | 5.52 m | 132.4 |
| 2 | 3.28 m | 37.1 | 17 | 5.36 m | 134.5 |
| 3 | 4.07 m | 74.0 | | | |
| 4 | 1.67 m | 24.9 | 18 | 2.27 m | 42.2 |
| | | | 19 | 3.58 d (9.2) | 81.7 |
| 5a | 1.28 m | 27.1 | 20 | | 132.3 |
| 5b | 1.40 m | | 21 | 5.20 d (8.9) | 136.6 |
| 6 | 1.52 m | 26.1 | 22 | 2.59 m | 26.1 |
| 7 | 3.77 d (8.9) | 76.1 | 23 | 0.97 d (6.6) | 22.9 |
| 8 | 2.04 m | 33.0 | 24 | 0.98 d (6.5) | 22.9 |
| 9 | 3.72 dd (9.6, 2.1) | 84.3 | 25 | 1.63 s | 10.1 |
| 10 | 2.01 m | 36.5 | 26 | 0.86 d (6.7) | 16.7 |
| 11 | 3.49 m | 83.3 | 27 | 1.15 d (7.2) | 10.8 |
| 12 | 1.75 m | 36.0 | 28 | 0.70 d (6.7) | 12.3 |
| | | | 29 | 1.11 d (7.0) | 11.3 |
| 13 | 4.06 m | 69.4 | 30 | 0.81 d (6.5) | 17.1 |
| 14a | 1.35 m | 34.2 | 31 | 1.18 d (7.1) | 15.4 |
| 14b | 1.77 m | | | | |
| 15a | 2.13 m | 28.6 | | | |
| 15b | 2.18 m | | | | |

### 3.4. Biological Activity

The cytotoxic activities of compounds 1–8 against K562, HCT-116, and HepG2 cancer cell lines are showed in Table 3. Eight compounds restrained proliferation of the tested cells and compound **1** showed the highest cytotoxic activity, and the average $IC_{50}$ values were lower than 10.0 μg/mL.

**Table 3.** The cytotoxicity of compounds 1–8.

| Compound | $IC_{50}$ (μg/mL) | | |
|---|---|---|---|
| | K562 | HCT-116 | HepG2 |
| 1 | 8.8 ± 1.5 | 9.5 ± 0.8 | 9.6 ± 5.6 |
| 2 | 57.1 ± 7.3 | 75.42 ± 2.1 | — |
| 3 | — | 68.39 ± 3.3 | 53.78 ± 6.7 |
| 4 | — | 36.8 ± 5.6 | 17.5 ± 1.9 |
| 5 | 28.3 ± 1.1 | 14.3 ± 1.6 | 27.3 ± 5.8 |
| 6 | 36.6 ± 2.4 | 21.6 ± 4.1 | 79.7 ± 5.9 |
| 7 | — | 112.3 ± 5.7 | — |
| 8 | 11.42 ± 3.05 | 15.13 ± 1.76 | 10.83 ± 3.47 |
| Doxorubicin | 1.1 ± 0.1 | 0.9 ± 0.3 | 2.1 ± 0.2 |

The result of minimum inhibitory concentrations (MICs) showed that compound **1** showed good activities against Gram-positive bacteria *Staphylococcus aureus*, *Sarcina lutea*, and *Bacillus subtilis*, and the Gram-negative bacteria *Klebsiella pneumoniae* in vitro (Table 4). Compound **8** showed weak antibacterial activity against two Gram-positive bacterium and the minimum inhibitory concentrations (MICs) of compounds **2–7** were determined to be >10 mg/mL, so they had no activity against these tested pathogens.

**Table 4.** The antibacterial activity of compounds 1-8.

| Compounds | MIC (μg/mL) | | | | |
|---|---|---|---|---|---|
| | Gram-Positive Bacteria | | | Gram-Negative Bacteria | |
| | *Staphylococcus aureus* | *Sarcina lutea* | *Bacillus subtilis* | *Klebsiella pneumonie* | *Escherichia coli* |
| **1** | 31.0 ± 2.5 | 44.0 ± 5.8 | 3.5 ± 0.5 | 25.0 ± 1.5 | — |
| **2–7** | — | — | — | — | — |
| **8** | 210.0 ± 20.0 | 190.0 ± 15.0 | — | — | — |

## 4. Discussion

In this research, the results of morphological, physiological, and biochemical tests showed that strain NEAU-wh3-1 has typical characteristics of the genus *Embleya* [41]. Such as containing LL-diaminopimelic acid as the cell wall peptidoglycan, MK-9($H_4$), and MK-9($H_6$) as the major menaquinones, phosphatidylethanolamine (PE) as the predominant phospholipid and arabinose in the whole sugars. Moreover, strain NEAU-wh3-1 formed spiral spore chains and the spore surface was rough, which are consistent with *E. scabrispora* DSM 41855$^T$ and *E. hyalina* MB891-A1$^T$ [39,41]. In addition, the phylogenetic trees constructed from the 16S rRNA gene sequences and the concatenated sequences alignment (1979 bp) of five housekeeping genes all suggested that the isolate should be assigned to the genus *Embleya*.

However, some obvious differences could also be found between strain NEAU-wh3-1 and its closely related strains regarding several phenotypic and chemotaxonomic characteristics (Table 1). The isolate was able to grow at 4 °C, in contrast to its closely related strains, which were not. The composition of phospholipids and menaquinones of strain NEAU-wh3-1 was also different from its related species, *E. scabrispora* DSM 41855$^T$, *E. hyalina* MB891-A1$^T$ and *S. lasii* 5H-CA11$^T$. Most notably, the whole-cell sugars of strain NEAU-wh3-1 was found to contain arabinose, glucose, and ribose, while *E. scabrispora* DSM 41855$^T$ only contains arabinose, *E. hyalina* MB891-A1$^T$ contains arabinose and glucose and *S. lasii* 5H-CA11$^T$ contains glucose and ribose, which also could distinguish the strain from its closely related strains. Other phenotypic differences include the temperature and pH range of growth, patterns of carbon and nitrogen utilization, hydrolysis of cellulose, starch, and Tweens (20, 40, and 80), liquefaction of gelatin, peptonization of milk, production of $H_2S$, and urease and reduction of nitrate. Therefore, it is evident from the phenotypic, genotypic, and chemotaxonomic data that strain NEAU-wh3-1 may represent a novel species of the genus *Embleya*.

The genus *Embleya*, recently transferred from genus *Streptomyces*, is a new member of the family *Streptomycetaceae* [39,40]. At present, it contains only two species: *Embleya scabrispora*, could produce hitachimycin with antitumor, antibacterial, and antiprotozoal activities [44–46]; and *Embleya hyaline*, could produce nybomycin which is an effective agent against antibiotic-resistant *Staphylococcus aureus* and is called a reverse antibiotic [48]. During the study of the chemical properties of the active ingredients of strain NEAU-wh3-1, eight active compounds were obtained, including one macrolide dilactone, five piericidins, one β-hydroxy acetamides, an analogue of monocarboxylic acid ionophore, which were observed to fit into at least three types based on their molecular skeletons. This shows to some extent that this strain has the ability to produce metabolites with a wide variety of different skeletal structures. Out of these compounds, Piericidins (**3-7**) were a class of polyene alpha-pyridone heterocyclic antibiotics, among them, Piericidin A1(**5**) was first reported [81], which was isolated from *Streptomyces mobaraensis*. Piericidins exhibit interesting biological activities, in particular antitrypanosomal [82]. In our research, compounds **3-7** exhibited different degrees of cytotoxicity on three types of tumor cells, but they did not show any antibacterial activity, which was consistent with previous reports [78,81]. As a Piericidins-producing strain, NEAU-wh3-1 has certain application potential in pest control. Conglobatin (**2**) is an unusual 16-membered macrocyclic diolide originally isolated from a polyether-producing strain of *Streptomyces conglobatus* ATCC 31005$^T$ and was reported to be essentially devoid of antifungal, antibacterial, antitumor, and antiprotozoal activity at that time [75]. However, in recent research, FW-04-806 is identical in structure to conglobatin, and it has been reported to inhibit the growth of a human chronic myelocytic leukemia K562 cell line with an $IC_{50}$ of 6.66 μg/mL, further study also investigated the effects of FW-04-806 on SKBR3 and MCF-7, respectively [83]. Its mechanism of action appears to be novel, via direct binding to the N-terminal domain of Hsp90 and disruption of its interaction with co-chaperone Cdc37 [84]. In our antitumor activity test, Conglobatin (**2**) showed good bioactivity against two tumor cell lines, supporting it at least partially accounted for the cytotoxic activity of the strain NEAU-wh3-1 extract. BE-52211 D (**8**) was a cytotoxic metabolite from a strain of *Streptomyces* and had been reported to have moderate cytotoxicity against human hepatocellular liver carcinoma cells HepG2, human leukemia cells K562, and human colon carcinoma cells HCT-116 with the $IC_{50}$ values of >10 μg/mL [80], which is consistent with the result in the present study. Compound **1** structurally related to zincophorin, which was also referred to as M144255 or griseochellin and is a polyoxygenated ionophoric antibiotic isolating from *Streptomyces griseus* in 1984 [17]. It has been reported to possess strong in vivo activity against Gram-positive bacteria and have strong cytotoxicity against human lung carcinoma cells A549 and Madin-Darby canine kidney cells MDCK [18]. No biological activity has been reported against Gram-negative bacteria, yeasts, and fungi [85]. The second member in the zincophorin family named CP-78545, was found in the culture broth of *Streptomyces* sp. N731-45. The structural difference between them is that CP-78545 has an extra terminal double bond; but they have similar spectrum and potency on biological properties except for the antitumor activity (no reports) [86]. In our antitumor and antimicrobial assays, compound **1** showed the highest antitumor activity against three human cell lines and good antibacterial activity against Gram-negative bacteria. To our knowledge, this is the first report of this kind of compound with antibacterial activity against Gram-negative bacteria. This study has enriched the activity spectrum of Zincophorins.

## 5. Conclusions

Strain producing a new compound with strong antitumor activity, isolated from the rhizosphere soil of wheat (*Triticum aestivum* L.) in HeBei province, China. Morphological and chemotaxonomic features together with phylogenetic analysis suggested that strain NEAU-wh3-1 belonged to the genus *Embleya*. Cultural and biochemical characteristics combined with multilocus sequence analysis clearly revealed that strain NEAU-wh3-1 may represent a novel species of the genus *Embleya*. Moreover, eight compounds, including one new compound with higher antitumor activities against three human cell lines, were isolated from the strain.

**Supplementary Materials:**
Table S1: GenBank accession numbers of the sequences used in MLSA; Table S2: MLSA distance values for selected strains in this study; Figure S1: The polar lipids of strain NEAU-wh3-1; Figure S2: Maximum-likelihood tree based on 16S rRNA gene sequences showing relationship between strain NEAU-wh3-1 and related taxa; Figure S3: Maximum-likelihood tree based on multilocus sequence analysis (MLSA)analysis of the concatenated partial sequences (1979 bp) from five housekeeping genes (*atpD*, *gyrB*, *recA*, *rpoB*, and *trpB*) of strain NEAU-wh3-1 (in bold) with related taxa; Figure S4: $^{1}$H NMR (400 MHz) spectrum of compound **1** in $CDCl_3$; Figure S5: $^{13}$C NMR (150 MHz) spectrum of compound **1** (in $CDCl_3$); Figure S: $^{1}$H-$^{1}$H COSY spectrum (400 MHz) of compound **1** (in $CDCl_3$); Figure S7: HSQC spectrum (400 MHz) of compound **1** (in $CDCl_3$); Figure S8: HMBC spectrum (400 MHz) of compound **1** (in $CDCl_3$); Figure S9: IR spectrum of compound **1**(in EtOH); Figure S10: UV spectrum of compound **1** (in EtOH); Figure S11: The HRESIMS spectrum of compound **1**; Figure S12: $^{1}$H NMR (400 MHz) spectrum of compound **2** in $CDCl_3$; Figure S13: $^{13}$C NMR (150 MHz) spectrum of compound **2** in $CDCl_3$; Figure S14: The ESI-MS spectrum of compound **2**; Figure S15: $^{1}$H NMR (400 MHz) spectrum of compound **3** in $CDCl_3$; Figure S16: The ESI-MS spectrum of compound **3**; Figure S17: $^{1}$H NMR (400 MHz) spectrum of compound **4** in $CDCl_3$; Figure S18: The ESI-MS spectrum of compound **4**; Figure S19: $^{1}$H NMR (400 MHz) spectrum of compound **5** in $CDCl_3$; Figure S20: The ESI-MS spectrum of compound **5**; Figure S21: $^{1}$H NMR (400 MHz) spectrum of compound **6** in $CDCl_3$; Figure S22: The ESI-MS spectrum of compound **6**; Figure S23: $^{1}$H NMR (400 MHz) spectrum of compound **7** in $CDCl_3$; Figure S24: ESI-MS spectrum of compound of compound **7**; Figure S25: $^{1}$H NMR (400 MHz) spectrum of compound **8** in $CDCl_3$; Figure S26: $^{13}$C NMR (150 MHz) spectrum of compound **8** in $CDCl_3$; Figure S27: The HRESIMS spectrum of compound **8**.

**Author Contributions:** H.W., T.S., and W.S. performed the experiments. X.G. prepared the figures and tables. P.C. and X.X. analyzed the data. Y.S. and J.Z. designed the experiments and reviewed the manuscript. All authors have read and agreed to the published version of the manuscript.

## References


1. Torre, L.A.; Bray, F.; Siegel, R.L.; Ferlay, J.; Lortet-Tieulent, J.; Jemal, A. Global cancer statistics, 2012. *CA. Cancer J. Clin.* **2015**, *65*, 87–108. [CrossRef]
2. Commander, H.; Whiteside, G.; Perry, C. Vandetanib: First global approval. *Drugs* **2011**, *71*, 1355–1365. [CrossRef]
3. Zhu, F.; Zhao, X.; Li, J.; Guo, L.; Bai, L.; Qi, X. A new compound Trichomicin exerts antitumor activity through STAT3 signaling inhibition. *Biomed. Pharmacother.* **2020**, *121*, 109608. [CrossRef]
4. Zhang, Z.; Yu, X.; Wang, Z.; Wu, P.; Huang, J. Anthracyclines potentiate anti-tumor immunity: A new opportunity for chemoimmunotherapy. *Cancer Lett.* **2015**, *369*, 331–335. [CrossRef]
5. Sznarkowska, A.; Kostecka, A.; Meller, K.; Bielawski, K.P. Inhibition of cancer antioxidant defense by natural compounds. *Oncotarget* **2017**, *8*, 15996–16016. [CrossRef]
6. Newman, D.J.; Cragg, G.M. Natural products as sources of new drugs over the last 25 years. *J. Nat. Prod.* **2007**, *70*, 461–477. [CrossRef]
7. Cragg, G.M.; Newman, D.J.; Snader, K.M. Natural products in drug discovery and development. *J. Nat. Prod.* **1997**, *60*, 52–60. [CrossRef]
8. Chin, Y.W.; Balunas, M.J.; Chai, H.B.; Kinghorn, A.D. Drug discovery from natural sources. *Drug Addict. From Basic Res. Ther.* **2008**, *8*, 17–39.

9. Pettit, R.K. Culturability and Secondary Metabolite Diversity of Extreme Microbes: Expanding Contribution of Deep Sea and Deep-Sea Vent Microbes to Natural Product Discovery. *Mar. Biotechnol.* **2011**, *13*, 1–11. [CrossRef]
10. Qin, S.; Li, J.; Chen, H.H.; Zhao, G.Z.; Zhu, W.Y.; Jiang, C.L.; Xu, L.H.; Li, W.J. Isolation, diversity, and antimicrobial activity of rare actinobacteria from medicinal plants of tropical rain forests in Xishuangbanna China. *Appl. Environ. Microbiol.* **2009**, *75*, 6176–6186. [CrossRef]
11. Parkinson, D.R.; Arbuck, S.G.; Moore, T.; Pluda, J.M.; Christian, M.C. Clinical development of anticancer agents from natural products. *Stem Cells* **1994**, *12*, 30–43. [CrossRef]
12. Berdy, J. Bioactive microbial metabolites. *J Antibiot.* **2005**, *58*, 1–26. [CrossRef]
13. Ventura, M.; Canchaya, C.; Tauch, A.; Chandra, G.; Fitzgerald, G.F.; Chater, K.F.; van Sinderen, D. Genomics of Actinobacteria: Tracing the Evolutionary History of an Ancient Phylum. *Microbiol. Mol. Biol. Rev.* **2007**, *71*, 495–548. [CrossRef]
14. Goodfellow, M.; Williams, S.T. Ecology of *actinomycetes*. *Annu. Rev. Microbiol.* **1983**, *37*, 189–216. [CrossRef]
15. Rabbani, A.; Finn, R.M.; Ausió, J. The anthracycline antibiotics: Antitumor drugs that alter chromatin structure. *BioEssays* **2005**, *27*, 50–56. [CrossRef]
16. Demain, A.L.; Sanchez, S. Microbial drug discovery: 80 Years of progress. *J. Antibiot.* **2009**, *62*, 5–16. [CrossRef]
17. Brooks, H.A.; Gardner, D.; Poyser, J.P.; King, T.J. The structure and absolute stereochemistry of zincophorin (antibiotic m144255): a monobasic carboxylic acid ionophore having a remarkable specificity for divalent cations. *J. Antibiot.* **1984**, *37*(11), 1501–1504. [CrossRef]
18. Walther, E.; Boldt, S.; Kage, H.; Lauterbach, T.; Martin, K.; Roth, M.; Hertweck, C.; Sauerbrei, A.; Schmidtke, M.; Nett, M. Zincophorin-biosynthesis in *Streptomyces griseus* and antibiotic properties. *GMS Infect. Dis.* **2016**, *4*, Doc08.
19. Yang, S.X.; Gao, J.M.; Zhang, A.L.; Laatsch, H.S. A novel toxic macrolactam polyketide glycoside produced by actinomycete *Streptomyces sannanensis*. *Bioorg. Med. Chem. Lett.* **2011**, *21*, 3905–3908. [CrossRef]
20. Subramani, R.; Aalbersberg, W. Culturable rare Actinomycetes: Diversity, isolation and marine natural product discovery. *Appl. Microbiol. Biotechnol.* **2013**, *97*, 9291–9321. [CrossRef]
21. Heidari, B.; Mohammadipanah, F. Isolation and identification of two alkaloid structures with radical scavenging activity from *Actinokineospora* sp. UTMC 968, a new promising source of alkaloid compounds. *Mol. Biol. Rep.* **2018**, *45*, 2325–2332. [CrossRef] [PubMed]
22. Gao, M.Y.; Qi, H.; Li, J.S.; Zhang, H.; Zhang, J.; Wang, J.D.; Xiang, W.S. A new polysubstituted cyclopentene derivative from *Streptomyces* sp. HS-NF-1046. *J. Antibiot.* **2017**, *70*, 216–218. [CrossRef] [PubMed]
23. Peláez, F. The historical delivery of antibiotics from microbial natural products - Can history repeat? *Biochem. Pharmacol.* **2006**, *71*, 981–990. [CrossRef]
24. Kekuda, P.; Onkarappa, R.; Jayanna, N. Characterization and Antibacterial Activity of a Glycoside Antibiotic from *Streptomyces variabilis* PO-178. *Sci. Technol. Arts. Res. J.* **2015**, *3*, 116. [CrossRef]
25. Miao, V.; Davies, J. Actinobacteria: The good, the bad, and the ugly. *Antonie van Leeuwenhoek* **2010**, *98*, 143–150. [CrossRef]
26. Subramani, R.; Aalbersberg, W. Marine actinomycetes: An ongoing source of novel bioactive metabolites. *Microbiol. Res.* **2012**, *167*, 571–580. [CrossRef]
27. Clardy, J.; Fischbach, M.A.; Walsh, C.T. New antibiotics from bacterial natural products. *Nat. Biotechnol.* **2006**, *24*, 1541–1550. [CrossRef]
28. Ayed, A.; Slama, N.; Mankai, H.; Bachkouel, S.; ElKahoui, S.; Tabbene, O.; Limam, F. *Streptomyces tunisialbus* sp. nov., a novel *Streptomyces* species with antimicrobial activity. *Antonie van Leeuwenhoek* **2018**, *111*, 1571–1581. [CrossRef]
29. Mccully, M.; Harper, J.D.I.; An, M.; Kent, J.H. The rhizosphere: the key functional unit in plant soil microbial interactions in the field implications for the understanding of allelopathic effects. *Pol. J. Vet. Sci.* **2005**, *15*, 493–498.
30. Hiltner, L. Uber neuer erfahrungen und probleme auf dem gebiet der berücksichtigung unter besonderer berücksichtigung der gründüngung und brache. *Arbeiten der Deustchen Landwirtschaftsgesellesschaft* **1904**, *32*, 1405–1417.
31. Berendsen, R.L.; Pieterse, C.M.J.; Bakker, P.A.H.M. The rhizosphere microbiome and plant health. *Trends. Plant Sci.* **2012**, *17*, 478–486. [CrossRef]

32. Loper, J.E.; Gross, H. Genomic analysis of antifungal metabolite production by *Pseudomonas fluorescens* Pf-5. *Eur. J. Plant Pathol.* **2007**, *119*, 265–278. [CrossRef]
33. Reiter, B.; Sessitsch, A.; Nowak, J.; Cle, C. Endophytic colonization of *Vitis vinifera* L. by plant growth-promoting bacterium *Burkholderia* sp. strain PsJN. *Society* **2005**, *71*, 1685–1693.
34. Lanteigne, C.; Gadkar, V.J.; Wallon, T.; Novinscak, A.; Filion, M. Production of DAPG and HCN by Pseudomonas sp. LBUM300 contributes to the biological control of bacterial canker of tomato. *Phytopathology* **2012**, *102*, 967–973. [CrossRef]
35. Osei, E.; Kwain, S.; Mawuli, G.T.; Anang, A.K.; Owusu, K.B.A.; Camas, M.; Camas, A.S.; Ohashi, M.; Alexandru-Crivac, C.N.; Deng, H. Paenidigyamycin A, potent antiparasitic imidazole alkaloid from the ghanaian *Paenibacillus* sp. De2Sh. *Mar. Drugs* **2019**, *17*, 9. [CrossRef]
36. Mukhtar, S.; Mehnaz, S.; Mirza, M.S.; Malik, K.A. Isolation and characterization of bacteria associated with the rhizosphere of halophytes (Salsola stocksii and Atriplex amnicola) for production of hydrolytic enzymes. *Brazilian J. Microbiol.* **2019**, *50*, 85–97. [CrossRef]
37. Orfali, R.; Perveen, S. Secondary metabolites from the *Aspergillus* sp. in the rhizosphere soil of *Phoenix dactylifera* (Palm tree). *BMC Chem.* **2019**, *13*, 1–6. [CrossRef]
38. Wang, R.J.; Zhang, S.Y.; Ye, Y.H.; Yu, Z.; Qi, H.; Zhang, H.; Xue, Z.L.; Wang, J.D.; Wu, M. Three new isoflavonoid glycosides from the mangrove-derived actinomycete micromonospora aurantiaca 110B. *Mar. Drugs* **2019**, *17*, 294. [CrossRef]
39. Nouioui, I.; Carro, L.; García-López, M.; Meier-Kolthoff, J.P.; Woyke, T.; Kyrpides, N.C.; Pukall, R.; Klenk, H.-P.; Goodfellow, M.; Göker, M. Genome-Based Taxonomic Classification of the Phylum Actinobacteria. *Front. Microbiol.* **2018**, *9*, 2007. [CrossRef]
40. Oren, A.; Garrity, G.M. List of new names and new combinations previously effectively, but not validly, published. *Int. J. Syst. Evol. Microbiol.* **2018**, *68*, 2707–2709. [CrossRef]
41. Komaki, H.; Hosoyama, A.; Kimura, A.; Ichikawa, N.; Igarashi, Y.; Tamura, T. Classification of '*Streptomyces hyalinum*' Hamada and Yokoyama as *Embleya hyalina* sp. nov., the second species in the genus *Embleya*, and emendation of the genus *Embleya*. *Int. J. Syst. Evol. Microbiol.* **2020**, 1–5. [CrossRef]
42. Kämpfer, P.; Genus, I. Streptomyces Waksman and Henrici 1943, 339 $^{AL}$. In *Bergey's Manual of Systematic Bacteriology*, 2nd ed.; Springer: New York, NY, USA, 2012; pp. 1679–1680.
43. Ping, X.; Takahashi, Y.; Seino, A.; Iwai, Y.; Omura, S. *Streptomyces scarbrisporus* sp. nov. *Int. J. Syst. Evol. Microbiol.* **2004**, *54*, 577–581. [CrossRef]
44. Omura, S.; Nakagawa, A.; Shibata, K.; Sano, H. The structure of hitachimycin, a novel macrocyclic lactam involving β-phenylalanine. *Tetrahedron Lett.* **1982**, *23*, 4713–4716. [CrossRef]
45. Komiyama, K.; Edanami, K.I.; Yamamoto, H.; Umezawa, I. Antitumor activity of a new antitumor antibiotic, stubomycin. *J. Antibiot.* **1982**, *35*, 703–706. [CrossRef]
46. Shibata, K.; Satsumabayashi, S.; Sano, H.; Komiyama, K.; Nakagawa, A.; Omura, S. Chemical modification of hitachimycin. Synthesis, antibacterial, cytocidal and in vivo antitumor activities of hitachimycin derivatives. *J. Antibiot.* **1988**, *41*, 614–623. [CrossRef]
47. Naganawa, H.; Wakashiro, T.; Yagi, A.; Kondo, S.; Takita, T.; Hamada, M.; Maeda, K.; Umezawa, H. Deoxynybomycin from a *streptomyces*. *J. Antibiot.* **1970**, *23*, 365–368. [CrossRef]
48. Hiramatsu, K.; Igarashi, M.; Morimoto, Y.; Baba, T.; Umekita, M.; Akamatsu, Y. Curing bacteria of antibiotic resistance: Reverse antibiotics, a novel class of antibiotics in nature. *Int. J. Antimicrob. Agents* **2012**, *39*, 478–485. [CrossRef]
49. Hayakawa, M.; Nonomura, H. Humic acid-vitamin agar, a new medium for the selective isolation of soil actinomycetes. *J. Ferment. Technol.* **1987**, *65*, 501–509. [CrossRef]
50. Shirling, E.B.; Gottlieb, D. Methods for characterization of *Streptomyces* species. *Int. J. Syst. Bacteriol.* **1966**, *16*, 313–340. [CrossRef]
51. Vichai, V.; Kirtikara, K. Sulforhodamine B colorimetric assay for cytotoxicity screening. *Nat. Protoc.* **2006**, *1*, 1112–1116. [CrossRef]
52. Jin, L.; Zhao, Y.; Song, W.; Duan, L.; Jiang, S.; Wang, X.; Zhao, J.; Xiang, W. *Streptomyces inhibens* sp. Nov., a novel actinomycete isolated from rhizosphere soil of wheat (*Triticum aestivum* L.). *Int. J. Syst. Evol. Microbiol.* **2019**, *69*, 688–695. [CrossRef] [PubMed]

53. Zhao, J.W.; Han, L.Y.; Yu, M.Y.; Cao, P.; Li, D.M.; Guo, X.W.; Liu, Y.Q.; Wang, X.J.; Xiang, W.S. Characterization of *Streptomyces sporangiiformans* sp. nov., a Novel Soil Actinomycete with Antibacterial Activity against *Ralstonia solanacearum*. *Microorganisms* **2019**, *7*, 360. [CrossRef] [PubMed]
54. Smibert, R.M.; Krieg, N.R. Phenotypic characterization. In *Methods for General and Molecular Bacteriology*, 2nd ed.; Gerhardt, P., Murray, R.G.E., Wood, W.A., Krieg, N.R., Eds.; American Society for Microbiology: Washington, DC, USA, 1994; pp. 607–654.
55. Gordon, R.E.; Barnett, D.A.; Handerhan, J.E.; Pang, C. Nocardia coeliaca, Nocardia autotrophica, and the nocardin strain. *Int. J. Syst. Bacteriol.* **1974**, *24*, 54–63. [CrossRef]
56. Yokota, A.; Tamura, T.; Hasegawa; Huang, L.H. *Catenuloplanes japonicas* gen. nov., sp. nov., nom. rev., a new genus of the order *Actinomycetales*. *Int. J. Syst. Bacteriol.* **1993**, *43*, 805–812. [CrossRef]
57. Lechevalier, M.P.; Lechevalier, H.A. The chemotaxonomy of actinomycetes. In *Actinomycete taxonomy*, 2nd ed.; Dietz, A., Thayer, D.W., Eds.; special publication for Society of Industrial Microbiology: Arlington, TX, USA, 1980; pp. 227–291.
58. Minnikin, D.E.; O'Donnell, A.G.; Goodfellow, M.; Alderson, G.; Athalye, M.; Schaal, A.; Parlett, J.H. An integrated procedure for the extraction of bacterial isoprenoid quinones and polar lipids. *J. Microbiol. Methods* **1984**, *2*, 233–241. [CrossRef]
59. Collins, M.D. Isoprenoid quinone analyses in bacterial classification and identification. In *Chemical methods in bacterial systematics*; Goodfellow, M., Minnikin, D.E., Eds.; Academic Press: London, UK, 1985; pp. 267–284.
60. Shen, Y.; Sun, T.; Jiang, S.; Mu, S.; Li, D.; Guo, X.; Zhang, J.; Zhao, J.; Xiang, W. *Streptomyces lutosisoli* sp. nov., a novel actinomycete isolated from muddy soil. *Antonie Van Leeuwenhoek, Int. J. Gen. Mol. Microbiol.* **2018**, *111*, 2403–2412. [CrossRef]
61. Wu, C.; Lu, X.; Qin, M.; Wang, Y.; Ruan, J. Analysis of menaquinone compound in microbial cells by HPLC. *Microbiology* **1989**, *16*, 176–178.
62. Kim, S.B.; Brown, R.; Oldfield, C.; Gilbert, S.C.; Iliarionov, S.; Goodfellow, M. Gordonia amicalis sp. nov., a novel dibenzothiophene-desulphurizing actinomycete. *Int. J. Syst. Evol. Microbiol.* **2000**, *50*, 2031–2036. [CrossRef]
63. Yoon, S.H.; Ha, S.M.; Kwon, S.; Lim, J.; Kim, Y.; Seo, H.; Chun, J. Introducing EzBioCloud: A taxonomically united database of 16S rRNA gene sequences and whole-genome assemblies. *Int. J. Syst. Evol. Microbiol.* **2017**, *67*, 1613–1617. [CrossRef]
64. Saitou, N.; Nei, M. The neighbor-joining method: a new method for reconstructing phylogenetic trees. *Mol. Biol. Evol.* **1987**, *4*, 406–425.
65. Felsenstein, J. Evolutionary trees from DNA sequences: A maximum likelihood approach. *J. Mol. Evol.* **1981**, *17*, 368–376. [CrossRef] [PubMed]
66. Kumar, S.; Stecher, G.; Tamura, K. Mega7: molecular evolutionary genetics analysis version 7.0 for bigger datasets. *Mol. Biol. Evol.* **2016**, *33*, 1870–1874. [CrossRef] [PubMed]
67. Felsenstein, J. Confidence Limits on Phylogenies: an Approach Using the Bootstrap. *Evolution (N. Y).* **1985**, *39*, 783–791.
68. Kimura, M. A simple method for estimating evolutionary rates of base substitutions through comparative studies of nucleotide sequences. *J. Mol. Evol.* **1980**, *16*, 111–120. [CrossRef]
69. Hatano, K.; Nishii, T.; Kasai, H. Taxonomic re-evaluation of whorl-forming *Streptomyces* (formerly *Streptoverticillium*) species by using phenotypes, DNA-DNA hybridization and sequences of gyrB, and proposal of *Streptomyces luteireticuli* (ex Katoh and Arai 1957) corrig., sp. nov., nom. rev. *Int. J. Syst. Evol. Microbiol.* **2003**, *53*, 1519–1529. [CrossRef]
70. Guo, Y.P.; Zheng, W.; Rong, X.Y.; Huang, Y. A multilocus phylogeny of the *Streptomyces griseus* 16S rRNA gene clade: Use of multilocus sequence analysis for streptomycete systematics. *Int. J. Syst. Evol. Microbiol.* **2008**, *58*, 149–159. [CrossRef]
71. Wang, J.; Zhang, H.; Ying, L.; Wang, C.; Jiang, N.; Zhou, Y.; Wang, H.; Bai, H. Five new epothilone metabolites from *Sorangium cellulosum* strain So0157-2. *J. Antibiot.* **2009**, *62*, 483–487. [CrossRef]
72. Wayne, P.A. Clinical and Laboratory Standards Institute. Methods for dilution antimicrobial susceptibility tests for bacteria that grow aerobically; approved standard—ninth edition. *Clin. Lab. Standards Institute* **2012**, *32*, 2.

73. Liu, C.; Han, C.; Jiang, S.; Zhao, X.; Tian, Y.; Yan, K.; Wang, X.; Xiang, W. *Streptomyces lasii* sp. nov., a Novel Actinomycete with Antifungal Activity Isolated from the Head of an Ant (*Lasius flavus*). *Curr. Microbiol.* **2018**, *75*, 353–358. [CrossRef]
74. Rong, X.; Huang, Y. Taxonomic evaluation of the *Streptomyces hygroscopicus* clade using multilocus sequence analysis and DNA-DNA hybridization, validating the MLSA scheme for systematics of the whole genus. *Syst. Appl. Microbiol.* **2012**, *35*, 7–18. [CrossRef]
75. Westley, J.W.; Liu, C.M.; Evans, R.H.; Blount, J.F. Conglobatin, a novel macrolide dilactone from *Streptomyces conglobatus* ATCC 31005. *J. Antibiot.* **1979**, *32*, 874–877. [CrossRef]
76. Yoshida, S.; Yoneyama, K.; Shiraishi, S.; Watanabe, A.; Takahashi, N. Chemical structures of new piericidins produced by *Streptomyces pactum*. *Agric. Biol. Chem.* **1977**, *41*, 855–862. [CrossRef]
77. Kubota, N.K.; Ohta, E.; Ohta, S.; Koizumi, F.; Suzuki, M.; Ichimura, M.; Ikegami, S. Piericidins C5 and C6: New 4-pyridinol compounds produced by *Streptomyces* sp. and *Nocardioides* sp. *Bioorganic Med. Chem.* **2003**, *11*, 4569–4575. [CrossRef]
78. Tamura, S.; Takahashi, N.; Miyamoto, S.; Mori, R.; Suzuki, S.; Nagatsu, J. Isolation and physiological activities of piericidin A, a natural insecticide produced by *Streptomyces*. *Agr. Biol. Chem.* **1963**, *27*, 576–582. [CrossRef]
79. Kominato, K.; Watanabe, Y.; Hirano, S.I.; Kioka, T.; Tone, H. Mer-a2026a and b, novel piericidins with vasodilating effect. ii. physico-chemical properties and chemical structures. *J. Antibiot.* **1994**, *48*, 103–105. [CrossRef]
80. Wang, H.; Zhao, X.L.; Gao, Y.H.; Qi, H.; Zhang, H.; Xiang, W.S.; Wang, J.D.; Wang, X.J. Two new cytotoxic metabolites from *Streptomyces* sp. HS-NF-813. *J. Asian. Nat. Prod. Res.* **2018**, *22*, 249–256. [CrossRef]
81. Hall, C.; Wu, M.; Crane, F.L.; Takahashi, H.; Tamura, S.; Folkers, K. Piericidin A: A new inhibitor of mitochondrial electron transport. *Biochem. Biophys. Res. Commun.* **1966**, *25*, 373–377. [CrossRef]
82. Shaaban, K.A.; Helmke, E.; Kelter, G.; Fiebig, H.H.; Laatsch, H. Glucopiericidin C: A cytotoxic piericidin glucoside antibiotic produced by a marine *Streptomyces* isolate. *J. Antibiot.* **2011**, *64*, 205–209. [CrossRef]
83. Huang, W.; Ye, M.; Zhang, L.; Wu, Q.; Zhang, M.; Xu, J.; Zheng, W. FW-04-806 inhibits proliferation and induces apoptosis in human breast cancer cells by binding to N-terminus of Hsp90 and disrupting Hsp90-Cdc37 complex formation. *Mol. Cancer* **2014**, *13*, 1–13. [CrossRef]
84. Huang, W.; Wu, Q.D.; Zhang, M.; Kong, Y.L.; Cao, P.R.; Zheng, W.; Xu, J.H.; Ye, M. Novel Hsp90 inhibitor FW-04-806 displays potent antitumor effects in HER2-positive breast cancer cells as a single agent or in combination with lapatinib. *Cancer Lett.* **2015**, *356*, 862–871. [CrossRef]
85. Song, Z.; Lohse, A.G.; Hsung, R.P. Challenges in the synthesis of a unique mono-carboxylic acid antibiotic, (+)-zincophorin. *Cheminform* **2009**, *26*, 560–571. [CrossRef] [PubMed]
86. Dirlam, J.P.; Belton, A.M.; Chang, S.P.; Cullen, W.P.; Huang, L.H.; Kojima, Y.; Maeda, H.; Nishiyama, S.; Oscarson, J.R.; Sakakibara, T.; et al. Cp-78,545, a new monocarboxylic acid ionophore antibiotic related to zincophorin and produced by a *streptomyces*. *J. Antibiot.* **1989**, *42*, 1213–1220. [CrossRef] [PubMed]

# 14

# Diversity and Bioactive Potential of Actinobacteria from Unexplored Regions of Western Ghats, India

**Saket Siddharth [1], Ravishankar Rai Vittal [1,*], Joachim Wink [2] and Michael Steinert [3]**

[1] Department of Studies in Microbiology, University of Mysore, Manasagangotri, Mysore 570 006, India; saketsiddharth@gmail.com

[2] Microbial Strain Collection, Helmholtz Centre for Infection Research GmbH (HZI), Inhoffenstrasse 7, 38124 Braunschweig, Germany; joachim.wink@helmholtz-hzi.de

[3] Institute of Microbiology, Technische Universität Braunschweig, 38106 Braunschweig, Germany; m.steinert@tu-bs.de

* Correspondence: raivittal@gmail.com

**Abstract:** The search for novel bioactive metabolites continues to be of much importance around the world for pharmaceutical, agricultural, and industrial applications. Actinobacteria constitute one of the extremely interesting groups of microorganisms widely used as important biological contributors for a wide range of novel secondary metabolites. This study focused on the assessment of antimicrobial and antioxidant activity of crude extracts of actinobacterial strains. Western Ghats of India represents unique regions of biologically diverse areas called "hot spots". A total of 32 isolates were obtained from soil samples of different forest locations of Bisle Ghat and Virjapet situated in Western Ghats of Karnataka, India. The isolates were identified as species of *Streptomyces*, *Nocardiopsis*, and *Nocardioides* by cultural, morphological, and molecular studies. Based on preliminary screening, seven isolates were chosen for metabolites extraction and to determine antimicrobial activity qualitatively (disc diffusion method) and quantitatively (micro dilution method) and scavenging activity against DPPH (2,2-diphenyl-1-picrylhydrazyl) and ABTS (2,2′-Azino-bis(3-ethylbenzothiazoline-6-sulfonic acid) radicals. Crude extracts of all seven isolates exhibited fairly strong antibacterial activity towards MRSA strains (MRSA ATCC 33591, MRSA ATCC NR-46071, and MRSA ATCC 46171) with MIC varying from 15.62 to 125 μg/mL, whereas showed less inhibition potential towards Gram-negative bacteria *Salmonella typhi* (ATCC 25241) and *Escherichia coli* (ATCC 11775) with MIC of 125–500 μg/mL. The isolates namely S1A, SS5, SCA35, and SCA 11 inhibited *Fusarium moniliforme* (MTCC 6576) to a maximum extent with MIC ranging from 62.5 to 250 μg/mL. Crude extract of SCA 11 and SCA 13 exhibited potent scavenging activities against DPPH and ABTS radicals. The results from this study suggest that actinobacterial strains of Western Ghats are an excellent source of natural antimicrobial and antioxidant compounds. Further research investigations on purification, recovery, and structural characterization of the active compounds are to be carried out.

**Keywords:** Western Ghats; diversity; actinobacteria; antimicrobial; MRSA; antioxidants

## 1. Introduction

The quest for novel biologically active secondary metabolites from microorganisms continues to rise due to emergence of drug resistance in pathogens causing life threatening diseases around the globe [1]. Particularly, methicillin resistant *Staphylococcus aureus* (MRSA) and methicillin resistant *Staphylococcus epidermidis* (MRSE) strains are not only exposed to hospital-acquired infections but also to community-acquired infections [2]. The mortality and morbidity associated with these infections are largely affecting economic conditions of patients and hospitals [3]. Therefore, there is an urgent need for developing novel and effective antimicrobial agents to overcome or delay acquired resistance to existing drugs. Reactive oxygen species (ROS) play an important role as signaling molecules involved in mitogenesis. However, high generation of ROS during aerobic metabolism creates oxidative stress within the intracellular milieu causing oxidative damage to cells [4]. The oxidative stress caused is often associated with many human diseases including cancer, diabetes [5], cardiovascular [6], and neurodegenerative diseases [7]. In order to withstand the oxidative stress caused, cells or organisms make use of antioxidants, that are able to block or delay the damage caused by several possible mechanisms such as halting chain reactions, preventing the formation of free radicals, neutralizing the singlet oxygen molecule, promoting anti-oxidant enzymes, and inhibiting pro-oxidative enzymes [8]. Formation of free radicals can be prevented by antioxidant systems present within the cells. However, these defense mechanisms are insufficient to prevent the damages that arise, therefore exogenous antioxidants through dietary intake and supplements are required [9]. Natural antioxidants are found abundantly in metabolites produced by microorganisms. These products have consistently been considered as mainstay for drugs with various interesting biological activities [10]. They are considered to be an excellent scaffold for the formulation and development of antibiotics, antioxidants, immunomodulators, enzyme inhibitors, anticancer agents, plant growth hormones, and insect control agents [11]. With many improved techniques under combinatorial chemistry for high throughput findings of novel compounds, natural products from microbial sources have been screened extensively and gained much attention owing to their massive chemical and biological diversity [12]. Under various screening strategies, the rate of discovery of natural products has increased many folds, of which around 22,250 bioactive compounds are of microbial origin [13]. Among microorganisms, actinomycetes have contributed nearly 45% of all the reported metabolites [14].

A major group of natural products from microbial origin have been identified from organisms that inhabit the soil. Since soil itself is a mixture of minerals and organic matter, the filamentous bacteria are predominantly more present in the gaps between the soil particles than their unicellular counterparts [15]. Actinobacteria are ubiquitous in soils. They are responsible for biodegradation and biodeterioration processes in nature. Their flexile and proven abilities have prompted biologists to screen these organisms from unexplored niche habitats in order to obtain novel molecules [16].

Western Ghats of India is considered as one of the global biodiversity hotspots covering an area of 180,000 km$^2$ and harbors numerous species of plants, animals, and microbes [17]. The unique biodiversity of Western Ghats is conserved and protected by wildlife sanctuaries, national parks, and biosphere reserves situated in states where hill ranges run through, like Karnataka, Gujarat, Tamil Nadu, Maharashtra, and Kerala [18]. The forest regions in Western Ghats are largely underexplored, though in recent times few studies were carried out for bioprospection. Ganesan et al. [19] reported larvicidal, ovicidal, and repellent activities of *Streptomyces enissocaesilis* (S12–17) isolated from Western Ghats of Tamil Nadu, India. Actinobacterial strains isolated from Western Ghats soil of Tamil Nadu were reported to produce antimicrobial compounds against range of pathogens [20]. In the present study, the forest range in Western Ghats of Karnataka was studied for microbial population and taxonomical identification of potential actinobacteria. An attempt was also made to characterize microbial diversity for the potential to produce antioxidants and antimicrobial compounds.

## 2. Materials and Methods

### *2.1. Site, Sampling, Pre-Treatment, and Selective Isolation*

Soil samples were collected from different forest locations of Bisle Ghat and Virajpet in Western Ghats regions of Karnataka, India. The samples were collected from a depth of 15–25 cm in dry sterile insulated containers and stored aseptically at 4 °C until subjected to plating. Samples were air dried in a hot air oven (Equitron, India) at 50 °C for 72 h. Pre-treated samples were ground aseptically with mortar and pestle and serially diluted up to $10^{-6}$ in 10-fold dilution. The aliquots of each dilution (100 µL) were spread evenly on starch casein agar (SCA, Himedia, India) plates in triplicates supplemented with cycloheximide (30 µg/mL) and nalidixic acid (25 µg/mL). The plates were incubated at 28 ± 2 °C for 14 days. Emerging colonies with different morphological characters were selected and the purified strains were maintained on International *Streptomyces* Project (ISP-2, Himedia, India) agar slants and stored at 4 °C as stock for further use.

### *2.2. Morphological Characterization of Isolates*

Morphological characteristics of isolates were assessed by scanning electron microscopy (SEM). Bacterial colonies were inoculated in ISP-2 medium and incubated at 28 ± 2 °C for 7 days. Cells were centrifuged (Eppendorf, USA) at 8000× *g* for 10 min and pellet was resuspended in 2%–5% gluteraldehyde (Sigma, Burlington, VT, USA) prepared in 0.1M phosphate buffer, pH 7.2. After incubating samples for 30 min, supernatant was discarded and pellet was resuspended in 1% osmium tetraoxide (Sigma, Burlington, VT, USA), incubated for 1 h and centrifuged at 5000 × *g*. To the pellet, sterile water was added and centrifuged twice for 10 min at 5000 × *g*. For dehydration, the pellet was resuspended in 35% ethanol for 10 min, 50% ethanol for 10 min, 75% ethanol for 10 min, 95% ethanol for 10 min, and a final wash with 100% ethanol for 10 min. For SEM analysis, sterilized aluminum stubs and cover slips were inserted into the SCA plates at an angle of about 45 °C. The plates with stubs and coverslips were incubated at 37 °C for 24 h to check any contamination. After 24 h, isolates were introduced along the line where the surface of the stub met the agar medium and incubated at 28 ± 2 °C for 7 days. The stubs were then carefully removed and coated under vacuum, with a film of gold for 25–30 min and viewed on the scanning electron microscope (Zeiss Evo 40 EP, Germany).

### *2.3. Molecular Identification and Phylogenetic Analysis*

The total genomic DNA of bacteria was extracted by phenol-chloroform method, quality checked by agarose gel electrophoresis and quantified using NanoDrop1000 (Thermo-Scientific, USA). The PCR amplification of 16S rRNA gene was carried out with universal primers: 27F (5′-AGA GTT TGA TCC TGG CTC AG-3′) and 1492R (5′- ACG GCT ACC TTG TTA CGA CTT-3′) using the following conditions: initial denaturation temperature was set at 95 °C for 5 min, followed by 35 cycles at same temperature for 1 min, primer annealing at 54 °C for 1 min, and primer extension at 72 °C for 2 min. The reaction mixture was kept at 72 °C for 10 min subsequently and then cooled to 4 °C. The PCR products were checked in 1.5% agarose gel and visualized in a UV transilluminator (Tarsons, India) and the gel imaging was done using a Gel documentation system (Bio-Rad, USA). The amplified PCR products were sequenced using same set of primers (27F′ and 1492R′) on Applied Biosystems 3130 Genetic Analyzer (Applied Biosystems, USA). The genetic relationship between the strains was determined by neighbor-joining tree algorithm method. The phylogenetic tree was constructed with a bootstrapped database containing 1000 replicates in MEGA 7.0 software (Mega, Raynham, MA, USA). The nearly complete 16S rRNA consensus sequences were deposited in the GenBank database.

### *2.4. Isolates Cultivation and Metabolites Extraction*

Pure isolates were subcultured in Tryptone Yeast Extract broth as seed medium (ISP-1 medium, Himedia, India) at 28 °C for 2 weeks prior to fermentation process. The production medium, ISP-2 was autoclaved at 121 °C and 1.5 atm for 15 min. Fermentation was carried out in 750 mL of (in 7 nos. conical flasks –1000 mL) ISP-2 medium, shaking at 140 rev $min^{-1}$ for 14 days at 28 °C, inoculated with 250 μL of seed medium. After incubation, the culture medium was split into mycelium and filtrate by centrifugation at 12,000× *g* for 15 min. The cell free supernatant from each flask was subjected to extraction thrice with equal volume of ethyl acetate (Qualigens Fine Chemicals Pvt. Ltd., San Diego, USA) and the organic phase was concentrated by rotary vacuum evaporator (Hahn-Shin, Bucheon, South Korea) at 50 °C. The crude concentrate was dried in a desiccator and suspended in methanol prior to bioactivity screening assays.

*2.5. Antimicrobial Susceptibility Test*

2.5.1. Disc Diffusion Assay

Antimicrobial susceptibility assay was carried out by disc diffusion method against methicillin-resistance *Staphylococcus aureus* (MRSA ATCC 33591, MRSA ATCC NR-46071 and MRSA ATCC 46171), Gram-negative bacteria (*Salmonella typhi* (ATCC 25241) and *Escherichia coli* (ATCC 11775)) and fungus (*Fusarium moniliforme* (MTCC 6576)). Gentamicin and Nystatin discs were used as positive control. The sterile discs (6mm, Himedia) were impregnated with 30 μL of crude extract dissolved in 0.5% DMSO. Discs impregnated with 0.5% DMSO (Qualigens Fine Chemicals Pvt. Ltd., San Diego, USA) were used as solvent control. The plates were left for 30 min at 4 °C to allow the diffusion of extracts before they were incubated for 24–48 h at 37 °C. The clear zones of inhibition observed around discs suggested antagonistic activity against test organisms and diameter of inhibition zones were measured subsequently. The test was performed in triplicate.

2.5.2. Minimum Inhibitory Concentration (MIC)

The minimum inhibitory concentration assay was determined by micro broth dilution method as previously reported by Siddharth and Rai [21]. The serially diluted fraction of extracts with sterile Mueller Hinton broth (Himedia, India) was added to pre coated microbial cultures in 96-well micro titer plates to give a final concentration of 1–3.8 μg/mL. The titer plate was incubated for 24 h at 37 °C. The lowest concentration of extract which completely inhibited the bacterial growth (no turbidity) was considered as MIC. Each test was done in triplicate.

*2.6. Antioxidant Assays*

2.6.1. 2,2-diphenyl-1-picrylhydrazyl Radical Scavenging activity (DPPH)

DPPH radical scavenging activity of crude extracts was examined based on a previously described method by Siddharth and Rai [22]. Crude extracts at varying concentrations (7.81–1000 μg/mL) were reacted with freshly prepared DPPH in methanol (60 mM, Sigma, USA). Reaction mixture was incubated at room temperature for 30 min in the dark prior to the measurement of absorbance at 520 nm. The radical scavenging activity was expressed as $IC_{50}$ (μg/mL). The percentage scavenging of DPPH radicals was computed by the following equation:

$$\text{DPPH scavenging effect (\%)} = [(A_o - A_1)/A_o] \times 100 \quad (1)$$

where, $A_o$ = Absorbance of control, $A_1$ = Absorbance of crude extracts. Trolox (Sigma) was used as a reference compound whereas methanol was used as blank.

2.6.2. 2,2′-Azino-bis(3-ethylbenzothiazoline-6-sulfonic acid) Radical Scavenging Activity (ABTS)

ABTS radical scavenging activity was performed according to the method developed by Ser et al. [23] with slight modifications. Crude extracts at concentrations (7.81–1000 μg/mL) were mixed with ABTS (Sigma, USA) cation complex and incubated in dark at room temperature for 30 min. The absorbance was measured at 415 nm. The radical scavenging activity was expressed as $IC_{50}$ (μg/mL). The percentage inhibition of $ABTS^{*+}$ radicals were computed by using following equation:

$$ABTS^{*+} \text{ scavenging effect (\%)} = [(A_o - A_1)/A_o] \times 100 \quad (2)$$

where, $A_o$ = Absorbance of control, $A_1$ = Absorbance of crude extracts. Trolox (Sigma) was used as a positive reference.

## 3. Results and Discussion

Soil is among the most productive habitat colonized by a large number of organisms. The rhizosphere soil in the vicinity of plant roots provides essential nutrients in the form of exudates which favors the growth of microbial communities [24]. Soil microbes are a major source of a number of natural products including clinical important antibiotics, immunomodulators, enzyme inhibitors, antioxidants, anti-tumor and anticancer agents. Actinobacteria are abundant in soil, species of *Streptomyces* in particular represent the dominance over other microbes present in soil and play a vital role in recycling of materials and production of important metabolites [25]. Rare actinobacteria genera such as *Nocardia*, *Nocardiopsis*, and *Nocardioides* are also encountered in soils, though their presence is subjected to conditions of soil such as salinity and alkalinity [26]. The rapid emergence of drug resistant pathogens urges the exploration of new niche habitats for the isolation of new microbial species which can contribute to the uncovering of novel, safe, effective, and broad spectrum bioactive compounds [27].

### *3.1. Isolation, Morphological, and Molecular Characterization of Actinobacterial Isolates*

In this study, from forest soil of Bisle Ghat and Virajpet of Western Ghats region of Karnataka, we targeted the isolation of different genera of actinobacteria in search of new natural products (Table 1). A total of 32 actinobacterial isolates were recovered on Starch casein agar and Actinomycetes isolation agar. Of them, seven isolates grown on starch casein agar medium showing marked antimicrobial activity against test organisms in primary screening were characterized on the basis of cultural, morphological, and molecular characteristics. The colonies of isolates revealed diverse morphological appearances with varied spore color, aerial and substrate mycelium, and colony morphology (Table 2 and Figure 1). Scanning electron microscope examination showed chains of smooth and spiny spores in oval, round, and spiral ornamentation (Figure 2). The molecular identification of isolates by amplification of 16S-rRNA gene was done by using universal primers 27F′ and 1492R′ (Figure S1). In 16S-rRNA sequencing, alignment of the nucleotide sequences of strain S1A, SS4, SS5, SS6, and SCA35 exhibited a similarity of 98.32%, 99.71%, 99.70%, 99.72%, and 99.35% with closely related *Streptomyces* species, respectively. The strain SCA11 was considered to represent a species of the genus *Nocardiopsis*, since it was closely related to *Nocardiopsis* species with 98.99% sequence similarity. The nucleotide sequence of strain SCA13 showed 98.28% similarity with the closely related *Nocardioides* sp. (Table 3). The phylogenetic relatedness of strains with their closely related species obtained by neighbor-joining method is shown in Figures 3–5.

**Table 1.** Actinobacterial isolates and sampling areas in Western Ghats regions of Karnataka, India.

| Name | Sample Code | Sampling Area | Latitude (N) | Longitude (E) | Elevation (m) |
|---|---|---|---|---|---|
| *Streptomyces* Sp. | S1A | Bisle Ghat, Hassan District | 12°71′88.04″ | 75°68′70.02″ | 802 |
| *Streptomyces* Sp. | SS4 | Bisle Ghat, Hassan District | 12°72′00.89″ | 75°68′41.42″ | 752 |
| *Streptomyces* Sp. | SS5 | Bisle Ghat, Hassan District | 12°71′24.30″ | 75°68′04.74″ | 710 |
| *Streptomyces* Sp. | SS6 | Virajpet, Madikeri District | 12°19′75.83″ | 75°79′52.93″ | 885 |
| *Streptomyces* Sp. | SCA35 | Virajpet, Madikeri District | 12°18′83.38″ | 75°83′06.79″ | 864 |
| *Nocardiopsis* Sp. | SCA11 | Virajpet, Madikeri District | 12°21′21.64″ | 75°80′24.84″ | 830 |
| *Nocardioides* Sp. | SCA13 | Virajpet, Madikeri District | 12°18′47.27″ | 75°76′24.17″ | 798 |

**Table 2.** Morphological characteristics of isolated actinobacterial strains.

| Isolate | Medium | Diffusible Pigment | Colony Morphology | Aerial Mycelium | Substrate Mycelium |
|---|---|---|---|---|---|
| *Streptomyces* Sp. S1A | SCA | None | Powdery | White | Cream |
| *Streptomyces* Sp. SS4 | SCA | None | Cottony | Dark Grey | Grey |
| *Streptomyces* Sp. SS5 | SCA | None | Cottony | Grey | Cream |
| *Streptomyces* Sp. SS6 | SCA | Pink | Rough | Pale red | Pink |
| *Streptomyces* Sp. SCA35 | SCA | None | Powdery | White | Cream |
| *Nocardiopsis* Sp. SCA11 | SCA | None | Powdery | Cream | Light brown |
| *Nocardioides* Sp. SCA13 | SCA | None | Raised | White | White |

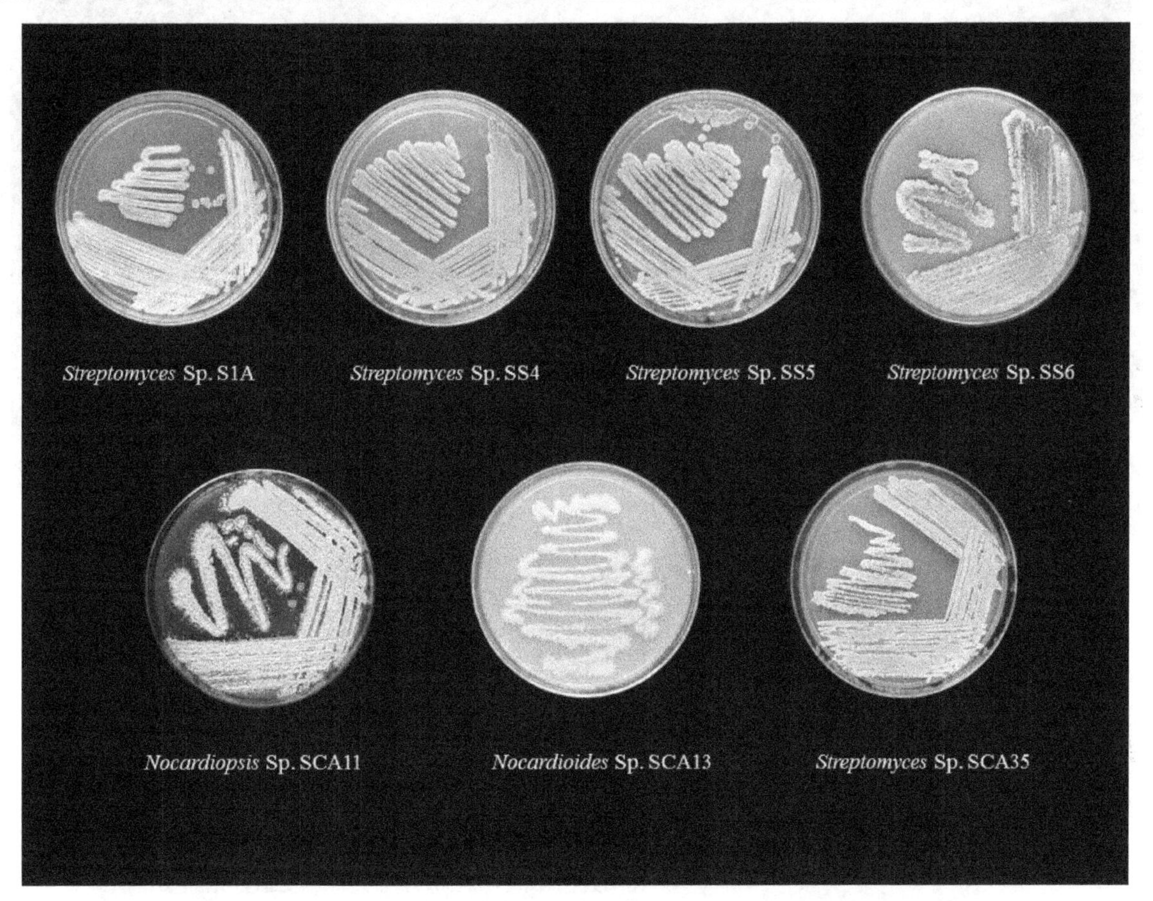


**Figure 1.** Morphological characterization of actinobacterial isolates on starch casein agar plates.

**Table 3.** Molecular identification (based on 16S rRNA amplification) of actinobacterial strains isolated from Western Ghats.

| Source | Organism | GenBank Accession No | % Similarity |
|---|---|---|---|
| Soil | *Streptomyces* Sp. S1A | KU921223 | 98.32% |
| Soil | *Streptomyces* Sp. SS4 | MF668120 | 99.71% |
| Soil | *Streptomyces* Sp. SS5 | MF925722 | 99.70% |
| Soil | *Streptomyces* Sp. SS6 | MF925723 | 99.72% |
| Soil | *Streptomyces* Sp. SCA35 | MN176654 | 99.35% |
| Soil | *Nocardiopsis* Sp. SCA11 | MG934272 | 98.99% |
| Soil | *Nocardioides* Sp. SCA13 | MG934273 | 98.28% |

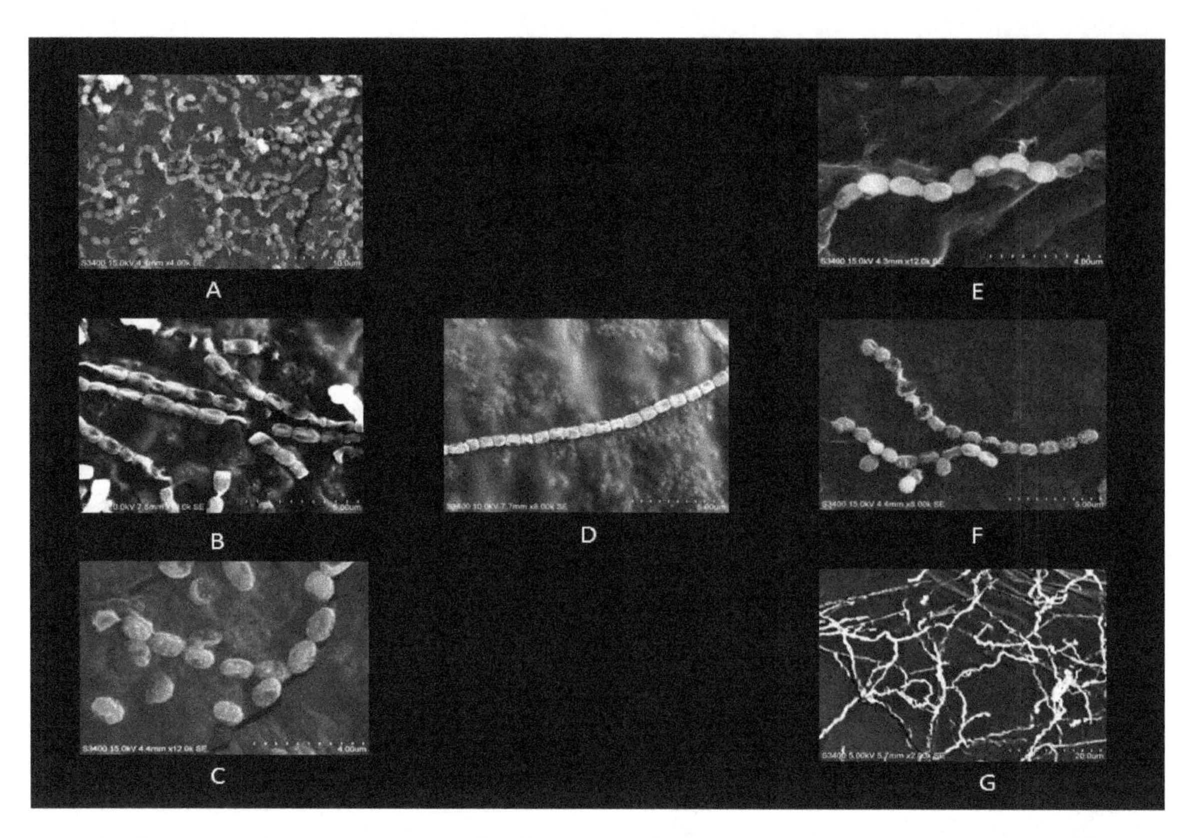


**Figure 2.** Scanning electron micrograph of strains (**A**) S1A, (**B**) SCA11, (**C**) SS4, (**D**) SS5, (**E**) SS6, (**F**) SCA35, (**G**) SCA13; scale bar represents 5 μm.

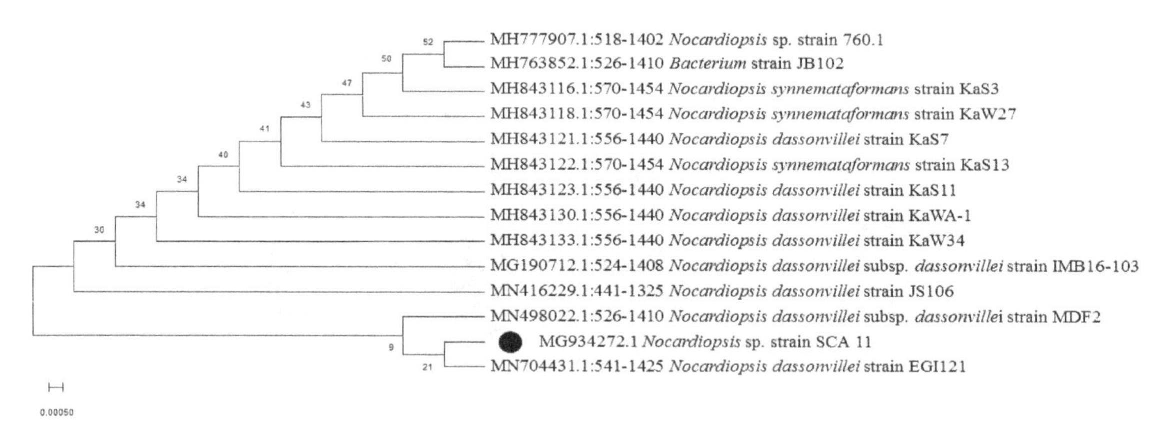


**Figure 3.** Phylogenetic analysis of isolate *Nocardiopsis* sp. strain SCA11. Neighbor-joining phylogenetic tree showing evolutionary relationship of selected isolate based on 16S r-RNA sequence alignments. Bootstrap values at the nodes indicate collated values based on 1000 resampled datasets. Bar indicates 0.0005 substitutions per nucleotide position.

### *3.2. Antimicrobial and Antioxidant Potential of Isolates*

In this study, all seven isolates showed antibacterial activity against at least one test bacterium. All isolates inhibited MRSA strains significantly (MRSA ATCC 33591, MRSA ATCC NR-46071, and MRSA ATCC 46171), whereas showed less inhibition potential against Gram-negative bacteria *Salmonella typhi* (ATCC 25241) and *Escherichia coli* (ATCC 11775). The isolates namely S1A, SS5, SCA35, and SCA 11 inhibited *Fusarium moniliforme* (MTCC 6576) to a maximum extent (Figure 6). The minimum inhibition concentration (MIC) ranges from 15.62 to 125 μg/mL for MRSA strains, 125–500 μg/mL for Gram-negative bacteria, and 62.5–250 μg/mL for fungi (Table 4). Numerous studies have reported antimicrobial activity of actinobacterial species. Sengupta et al. [28] reported potential antimicrobial activity of three isolates against *Pseudomonas aueroginosa, Enterobacter aueroginosa, Salmonella typhi, Salmonella typhimurium, Escherichia coli, Bacillus subtilis,* and *Vibrio cholera.* Satheeja and Jebakumar [29] reported isolation of *Streptomcyes* species from a mangrove ecosystem for antibacterial activity against clinical isolates of MRSA, methicillin-susceptible *Staphylococcus aureus* (MSSA), and *Salmonella typhi.* Vu et al. [30] reported antimicrobial activity of *Streptomyces cavourensis* YBQ59 against methicillin-resistant *Staphylococcus aureus* ATCC 33591 and methicillin-resistant *Staphylococcus epidermidis* ATCC 358984. Dashti et al. [31] reported the co-cultivation of *Actinokineospora* sp.EG49 and *Nocardiopsis* sp.RV163 and metabolites produced were tested for antimicrobial activity against the range of pathogens. The bioactive metabolite from *Streptomyces cyaneofuscatus* M-169 showed significant inhibition of Gram-positive bacteria with MIC value of 0.03 μg/mL [32]. It has been shown that actinobacterial isolates from Western Ghats exhibit antimicrobial activity. The *Streptomyces* species from Agumbe [33], *Streptomyces* sp. RAMPP-065 from Kudremukh [34], and *Streptomyces* sp. GOS1 isolated from Western Ghats of Agumbe, Karnataka [35] exhibited remarkable antimicrobial activity. In earlier studies, *Streptomyces* species isolated from Kodachadri were found to possess antifungal activity [33]. Each extract was evaluated for scavenging activity against DPPH and ABTS radicals for antioxidant activity. Crude extract of strain SCA11 showed potent scavenging activity against DPPH radicals with $IC_{50}$ (μg/mL) 30.91 ± 0.25, whereas SCA13 showed remarkable scavenging activity against ABTS radicals with $IC_{50}$ (μg/mL) 37.91 ± 0.17. Trolox as a standard showed significant activity with $IC_{50}$ (μg/mL) 11.07 ± 0.06 and 9.87 ± 0.01 against DPPH and ABTS radicals, respectively (Table 5). Similar studies were carried out for the detection of compounds with antioxidant activity from *Streptomyces* spp., dihydroherbimycin A [36], 5-(2,4-dimethylbenzyl)pyrrolidin-2-one [37]. Ser et al. [38] reported antioxidant activity of pyrrolo[1,2-a]pyrazine-1,4-dione, hexahydro extracted from *Streptomyces mangrovisoli,* a novel *Streptomyces* species isolated from a mangrove forest in Malaysia. Narendhran et al. [39] successfully reported antioxidant potential of phenol, 2,4-bis(1,1-dimethylethyl) in *Streptomyces cavouresis* KUV39 isolated from vermicompost samples. Tian et al. [40] reported antioxidant, antifungal, and antibacterial activity of *p*-Terphenyls isolated from halophilic actinobacteria *Nocardiopsis gilva* YIM 90087. Current findings indicated the bioactive potential of actinobacterial isolates from Western Ghats region of Karnataka. The isolates were found to possess significant antimicrobial activity against Gram-positive MRSA bacteria, Gram-negative bacteria, and fungal pathogens. They also exhibited potent scavenging activity against DPPH and ABTS radicals suggesting their antioxidant potential. It is anticipated that findings of the study will be useful in the discovery of novel species of actinobacteria for a potential source of bioactive compounds from underexplored environments.

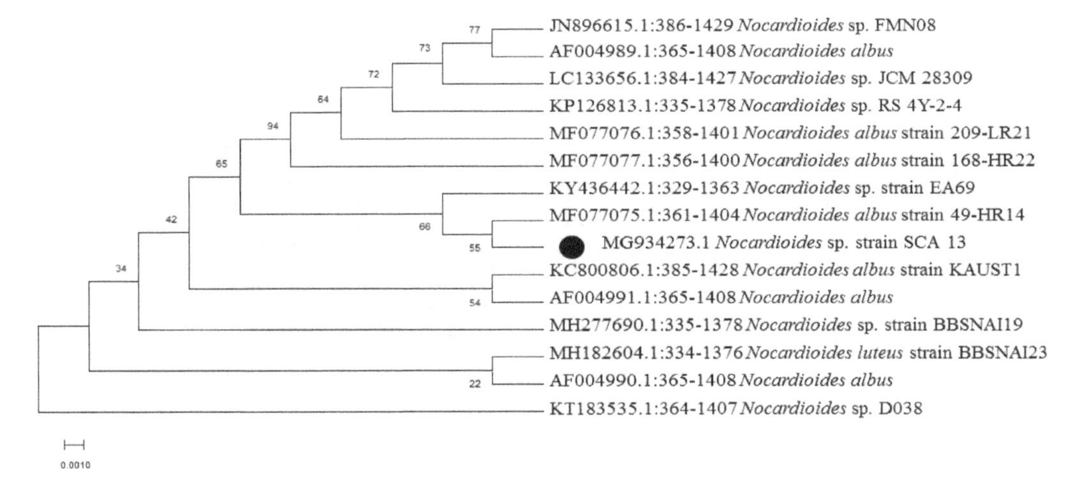


**Figure 4.** Phylogenetic analysis of isolate *Nocardioides* sp. strain SCA13. Neighbor-joining phylogenetic tree showing evolutionary relationship of selected isolate based on 16S r-RNA sequence alignments. Bootstrap values at the nodes indicate collated values based on 1000 resampleddatasets. Bar indicates 0.001 substitutions per nucleotide position.

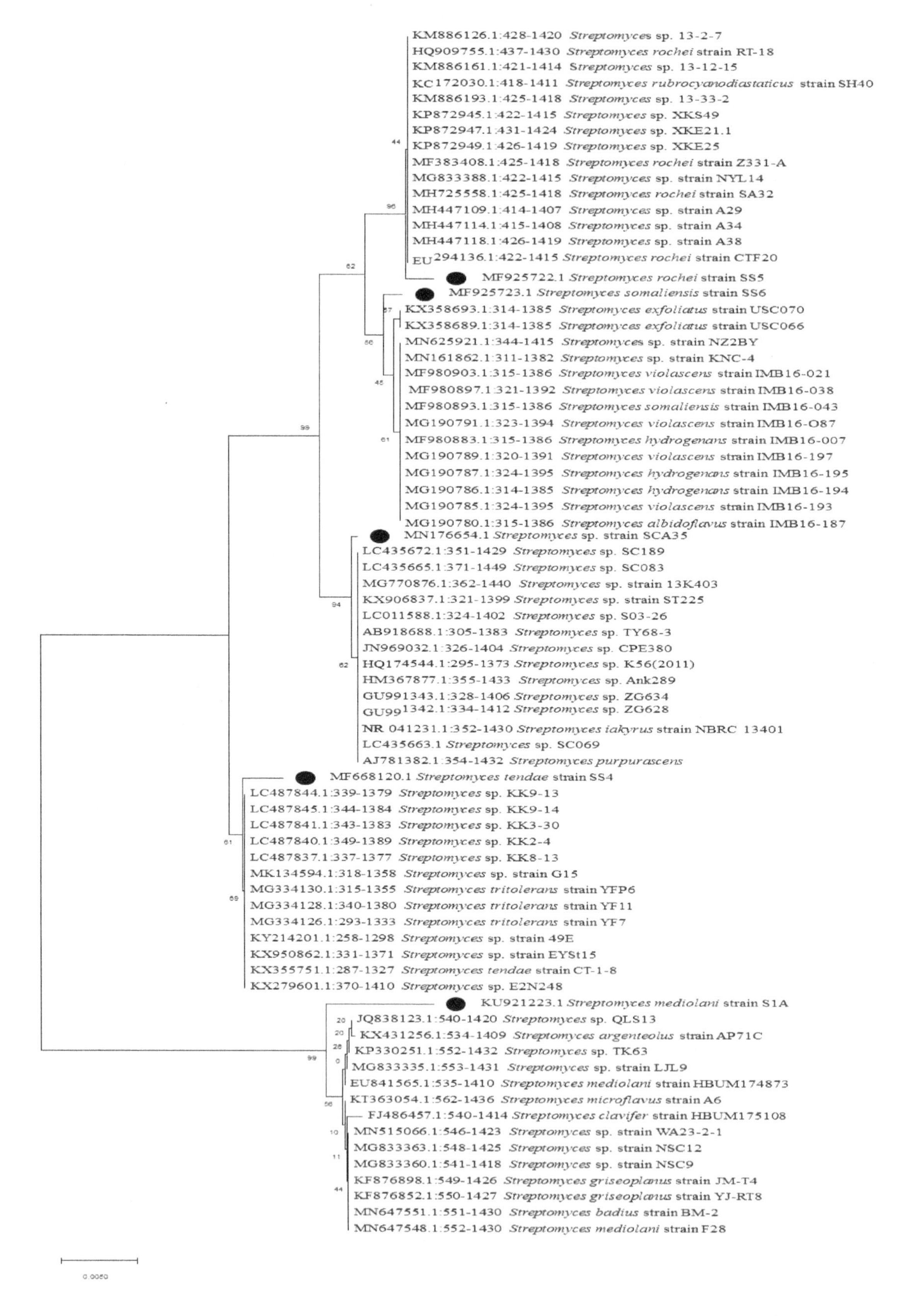


**Figure 5.** Neighbor-joining phylogenetic tree showing evolutionary relationship between isolates S1A, SS4, SS5, SS6, and SCA35 based on 16S r-RNA sequence alignments. Bootstrap values at the nodes indicate collated values based on 1000 resampled datasets. Bar indicates 0.005 substitutions per nucleotide position.

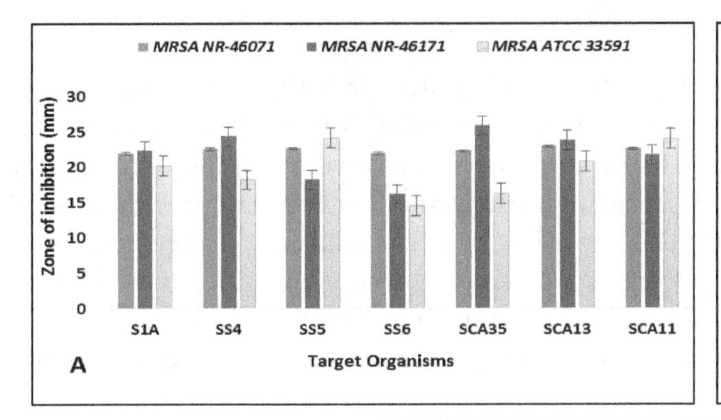


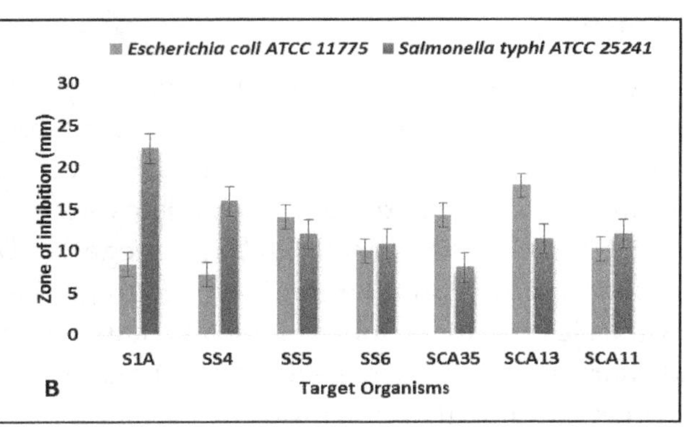


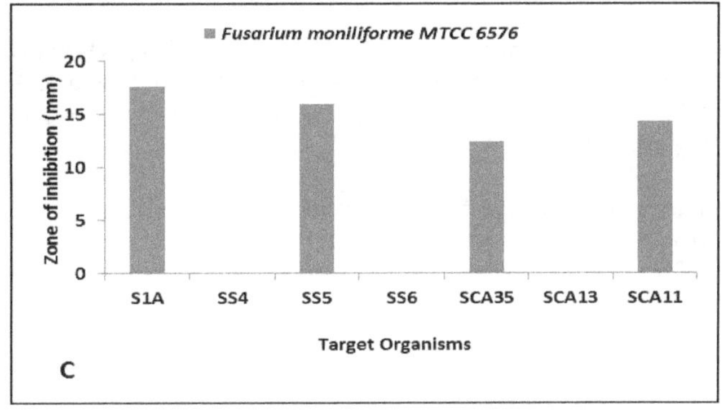


**Figure 6.** Antimicrobial activity of actinobacterial isolates against (**A**) MRSA ATCC NR 46071, MRSA ATCC NR 46171, and MRSA ATCC 33591 (**B**) *Escherichia coli* ATCC 11775 and *Salmonella typhi* ATCC 25241, (**C**) *Fusarium moniliforme* MTCC 6576.

**Table 4.** Minimum inhibitory concentration (MIC in μg/mL) of actinobacterial isolates against pathogenic test organisms.

| Organisms | Minimum Inhibition Concentration (μg/mL) | | | | | |
|---|---|---|---|---|---|---|
| | Test Organisms | | | | | |
| | MRSA ATCC NR-46071 | MRSA ATCC NR-46171 | MRSA ATCC 33591 | *Salmonella typhi* ATCC 25241 | *Escherichia coli* ATCC 11775 | *Fusarium moniliforme* MTCC 6576 |
| *Streptomyces* Sp. S1A | 15.62 | 15.62 | 62.5 | 125 | >500 | 62.5 |
| *Streptomyces* Sp. SS4 | 62.5 | 31.25 | 125 | 250 | >500 | - |
| *Streptomyces* Sp. SS5 | 31.25 | 125 | 15.62 | 250 | 250 | 125 |
| *Streptomyces* Sp. SS6 | 31.25 | 62.5 | 62.5 | 250 | 125 | - |
| *Streptomyces* Sp. SCA35 | 62.5 | 15.62 | 125 | >500 | 125 | 250 |
| *Nocardiopsis* Sp. SCA11 | 31.25 | 15.62 | 31.25 | 250 | >500 | 125 |
| *Nocardioides* Sp. SCA13 | 62.5 | 62.5 | 125 | 250 | 125 | - |

**Table 5.** Comparison of $IC_{50}$ (μg/mL) of crude extracts and trolox for DPPH and ABTS radical scavenging activity.

| Isolates | $IC_{50}$ (μg/mL) Crude Extracts | |
|---|---|---|
| | DPPH | ABTS |
| *Streptomyces* Sp. S1A | 189.40 ± 0.12 | 156.81 ± 0.06 |
| *Streptomyces* Sp. SS4 | 98.29 ± 0.32 | 123.48 ± 0.13 |
| *Streptomyces* Sp. SS5 | 86.45 ± 0.04 | 114.87 ± 0.29 |
| *Streptomyces* Sp. SS6 | 114.15 ± 0.03 | 164.04 ± 0.07 |
| *Streptomyces* Sp. SCA35 | 65.86 ± 0.49 | 49.11 ± 0.73 |
| *Nocardiopsis* Sp. SCA11 | 30.91 ± 0.25 | 48.24 ± 0.30 |
| *Nocardioides* Sp. SCA13 | 42.30 ± 0.10 | 37.91 ± 0.17 |
| **Trolox (Standard)** | 11.07 ± 0.06 | 9.87 ± 0.01 |

## 4. Conclusions

The present study was successful in determining the diversity and bioactive potential of actinobacterial isolates from Western Ghats region of Karnataka. Relatively underexplored forest regions of Western Ghats of Karnataka are found to be promising resources for the discovery of natural bioactive metabolites. The isolates showed significant antimicrobial activity against pathogenic Gram-positive MRSA bacteria, Gram-negative bacteria, and fungi. They were also found to possess antioxidant potential. Our studies encourage the exploration of diverse ecosystems for the isolation of new species for novel and biologically active compounds. Further studies are under progress to purify and characterize the crude extracts that may result in the economic production of bioactive compounds for pharmaceutical applications.

**Author Contributions:** Conceptualization, S.S., R.R.V., W.J. and M.S.; Methodology, S.S. and R.R.V.; Formal analysis, S.S. and R.R.V.; Investigation, S.S., Original Draft Preparation, S.S. and R.R.V.; Writing—Reviewing and Editing, S.S., R.R.V., W.J. and M.S.; Visualization, S.S. and R.R.V.; Supervision, R.R.V. All authors have read and agreed to the published version of the manuscript.

**Acknowledgments:** The authors would like to acknowledge the facilities provided by University of Mysore. The authors gratefully acknowledge the financial support grant to Mr. Saket Siddharth, ICMR-SRF from India Council of Medical Research, New Delhi (ICMR-SRF/AMR/FELLOWSHIPS/5/2019-ECD-II, dated: 27.06.2019).

## References

1. Liu, H.; Xiao, L.; Wei, J.; Schmitz, J.C.; Liu, M.; Wang, C.; Cheng, L.; Wu, N.; Chen, L.; Zhang, Y.; et al. Identification of *Streptomyces* sp. nov. WH26 producing cytotoxic compounds isolated from marine solar saltern in China. *World J. Microbiol. Biotechnol.* **2013**, *29*, 1271–1278. [CrossRef]
2. David, M.Z.; Daum, R.S. Community-associated methicillin-resistant *Staphylococcus aureus*: Epidemiology and clinical consequences of an emerging epidemic. *Clin. Microbiol. Rev.* **2010**, *23*, 616–687. [CrossRef] [PubMed]
3. Chalasani, A.G.; Dhanarajan, G.; Nema, S.; Sen, R.; Roy, U. An antimicrobial metabolite from *Bacillus* sp.: Significant activity against pathogenic bacteria including multidrug-resistant clinical strains. *Front. Microbiol.* **2015**, *6*, 1335. [CrossRef] [PubMed]
4. Torres, M.A.; Jones, J.D.; Dangl, J.L. Reactive oxygen species signalling in response to pathogens. *Plant. Physiol.* **2006**, *141*, 373–378. [CrossRef] [PubMed]
5. Giacco, F.; Brownlee, M. Oxidative stress and diabetic complications. *Circ. Res.* **2010**, *107*, 1058–1070. [CrossRef] [PubMed]
6. Fearon, I.M.; Faux, S.P. Oxidative stress and cardiovascular disease: Novel tools give (free) radical insight. *J. Mol. Cell. Cardiol.* **2009**, *47*, 372–381. [CrossRef] [PubMed]
7. Barnham, K.J.; Masters, C.L.; Bush, A.I. Neurodegenerative diseases and oxidative stress. *Nat. Rev. Drug Discov.* **2004**, *3*, 205–214. [CrossRef]
8. Zhang, H.; Tsao, R. Dietary polyphenols, oxidative stress and antioxidant and anti-inflammatory effects. *Curr. Opin. Food Sci.* **2016**, *8*, 33–42. [CrossRef]
9. Carocho, M.; Ferreira, I.C. A review on antioxidants, prooxidants and related controversy: Natural and synthetic compounds, screening and analysis methodologies and future perspectives. *Food Chem. Toxicol.* **2013**, *51*, 15–25. [CrossRef]
10. Thompson, C.C.; Kruger, R.H.; Thompson, F.L. Unlocking marine biotechnology in the developing world. *Trends Biotechnol.* **2017**, *35*, 1119–1121. [CrossRef]
11. Berdy, J. Thoughts and facts about antibiotics: Where we are now and where we are heading. *J. Antibiot.* **2012**, *65*, 385–395. [CrossRef] [PubMed]
12. Qin, S.; Xing, K.; Jiang, J.H.; Xu, L.H.; Li, W.J. Biodiversity, bioactive natural products and biotechnological potential of plant-associated endophytic actinobacteria. *Appl. Microbiol. Biotechnol.* **2011**, *89*, 457–473. [CrossRef] [PubMed]
13. Manivasagan, P.; Venkatesan, J.; Sivakumar, K.; Kim, S.K. Pharmaceutically active secondary metabolites of marine actinobacteria. *Microbiol. Res.* **2013**, *169*, 262–278. [CrossRef] [PubMed]

14. Nagpure, A.; Choudhary, B.; Kumar, S.; Gupta, R.K. Isolation and characterization of chitinolytic *Streptomyces* sp. MT7 and its antagonism towards wood-rotting fungi. *Ann. Microbiol.* **2014**, *64*, 531–541. [CrossRef]
15. Che, Q.; Zhu, T.; Qi, X.; Mandi, A.; Mo, X.; Li, J. Hybrid isoprenoids from a reed rhizosphere soil derived actinomycete *Streptomyces* sp. CHQ-64. *Org. Lett.* **2012**, *14*, 3438–3441. [CrossRef]
16. Ibeyaima, A.; Rana, J.; Dwivedi, A.K.; Saini, N.; Gupta, S.; Sarethy, I.P. *Pseudonocardiaceae* sp. TD-015 from the Thar Desert, India: Antimicrobial activity and identification of antimicrobial compounds. *Curr. Bioact. Compd.* **2017**, *14*, 112–118. [CrossRef]
17. Gautham, S.A.; Shobha, K.S.; Onkarappa, R.; Kekuda, P.T.R. Isolation, characterization and antimicrobial potential of *Streptomyces* species from Western Ghats of Karnataka, India. *Res. J. Pharm. Technol.* **2012**, *5*, 233–238.
18. Nampoothiri, M.K.; Ramkumar, B.; Pandey, A. Western Ghats of India: Rich source of microbial diversity. *J. Sci. Ind. Res.* **2013**, *72*, 617–623.
19. Ganesan, P.; Anand, S.; Sivanandhan, S.; David, R.H.A.; Paulraj, M.G.; Al-Dhabi, N.A.; Ignacimuthu, S. Larvicidal, ovicidal and repellent activities of *Streptomyces enissocaesilis* (S12-17) isolated from Western Ghats of Tamil Nadu, India. *J. Entomol.* **2018**, *6*, 1828–1835.
20. Ganesan, P.; Reegan, A.D.; David, R.H.A.; Gandhi, M.R.; Paulraj, M.G.; Al-Dhabi, N.A.; Ignacimuthu, S. Antimicrobial activity of some actinomycetes from Western Ghats of Tamil Nadu, India. *Alexandria J.* **2017**, *53*, 101–110. [CrossRef]
21. Siddharth, S.; Vittal, R.R. Evaluation of antimicrobial, enzyme inhibitory, antioxidant and cytotoxic activities of partially purified volatile metabolites of marine *Streptomyces* sp. S2A. *Microorganisms* **2018**, *6*, 72. [CrossRef]
22. Siddharth, S.; Vittal, R.R. Isolation and characterization of bioactive compounds with antibacterial, antioxidant and enzyme inhibitory activities from marine-derived rare actinobacteria, *Nocardiopsis* sp. SCA21. *Microb. Pathog.* **2019**, 103775. [CrossRef] [PubMed]
23. Ser, H.L.; Tan, L.T.; Palanisamy, U.D.; Abd Malek, S.N.; Yin, W.F.; Chan, K.G. *Streptomyces antioxidans* sp. nov., a novel mangrove soil actinobacterium with antioxidative and neuroprotective potentials. *Front. Microbiol.* **2016**, *7*, 899. [CrossRef] [PubMed]
24. Ceylan, O.; Okmen, G.; Ugur, A. Isolation of soil *Streptomyces* as source of antibiotics active against antibiotic-resistant bacteria. *EurAsian J. Biosci.* **2008**, *2*, 73–82.
25. Junaid, S.; Dileep, N.; Rakesh, K.N.; Kekuda, P.T.R. Antimicrobial and antioxidant efficacy of *Streptomyces* species SRDP-TK-07 isolated from a soil of Talakaveri, Karnataka, India. *Pharmanest* **2013**, *4*, 736–750.
26. Bennur, T.; Kumar, A.R.; Zinjarde, S.; Javdekar, V. *Nocardiopsis* species as potential sources of diverse and novel extracellular enzymes. *Appl Microbiol Biotechnol.* **2014**, *98*, 9173–9185. [CrossRef]
27. Fiedler, H.P.; Bruntner, C.; Bull, A.T. Marine actinomycetes as a source of novel secondary metabolites. *Ant. V. Leeuw.* **2004**, *87*, 37–42. [CrossRef]
28. Sengupta, S.; Pramanik, A.; Ghosh, A.; Bhattacharyya, M. Antimicrobial activities of actinomycetes isolated from unexplored regions of Sundarbans mangrove ecosystem. *BMC Microbiol.* **2015**, *15*, 170. [CrossRef]
29. Satheeja, S.; Jebakumar, S.R.D. Phylogenetic analysis and antimicrobial activities of *Streptomyces* isolates from mangrove sediment. *J. Basic Microbiol.* **2011**, *51*, 71–79. [CrossRef]
30. Vu, H.T.; Nguyen, D.T.; Nguyen, H.Q. Antimicrobial and cytotoxic properties of bioactive metabolites produced by *Streptomyces cavourensis* YBQ59 isolated from *Cinnamomum cassia Prels* in Yen Bai province of Vietnam. *Curr. Microbiol.* **2018**, *75*, 1247–1255. [CrossRef]
31. Dashti, Y.; Grkovic, T.; Abdelmohsen, U.R. Production of induced secondary metabolites by a co-culture of sponge-associated actinomycetes, *Actinokineospora* sp. EG49 and *Nocardiopsis* sp. RV163. *Mar. Drugs* **2014**, *12*, 3046–3059. [CrossRef] [PubMed]
32. Rodríguez, V.; Martín, J.; Sarmiento-Vizcaíno, A. Anthracimycin B, a potent Antibiotic against gram-positive bacteria isolated from cultures of the deep-sea actinomycetes *Streptomyces cyaneofuscatus* M-169. *Mar. Drugs* **2018**, *16*, 406. [CrossRef] [PubMed]
33. Shobha, K.S.; Onkarappa, R. In vitro susceptibility of C. albicans and C. neoformens to potential metabolites from *Streptomycetes*. *Indian J. Microbiol.* **2011**, *51*, 445–449. [CrossRef] [PubMed]
34. Manasa, M.; Poornima, G.; Abhipsa, V.; Rekha, C.; Prashith, K.T.R.; Onkarappa, R.; Mukunda, S. Antimicrobial and antioxidant potential of Streptomyces sp. RAMPP-065 isolated from Kudremukh soil, Karnataka, India. *Sci. Technol. Arts Res. J.* **2012**, *1*, 39–44. [CrossRef]

35. Gautham, S.A.; Onkarappa, R. Pharmacological activities of metabolite from *Streptomyces fradiae* strain GOS1. *Int. J. Chem.* **2013**, *11*, 583–590.
36. Chang, H.B.; Kim, J.-H. Antioxidant properties of dihydroherbimycin A from a newly isolated *Streptomyces* sp. *Biotechnol. Lett.* **2007**, *29*, 599–603. [CrossRef]
37. Saurav, K.; Kannabiran, K. Cytotoxicity and antioxidant activity of 5-(2, 4-dimethylbenzyl) pyrrolidin-2-one extracted from marine *Streptomyces* VITSVK5 spp. *Saudi J. Biol. Sci.* **2012**, *19*, 81–86. [CrossRef]
38. Ser, H.L.; Palanisamy, U.D.; Yin, W.F.; Malek, S.N.A.; Chan, K.G.; Goh, B.H. Presence of antioxidative agent, Pyrrolo[1,2-a] pyrazine-1,4-dione, hexahydro-in newly isolated *Streptomyces mangrovisoli* sp. nov. *Front. Microbiol.* **2015**, *6*, 854. [CrossRef]
39. Narendhran, S.; Rajiv, P.; Vanathi, P.; Sivaraj, R. Spectroscopic analysis of bioactive compounds from Streptomyces cavouresis kuv39: Evaluation of antioxidant and cytotoxicity activity. *Int. J. Pharm. Pharm. Sci.* **2014**, *6*, 319–322.
40. Tian, S.Z.; Pu, X.; Luo, G.; Zhao, L.X.; Xu, L.H.; Li, W.J.; Luo, Y. Isolation and characterization of new p-Terphenyls with antifungal, antibacterial, and antioxidant activities from halophilic actinomycete *Nocardiopsis gilva* YIM 90087. *J. Agric. Food Chem.* **2013**, *61*, 3006–3012. [CrossRef]

# New Antimicrobial Phenyl Alkenoic Acids Isolated from an Oil Palm Rhizosphere-Associated Actinomycete, *Streptomyces palmae* CMU-AB204$^T$

**Kanaporn Sujarit [1,2], Mihoko Mori [2,3,*], Kazuyuki Dobashi [2], Kazuro Shiomi [2,3], Wasu Pathom-aree [1,4] and Saisamorn Lumyong [1,4,5,*]**

1 Research Center of Microbial Diversity and Sustainable Utilization, Faculty of Science, Chiang Mai University, Chiang Mai 50200, Thailand; k.sujarit@gmail.com (K.S.); wasu215793@gmail.com (W.P.-a.)
2 Kitasato Institute for Life Sciences, Kitasato University, 5-9-1 Shirokane, Minato-ku, Tokyo 108-8641, Japan; dobashi.kazu@gmail.com (K.D.); shiomi@lisci.kitasato-u.ac.jp (K.S.)
3 Graduate School of Infection Control Sciences, Kitasato University, 5-9-1 Shirokane, Minato-ku, Tokyo 108-8641, Japan
4 Department of Biology, Faculty of Science, Chiang Mai University, Chiang Mai 50200, Thailand
5 Academy of Science, The Royal Society of Thailand, Bangkok 10300, Thailand
* Correspondence: morigon5454@gmail.com (M.M.); scboi009@gmail.com (S.L.).

**Abstract:** Basal stem rot (BSR), or *Ganoderma* rot disease, is the most serious disease associated with the oil palm plant of Southeast Asian countries. A basidiomycetous fungus, *Ganoderma boninense*, is the causative microbe of this disease. To control BSR in oil palm plantations, biological control agents are gaining attention as a major alternative to chemical fungicides. In the course of searching for effective actinomycetes as potential biological control agents for BSR, *Streptomyces palmae* CMU-AB204$^T$ was isolated from oil palm rhizosphere soil collected on the campus of Chiang Mai University. The culture broth of this strain showed significant antimicrobial activities against several bacteria and phytopathogenic fungi including *G. boninense*. Antifungal and antibacterial compounds were isolated by antimicrobial activity-guided purification using chromatographic methods. Their structures were elucidated by spectroscopic techniques, including Nuclear Magnetic Resonance (NMR), Mass Spectrometry (MS), Ultraviolet (UV), and Infrared (IR) analyses. The current study isolated new phenyl alkenoic acids **1–6** and three known compounds, anguinomycin A (**7**), leptomycin A (**8**), and actinopyrone A (**9**) as antimicrobial agents. Compounds **1** and **2** displayed broad antifungal activity, though they did not show antibacterial activity. Compounds **3** and **4** revealed a strong antibacterial activity against both Gram-positive and Gram-negative bacteria including the phytopathogenic strain *Xanthomonas campestris* pv. *oryzae*. Compounds **7–9** displayed antifungal activity against *Ganoderma*. Thus, the antifungal compounds obtained in this study may play a role in protecting oil palm plants from *Ganoderma* infection with the strain *S. palmae* CMU-AB204$^T$.

**Keywords:** actinomycetes; antimicrobial; phenyl alkenoic acid; rhizosphere; *Streptomyces palmae*

## 1. Introduction

Oil palm (*Elaeis guineensis* Jacq.) is an important economic crop in many tropical areas. In particular, Indonesia, Malaysia, and Thailand are the leading palm oil producing countries of this region. The oil palm plant typically has a productive life of 20 or more years, and oil can be harvested several times each year. Consequently, it holds an advantage over all other oil-producing crops [1]. However, the plant is often damaged by fungal infections, and these can cause a decrease of crop yields and result in the death of oil palm trees.

Fungal pathogens mainly infect the stems and leaves of oil palm trees during all stages of growth, from seedlings to the mature stage, and consequently can affect both the quality and quantity of palm oil. Basal stem rot (BSR), or *Ganoderma* rot disease, is the most severe disease of oil palm trees in Southeast Asian countries, especially Malaysia and Indonesia [2]. In addition to these countries, BSR has also destroyed oil palm plantations in Africa, Colombia [3], Papua New Guinea [4], and Thailand [5]. The causative fungus *Ganoderma boninense* is a basidiomycetous fungus and belongs to the order *Polyporales* and the family of *Ganodermataceae*. Fruiting bodies of *Ganoderma* typically form on the exterior of the oil palm trunk and then release and spread the spores to the soil. The usual method of controlling BSR in oil palm plantations is the use of chemical fungicides. Many fungicides, such as azoxystrobin, benomyl, carbendazim, carboxin, cycloheximide, cyproconazole, drazoxolone, hexaconazole, methfuroxam, nystatin, penconazole, thiram, triadimefon, triadimenol, tridemorph, and quintozene, could inhibit the growth of *Ganoderma* [6–10]. However, the fungicides cannot actually cure infected palm trees; they can only delay the spreading of the disease [9]. Furthermore, the applications of these chemical treatments have some worrying effects on human health and ecosystems. Examples of this would be toxicity to organisms and the suppression of beneficial microbes [9,11–13]. Nowadays, raising concerns about the high cost of chemicals, and the environmental problems they are associated with, have encouraged researchers to seek alternative strategies for BSR suppression.

The use of biological control agents represents a major alternative approach in the management of oil palm diseases. Fungal species, such as *Trichoderma harzianum, Trichoderma viride*, and *Gliocladium viride*, have been studied for their anti-*Ganoderma* activity, and their effectiveness against *Ganoderma* in a glasshouse and in a field trial [2,14–16]. Certain *Trichoderma* species are known as mycoparasites and have been utilized to control fungal pathogens. One of the biocontrol mechanisms of *Trichoderma* spp. is the release of glucanase and chitinase enzymes that are involved in the cell-wall degradation of *G. boninense*, and these can be elicitors in inducing a plant defense response [17,18]. Several strains of bacteria, especially *Pseudomonas aeruginosa, Pseudomonas syringae,* and *Burkholderia cepacia,* have also been studied for their potential to be applied as biological control agents [19–22]. Their potential abilities to inhibit the spread of *G. boninense* and to reduce the incidence of the disease have been documented [19–22]. Although the control mechanisms of these bacteria have not yet been clarified, they may control *Ganoderma* by producing antifungal secondary metabolites. In addition, several actinomycetes were screened for their antagonistic activity against *G. boninense.* Actinomycetes, especially the genus *Streptomyces*, are well known for their ability to produce a wide variety of bioactive metabolites [23–26]. Many *Streptomyces* species, such as *Streptomyces hygroscopicus, Streptomyces ahygroscopicus*, *Streptomyces abikoensis*, and *Streptomyces angustmyceticus*, were found to be promising biocontrol agents for BSR disease [27,28]. *Streptomyces violaceorubidus* released not only secondary metabolites towards *G. boninense* but also released cell-wall degrading enzymes involved in the control of this pathogen [29,30].

Actinomycetes associated with the oil palm rhizosphere may have an important role in protecting plants from *Ganoderma* infection by releasing antibiotics and enzymes. Thus, we isolated actinomycetes from the rhizosphere of healthy oil palm plants and screened the antifungal activities of their culture broth against *G. boninense*. One actinomycete strain, CMU-AB204$^T$, showed significant antimicrobial activities against, not only *G. boninense* but also phytopathogenic fungi and several bacteria. We had previously identified this strain and proposed that it could serve as a novel species, namely *Streptomyces palmae* CMU-AB204$^T$ [31]. This actinomycete was selected to investigate antimicrobial secondary metabolites. This report describes the results of the isolation, structural elucidation, and antimicrobial activities of six new compounds, AB204-A–F (**1**–**6**), and three known compounds, anguinomycin A (**7**), leptomycin A (**8**), and actinopyrone A (**9**), that were produced by *S. palmae* CMU-AB204$^T$.

## 2. Materials and Methods

### 2.1. Microbial Material

*Streptomyces palmae* CMU-AB204$^T$ was previously isolated from the rhizosphere of an oil palm tree collected from the oil palm plantation at Chiang Mai University, Chiang Mai Province, Thailand, in October 2012. This strain has been characterized using a polyphasic approach and was previously proposed as *S. palmae* (type strain CMU-AB204$^T$ = JCM 31289$^T$ = TBRC 1999$^T$) [31].

### 2.2. Culture Conditions

*S. palmae* CMU-AB204$^T$ was grown in the International Streptomyces Project medium 2 (ISP2) agar [32] at 28 °C. For seed culture, 100 mL of ISP2 medium, consisting of 0.4% yeast extract (Becton, Dickinson and Company, Sparks, MD, USA), 1.0% malt extract (Becton, Dickinson and Company, Sparks, MD, USA), and 0.4% glucose, was prepared in an Erlenmeyer flask and the pH was adjusted to 7.0 before sterilization. The slant culture of *S. palmae* was scraped by an inoculating loop and inoculated into ISP2 medium. The inoculated flask was incubated at 30 °C for three days on a rotary shaker at 150 rpm. Two mL portions of this seed culture were transferred into 500 mL Erlenmeyer flasks containing 150 mL of ISP2 medium, which was followed by fermentation using a rotary shaker at 150 rpm, 30 °C for seven days.

### 2.3. Compound Extraction and Isolation Procedure

The mycelia were separated from fermentation broth (40.0 L) by filtration. The culture filtrate and mycelium were separately extracted twice with an equal volume of EtOAc. The organic layer was evaporated using a rotary evaporator. Extracts from culture filtrate and mycelium were combined and concentrated to dryness in vacuo to obtain a crude extract as a brown oil. The active secondary metabolites were isolated by biological activity-guided purification. The crude extract (4.9 g) was separated using an open column with silica gel (silica gel 60, 0.063–0.200 mm, Merck, Darmstadt, Germany, 150 g of silica gel, Ø40 mm × 240 mm) and eluted with a stepwise gradient of $CHCl_3$/MeOH: 100:0, 99:1, 98:2, 95:5, 90:10, 80:20, 50:50 and 0:100 (v/v), with 1.0 L each. Each eluent was collected in two 500 mL Erlenmeyer flasks (S1–S16) and concentrated in vacuo. The components of each fraction were analyzed using thin-layer chromatography (TLC, silica gel F254, Merck, Darmstadt, Germany) plates with a thickness of 0.25 mm, developed with the $CHCl_3$/MeOH solvent system. Compounds were detected by UV light and phosphomolybdic acid reagent and followed by heating. The active fractions S3 (580.4 mg) and S4 (608.8 mg) eluted with 99:1 (v/v) of $CHCl_3$/MeOH were dissolved in a small amount of MeOH and then separately subjected to Sephadex LH-20 column chromatography (GE Healthcare Bio-Sciences, USA, Ø20 mm × 650 mm) with MeOH as the eluent. The eluate was automatically fractionated into 100 fractions (L1–L100) by a fraction collector (CHF100AA, Advantec, Tokyo, Japan). The active materials were detected from fractions L52–L64. From fractions S3 and S4, 59.6 mg of yellow semi-solid substance was obtained as an active material. Analytical and preparative HPLC of these fractions were carried out on a JASCO HPLC system (JASCO, Tokyo, Japan); pump, PU-2080 Plus; solvent mixer, LG-2808-04; UV detector, MD-1510. The HPLC columns included an analytical column (Pegasil ODS SP100, Ø4.6 mm × 250 mm; Senshu Scientific, Tokyo, Japan) and a preparative column (Pegasil ODS SP100, Ø20 mm × 250 mm; Senshu Scientific). This dried material (59.6 mg) was subjected to preparative HPLC developed with a gradient system of $CH_3CN$ aqueous solution containing 0.1% trifluoroacetic acid (60–90% $CH_3CN$ for 20 min, 90% $CH_3CN$ for 20 min) at flow rate of 7.0 mL/min, and detection was achieved at 254 nm. The eluates at retention times of 16, 21, 32, 33, and 34 min were collected and concentrated in vacuo to dryness in order to afford AB204-A (**1**), AB204-B (**2**), AB204-E (**5**), AB204-F (**6**), and a mixture of AB204-C (**3**) and D (**4**), respectively. Compound **9** was obtained from side fractions (L36–L49) of LH-20 column chromatography of S3. The combined fractions (L36–L49 of S3) were purified by preparative TLC (silica gel, Merck, Darmstadt, Germany) with a developing solvent of $CHCl_3$/MeOH (20:1) to obtain **9**. Compounds **7** and **8** were isolated from the active fraction that was eluted with 98:2 (v/v) of $CHCl_3$/MeOH. The fraction was subjected to silica gel column chromatography with the $CHCl_3$/MeOH solvent system, and active compounds were obtained from the 95:5 (v/v) fraction. This fraction was purified by preparative HPLC with a linear gradient system of 60–90% $CH_3CN$–$H_2O$ containing 0.1% trifluoroacetic acid for 30 min at a flow rate of 7 mL/min and at room temperature. Detection was achieved at 254 nm. Compounds **7** and **8** were eluted at 24 min and 27 min, respectively.

### *2.4. Analyses of the Chemical Structure and Physicochemical Properties*

The purified compounds were prepared at a concentration of 1 mg/mL in MeOH for the measurement of optical rotation, UV spectra, and IR spectra. An optical rotation $[\alpha]_D$ of the compound suspension was measured using a P-2200 polarimeter (JASCO, Tokyo, Japan). UV spectra of each compound were recorded with a U-2810 spectrophotometer (Hitachi High-Tech Science Co., Tokyo, Japan), and IR spectra (ATR) were measured using a FT–IR 4600 (JASCO, Tokyo, Japan). The isolated compounds were dissolved in chloroform-*d* ($CDCl_3$) or methanol-$d_4$ ($CD_3OD$) for NMR analyses. NMR spectra of each compound were obtained on a JNM ECP500 NMR spectrometer (JEOL, Tokyo, Japan) with 500 MHz for $^1$H NMR and 125 MHz for $^{13}$C NMR. Chemical shifts (ppm) of $CDCl_3$ ($\delta_H$ 7.26, $\delta_C$ 77.0) and $CD_3OD$ ($\delta_H$ 3.30, $\delta_C$ 49.0) were used as references. The accurate mass and molecular formulas of the isolated compounds were established by liquid chromatography–mass spectrometry (LC–MS) analyses. Spectra of electron ionization mass spectrometry (EI–MS) were analyzed using a JMS-AX505 HA spectrometer (JEOL, Tokyo, Japan), while the spectra of electrospray ionization mass spectrometry (ESI–MS) were obtained by a JMS-T100LP spectrometer (JEOL, Tokyo, Japan) equipped with an Agilent1100 HPLC system (Agilent, CA, USA).

### *2.5. Measurement of Antimicrobial Activity*

In the purification process of antimicrobial compounds, every fraction obtained from each fractionation step was tested with representative microbes, *Xanthomonas campestris* pv. *oryzae* KB88, *Kocuria rhizophila* ATCC 9341, *Mucor racemosus* IFO 4581, and *G. boninense* BCC 21330, by paper disk diffusion assay. The antimicrobial activity of the purified compounds was analyzed using the paper disk diffusion method (Ø8 mm disk, Advantec, Co., Ltd., Tokyo, Japan) against fourteen microorganisms. Cell suspensions of *Bacillus subtilis* ATCC 6633 ($5 \times 10^5$ cfu/mL), *Escherichia coli* NIHJ ($5 \times 10^5$ cfu/mL), *K. rhizophila* ATCC 9341 ($2 \times 10^5$ cfu/mL), *Mycobacterium smegmatis* ATCC 607 ($5 \times 10^5$ cfu/mL), *Staphylococcus aureus* ATCC 6538p ($5 \times 10^5$ cfu/mL), *Klebsiella pneumonia* ATCC 10031 ($5 \times 10^5$ cfu/mL), *Proteus vulgaris* NBRC 3167 ($5 \times 10^5$ cfu/mL), *Pseudomonas aeruginosa* IFO 3080 ($1 \times 10^6$ cfu/mL), and *X. campestris* pv. *oryzae* KB88 ($1 \times 10^6$ cfu/mL) were individually mixed into the medium containing 0.5% peptone, 0.5% meat extract, and 0.8% agar, while *Aspergillus niger* ATCC 6275 ($1 \times 10^6$ spores/mL), *Candida albicans* ATCC 64548 ($2 \times 10^5$ cfu/mL), *G. boninense* BCC 21330 ($2 \times 10^5$ cfu/mL), *Mu. racemosus* IFO 4581 ($2 \times 10^5$ spores/mL), and *Saccharomyces cerevisiae* ATCC 9763 ($1 \times 10^6$ cfu/mL) were individually mixed into the medium containing 1.0% glucose, 0.5% yeast extract, and 0.8% agar, and poured into Petri dishes. After that, paper disks containing the purified compounds at 50 µg/disk were put onto agar plates of each microorganism with three replicates. All bacterial plates, except *M. smegmatis* ATCC 607 and *X. campestris* pv. *oryzae* KB 88, were incubated at 37 °C for 24 h. *M. smegmatis* ATCC 607 was incubated at the same temperature for 48–72 h., whereas *X. campestris*, yeasts, and fungi were incubated at 27 °C for 24–48 h. The diameter of the inhibition zone was measured in mm units.

## 3. Results

### 3.1. *Biological Activity-Guided Purification of Active Components from Culture Broth of S. palmae CMU-AB204*$^T$ *and Structure Determination of Active Components*

*S. palmae* CMU-AB204$^T$ was cultured in 40 L of ISP2 medium at 28 °C for seven days, and the broth and mycelia were extracted with EtOAc. The active components in culture broth extract of strain CMU-AB204$^T$ were isolated by biological activities-guided purification using paper disk assay. The extract was purified by silica gel column chromatography, Sephadex LH-20 column chromatography, preparative TLC, and preparative HPLC. The eluates were concentrated in vacuo to yield nine compounds, AB204-A (**1**, 10.2 mg), AB204-B (**2**, 8.2 mg), AB204-E (**5**, 14.2 mg), AB204-F (**6**, 9.9 mg), a mixture (6.5 mg) of AB204-C (**3**) and AB204-D (**4**), **7** (7.0 mg), **8** (1.7 mg), and **9** (17.0 mg) as depicted in Scheme 1.

AB204-A (**1**) was obtained as a pale yellow amorphous solid. It was found to be readily soluble in acetonitrile, MeOH, $CHCl_3$, and was observed to be less soluble in water. As the HREIMS analysis showed *m/z* 190.1000 $[M]^+$, the molecular formula of **1** was elucidated as $C_{12}H_{14}O_2$ (calculated value of 190.0994, Figure S1). The intense band at 1706 $cm^{-1}$ of the IR spectrum in MeOH solution was assigned as C=O stretching frequency of dimeric carboxylic acid moiety (Figure S2). Based on $^1H$ NMR analysis, **1** revealed four aromatic protons stacked at 7.14–7.18 ppm and a pair of olefinic protons stacked at 5.70 ppm and 6.51 ppm (Figure S3). Coupling constants ($J$ = 11.5 Hz) of the two olefinic protons showed *Z*-configuration of the olefin moiety. Compound **1** had four additional methylene protons at 2.41–2.51 ppm and one methyl singlet signal at 2.24 ppm (Table 1). The $^{13}C$ NMR spectrum showed 12 carbon signals: one carbonyl carbon at 177.2 ppm that indicated a carboxylic acid, eight aromatic or olefinic carbons, two methylene carbons at 23.5 and 33.7 ppm, and one methyl carbon at 19.8 ppm (Table 1, Figure S4).

**Table 1.** NMR spectroscopic data of AB204-A (**1**) and B (**2**) ($\delta_H$, 500 MHz; $\delta_C$, 125 MHz).

| Carbon No. | 1 (in $CDCl_3$) | | | 2 (in $CDCl_3$) | | |
|---|---|---|---|---|---|---|
| | $\delta_C$, Type | $\delta_H$, Mult ($J$ in Hz) | HMBC | $\delta_C$, type | $\delta_H$, Mult ($J$ in Hz) | HMBC |
| 1 | 177.2, C | | | 177.6, C | | |
| 2 | 33.7, $CH_2$ | 2.41–2.44, m | C1, C3, C4 | 33.4, $CH_2$ | 2.30, t (7.5) | C1, C3, C4 |
| 3 | 23.5, $CH_2$ | 2.46–2.51, m | C1, C2, C5 | 24.2, $CH_2$ | 1.63, tt (7.5, 7.5) | C1, C2, C5 |
| 4 | 129.7 or 129.8, CH | 5.70, dt (11.5, 7.0) | C1′ | 29.1, $CH_2$ | 1.45, tt (7.5, 7.5) | C2, C5, C6 |
| 5 | 129.7 or 129.8, CH | 6.51, d (11.5) | C3, C2′, C6′ | 27.8, $CH_2$ | 2.17, dtd (7.5, 7.5, 1.5) | C3, C4, C6, C7 |
| 6 | | | | 132.0, CH | 5.69, dt (11.5, 7.5) | C1′ |
| 7 | | | | 128.5, CH | 6.45, br.d (11.5) | C5, C6′ |
| 1′ | 136.2, C | | | 136.7, C | | |
| 2′ | 136.2, C | | | 136.2, C | | |
| 3′ | 129.8, CH | 7.14–7.18 *, m | | 129.8, CH | 7.13–7.18 *, m | - |
| 4′ | 125.4 or 127.1, CH | 7.14–7.18 *, m | | 125.3 or 126.8, CH | 7.13–7.18 *, m | C2′ |
| 5′ | 125.4 or 127.1, CH | 7.14–7.18 *, m | C3′ | 125.3 or 126.8, CH | 7.13–7.18 *, m | C3′ |
| 6′ | 128.8, CH | 7.14–7.18 *, m | | 128.9, CH | 7.13–7.18 *, m | C2′ |
| 2′-Me | 19.8, $CH_3$ | 2.24, s | C1′, C2′, C3′ | 19.9, $CH_3$ | 2.25, s | C1′, C2′, C3′ |

* overlapped.

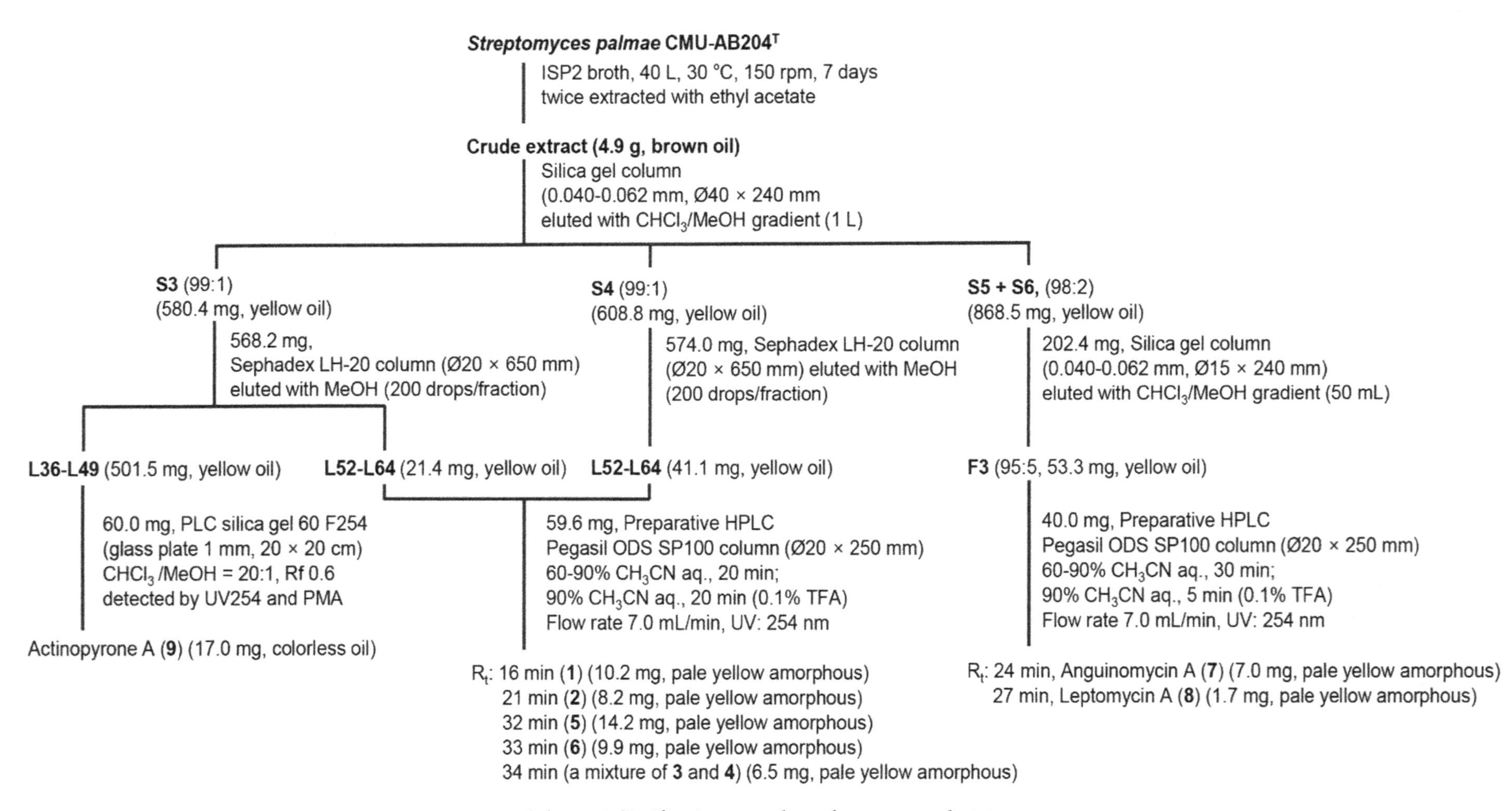


**Scheme 1.** Purification procedures for compounds **1–9**.

MS, HMBC, and HMQC analyses suggested **1** contained one disubstituted aromatic ring, one methyl, and pentenoic acid moieties (Figures S5 and S6). HMBC correlations were observed from two methylene protons (2.41–2.44 ppm and 2.46–2.51 ppm) to a carboxylic carbon at 177.2 ppm, and two olefinic carbons of C-4 and C-5 (129.7 and 129.8 ppm), as are given in Table 1. A correlation between the Z-olefinic proton at 6.51 ppm (H-5) and one methylene carbon (C-3) at 23.5 ppm was also observed; thus **1** was believed to possess 4,5-Z-pentenoic acid moiety in the structure. An HMBC correlation between one methyl proton at 2.24 ppm and three aromatic carbons of C-1′, C-2′, and C-3′ (136.2, 136.2, and 129.8 ppm, respectively), and between Z-olefinic protons and aromatic carbons, H-4 (5.70 ppm) and C-1′ (136.2 ppm), and H-5 (6.51 ppm) and C-6′ (128.8 ppm), indicated **1** was an *ortho*-methyl phenyl alkenoic acid compound, (*Z*)-5-(2-methylphenyl)-4-pentenoic acid (Figure 1). Differential NOE of **1** was observed between a methyl proton and both an aromatic 3′-proton and an olefinic proton of H-5 as well as between the two olefinic protons (Figure S7). The geometry of two substitutes of the aromatic ring was confirmed by NOE correlations, as is shown in Figure 2. From some *Streptomyces* strains, *E*-isomer of **1**, (*E*)-5-(2-methylphenyl)-4-pentenoic acid was identified [33–35]; however, there was no report on the *Z*-isomer (**1**) obtained from natural sources. Therefore, it was concluded that **1** was a novel natural product.

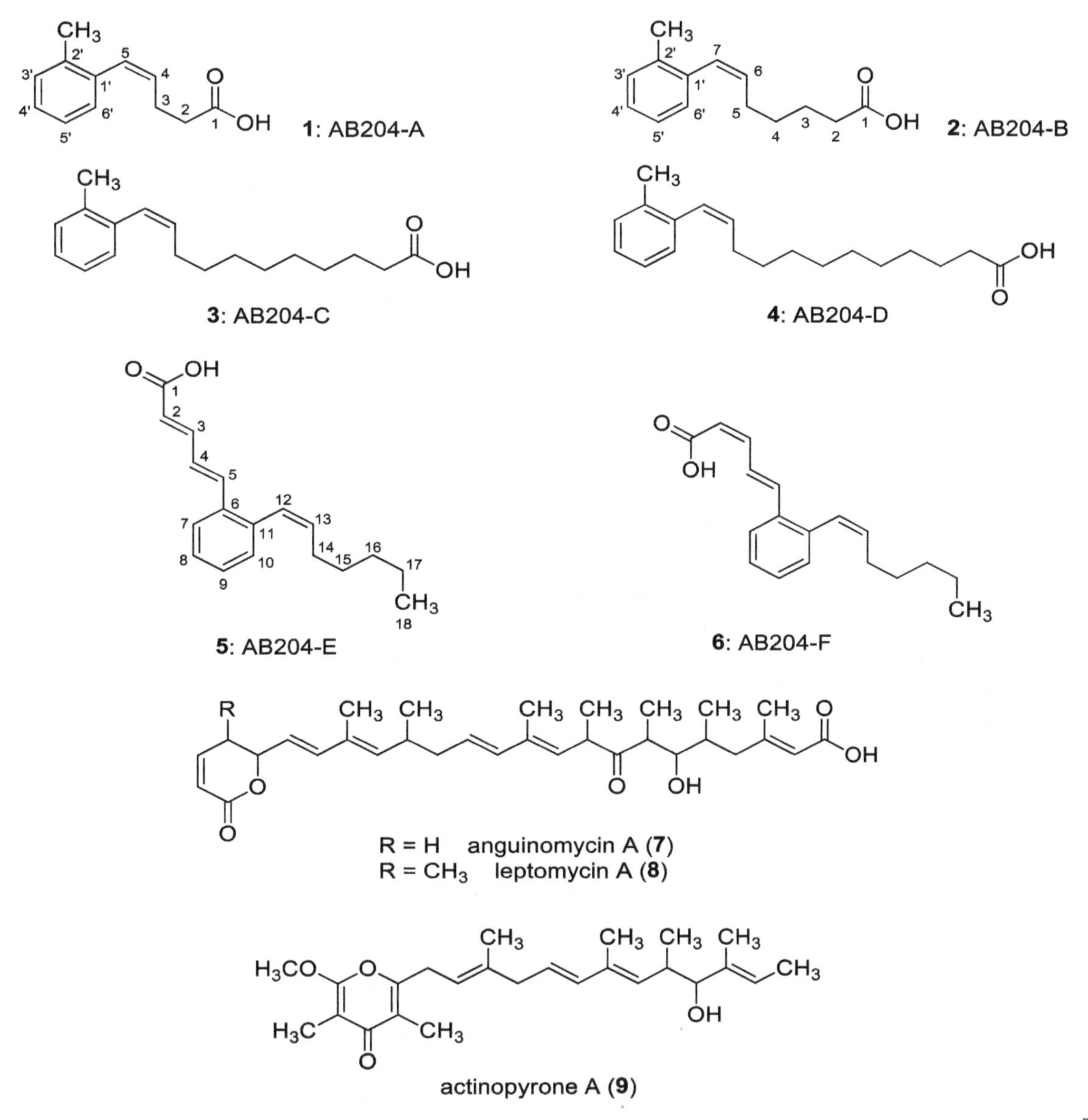


**Figure 1.** Antimicrobial compounds isolated from the broth extract of *Streptomyces palmae* CMU-AB204$^{T}$.

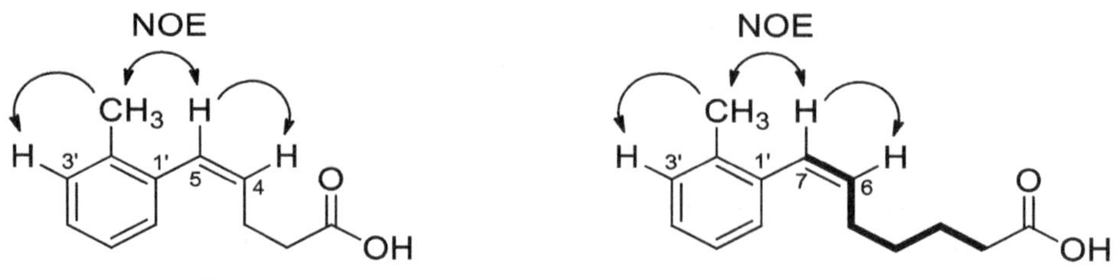

**Figure 2.** Observed COSY correlation (bold lines, in **2**) and NOE correlation in AB204-A (**1**) and B (**2**).

AB204-B (**2**) was isolated as a pale yellow amorphous solid. It was readily soluble in the same solvent as **1** and less soluble in water. The molecular formula of **2** was established as $C_{14}H_{18}O_2$ (calculated value of 218.1329) based on NMR data and the HREIMS ion peak of *m/z* 218.1301 $[M]^+$ (Figure S8), indicating compound **2** had one more $C_2H_4$ unit when compared to the molecular formula of **1**. The IR spectrum of **2** revealed the presence of C=O stretching frequency of dimeric carboxylic acid moiety at 1706 $cm^{-1}$, which was similar to the spectrum of **1** (Figure S9). The $^1H$ NMR spectrum of **2** revealed four aromatic protons at 7.14–7.16 ppm, a pair of olefinic protons at 5.69 and 6.45 ppm, eight methylene protons at 1.45, 1.63, 2.17, and 2.30 ppm and one methyl singlet signal at 2.25 ppm (Table 1, Figure S10). The coupling constant of two olefinic protons (11.5 Hz) indicated a *Z*-configuration. The $^{13}C$ NMR spectrum of **2** showed 14 carbon signals: one carbonyl carbon at 177.6 ppm, eight aromatic or olefinic carbons, four methylene carbons at 24.2, 27.8, 29.1, and 33.4 ppm, and one methyl carbon at 19.9 ppm (Table 1, Figure S11). These data support the conclusion that the compound had a closely related structure to **1**. Each methylene signal was assigned by COSY, as is shown in Figure 2. Eight methylene protons constructed a $C_4$ alkyl chain, and COSY correlation confirmed the connection between this $C_4$ alkyl chain and *Z*-olefin (Figure S12). This connection was supported by HMBC and HMQC spectra (Figures S13 and S14). HMBC correlations were observed from one olefinic proton H-6 (5.69 ppm) to an aromatic carbon at C-1′ (136.7 ppm) and from singlet methyl proton at 2.25 ppm to aromatic carbons at C-1′, C-2′, and C-3′. Therefore, one methyl moiety and one alkene chain were substituted for an aromatic ring in the *ortho* position. HMBC correlation from two methylene protons of the alkene chain at 1.63 and 2.30 ppm to the carbonyl carbon at 177.6 ppm, and a molecular formula of **2**, suggested that this compound had a carboxylic acid at the end of the alkene chain. The same NOE correlation was observed in **1** and **2** (Figure 2 and Figure S15). Thus, the structure of **2** was assigned as (Z)-7-(2-methylphenyl)-6-heptenoic acid, as is shown in Figure 1.

The structures of AB204-C (**3**) and AB204-D (**4**) were elucidated as a mixture of both compounds because of the difficulty associated with further purification. MS spectra showed the molecular formulas of **3** and **4** were $C_{18}H_{26}O_2$ and $C_{19}H_{28}O_2$, respectively (Figure S16). $^1H$ NMR data suggested compounds **3** and **4** were analogs of **1** and **2**, thus **3** and **4** might be (*Z*)-11-(2-methylphenyl)-10-undecenoic acid and (Z)-12-(2-methylphenyl)-11-dodecenoic acid, respectively (Figure 1 and Figure S17).

AB204-E (**5**) and AB204-F (**6**) were obtained as a pale yellow amorphous solid. The accurate mass and molecular formula of compounds **5** and **6** were analyzed by both HRESIMS and HREIMS. Molecular ion peaks were exhibited at *m/z* 271.1705 $[M + H]^+$ and 270.1616 $[M]^+$ for **5**, and *m/z* 271.1689 $[M + H]^+$ and 270.1633 $[M]^+$ for **6** (Figures S18–S21). These data suggest that both compounds had the same molecular formula as $C_{18}H_{22}O_2$ (calcd. 270.1620 for $C_{18}H_{22}O_2$ and calcd. 271.1698 for $C_{18}H_{23}O_2$). The C=O stretching frequency band at 1685 $cm^{-1}$ in the IR spectrum of **5** and 1684 $cm^{-1}$ in the spectrum of **6** was assigned as a carboxylic acid moiety (Figures S22 and S23). NMR spectra of both compounds demonstrated a structural similarity. The assignment of $^1H$ and $^{13}C$ NMR spectra of **5** and **6** are given in Table 2.

**Table 2.** NMR spectroscopic data of AB204-E (**5**) and F (**6**) ($\delta_H$, 500 MHz; $\delta_C$, 125 MHz).

| Carbon No. | 5 (in $CD_3OD$) | | | 6 (in $CDCl_3$) | | |
|---|---|---|---|---|---|---|
| | $\delta_C$, Type | $\delta_H$, Mult (*J* in Hz) | HMBC | $\delta_C$, Type | $\delta_H$, Mult (*J* in Hz) | HMBC |
| 1 | 170.5, C | | | 169.5, C | | |
| 2 | 122.3, CH | 5.99, d (15.5) | C1, C4 | 115.8, CH | 5.74, d (11.5) | C1, C4 |
| 3 | 146.8, CH | 7.41, dd (15.5, 11.0) | C1, C5 | 147.3, CH | 6.85, t (11.5) | C1, C5 |
| 4 | 128.2, CH | 6.95, dd (15.5, 11.0) | C2, C5, C6 | 125.4, CH | 8.04, dd (15.5, 11.5) | C6 |
| 5 | 140.0, CH | 7.13, d (15.5) | C3, C7, C11 | 140.6, CH | 7.08, d (15.5) | C3, C7, C11 |
| 6 | 135.6, C | | | 134.3, C | | |
| 7 | 126.7, CH | 7.67, m | C9, C11 | 126.2, CH | 7.72, m | C9, C11 |
| 8 | 128.4, CH | 7.26 *, m | | 127.1, CH | 7.26 *, m | C10 |
| 9 | 129.5, CH | 7.26 *, m | | 128.6, CH | 7.26 *, m | C11 |
| 10 | 131.0, CH | 7.15, m | C8, C12 | 129.9, CH | 7.17, m | C6, C8, C12 |
| 11 | 138.7, C | | | 137.5, C | | |
| 12 | 128.3, CH | 6.54, br.d (11.5) | C6, C10, C14 | 127.0, CH | 6.51, br.d (11.5) | C10, C14 |
| 13 | 135.7, CH | 5.83, dt (11.5, 7.5) | C11 | 134.9, CH | 5.81, dt (11.5, 7.5) | C11 |
| 14 | 29.4, $CH_2$ | 2.02, dtd (7.5, 7.5, 1.5) | C12, C13, C15, C16 | 28.4, $CH_2$ | 2.04, dtd (7.5, 7.5, 1.0) | C12, C13, C15, C16 |
| 15 | 30.2, $CH_2$ | 1.38, m | C13, C14, C16, C17 | 29.2, $CH_2$ | 1.37, tt (7.5, 7.5) | C14, C16, C17 |
| 16 | 32.6, $CH_2$ | 1.22, m | C17 | 31.4, $CH_2$ | 1.23 *, m | C17 |
| 17 | 23.5, $CH_2$ | 1.21, m | C16 | 22.5, $CH_2$ | 1.23 *, m | C16 |
| 18 | 13.3, $CH_3$ | 0.83, t (7.0) | C16, C17 | 14.0, $CH_3$ | 0.84, t (7.0) | C16, C17 |

* overlapped.

$^1$H NMR spectrum of **5** showed ten aromatic/olefinic protons, eight methylene protons, and one methyl triplet proton at 0.83 ppm (Figure S24). In $^{13}$C NMR spectrum of **5**, one calboxylic carbon at 170.5 ppm, twelve olefinic/aromatic carbons, four methylene carbons, and one methyl carbon were measured (Figure S25). $^1$H NMR spectrum of **5** suggested the existence of two pairs of *E*-olefin assigned by large coupling constants (15.5 Hz for each) and one pair of *Z*-olefin whose coupling constant was 11.5 Hz. Three partial structures were assigned by COSY correlation; one 1,2-substituted aromatic ring, one 1,2-*Z*-heptene group, and one diene group (Figure S26). HMBC correlations between diene protons of 5.99 and 7.41 ppm and a carbonyl carbon at 170.5 ppm, and between diene protons of 6.95 and 7.13 ppm and three aromatic carbons at positions C-6, C-7, and C-11 (135.6, 126.7, and 138.7 ppm, respectively) indicated that one end of the diene was connected to a carboxylic carbon and the other end of the diene was attached to an aromatic ring at position 6 (Figures S27 and S28). HMBC correlation between *Z*-olefinic protons (5.83 and 6.54 ppm) of 1,2-*Z*-heptene and aromatic ring carbons at C-10 and C-11 (131.0 and 138.7 ppm, respectively) suggested 1,2-*Z*-heptene moiety was connected to the aromatic ring at C-11. These data support the structure of **5** as (2*E*,4*E*)-5-(2-(1Z)-heptenylphenyl)-2,4-pentadienoic acid (Figure 1).

$^1$H NMR and $^{13}$C NMR spectra of **6** suggested that the structure was almost the same as **5** (Figures S29 and S30). However, the $^1$H NMR spectrum clarified that only one pair of *E*-olefin existed, while the other two pairs of olefin were identified as *Z*-configuration by analysis of coupling constants. These data indicated that **6** was a stereoisomer of **5**. COSY correlations revealed that two partial structures of **6**, one 1,2-substituted aromatic ring and one 1,2-*Z*-heptene moiety, were identical to those of **5**; however, a diene structure was constituted of both *E* and *Z*-olefins (Figure S31). The connection of 1,2-*Z*-heptene moiety to the aromatic ring at C-11 was confirmed by the HMBC spectrum (Figure 3, Figures S32 and S33). HMBC correlations between *Z*-olefinic protons of the diene moiety and a carboxylic carbon at 169.5 ppm and between one *E*-olefinic proton (7.08 ppm) of the diene moiety and aromatic ring carbons at C-7 and C-11 (126.2 and 137.5 ppm, respectively) established the structure of **6** as (2*Z*,4*E*)-5-(2-(1*Z*)-heptenylphenyl)-2,4-pentadienoic acid, as is depicted in Figure 1.

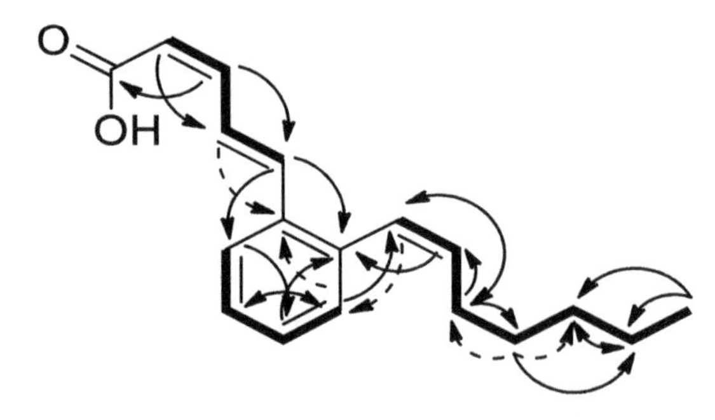


**Figure 3.** Observed COSY (bold lines) and HMBC correlations (arrows) in AB204-F (**6**). The same correlations were observed in AB204-E (**5**).

To clarify the geometry of two substituted chains of **5** and **6**, the differential NOE experiment was conducted. NOE correlations between H-5 and H-12 were observed in both compounds (Figures S34 and S35), which supported the geometry of two side chains in **5** and **6**, as is shown in Figure 4. This is the first report of **5** and **6** obtained from natural sources. Qureshi et al. [36] found structurally related compounds, MF-EA-705a and b, along with actinopyrone A from a broth extract of *Streptomyces* MF-EA-705. The most related structures were found as cinnamoyl moieties of rare peptide compounds, pepticinnamin E, WS9326A, and RP-1776 (skyllamycin A) [37–39].

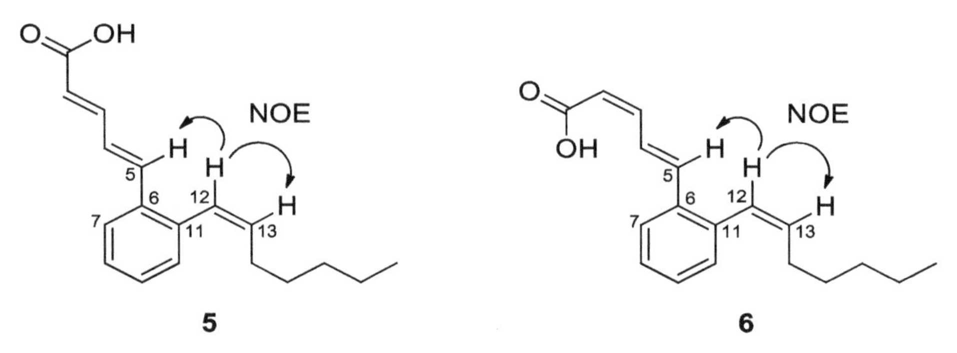


**Figure 4.** NOE correlations in AB204-E (**5**) and AB204-F (**6**).

Compounds **7** and **8** were pale-yellow amorphous solids that were determined by HRESI-MS analyses to have molecular formulas of $C_{31}H_{44}O_6$ and $C_{32}H_{46}O_6$, respectively (Figures S36 and S37). The planar structures of **7** and **8** (Figure 1) were confirmed by NMR spectra (Figures S38 and S39) as known compounds, anguinomycin A (**7**) [40] and leptomycin A (**8**) [41], which were known to be relative structures. Compound **9** was a colorless oil, and the molecular formula of **9** was determined to be $C_{25}H_{36}O_4$ on the basis of HRESI-MS data and the signals of $^{1}$H NMR spectrum (Figures S40 and S41). These data supported the structure of **9** to be actinopyrone A (**9**) as is shown in Figure 1 [42].

Physicochemical Properties of **1**, **2**, **5**, and **6**

Compound **1** (AB204-A): pale yellow amorphous solid; $[\alpha]_D$ –3.4 (*c* 0.1, MeOH); UV (MeOH) $\lambda_{max}$ (log $\varepsilon$) 204.5 (4.09), 235.5 (3.63) nm; IR (ATR) $\nu_{max}$ 3014, 2923, 1706, 1654, 1409 cm$^{-1}$; $^{1}$H and $^{13}$C NMR (chloroform-*d*) see Table 1; HREIMS *m/z* 190.1000 $[M]^+$ (calcd for $C_{12}H_{14}O_2$, 190.0994).

Compound **2** (AB204-B): pale yellow amorphous solid; $[\alpha]_D$ –0.3 (*c* 0.1, MeOH); UV (MeOH) $\lambda_{max}$ (log $\varepsilon$) 204.5 (3.98), 235.5 (3.53) nm; IR (ATR) $\nu_{max}$ 3011, 2925, 1706, 1653, 1457 cm$^{-1}$; $^{1}$H and $^{13}$C NMR (chloroform-*d*) see Table 1; HREIMS *m/z* 218.1301 $[M]^+$ (calcd for $C_{14}H_{18}O_2$, 218.1329).

Compound **5** (AB204-E): pale yellow amorphous solid; $[\alpha]_D$ –13.4 (*c* 0.1, MeOH); UV (MeOH) $\lambda_{max}$ (log $\varepsilon$) 204 (4.04), 251 (3.90), 309.5 (4.26) nm; IR (ATR) $\nu_{max}$ 2956, 2924, 2857, 1684, 1617, 1277, 1000 cm$^{-1}$; $^{1}$H and $^{13}$C NMR (methanol-$d_4$) see Table 2; HREIMS *m/z* 270.1616 $[M]^+$ (calcd for $C_{18}H_{22}O_2$, 270.1620).

Compound **6** (AB204-F): pale yellow amorphous solid; $[\alpha]_D$ –14.1 (*c* 0.1, MeOH); UV (MeOH) $\lambda_{max}$ (log $\varepsilon$) 204 (3.76), 251 (3.51), 309.5 (3.86) nm; IR (ATR) $\nu_{max}$ 2955, 2924, 2855, 1684, 1615, 1244, 958 cm$^{-1}$; $^{1}$H and $^{13}$C NMR (chloroform-*d*) see Table 2; HREIMS *m/z* 270.1633 $[M]^+$ (calcd for $C_{18}H_{22}O_2$, 270.1620).

### *3.2. Antimicrobial Activities of Isolated Compounds*

Antimicrobial activities of the purified compounds, except for the mixture of AB204-C (**3**) and AB204-D (**4**), were tested against four Gram-positive bacteria, five Gram-negative bacteria, two yeasts, and three fungi using paper disk diffusion assay with an equal amount of each compound at 50 μg/disk. The results are shown in Table 3. All compounds did not show activity against *E. coli* NIHJ, *K. pneumonia* ATCC 10031, *P. vulgaris* NBRC 3167, *Ps. aeruginosa* IFO 3080, and *Sa. cerevisiae* ATCC 9763. AB204-A (**1**) and B (**2**) displayed a weak activity towards *C. albicans* ATCC 64548, *A. niger* ATCC 6275, and *G. boninense* BCC 21330, with a clear zone range from 10.4 to 13.2 mm. However, they did not affect the Gram-positive and Gram-negative bacteria. AB204-E (**5**) and AB204-F (**6**) displayed no antifungal and antiyeast activities but showed good antibacterial activity against Gram-positive bacteria and weak activity against the Gram-negative bacterium *X. campestris* pv. *oryzae* KB88, which is a phytopathogenic strain. AB204-E (**5**) strongly inhibited *K. rhizophila* ATCC 9341, *B. subtilis* ATCC 6633, *M. smegmatis* ATCC 607, and *S. aureus* ATCC 6538p with inhibition zones of 41.3, 35.3, 32.7, and 26.0 mm, respectively, while AB204-F (**6**) showed a slightly lower activity against the same pathogens as is presented in Table 3. Anguinomycin A (**7**) revealed potent inhibitory activity against the Gram-positive bacterium *K. rhizophila* ATCC 9341 (19.1 mm), and two fungi, *Mu. racemosus* IFO 4581 (16.9 mm), and *G. boninense* BCC 21330 (19.6 mm), while leptomycin A (**8**) showed stronger activity against these pathogens at 30.6, 49.0, and 21.2 mm, respectively. Actinopyrone A (**9**) exhibited potent antifungal activity against *C. albicans* ATCC 64548 and three fungal strains with the inhibition zone in a range of 11.9 to 23.9 mm (Table 3). These results suggest that the antibacterial and antifungal activities of the *S. palmae* CMU-AB204$^{T}$ may have been displayed as a consequence of the contribution of all these antimicrobial secondary metabolites.

**Table 3.** Antimicrobial activities of the pure compounds against fourteen microorganisms using an equal amount of each compound at 50 μg/disk. Inhibition zone (mm) (Mean ± SD; *n* = 3) including the diameter of the paper disk (8 mm) was measured after 24 and 48 h of incubation. **1**, AB204-A; **2**, AB204-B; **5**, AB204-E; **6**, AB204-F; **7**, anguinomycin A; **8**, leptomycin A; **9**, actinopyrone A.

| Microorganism | Inhibition Zone (mm) of Seven Pure Compounds | | | | | | |
|---|---|---|---|---|---|---|---|
| | 1 | 2 | 5 | 6 | 7 | 8 | 9 |
| **Gram-positive bacteria** | | | | | | | |
| *Bacillus subtilis* ATCC 6633 | - | - | 35.3 ± 1.4 | 12.3 ± 1.8 | - | - | - |
| *Kocuria rhizophila* ATCC 9341 | - | - | 41.3 ± 2.0 | 17.5 ± 1.5 | 19.1 ± 1.9 | 30.6 ± 2.3 | - |
| *Mycobacterium smegmatis* ATCC 607 | - | - | 32.7 ± 1.2 | 14.0 ± 2.2 | - | - | - |
| *Staphylococcus aureus* ATCC 6538p | - | - | 26.0 ± 1.6 | 13.2 ± 1.9 | - | - | - |
| **Gram-negative bacteria** | | | | | | | |
| *Escherichia coli* NIHJ | - | - | - | - | - | - | - |
| *Klebsiella pneumonia* ATCC 10031 | - | - | - | - | - | - | - |
| *Proteus vulgaris* NBRC 3167 | - | - | - | - | - | - | - |
| *Pseudomonas aeruginosa* IFO 3080 | - | - | - | - | - | - | - |
| *Xanthomonas campestris* pv. *oryzae* KB 88 | - | - | 10.6 ± 1.3 | 11.0 ± 2.2 | | - | - |
| **Yeasts** | | | | | | | |
| *Candida albicans* ATCC 64548 | 13.1 ± 1.6 | 10.4 ± 0.9 | - | - | - | - | 20.8 ± 1.5 |
| *Saccharomyces cerevisiae* ATCC 9763 | - | - | - | - | - | - | - |
| **Fungi** | | | | | | | |
| *Mucor racemosus* IFO 4581 | - | - | - | - | 16.9 ± 1.7 | 49.0 ± 2.3 | 23.1 ± 1.6 |
| *Aspergillus niger* ATCC 6275 | 11.5 ± 1.1 | 11.1 ± 1.2 | - | - | - | - | 23.9 ± 1.8 |
| *Ganoderma boninense* BCC 21330 | 11.0 ± 1.4 | 13.2 ± 2.0 | - | - | 19.6 ± 1.6 | 22.1 ± 1.7 | 11.9 ± 1.0 |

## 4. Discussion

Several mechanisms have been proposed to control *G. boninense* causing BSR disease in oil palm trees. However, none of them have successfully been treated or been shown to suppress the disease [9]. The search for antifungal alternatives is representative of a potential solution that has drawn significant interest. In this study, new compounds were identified during the isolation of anti-*Ganoderma* substances from *S. palmae* CMU-AB204$^{T}$. The assessment of antimicrobial activity of four new phenyl alkenoic acids showed that AB204-A (**1**) and B (**2**) mildly inhibited the growth of fungi, *G. boninense*

BCC 21330, *Mu. racemosus* IFO 4581, and *A. niger* ATCC 6275, while AB204-E (**5**) and F (**6**) displayed a positive degree of activity against Gram-positive bacteria, *B. subtilis* ATCC 6633, *K. rhizophila* ATCC 9341, *M. smegmatis* ATCC 607, and *S. aureus* ATCC 6538p. New antifungal compounds, AB204-A (**1**) and B (**2**), possessed similar structures to phenylethyl alcohol (PEA), an antifungal aromatic compound that was obtained from *Trichoderma virens* 7b, which had significant potential as a biological control agent for BSR [43]. These compounds may exhibit a mechanism in inhibiting fungi similar to PEA which inhibits protein, DNA, RNA, and aminoacyl tRNA syntheses of fungi [44,45]. However, a mode of action of novel compounds in controlling fungi should be confirmed in the future.

AB204-B (**2**) contained more $C_2H_4$ units than AB204-A (**1**), but it displayed antimicrobial activity against the same pathogenic strains with similar inhibition zone sizes. This result indicated that the presence of a longer chain of the carboxylic group in **2** had not been involved in the antimicrobial activity. Biological activities of the *E*-isomer of AB204-A (**1**), (*E*)-5-(2-methylphenyl)-4-pentenoic acid was previously reported to be an inactive compound against bacteria and fungi but the tested concentration and the strain of tested microorganisms have not been indicated [33]. However, the difference of an antimicrobial activity between (*E*)-5-(2-methylphenyl)-4-pentenoic acid and new compounds, AB204-A (**1**) and AB204-B (**2**), revealed that the existence of *Z*-olefin in **1** and **2** had been involved in their antifungal activity. The structures of a mixture of AB204-C (**3**) and AB204-D (**4**) were predicted based on HREI-MS and $^1$H NMR spectra. In the future, the mixture should be reseparated using other techniques, and additional data is needed to confirm their structures and antimicrobial activities. AB204-E (**5**) and F (**6**) have an 1,2-*Z*-heptene moiety connected to the aromatic ring. These metabolites have not shown antifungal activity but exhibited strong antibacterial activity when associated with this moiety. Moreover, the existence of one pair of *Z*-olefin in the chain of the carboxylic group of AB204-E (**5**), instead of the *E* and *Z*-olefins of AB204-F (**6**), increased the antibacterial activity of this compound.

Previously, Thong et al. [35] found two closely related compounds of **1–4**, and they were isolated from a *Streptomyces* that had been spontaneously acquired rifampicin resistance. These compounds contained *E*-olefins and have both a methylbenzene unit and a 2-amino-3-hydroxycyclopent-2-enone ($C_5N$) moiety. The phenyl alkenoic acid-associated metabolites discovered by Thong et al. did not display antibacterial activity against *E. coli*, *M. luteus*, *S. aureus*, and *B. subtilis* in testing with a microplate assay at 100 μM or approximately 28.5 and 33.5 mg/mL, thus revealing similar results to AB204-A (**1**) and B (**2**). Notably, the presence of a carboxylic acid moiety in the novel compounds, and a $C_5N$ moiety in the known compounds, may not be involved in the antimicrobial activity. Based on draft genome sequences of the rifampicin-resistant mutant (TW-R50–13), the methylbenzene moiety may be biosynthesized by the expression of polyketide synthase (PKS) genes that are located at a different locus from the biosynthetic genes for the $C_5N$ moiety [35]. The genes encoding for PKS have been disclosed to complex biosynthetic mechanisms, which were involved in the production of many metabolites in microorganisms [46]. Genome sequences would provide the data of potential gene clusters to understand the metabolic pathways of *S. palmae* CMU-AB204$^T$. Thus, the genome sequences of this strain should be further studied to determine the presence of both silent and cryptic secondary metabolite biosynthetic gene clusters that are able to synthesize the corresponding novel natural products.

In addition to **1** and **2**, other antifungal compounds, anguinomycin A (**7**), leptomycin A (**8**), and actinopyrone A (**9**) obtained from the same broth of *S. palmae* CMU-AB204$^T$, also displayed anti-*Ganoderma* activity. The ability of *S. palmae* to produce a variety of antifungal compounds was proven. This strain might produce each antifungal secondary metabolite depending on the prevailing environmental conditions, such as nutritional source, incubation period, pH value, and temperature [47,48]. Hence, the optimization of culture conditions should be studied in order to obtain high yields of the antifungal metabolites. The protecting effect of *S. palmae* CMU-AB204$^T$ against BSR has also been confirmed in a glasshouse experiment [49]. The results obtained from this study strongly suggest that the antimicrobial secondary metabolites were involved in the mechanism exhibiting anti-BSR effects by this *Streptomyces* strain.

Although the new compounds obtained in this study showed moderate activity towards *G. boninense*, they inhibited clinical bacterial pathogens and other phytopathogenic fungi, suggesting a possible utility of the four new antimicrobial substances in both agricultural and medical treatments. However, cytotoxicity to mammalian cell of both new compounds and three known compounds should be tested before being applied to these compounds. The recovery of novel actinomycetes species, especially the genus *Streptomyces*, has the potential to be a rich source of both new and known natural products [50,51]. Notably, *S. palmae* CMU-AB204$^{T}$ was found to produce various bioactive metabolites.

**Author Contributions:** The project approach was designed by K.S., W.P.-a., S.L.; *Streptomyces* strain was selected by K.S.; activity-guided purification was conducted by K.S., M.M., K.D.; structural determination was conducted by K.S., M.M., K.D.; antimicrobial activity was measured by K.S.; the research was supervised by K.S., W.P.-a., S.L.; funding was acquired by M.M. and S.L. All authors have read and agreed to the published version of the manuscript.

**Acknowledgments:** We thank Kenichiro Nagai, School of Pharmacy, Kitasato University, for help in obtaining mass data, and Toshiyuki Tokiwa, Kitasato Institute for Life Sciences, Kitasato University, for providing the microbes used to measure antimicrobial activity.

## References

1. Rees, R.W.; Flood, J.; Hasan, Y.; Cooper, R.M. Effects of inoculum potential, shading and soil temperature on root infection of oil palm seedlings by the basal stem rot pathogen *Ganoderma boninense*. *Plant. Pathol.* **2007**, *56*, 862–870. [CrossRef]
2. Susanto, A.; Sudharto, P.S.; Purba, R.Y. Enhancing biological control of basal stem rot disease (*Ganoderma boninense*) in oil palm plantations. *Mycopathologia* **2005**, *159*, 153–157. [CrossRef] [PubMed]
3. Nieto, L. Incidence of oil palm stem rots in Colombia. *Palmas* **1995**, *16*, 227–232.
4. Turner, P.D. *Oil Palm Diseases and Disorders*; Oxford University Press: Oxford, UK, 1981; pp. 88–110.
5. Likhitekaraj, S.; Tummakate, A. Basal stem rot of oil palm in Thailand caused by *Ganoderma*. In *Ganoderma Diseases of Perennial Crops*; Flood, J., Bridge, P.D., Holderness, M., Eds.; CABI Publishing: Wallingford, UK, 2000; pp. 66–70.
6. Jollands, P. Laboratory investigations on fungicides and biological agents to control three diseases of rubber and oil palm and their potential applications. *Trop. Pest. Manag.* **1983**, *29*, 33–38. [CrossRef]
7. Idris, A.S.; Ismail, S.; Ariffin, D.; Ahmed, D. Control of *Ganoderma* infected palm-development of pressure injection and field application. *MPOB Info. Ser.* **2002**, *148*, 2.
8. Idris, A.S.; Arifurrahman, R. Determination of 50% effective concentration ($EC_{50}$) of fungicides against pathogenic *Ganoderma*. *MPOB Info. Ser.* **2008**, *449*, 2.
9. Naher, L.; Siddiquee, S.; Yusuf, U.K.; Mondal, M.M.A. Issues of *Ganoderma* spp. and basal stem rot disease management in oil palm. *Am. J. Agri. Sci.* **2015**, *2*, 103–107.
10. Shariffah-Muzaimah, S.; Idris, A.; Madihah, A.; Dzolkhifli, O.; Kamaruzzaman, S.; Cheong, P. Isolation of actinomycetes from rhizosphere of oil palm (*Elaeis guineensis* Jacq.) for antagonism against *Ganoderma boninense*. *J. Oil Palm Res.* **2015**, *27*, 19–29.
11. Wightwick, A.; Walters, R.; Allinson, G.; Reichman, S.; Menzies, N. Environmental risks of fungicides used in horticultural production systems. In *Fungicides*; Carisse, O., Ed.; IntechOpen: London, UK, 2010; pp. 273–304.
12. Ji, X.Y.; Zhong, Z.J.; Xue, S.T.; Meng, S.; He, W.Y.; Gao, R.M.; Li, Y.H.; Li, Z.R. Synthesis and antiviral activities of synthetic Glutarimide derivatives. *Chem. Pharm. Bull.* **2010**, *58*, 1436–1441. [CrossRef]
13. Gupta, P.K. Toxicity of fungicides. In *Veterinary Toxicology, Basic and Clinical Principles*, 3rd ed.; Gupta, R.C., Ed.; Academic Press: London, UK, 2018; pp. 569–580.
14. Nur Ain Izzati, M.Z.; Abdullah, F. Disease suppression in *Ganoderma*-infected oil palm seedlings treated with *Trichoderma harzianum*. *Plant. Protect. Sci.* **2008**, *44*, 101–107. [CrossRef]
15. Sundram, S.; Abdullah, F.; Ahmad, Z.A.M.; Yusuf, U.K. Efficacy of single and mixed treatments of *Trichoderma harzianum* as biocontrol agents of *Ganoderma* basal stem rot in oil palm. *J. Oil Palm Res.* **2008**, *20*, 470–483.
16. Naher, L.; Tan, S.G.; Yusuf, U.K.; Ho, C.-L.; Abdullah, F. Biocontrol agent *Trichoderma harzianum* strain FA 1132 as an enhancer of oil palm growth. *Pertanika J. Trop. Agric. Sci.* **2012**, *35*, 173–182.

17. Harman, G.E.; Howell, C.R.; Viterbo, A.; Chet, I.; Lorito, M. *Trichoderma* species-opportunistic, avirulent plant symbionts. *Nat. Rev. Microbiol.* **2004**, *2*, 43–56. [CrossRef]
18. Naher, L.; Ho, C.-L.; Tan, S.G.; Yusuf, U.K.; Abdullah, F. Cloning of transcripts encoding chitinases from *Elaeis guineensis* Jacq. and their expression profiles in response to fungal infections. *Physiol. Mol. Plant. Pathol.* **2011**, *76*, 96–103. [CrossRef]
19. Sapak, Z.; Meon, S.; Ahmad, Z.A.M. Effect of endophytic bacteria on growth and suppression of *Ganoderma* infection in oil palm. *Int. J. Agric. Biol.* **2008**, *10*, 127–132.
20. Bivi, M.R.; Farhana, M.; Khairulmazmi, A.; Idris, A. Control of *Ganoderma boninense*: A causal agent of basal stem rot disease in oil palm with endophyte bacteria in vitro. *Int. J. Agric. Biol.* **2010**, *12*, 833–839.
21. Sundram, S.; Meon, S.; Seman, I.A.; Othman, R. Symbiotic interaction of endophytic bacteria with arbuscular mycorrhizal fungi and its antagonistic effect on *Ganoderma boninense*. *J. Microbiol.* **2011**, *49*, 551–557. [CrossRef]
22. Nurrashyeda, R.; Maizatul, S.; Idris, A.; Madihah, A.; Nasyaruddin, M. The potential of endophytic bacteria as a biological control agent for *Ganoderma* disease in oil palm. *Sains Malaysiana* **2016**, *45*, 401–409.
23. Katz, L.; Baltz, R.H. Natural product discovery: past, present, and future. *J. Ind. Microbiol. Biotechnol.* **2016**, *43*, 155–176. [CrossRef]
24. Procópio, R.E.; Silva, I.R.; Martins, M.K.; Azevedo, J.L.; Araújo, J.M. Antibiotics produced by *Streptomyces*. *Braz. J. Infect. Dis.* **2012**, *16*, 466–471. [CrossRef]
25. Ōmura, S.; Ikeda, H.; Ishikawa, J.; Hanamoto, A.; Takahashi, C.; Shinose, M.; Takahashi, Y.; Horikawa, H.; Nakazawa, H.; Osonoe, T.; et al. Genome sequence of an industrial microorganism *Streptomyces avermitilis*: Deducing the ability of producing secondary metabolites. *PNAS* **2001**, *98*, 12215–12220.
26. Watve, M.G.; Tickoo, R.; Jog, M.M.; Bhole, B.D. How many antibiotics are produced by the genus *Streptomyces*? *Arch. Microbiol.* **2001**, *176*, 386–390. [CrossRef]
27. Pithakkit, S.; Petcharat, V.; Chuenchit, S.; Pornsuriya, C.; Sunpapao, A. Isolation of antagonistic actinomycetes species from rhizosphere as effective biocontrol against oil palm fungal diseases. *Walailak J. Sci. Tech.* **2014**, *12*, 481–490.
28. Shariffah-Muzaimah, S.; Idris, A.; Madihah, A.; Dzolkhifli, O.; Kamaruzzaman, S.; Maizatul-Suriza, M. Characterization of *Streptomyces* spp. isolated from the rhizosphere of oil palm and evaluation of their ability to suppress basal stem rot disease in oil palm seedlings when applied as powder formulations in a glasshouse trial. *World J. Microbiol. Biotech.* **2018**, *34*, 15. [CrossRef]
29. Ting, A.S.Y.; Hermanto, A.; Peh, K.L. Indigenous actinomycetes from empty fruit bunch compost of oil palm: evaluation on enzymatic and antagonistic properties. *Biocat. Agric. Biotech.* **2014**, *3*, 310–315. [CrossRef]
30. Pal, K.K.; Gardener, B.M. Biological control of plant pathogens. *Plant. Health Instructor* **2006**, *2*, 1117–1142. [CrossRef]
31. Sujarit, K.; Kudo, T.; Ohkuma, M.; Pathom-Aree, W.; Lumyong, S. *Streptomyces palmae* sp. nov., isolated from oil palm (*Elaeis guineensis*) rhizosphere soil. *Int. J. Syst. Evol. Microbiol.* **2016**, *66*, 3983–3988. [CrossRef]
32. Shirling, E.B.; Gottlieb, D. Methods for characterization of *Streptomyces* species. *Int. J. Syst. Bacteriol.* **1966**, *16*, 313–340. [CrossRef]
33. Mukku, V.J.R.V.; Maskey, R.P.; Monecke, P.; Grün-Wollny, I.; Laatsch, H. 5-(2-Methylphenyl)-4-pentenoic acid from a terrestrial Streptomycete. *Z. Naturforsch.* **2002**, *57b*, 335–337. [CrossRef]
34. Shaaban, K.A.; Helmke, E.; Kelter, G.; Fiebig, H.H.; Laatsch, H. Glucopiericidin C: A cytotoxic piericidin glucoside antibiotics produced by a marine *Streptomyces* isolate. *J. Antibiot.* **2011**, *64*, 205–209. [CrossRef]
35. Thong, W.L.; Shin-ya, K.; Nishiyama, M.; Kuzuyama, T. Methylbenzene-containing polyketides from a *Streptomyces* that spontaneously acquired rifampicin resistance: Structural elucidation and biosynthesis. *J. Nat. Prod.* **2016**, *79*, 857–864. [CrossRef] [PubMed]
36. Qureshi, A.; Mauger, J.B.; Cano, R.J.; Galazzo, J.L.; Lee, M.D. MF-EA-705a & MF-EA-705b, new metabolites from microbial fermentation of a *Streptomyces* sp. *J. Antibiot.* **2001**, *54*, 1100–1103. [PubMed]
37. Shiomi, K.; Yang, H.; Inokoshi, J.; Van der Pyl, D.; Nakagawa, A.; Takeshima, H.; Ōmura, S. Pepticinnamins, new farnesyl-protein transferase inhibitors produced by an actinomycete. II. Structural elucidation of pepticinnamin E. *J. Antibiot.* **1993**, *46*, 229–234. [CrossRef] [PubMed]
38. Shigematsu, N.; Hayashi, K.; Kayakiri, N.; Takase, S.; Hashimoto, M.; Tanaka, H. Structure of WS9326A, a novel tachykinin antagonist from a *Streptomyces*. *J. Org. Chem.* **1993**, *58*, 170–175. [CrossRef]

39. Toki, S.; Agatsuma, T.; Ochiai, K.; Saitoh, Y.; Ando, K.; Nakanishi, S.; Lokker, N.; Giese, N.A.; Matsuda, Y. RP-1776, a novel cyclic peptide produced by *Streptomyces* sp., inhibits the binding of PDGF to the extracellular domain of its receptor. *J. Antibiot.* **2001**, *54*, 405–414. [CrossRef]
40. Hayakawa, Y.; Adachi, K.; Komeshima, N. New antitumor antibiotics, anguinomycins A and B. *J. Antibiot.* **1987**, *40*, 1349–1352. [CrossRef]
41. Hamamoto, T.; Seto, H.; Beppu, T. Leptomycins A and B, new antifungal antibiotics. II. Structure elucidation. *J. Antibiot.* **1983**, *36*, 646–650. [CrossRef]
42. Yano, K.; Yokoi, K.; Sato, J.; Oono, J.; Kouda, T.; Ogawa, Y.; Nakashima, T. Actinopyrones A, B, and C, new physiologically active substances. II. Physicochemical properties and chemical structures. *J. Antibiot.* **1986**, *39*, 38–43. [CrossRef]
43. Angel, L.P.L.; Yusof, M.T.; Ismail, I.S.; Ping, B.T.Y.; Mohamed Azni, I.N.A.; Kamarudin, N.H.; Sundram, S. An in vitro study of the antifungal activity of *Trichoderma virens* 7b and a profile of its non-polar antifungal components related against *Ganoderma boninense*. *J. Microbiol.* **2016**, *54*, 732–744. [CrossRef]
44. Lester, G. Inhibition of growth, synthesis, and permeability in *Neurospora crassa* by phenethyl alcohol. *J. Bacteriol.* **1965**, *90*, 29–37. [CrossRef]
45. Liu, P.; Cheng, Y.; Yang, M.; Liu, Y.; Chen, K.; Long, C.A.; Deng, X. Mechanisms of action for 2-phenylethanol isolated from *Kloeckera apiculata* in control of *Penicillium* molds of citrus fruits. *BMC Microbiol.* **2014**, *14*, 242. [CrossRef]
46. Staunton, J.; Weissman, K.J. Polyketide biosynthesis: a millennium review. *Nat. Prod. Rep.* **2001**, *18*, 380–416. [CrossRef]
47. Banga, J.; Praveen, V.; Singh, V.; Tripathi, C.K.M.; Bihari, V. Studies on medium optimization for the production of antifungal and antibacterial antibiotics from a bioactive soil actinomycete. *Med. Chem. Res.* **2008**, *17*, 425–436. [CrossRef]
48. Ruiz, B.; Chávez, A.; Forero, A.; García-Huante, Y.; Romero, A.; Sánchez, M.; Rocha, D.; Sánchez, B.; Rodríguez-Sanoja, R.; Sánchez, S.; et al. Production of microbial secondary metabolites: Regulation by the carbon source. *Crit. Rev. Microbiol.* **2010**, *36*, 146–167. [CrossRef]
49. Sujarit, K. Selection and Characterization of Actinomycetes for Growth Promotion of Oil Palm and Biological Control of Oil Palm Diseases. Ph.D. Thesis, Chiang Mai University, Chiang Mai, Thailand, 2018.
50. Bérdy, J. Thoughts and facts about antibiotics: where we are now and where we are heading. *J. Antibiot.* **2012**, *65*, 385–395. [CrossRef]
51. Lucas, X.; Senger, C.; Erxleben, A.; Grüning, B.A.; Döring, K.; Mosch, J.; Flemming, S.; Günther, S. StreptomeDB: A resource for natural compounds isolated from *Streptomyces* species. *Nucleic Acids Res.* **2013**, *41*, D1130–D1136. [CrossRef]

# Permissions

 Every chapter published in this book has been scrutinized by our experts. Their significance has been extensively debated. The topics covered herein carry significant findings which will fuel the growth of the discipline. They may even be implemented as practical applications or may be referred to as a beginning point for another development.

The contributors of this book come from diverse backgrounds, making this book a truly international effort. This book will bring forth new frontiers with its revolutionizing research information and detailed analysis of the nascent developments around the world.

We would like to thank all the contributing authors for lending their expertise to make the book truly unique. They have played a crucial role in the development of this book. Without their invaluable contributions this book wouldn't have been possible. They have made vital efforts to compile up to date information on the varied aspects of this subject to make this book a valuable addition to the collection of many professionals and students.

This book was conceptualized with the vision of imparting up-to-date information and advanced data in this field. To ensure the same, a matchless editorial board was set up. Every individual on the board went through rigorous rounds of assessment to prove their worth. After which they invested a large part of their time researching and compiling the most relevant data for our readers.

The editorial board has been involved in producing this book since its inception. They have spent rigorous hours researching and exploring the diverse topics which have resulted in the successful publishing of this book. They have passed on their knowledge of decades through this book. To expedite this challenging task, the publisher supported the team at every step. A small team of assistant editors was also appointed to further simplify the editing procedure and attain best results for the readers.

Apart from the editorial board, the designing team has also invested a significant amount of their time in understanding the subject and creating the most relevant covers. They scrutinized every image to scout for the most suitable representation of the subject and create an appropriate cover for the book.

The publishing team has been an ardent support to the editorial, designing and production team. Their endless efforts to recruit the best for this project, has resulted in the accomplishment of this book. They are a veteran in the field of academics and their pool of knowledge is as vast as their experience in printing. Their expertise and guidance has proved useful at every step. Their uncompromising quality standards have made this book an exceptional effort. Their encouragement from time to time has been an inspiration for everyone.

The publisher and the editorial board hope that this book will prove to be a valuable piece of knowledge for researchers, students, practitioners and scholars across the globe.

# List of Contributors

**Manar Ibrahimi**
Laboratory of Microbial Biotechnologies, Agrosciences and Environment (BioMAgE), Faculty of Sciences Semlalia, Cadi Ayyad University, Marrakesh, Morocco
Institut de Chimie des Milieux et Matériaux de Poitiers (IC2MP - CNRS UMR 7285), Université de Poitiers, 4 rue Michel Brunet - TSA 51106, 86073 Poitiers Cedex 9, France
Laboratory of Microbiology and Virology, Faculty of Medicine and Pharmacy, Cadi Ayyad University, Marrakesh, Morocco

**Laurent Lemee**
Institut de Chimie des Milieux et Matériaux de Poitiers (IC2MP - CNRS UMR 7285), Université de Poitiers, 4 rue Michel Brunet - TSA 51106, 86073 Poitiers Cedex 9, France

**Souad Loqman**
Laboratory of Microbiology and Virology, Faculty of Medicine and Pharmacy, Cadi Ayyad University, Marrakesh, Morocco

**Wassila Korichi**
Laboratory of Microbial Biotechnologies, Agrosciences and Environment (BioMAgE), Faculty of Sciences Semlalia, Cadi Ayyad University, Marrakesh, Morocco
Laboratory of Microbiology and Virology, Faculty of Medicine and Pharmacy, Cadi Ayyad University, Marrakesh, Morocco

**Mohamed Hafidi and Yedir Ouhdouch**
Laboratory of Microbial Biotechnologies, Agrosciences and Environment (BioMAgE), Faculty of Sciences Semlalia, Cadi Ayyad University, Marrakesh, Morocco
Agro Bio Sciences Program, Mohammed VI Polytechnic University (UM6P), Benguerir, 43150, Morocco

**Nadja A. Henke, Susanne Götker, Petra Peters-Wendisch and Volker F. Wendisch**
Faculty of Biology & CeBiTec, Bielefeld University, 33615 Bielefeld, Germany

**Sophie Austermeier**
Faculty of Biology & CeBiTec, Bielefeld University, 33615 Bielefeld, Germany
Department of Microbial Pathogenicity Mechanisms, Leibniz Institute for Natural Product Research and Infection Biology (HKI), 07745 Jena, Germany

**Isabell L. Grothaus**
Faculty of Biology & CeBiTec, Bielefeld University, 33615 Bielefeld, Germany
Faculty of Production Engineering, Bremen University, 28359 Bremen, Germany

**Marcus Persicke**
Faculty of CeBiTec, Bielefeld University, 33615 Bielefeld, Germany

**Xiaoxin Zhuang**
Heilongjiang Provincial Key Laboratory of Agricultural Microbiology, Northeast Agricultural University, Harbin 150030, China

**Chongxi Liu and Zhiyin Yu**
Heilongjiang Provincial Key Laboratory of Agricultural Microbiology, Northeast Agricultural University, Harbin 150030, China
State Key Laboratory of Phytochemistry and Plant Resources in West China, Kunming Institute of Botany, Chinese Academy of Sciences, Kunming 650201, China

**Xiaowei Guo**
Key Laboratory of Agricultural Microbiology of Heilongjiang Province, Northeast Agricultural University, No. 600 Changjiang Road, Xiangfang District, Harbin 150030, China
State Key Laboratory of Phytochemistry and Plant Resources in West China, Kunming Institute of Botany, Chinese Academy of Sciences, Kunming 650201, China

**Zhiyan Wang and Yongjiang Wang**
State Key Laboratory of Phytochemistry and Plant Resources in West China, Kunming Institute of Botany, Chinese Academy of Sciences, Kunming 650201, China

**Joachim Wink**
Microbial Strain Collection (MISG), Helmholtz Centre for Infection Research (HZI), 38124 Braunschweig, Germany

**Chandra Risdian**
Microbial Strain Collection (MISG), Helmholtz Centre for Infection Research (HZI), 38124 Braunschweig, Germany
Research Unit for Clean Technology, Indonesian Institute of Sciences (LIPI), Bandung 40135, Indonesia

**Tjandrawati Mozef**
Research Center for Chemistry, Indonesian Institute of Sciences (LIPI), Serpong 15314, Indonesia

**Chuanyu Han and Bing Yu**
Key Laboratory of Agricultural Microbiology of Heilongjiang Province, Northeast Agricultural University, Harbin 150030, China

**Yijun Yan and Shengxiong Huang**
State Key Laboratory of Phytochemistry and Plant Resources in West China, Kunming Institute of Botany, Chinese Academy of Sciences, Kunming 650201, China

**Radha Singh and Ashok K. Dubey**
Department of Biological Sciences & Engineering, Netaji Subhas Institute of Technology, New Delhi 110078, India

**Liyuan Han, Mingying Yu, Peng Cao, Dongmei Li, Yongqiang Liu and Xiangjing Wang**
Key Laboratory of Agricultural Microbiology of Heilongjiang Province, Northeast Agricultural University, No. 59 Mucai Street, Xiangfang District, Harbin 150030, China

**Wensheng Xiang**
Key Laboratory of Agricultural Microbiology of Heilongjiang Province, Northeast Agricultural University, No. 59 Mucai Street, Xiangfang District, Harbin 150030, China
State Key Laboratory for Biology of Plant Diseases and Insect Pests, Institute of Plant Protection, Chinese Academy of Agricultural Sciences, Beijing 100193, China

**Saket Siddharth and Ravishankar Rai Vittal**
Department of Studies in Microbiology, University of Mysore, Manasagangotri, Mysore 570006, India

**Ling Ling, Xiaoyang Han, Xiao Li, Xue Zhang, Han Wang, Lida Zhang and Yutong Wu**
Key Laboratory of Agricultural Microbiology of Heilongjiang Province, Northeast Agricultural University, No. 59 Mucai Street, Xiangfang District, Harbin 150030, China

**Eduardo L. Almeida and Andrés Felipe Carrillo Rincón**
School of Microbiology, University College Cork, T12 YN60 Cork, Ireland

**Navdeep Kaur, Laurence K. Jennings and Olivier P. Thomas**
Marine Biodiscovery, School of Chemistry and Ryan Institute, National University of Ireland Galway (NUI Galway), University Road, H91 TK33 Galway, Ireland

**Stephen A. Jackson and Alan D.W. Dobson**
School of Microbiology, University College Cork, T12 YN60 Cork, Ireland
Environmental Research Institute, University College Cork, T23 XE10 Cork, Ireland

**Katherine Gregory, Laura A. Salvador, Shukria Akbar, Barbara I. Adaikpoh and D. Cole Stevens**
Department of BioMolecular Sciences, School of Pharmacy, University of Mississippi, University, MS 38677, USA

**Maksym Myronovskyi, Birgit Rosenkränzer and Marc Stierhof**
Pharmazeutische Biotechnologie, Universität des Saarlandes, 66123 Saarbrücken, Germany

**Lutz Petzke and Tobias Seiser**
BASF SE, 67056 Ludwigshafen, Germany

**Andriy Luzhetskyy**
Pharmazeutische Biotechnologie, Universität des Saarlandes, 66123 Saarbrücken, Germany
Helmholtz-Institut für Pharmazeutische Forschung Saarland, 66123 Saarbrücken, Germany

**Han Wang, Tianyu Sun, Wenshuai Song, Peng Cao, Xi Xu and Junwei Zhao**
Key Laboratory of Agricultural Microbiology of Heilongjiang Province, Northeast Agricultural University, No. 600 Changjiang Road, Xiangfang District, Harbin 150030, China

**Yue Shen**
Key Laboratory of Agricultural Microbiology of Heilongjiang Province, Northeast Agricultural University, No. 600 Changjiang Road, Xiangfang District, Harbin 150030, China
College of Science, Northeast Agricultural University, No. 600 Changjiang Road, Xiangfang District, Harbin 150030, China

**Michael Steinert**
Institute of Microbiology, Technische Universität Braunschweig, 38106 Braunschweig, Germany

**Kanaporn Sujarit**
Research Center of Microbial Diversity and Sustainable Utilization, Faculty of Science, Chiang Mai University, Chiang Mai 50200, Thailand
Kitasato Institute for Life Sciences, Kitasato University, 5-9-1 Shirokane, Minato-ku, Tokyo 108-8641, Japan

**Kazuyuki Dobashi**
Kitasato Institute for Life Sciences, Kitasato University, 5-9-1 Shirokane, Minato-ku, Tokyo 108-8641, Japan

**Mihoko Mori and Kazuro Shiomi**
Kitasato Institute for Life Sciences, Kitasato University, 5-9-1 Shirokane, Minato-ku, Tokyo 108-8641, Japan
Graduate School of Infection Control Sciences, Kitasato University, 5-9-1 Shirokane, Minato-ku, Tokyo 108-8641, Japan

**Wasu Pathom-aree**
Research Center of Microbial Diversity and Sustainable Utilization, Faculty of Science, Chiang Mai University, Chiang Mai 50200, Thailand
Department of Biology, Faculty of Science, Chiang Mai University, Chiang Mai 50200, Thailand

**Saisamorn Lumyong**
Research Center of Microbial Diversity and Sustainable Utilization, Faculty of Science, Chiang Mai University, Chiang Mai 50200, Thailand
Department of Biology, Faculty of Science, Chiang Mai University, Chiang Mai 50200, Thailand
Academy of Science, The Royal Society of Thailand, Bangkok 10300, Thailand

# Index